AF685279

Valorization of Food Processing By-Products

Valorization of Food Processing By-Products

Editors

Marco Poiana
Francesco Caponio
Antonio Piga

MDPI • Basel • Beijing • Wuhan • Barcelona • Belgrade • Manchester • Tokyo • Cluj • Tianjin

MDPI

Editors
Marco Poiana
Department of AGRARIA,
University Mediterranea of
Reggio Calabria,
Reggio Calabria, Italy

Francesco Caponio
Department of Soil, Plant and
Food Sciences (DiSSPA),
University of Bari "Aldo Moro",
Bari, Italy

Antonio Piga
Department of Agriculture,
University of Sassari,
Sassari, Italy

Editorial Office
MDPI
St. Alban-Anlage 66
4052 Basel, Switzerland

This is a reprint of articles from the Special Issue published online in the open access journal *Foods* (ISSN 2304-8158) (available at: https://www.mdpi.com/journal/foods/special_issues/valorization_food_byproducts).

For citation purposes, cite each article independently as indicated on the article page online and as indicated below:

LastName, A.A.; LastName, B.B.; LastName, C.C. Article Title. *Journal Name* **Year**, *Volume Number*, Page Range.

ISBN 978-3-0365-5949-0 (Hbk)
ISBN 978-3-0365-5950-6 (PDF)

© 2022 by the authors. Articles in this book are Open Access and distributed under the Creative Commons Attribution (CC BY) license, which allows users to download, copy and build upon published articles, as long as the author and publisher are properly credited, which ensures maximum dissemination and a wider impact of our publications.
The book as a whole is distributed by MDPI under the terms and conditions of the Creative Commons license CC BY-NC-ND.

Contents

About the Editors

Marco Poiana

Marco Poiana is currently working as a full professor of food technology at the Mediterranean University of Reggio Calabria, Italy. His research interests include food processing and food valorisation, with a particular focus on olive oil production, table olives, and transformed vegetables foods. In view of a circular economy, his research investigates the exploitation of food by-products in terms of the extraction of active fractions and their synthesis with foodstuffs to improve their quality. He has published over 100 scientific papers in international peer-reviewed journals on food science and technology and several book chapters. He has held the roles of PI and research unit leader in a number of regional and national projects.

Francesco Caponio

Francesco Caponio (Full Professor) graduated in agrarian sciences from the University of Bari (Italy). Since 2017, he has been working as a full professor in food science and technology. He is the principal investigator of several competitive research projects and has been an invited speaker at many national and regional conferences. He is the author of over 245 works published in cited international journals. He has worked as the coordinator of the bachelor's and master's degree programmes in "Food Science and Technology" at the University of Bari Aldo Moro and is a member of the Ph.D. School in "Soil and Food Sciences". His research interests cover olive oil technology; the setup of analytical techniques designed to ensure the quality and genuineness of foods; the extraction, characterization, and valorisation of bioactive compounds from the processing of wastes and their application to increase the shelf-life of foods; and the formulation of innovative and functional foods.

Antonio Piga

Antonio Piga has held the role of full professor since 31 December 2016, in the Food Science and Technology group and teaches food technology processing. He is the author of more than 200 scientific papers published in national and international journals, as well as symposia proceedings, and he has participated in numerous seminars and national and international congresses. At the Dipartimento di Agraria (formerly Disaaba), Antonio Piga carries out research on topics related to food technology. His main research interests include the storage, processing, and stabilisation of foods, with a particular emphasis on the shelf-life of baked foods, drying of fruit and vegetables, olive processing technologies for the production of extra-virgin olive oils and table olives, the conditioning and minimal processing of fruit, and the changes in the antioxidant activities of different foods subjected to processing and storage.

foods

MDPI

Editorial

Valorization of Food Processing By-Products

Francesco Caponio [1], Antonio Piga [2] and Marco Poiana [3,*]

1 Department of Soil, Plant and Food Science (DISSPA), University of Bari Aldo Moro, Via Amendola 165/a, 70126 Bari, Italy
2 Dipartimento di Agraria, Università degli Studi di Sassari, Viale Italia 39/A, 07100 Sassari, Italy
3 Department of AGRARIA, University Mediterranea of Reggio Calabria, Feo di Vito, 89124 Reggio Calabria, Italy
* Correspondence: mpoiana@unirc.it

Nowadays, the valorization of by-products of the food industry is a priority linked to the need to release the smallest amount of products from processes. The sustainable development goal provides for the creation of a circular economy in order to reduce the environmental impact of production processes and increase the incomes from it. In one of the several definitions of a circular economy, it could be considered as an economic model that aims to the maximum reuse and recycling of materials, goods, and components in order to decrease waste generation [1] Within this economic model, the valorization of the food industry by-products becomes very important considering the quantities that are released by the different transformation processes. In addition, the food by-products contain important amounts of biologically active compounds that could be used in various sectors, including food manufacturing itself [2]. The scientific research is called to study and provide new solutions for the exploitation of the food by-products, in fact, in recent years, the publications that contain in the title, abstract or key-words the "food by products" exceed a thousand a year with a great increase in recent years (source Scopus).

Food by-products derive from different sources, among them vegetable, meat, brewery, winery, dairy, and fish processing and industries provide very interesting quantities and high-activity molecules for their reuse. The purpose of this Special Issue is to bring together some of the recent studies that can allow the development of new knowledge for the exploitation of the by-products of the food industry.

The valorization of by-products requires an articulated strategy. In fact, it becomes important to the management of the by-product immediately after its production and until its use, its transformation or extraction of the active fractions, the treatment of the active fractions. When the extracted fractions are added to foods it is important to know the interactions between compounds (or fractions) and foodstuff. In view of a food utilization of the obtained extract, the choice of the correct solvent and extraction process are very important. In the same way, the other operations involved should be carefully evaluated. Some manuscripts of this Special Issues discuss pretreatments applied to different raw materials. Drying is applied to reduce water content and improve extraction of active substances, but the process could have a thermal impact on the molecules. In addition to conventional air drying of vine shoots [3], freeze-drying, heat drying, and non-thermal air-drying to reduce water content of the tomato pomace [4] were studied. Considering the potential food utilization of the lycopene extracts from tomato pomace, green deep eutectic solvent mixtures were used and compared with traditional ones (hexane). The study has demonstrated that the green deep eutectic solvents coupled with a preliminary non-thermal drying could be an effective alternative to traditional processes. Another study has studied the grape pomace extraction and a large scale solid–liquid extraction process was set up to optimize the process and obtain catechin enriched extracts [5].

A problem of active molecules extraction is their concentration in the extract. Some techniques could be applied to enriched (or concentrated) items, the use of macroporous

Citation: Caponio, F.; Piga, A.; Poiana, M. Valorization of Food Processing By-Products. *Foods* **2022** *11*, 3246. https://doi.org/10.3390/foods11203246

Received: 5 October 2022
Accepted: 8 October 2022
Published: 18 October 2022

Publisher's Note: MDPI stays neutral with regard to jurisdictional claims in published maps and institutional affiliations.

Copyright: © 2022 by the authors. Licensee MDPI, Basel, Switzerland. This article is an open access article distributed under the terms and conditions of the Creative Commons Attribution (CC BY) license (https://creativecommons.org/licenses/by/4.0/).

resins is described and tested on hazelnut skin extracts [6]. With this operation, a great concentration of phenols and anti-oxidant activity was obtained.

Many agricultural and livestock breeding products show important biological activities; the molecules responsible for these properties are also present in the anatomical parts that the industrial process of transformation does not use. The resulting by-products, if not used and consequently treated as waste, would result in a loss of usable components. Research on the valorization of by-products has involved various raw materials that show interesting properties and their application in different sectors not only as food ingredients.

Often the by-products from the vegetable oil industry contain high value-added molecules, such as antioxidants. A wide study, included in this volume, reports the content of total phenols and flavonoid, as well as evaluated the antioxidant activity by different essays (DPPH, FRAP, CUPRAC, ABTS, and by photochemiluminescence method-PCL-ACL) of by-products obtained from different vegetable oil industry (Sea buckthorn flour, hemp flour, walnut flour, grape seed flour, rapeseed meals, sunflower meals, black sesame meals, red grape seed meals, golden flax meals, thistle meals, sesame groats, thistle groats, coriander groats, and sunflower groats) [7]. Although the molecules and biological activities are influenced by the variety of the raw material and the extraction method, the study showed that these by-products could be an interesting source of phenolic compounds, especially flavonoids, with antioxidant properties that could be valuable ingredients for functional foods.

An environmentally impacting food process is the extraction of oil from olives. Olive mill pomace and wastewater represent a significant amount of the whole raw material: less than 15–20% of olives is oil, the remaining part is the abovementioned waste. Due to the high content of polyphenols and other compounds, olive oil wastewater is considered highly polluting, but, on the other hand, the presence of many of these compounds could be a resource. The importance of polyphenols is well known and, among these, their antioxidant activity can be used in different food formulations with the aim of preserving the food itself, but also providing a greater supply of these substances to consumers. During extraction operations, only a small amount of olives' phenolic components is transferred to oil while the remaining portion could be lost. Many studies appear in scientific literature about extraction and purification methods, and utilization of the phenolic fraction derived from olive mill by-products. In this Special Issue, there are interesting applications on beef burgers (and minced meat in general) shelf life extension, adding polyphenols derived from olive mill waste water (OMWW) concentrated by mean of selective membranes [8]. The addition of the important substances obtained from by-products led to a great reduction in meat oxidation and an increase in its shelf life. The mayonnaise, as an O/W emulsion model, was studied in another research that add phenol substances from OMWW [9], the added phenolic molecules play an important role in the nutritional parameters of mayonnaise. The concentration of these compounds transferred in the samples allows slowing down of oxidation processes with a consequent shelf life extension of the food. Furthermore, it could have potential health-properties for consumers even if the color and taste appear slightly different from no-added mayonnaise.

The mayonnaise is a very heterogeneous and unstable physical system both for the formation of phase separation and the oxidation that some components can undergo. For this reason, it has been widely studied and the manuscripts in this Special Issue report different research about this matter. The research of Włodarczyk et al. focused on the evaluation of radical scavenging characteristics, oxidation stability, microstructures, and optical properties of vegan mayonnaises containing aquafaba from chickpeas and blends of refined rapeseed [10]. Aquafaba is derived from cooking legumes and, on an industrial-scale, huge quantities are produced that are unutilized. The uses for aquafaba have expanding in the recent years due to its evident properties, including those that are cited in this Special Issue.

The studies about dispersed systems, such as mayonnaise, are numerous, especially in regards to search for emulsifying and fat-replacer components to substitute the egg yolk in mayonnaise formulations and obtain a stable semi-solid product. The application of by-products to the production of mayonnaise has also been studied in another work of the Special Issue. Hijazi et al. have studied rheological properties of gums obtained from the cold-pressed seed oil by-products of chia (*Salvia hispanica* L.), flaxseed (*Linum usitatissimum* L.), and rocket (*Eruca sativa*) [11]. The obtained emulsion (O/W) is a mayonnaise with a good structure and an improvement of the oxidative stability, as already reported in other manuscripts.

Chia seed oil is very interesting due to the high content of oil (near 30% in weight) and the concentration of some polyunsaturated fatty acids, mainly linoleic (C18:2–ω-6) and linolenic acid (C18:3–ω-3), that give to it a nutritional function as an integrator of essential fatty acids. After oil extraction, the defatted seeds of chia contain proteins, carbohydrates, and dietary fiber with emulsifying properties that can be used to produce ice cream with low fat content and rheological properties, similar to those obtained with commercial stabilizers, studied by a work inside this Special Issue [12].

Among the agricultural crops, tomato (*Solanum lycopersicum* L.) is one of the most worldwide cultivated and a large quantity is sent to industrial transformation. Skin and seeds are undesirable anatomical parts and so they are removed. Even though these parts represent a small percentage of the entire fruit (1–5%), they are interesting due to the content of fiber and other compounds, such as sugars, proteins, pectins, fats, and vitamins, together with an important amount of carotenoids, mainly lycopene. The study previously reported described the possibility to obtain enriched fractions of lycopene and carotene for subsequent applications in food and pharmaceutical industries [4].

The process of grapes into wine, from the vineyard to the cellar, presents a series of interesting by-products. An interesting residue is grape pomace, its importance is due to the large quantity of grapes used worldwide for winemaking. Inside this by-product, interesting molecules with different biological activities are present. The goal of improving the extraction of these substances is the subject of many studies, including one within this Special Issue [5]. By-products with important functionality do not come only from winemaking; important molecules are also present in the vine plant. The stilbene molecules show beneficial effects for human health, due to these properties, they are highly studied for a wide range of applications, among food and pharmaceutical fields, which are described in this Special Issue. Their sources of production are also studied. Winery wastes and by-products are highly rich in stilbenes and the whole vine plant contains these molecules. Vine shoots from pruning have a low value, they are burned or cut into small pieces and put inside soil. An interesting study reported in this Special Issue describes the importance of the grape variety as a source of stilbene, such as trans-resveratrol and viniferin [3].

Hazelnut is a wide transformed nut, nearly 50% of the whole weight is discarded, as shell skin and other parts of the fruit are rich in bioactive compounds, such as polyphenols. With the adequate extraction and concentration of anti-oxidant fractions, they could be obtained to add in to food and for cosmetic and pharmaceutical purposes [6].

Pomace from different berries could be an interesting source of dietary fiber. The enzymatic treatment (with different commercially available formulates of hydrolases) of cranberry pomace and the evaluation of the product obtained in terms of technological properties and probiotic potential were carried out by Jagelaviciute et al. [13].

Another rich source of biocomponents is kiwi processing waste, such as peels and pomace. An application of the valorization of this by-product was reported by Ille et al. [14]. After freeze drying, these substances and the obtained powder was added with black rice and buckwheat flour, inoculated with Lactobacillus spp. and tested as probiotic food. The study also appears very interesting for future applications, not only in food but also in cosmetic and pharmaceutical industries.

Artichoke roots represent a source rich in inulin, a functional compound with prebiotic, and technological properties. To valorize this waste, Difonzo et al. exploit inulin with high degree of polymerization, by partially replacing durum wheat semolina for the production of functional fresh pasta [15]. The results highlighted the nutritional improvement of the glycemic index and prebiotic activity of pasta without compromising the sensory properties, although some structural features were affected.

The cheesemaking is characterized for a high amount of output resources: whey proteins are recovered by thermal flocculation and ricotta cheese is produced. The liquid residue, ricotta cheese exhaust whey (scotta), is often unutilized and becomes a pollutant waste. In recent years, many utilizations of scotta were studied due to its content of lactose, minerals, vitamins, soluble peptides, and free amino acids. The hydrolyzed whey peptides could be inhibitors of dipeptidyl peptidase-4 (DPP-IV) and could play an interesting role in diabetes type II treatment. Cabizza et al. evaluate the DPP-IV inhibition, antioxidant, and antibacterial activities from ovine scotta enzymatically hydrolyzed [16].

Most of the slaughter of pork is used in dishes of many countries. Pig brain is an important part, but it is not completely utilized. Chanted et al. reports the presence in the pig brain of important nutrients, such as phospholipids, and essential amino acids [17]. The research showed interesting amounts of umami-taste and functional amino acids which give a potential food use of this part of the pig.

Fish processing by-products have an environmental impact, they are near the 50% of the whole weight. For many years, studies have been researching their use and treatment, and most of them have high value applications. The omega-3 (ω-3)-rich fish oils represent an interesting valorization of fish by-products. Some methods were applied to evaluate the fatty acid content and their profile. One of them, which utilized the near infrared reflectance (NIR), was applied to fish by-products classification [18].

Other by-products derive from defatted *Aurantiochytrium* spp., an algae from which an interesting oil rich in polyunsaturated fatty acids (PUFA) is obtained, but a high amount of by-product residues cause some environmental troubles. After defatting, the residual biomass of *Aurantiochytrium* spp. contains a highly concentrated mixture of proteins and carbohydrates that could have nutritional and functional properties. The presence of interesting amounts of glutamic and aspartic acids give to the defatted and dried residue a flavor-enhancing action [19].

Food waste is a huge problem for our society. Of these, wasted bread is one of the most abundant. By means of bioprocessing bread was transformed and experimentally tested as alkaline soil amendment [20].

In conclusion, the overview of studies reported in this Special Issue shows how processing waste from the food industries can become an important economic resource. The Special Issue title as "valorization" is demonstrated not only for an economic point of view, but also from an environmental one. In fact, many applications derive from by-products with a high level of pollution. Their recovery could reduce this trouble.

Author Contributions: Conceptualization, F.C., A.P. and M.P.; writing—original draft preparation, M.P.; writing—review and editing, F.C. and A.P. All authors have read and agreed to the published version of the manuscript.

Funding: This research received no external funding.

Acknowledgments: We would like to thank Dana Min for technical support.

Conflicts of Interest: The authors declare no conflict of interest.

References

1. Ghisellini, P.; Ripa, M.; Ulgiati, S. Exploring Environmental and Economic Costs and Benefits of a Circular Economy Approach to the Construction and Demolition Sector. A Literature Review. *J. Clean. Prod.* **2018**, *178*, 618–643. [CrossRef]
2. Faustino, M.; Veiga, M.; Sousa, P.; Costa, E.M.; Silva, S.; Pintado, M. Agro-Food Byproducts as a New Source of Natural Food Additives. *Molecules* **2019**, *24*, 1056. [CrossRef] [PubMed]
3. Noviello, M.; Caputi, A.F.; Squeo, G.; Paradiso, V.M.; Gambacorta, G.; Caponio, F. Vine Shoots as a Source of Trans-Resveratrol and #-Viniferin: A Study of 23 Italian Varieties. *Foods* **2022**, *11*, 553. [CrossRef] [PubMed]
4. Lazzarini, C.; Casadei, E.; Valli, E.; Tura, M.; Ragni, L.; Bendini, A.; Gallina Toschi, T. Sustainable Drying and Green Deep Eutectic Extraction of Carotenoids from Tomato Pomace. *Foods* **2022**, *11*, 405. [CrossRef] [PubMed]

5. Lo, S.; Pilkington, L.I.; Barker, D.; Fedrizzi, B. Attempts to Create Products with Increased Health-Promoting Potential Starting with Pinot Noir Pomace: Investigations on the Process and Its Methods. *Foods* **2022**, *11*, 1999. [CrossRef] [PubMed]
6. Seif Zadeh, N.; Zeppa, G. Recovery and Concentration of Polyphenols from Roasted Hazelnut Skin Extract Using Macroporous Resins. *Foods* **2022**, *11*, 1969. [CrossRef] [PubMed]
7. Multescu, M.; Marinas, I.C.; Susman, I.E.; Belc, N. Byproducts (Flour, Meals, and Groats) from the Vegetable Oil Industry as a Potential Source of Antioxidants. *Foods* **2022**, *11*, 253. [CrossRef] [PubMed]
8. Roila, R.; Sordini, B.; Esposto, S.; Ranucci, D.; Primavilla, S.; Valiani, A.; Taticchi, A.; Branciari, R.; Servili, M. Effect of the Application of a Green Preservative Strategy on Minced Meat Products: Antimicrobial Efficacy of Olive Mill Wastewater Polyphenolic Extract in Improving Beef Burger Shelf-Life. *Foods* **2022**, *11*, 2447. [CrossRef] [PubMed]
9. De Bruno, A.; Romeo, R.; Gattuso, A.; Piscopo, A.; Poiana, M. Functionalization of a Vegan Mayonnaise with High Value Ingredient Derived from the Agro-Industrial Sector. *Foods* **2021**, *10*, 2684. [CrossRef] [PubMed]
10. Włodarczyk, K.; Zienkiewicz, A.; Szydłowska-Czerniak, A. Radical Scavenging Activity and Physicochemical Properties of Aquafaba-Based Mayonnaises and Their Functional Ingredients. *Foods* **2022**, *11*, 1129 [CrossRef] [PubMed]
11. Hijazi, T.; Karasu, S.; Tekin-Çakmak, Z.H.; Bozkurt, F. Extraction of Natural Gum from Cold-Pressed Chia Seed, Flaxseed, and Rocket Seed Oil By-Product and Application in Low Fat Vegan Mayonnaise. *Foods* **2022**, *11*, 363. [CrossRef] [PubMed]
12. Atik, I.; Tekin Cakmak, Z.H.; Avcı, E.; Karasu, S. The Effect of Cold Press Chia Seed Oil By-Products on the Rheological, Microstructural, Thermal, and Sensory Properties of Low-Fat Ice Cream. *Foods* **2021**, *10*, 2302. [CrossRef] [PubMed]
13. Jagelaviciute, J.; Basinskiene, L.; Cizeikiene, D.; Syrpas, M. Technological Properties and Composition of Enzymatically Modified Cranberry Pomace. *Foods* **2022**, *11*, 2321. [CrossRef] [PubMed]
14. Ilie, G.-I.; Milea, S.A.; Râpeanu, G.; Cîrciumaru, A.; Stănciuc, N. Sustainable Design of Innovative Kiwi Byproducts-Based Ingredients Containing Probiotics. *Foods* **2022**, *11*, 2334. [CrossRef] [PubMed]
15. Difonzo, G.; de Gennaro, G.; Caponio, G.R.; Vacca, M.; dal Poggetto, G.; Allegretta, I.; Immirzi, B.; Pasqualone, A. Inulin from Globe Artichoke Roots: A Promising Ingredient for the Production of Functional Fresh Pasta. *Foods* **2022**, *11*, 3032. [CrossRef] [PubMed]
16. Cabizza, R.; Fancello, F.; Petretto, G.L.; Addis, R.; Pisanu, S.; Pagnozzi, D.; Piga, A.; Urgeghe, P.P. Exploring the DPP-IV Inhibitory, Antioxidant and Antibacterial Potential of Ovine "Scotta" Hydrolysates. *Foods* **2021**, *10*, 3137. [CrossRef] [PubMed]
17. Chanted, J.; Panpipat, W.; Panya, A.; Phonsatta, N.; Cheong, L.-Z.; Chaijan, M. Compositional Features and Nutritional Value of Pig Brain: Potential and Challenges as a Sustainable Source of Nutrients. *Foods* **2021**, *10*, 2943. [CrossRef] [PubMed]
18. Nieto-Ortega, S.; Olabarrieta, I.; Saitua, E.; Arana, G.; Foti, G.; Melado-Herreros, Á. Improvement of Oil Valorization Extracted from Fish By-Products Using a Handheld near Infrared Spectrometer Coupled with Chemometrics. *Foods* **2022**, *11*, 1092. [CrossRef] [PubMed]
19. Reboleira, J.; Félix, R.; Félix, C.; de Melo, M.M.R.; Silva, C.M.; Saraiva, J.A.; Bandarra, N.M.; Teixeira, B.; Mendes, R.; Paulo, M.C.; et al. Evaluating the Potential of the Defatted By-Product of *Aurantiochytrium* sp. Industrial Cultivation as a Functional Food. *Foods* **2021**, *10*, 3058. [CrossRef] [PubMed]
20. Cacace, C.; Rizzello, C.G.; Brunetti, G.; Verni, M.; Cocozza, C. Reuse of Wasted Bread as Soil Amendment: Bioprocessing, Effects on Alkaline Soil and Escarole (*Cichorium endivia*) Production. *Foods* **2022**, *11*, 189. [CrossRef] [PubMed]

foods

MDPI

Article

Inulin from Globe Artichoke Roots: A Promising Ingredient for the Production of Functional Fresh Pasta

Graziana Difonzo [1,*], Giuditta de Gennaro [1], Giusy Rita Caponio [1], Mirco Vacca [1], Giovanni dal Poggetto [2], Ignazio Allegretta [1], Barbara Immirzi [2] and Antonella Pasqualone [1]

[1] Department of the Soil, Plant and Food Science (DiSSPA), University of Bari Aldo Moro, Via Amendola, 165/A, I-70126 Bari, Italy

[2] Istituto per i Polimeri, Compositi e Biomateriali, Consiglio Nazionale delle Ricerche, Via Campi Flegrei, 34-80078 Pozzuoli, Italy

* Correspondence: graziana.difonzo@uniba.it

Abstract: Globe artichoke roots represent an alternative and sustainable source for inulin extraction and are well-noted for their technological and functional properties. Therefore, the aim of our study was to exploit inulin with high degree of polymerization as a replacement of durum wheat semolina for the production of functional fresh pasta. The effect of increased level of substitution (5, 10, 15%) on cooking, structural, sensory, and nutritional properties were evaluated and compared with a control sample consisting exclusively of durum wheat semolina. Inulin addition caused changes to internal structure as evaluated by scanning electron microscopy. The enriched samples showed a lower swelling index, an increasing cooking time, and values of cooking loss (2.37–3.62%), mainly due to the leaching of inulin into the cooking water. Cooked and raw enriched pasta was significantly darker and firmer than the control, but the sensory attributes were not negatively affected, especially at 5 and 10% of substitution levels The increase of dietary fiber content in enriched pasta (3.44–12.41 g/100 g) resulted in a significant reduction of glycaemic index (pGI) and starch hydrolysis (HI). After gastrointestinal digestion, inulin-enriched pasta increased prebiotic growth able to significantly reduce *E. coli* cell density.

Keywords: waste reuse; inulin; high polymerization degree; functional pasta; glycemic index; prebiotics growth

Citation: Difonzo, G.; de Gennaro, G.; Caponio, G.R.; Vacca, M.; dal Poggetto, G.; Allegretta, I.; Immirzi, B.; Pasqualone, A. Inulin from Globe Artichoke Roots: A Promising Ingredient for the Production of Functional Fresh Pasta. *Foods* **2022**, *11*, 3032. https://doi.org/10.3390/foods11193032

Academic Editor: Saroat Rawdkuen

Received: 17 August 2022
Accepted: 22 September 2022
Published: 30 September 2022

Publisher's Note: MDPI stays neutral with regard to jurisdictional claims in published maps and institutional affiliations.

Copyright: © 2022 by the authors. Licensee MDPI, Basel, Switzerland. This article is an open access article distributed under the terms and conditions of the Creative Commons Attribution (CC BY) license (https://creativecommons.org/licenses/by/4.0/).

1. Introduction

Globe artichoke (*Cynara cardunculus* L. subsp. *scolymus* (L.)) is a perennial plant belonging to the Asteraceae family that is cultivated as a polyannual crop with vegetative propagation. Nevertheless, the length of the crop cycle, which negatively influences yields and the quality of the heads, has led artichoke growers to take an interest in the development of new seed-propagated cultivars for annual crops [1]. Italy represents one of the major producers, accounting for 33% of global production, followed by Spain, France, and Greece [2–5]. The edible part of the globe artichoke, commonly known as the "heart", is highly appreciated worldwide and it is consumed fresh, canned, or frozen [6]. However, from the harvesting phase to the processing one, a large amount of waste and by-products (80–85% of the biomass) are produced: roots, stems, bracts, and leaves, which have shown to be a rich source of antioxidant compounds and dietary fiber, mainly as pectin and inulin [6–8].

Artichoke roots are an important agricultural waste that, at the end of the harvesting period, remains unexploited in the field [1]. At the same time, their richness in inulin, characterized by a high degree of polymerization [9], makes them an alternative and sustainable source, considering that chicory roots, Jerusalem artichoke, and dahlia are commonly used together to produce commercial inulin [10–12]. This fiber is the most abundant reserve polysaccharide after starch, and its structure is characterized by a mixture of oligo and/or polysaccharides consisting of a variable number of d-fructose units bonded with a β-(2 → 1) linkages with a terminal glucose residue. The degree of polymerization can range from 2 to 60 units, which directly influences its physicochemical and nutritional properties and its application as a food ingredient [9,12–14]. Generally, short-chain inulin is used as an alternative low-calorie sweetener, with its good solubility contributing to improving the mouthfeel, while long-chain inulin mixed with water or aqueous solution forms a particle gel network that can act as a fat replacer or texture modifier due to its lower solubility and good viscosity stability [12,15].

From the nutritional point of view, the peculiar bond configuration β-(2 → 1) confers to inulin a prebiotic character [16]. Inulin, indeed, reaches the colon unaltered, where it is fermented by beneficial bacteria such as *Lactobacillus* and *Bifidobacterium*, determining the release of short-chain fatty acids in the gut and a pH lowering which, in turn, enhances the absorption of minerals (Ca^{2+} and Mg^{2+}), nutrients [4], and increases the functionality of colonocytes. Moreover, regular consumption of prebiotics has shown several health benefits like modulation of hyperglycemia, reduction of LDL cholesterol and serum lipids, prevention of colorectal cancer, and enhancement of immune system efficiency [17–19].

The consumer awareness about the relationship between health and food has led to the spread of functional products, i.e., those foods which contain bioactive compounds that, if taken in suitable quantities, have been shown to be able to prevent disease in addition to their nutritional functions [20,21]. In this regard, pasta seems to be particularly suitable for functional ingredients integration due to its easy preparation and its widespread consumption. Several studies have tried to improve the nutritional profile of pasta by adding soluble and insoluble fiber and evaluating their effect on quality properties [22–30].

However, the inclusion of dietary fiber could cause a deterioration of pasta quality in terms of cooking properties and sensory features due to the alteration of protein–starch network integrity, as observed by Foschia et al. [31]. Aravind et al. [32] found that the use of commercial inulin with a low degree of polymerization (DP) had a negative impact on firmness, cooking loss, and sensory acceptability of pasta; instead, the inulin with higher DP provided minimal impact. Similarly, Padalino et al. [33] added inulin at different DP and at two different concentrations (2 and 4%). In particular, the authors found interesting results in that the addition of inulin with a high DP, compared to the low one, had a greater disruptive effect on the starch–protein matrix. Peressini et al. [30] found an increase of firmness in pasta added with Barley Balance (BB), Psyllium seed husk (P) and BB–P, while a lower value than the control when inulin with high DP was used. Moreover, during the in vitro starch digestion, the pasta enriched with both inulin with high and low DP showed the highest reducing sugar release at 20 min compared to other fibers and control sample without fiber. Hence, Garbetta et al. [28] studied the effect of 4% addition of two different types of inulin, namely artichoke roots with high DP and chicory roots with low DP, observing good results in terms of sensory acceptability after the addition of inulin with high DP, suggesting the potential use of inulin-enriched spaghetti as a prebiotic food.

Therefore, an analysis of the literature revealed the influence of the DP of inulin on the technological and functional properties of foods, which needs to be further investigated based on the conflicting results in the literature.

In this framework, our study aimed to enhance the value of globe artichoke roots using them as an alternative source to extract inulin with a high degree of polymerization and their use as a functional ingredient for fresh pasta preparation. The effect of increasing the amount of the extracted inulin on structural, nutritional, and sensory properties of fresh pasta samples was evaluated.

2. Materials and Methods

2.1. Extraction and Characterzsation of Artichoke Root Inulin

2.1.1. Extraction Process

Inulin extraction from artichoke roots (AR) was carried out according to Castellino et al. [10], with slight modifications. Artichoke roots powder (ARP) was mixed with water at a pH of

6.8 with a ratio solid to water of 1:16 (*w/w*). The extraction took place in a thermostated bath at 80 °C for 2 h, with periodic stirring every 15 min. Afterwards, the sample was filtered with a Buchner funnel using Whatman™ (Darmstadt, Germany) filters with 11 µm of porosity, and then the filtrate was collected and submitted to precipitation phase through two cycles of freezing and thawing. The precipitate was centrifugated at 7500× *g* 15 min at 10 °C and the resulting pellet was washed by adding 10 mL of ethanol, centrifugated, dried overnight, and weighted.

Inulin yield was expressed in g per 100 g of ARP.

$$\text{Yield (\%)} = \left(\frac{\text{weight (g) of dried inulin}}{\text{weight (g) of ARP}}\right) \times 100$$

2.1.2. Moisture and Water Activity of AR Inulin Powder

Moisture was determined with a moisture analyzer (Mod. MAC 110/NP, Rodwang Wagi Elektroniczne, Radom, Poland) at 120 °C until constant weight. Water activity was measured using a hygrometer (Aqua Lab 100–240 V AC, Pullman, WA, USA). Each measurement was carried out in triplicate.

2.1.3. Identification and Quantification of AR Inulin

Identification and quantification of AR inulin was conducted via high-performance liquid chromatography (HPLC) analysis using a 1260 infinity series chromatograph (Agilent Technologies, Santa Clara, CA, USA) equipped with a refractive index detector (RID) and a cationic exchange column 300 × 7.8 mm (Rezex™ RCM-Monosaccharide Ca^{2+}, 8 µm, Torrance, CA, USA). The analysis was conducted isocratically using Milli-Q water as a mobile phase with a flow of 0.6 mL/min, column temperature of 80 °C, and RID of 35 °C. For calibration, commercial inulin with a high degree of polymerization and purity ≥98.5% (FibrulineTM XL chicory root fibre, COSUCRA Groupe Warcoing SA, Warcoing, Belgium) was used as standard. Standard inulin solutions were filtered through a 0.45 µm nylon filter and injected in triplicate at different concentrations (0.25 mg/mL, 0.5 mg/mL, and 1.0 mg/mL). AR inulin was properly diluted, filtered, and injected. A calibration curve was obtained for concentration versus area, and AR inulin was identified by coincidences of retention time with standard inulin and quantified through the corresponding calibration curve.

2.1.4. Polymerization Degree and Molecular Weight

Gel permeation chromatography was performed on extracted inulin. The analysis was carried out by using a GPC Max (Viscotek, Malvern, UK) system equipped with a TDA 305 detector (Refractive Index, Low Angle Light Scattering, Right Angle Light Scattering and Viscometer) and UV detector. We used a pre-column TSK PW_{XL} and TSK Gel GMPWXL (Tosoh Bioscience, Griesheim, Germany). The sample was dissolved at 3 mg/mL concentration and eluted in MilliQ water containing 0.2% NaN_3. After complete dissolution, the sample was filtered on a membrane having 0.22 µm porosity. The injection volume was 100 µL, and the flow rate was 0.8 mL/min. The chosen method of analysis was universal calibration, using polyethyleneoxide (PEO) standards, ranging from 21.300 kDa to 420 Da. The measurements, performed at 35 °C according to the temperatures of columns and detectors, were ran for 60 min in duplicate.

The polymerization degree (DP) was calculated as follows [34]:

$$M_n = 180 + 162 \times (DP_n - 1)$$

and

$$M_w = 180 + 162 \times (DP_w - 1)$$

where M_n and M_w are number average molecular weight and weight average molecular weight, respectively.

2.2. Fresh Pasta Preparation and Characterization

2.2.1. Experimental Design

Inulin-enriched fresh pasta was produced by substituting 5% (P5), 10% (P10), and 15% (P15) of durum wheat semolina (Mulino Martimucci, Altamura, Italy) with AR inulin powder. A control sample was produced with 100% durum wheat semolina. Durum wheat semolina and durum-wheat-semolina–inulin were mixed with an adequate amount of water and manually kneaded until obtaining a homogenous dough. The dough was left to rest for 30 min and then laminated (2 mm thickness) and cut with the iPasta (Imperia, Moncalieri, Italy) moulder machine to produce tagliatelle pasta. Afterwards, fresh tagliatelle were put on a wooden vessel, covered with a cotton cloth, and left to dry at room temperature until reaching a moisture content in a range of 26–28% and a_w values ranging from 0.92 and 0.97, according with Italian legal requirements for fresh pasta. Two different batches were produced.

2.2.2. Cooking Properties

Pasta samples were cooked in boiling distilled water at 1:10 (*w/v*) pasta-to-water ratio, without the addition of salt, as reported by Pasqualone et al. [35]. The optimum cooking time (OCT) was determined by removing tagliatelle from boiling water every 30 sec, cutting them, and checking for the white and opaque core to disappear, according to the AACC 16–50 official method [36]. After cooking, pasta samples were drained and rinsed with distilled water and allowed to rest for 5 min. Cooking loss was evaluated by combining cooking and rinsing water, measuring total volume, putting 20 mL in tarred Petri dishes, and evaporating in an air-oven at 105 °C until reaching constant weight. The residue, scaled up to total volume, was expressed as a percentage of the original pasta sample weight. After cooking at their OCT, the samples were analyzed for their water absorption and swelling index (grams of water per gram of dry pasta) according with Bustos et al. [37]. Each determination was carried out in triplicate.

$$\text{Water absorption } (\%) = \frac{\text{Wc} - \text{Wr}}{\text{Wr}} \times 100 \qquad \text{Swelling index} = \frac{\text{Wc} - \text{Wd}}{\text{Wd}}$$

W_r = weight of raw pasta (g)

W_c = weight of cooked pasta (g)

W_d = weight of dried pasta (g)

2.2.3. Inulin Loss in Cooking Water

P5, P10, and P15 were cooked at their OCT with a pasta-to-water ratio of 1:10 (*w/v*). Afterwards, 15 mL of cooking water was collected and analyzed for its inulin content by a cationic exchange HPLC following the same method described in Section 2.1.3. The inulin cooking water content was expressed as grams of inulin loss during cooking at OCT of 100 g of fresh pasta. The analysis was carried out in triplicate.

2.2.4. Color and Firmness Evaluation

Color of raw and cooked fresh pasta sample at its OCT (red index, corresponding to a*, yellow index, corresponding to b*, and luminosity index, corresponding to L*) was evaluated using a colorimeter CM-600d (Konica Minolta, Osaka, Japan) equipped with SpectraMagicX software. Six measurements were recorded for each pasta sample.

Firmness evaluation of raw and cooked pasta at its OCT was carried out using a Z1.0 TN (Zwick Roell, Ulm, Germany) equipped with a blade set with a guillotine and 50 N load cell. Pasta firmness was evaluated as the force required to cut 5 strands of pasta at a speed of 0.17 mm/s and was expressed as the maximum force (N) required to cut pasta strands. Eight measurements were recorded for each sample.

2.2.5. Microstructure Determination

The microstructure and the surface characteristics of the pasta were studied with a Zeiss Sigma 300 VP (Carl Zeiss NTS GmbH, Oberkochen, Germany) field-emission gun scanning electron microscope (FEG-SEM) equipped with a secondary electrons detector (SE). The analyses were done under vacuum (<10^{-4} Pa), using an accelerating voltage of 20 kV, an aperture of 30 μA, and a working distance between 4 and 5 mm and magnification of 1000×. Fragments of the different pasta samples were glued onto an aluminium stub with carbon tape. Before the analysis, all the samples were carbon-coated in order to make the surface of the specimen conductive.

2.2.6. Quantitative Descriptive Sensory Analysis

Sensory analyses were conducted by a panel of eight trained tasters (6 females and 2 males) to evaluate the sensory attributes according to Pasqualone et al. [35]. Pasta samples were coded by a three-digit number, cooked at their OCT, and evaluated for color, smell, taste, bulkiness, adhesiveness, hardness, and overall acceptability, using a structured scale ranging from 1 to 10. Color refers to the typical color of durum wheat pasta, where 1 = yellow color and 10 = brown; smell as perceived by olfaction and taste as perceived during mastication, which refer to the typical odor and taste of durum wheat pasta without anomalies, where 1 = very low and 10 = very high; bulkiness refers to adhesion of tagliatelle strands to each other, evaluated both visually and manually by pressing two tagliatelle together and determining the force required for detachment; stickiness, related to the organic matter released during cooking and still adhering to the surface of pasta, evaluated by pressing a single strand against the palate and determining the force require to remove it with the tongue; hardness, which is the resistance of cooked pasta to chewing measured while cutting the spaghetti strand using the front teeth. The analysis was carried out according to Pasqualone et al. [38], following the ethical guidelines of the Laboratory of Food Science and Technology of the Department of Plant, Soil, Food Science of University of Bari, Italy.

2.3. Functional Properties Determination

2.3.1. Proximate Composition of Fresh Pasta

Protein (total nitrogen × 5.7), ash, and lipids content were determined using the AOAC method 979.09, 923.03, and 945.38 F, respectively [39]. Total dietary fiber content was determined by enzymatic gravimetric as described by the AOAC Official Method 991.43. Moisture content was determined by a moisture analyzer (Mod. MAC 110/NP, Radwag Wagi Elektroniczne, Radom, Poland) at 120 °C. The carbohydrate content was determined as difference. The determinations were carried out in triplicate.

2.3.2. In Vitro Starch Hydrolysis

In vitro gastrointestinal digestion of fresh pasta and starch hydrolysis was determined according to Liljeberg et al. [40], simulating the in vivo digestion of starch. Briefly, aliquots of fresh pasta samples, cooked until the optimal cooking time and containing 1 g of starch (determined in cooked fresh pasta), were subjected to an enzymatic process (pancreatic amylase and pepsin-HCl), and the released glucose content was measured with D-fructose/D-glucose Assay Kit (Megazyme, Wicklow, Ireland). Simulated digests were dialyzed (cut-off of the membrane: 12,400 Da) for 180 min. Aliquots of dialysate, containing free glucose, and partially hydrolyzed starch were sampled every 30 min and further treated with amyloglucosidase. Then, free glucose was determined using the above-mentioned enzyme-based kit and finally converted into hydrolyzed (digested) starch in pasta. Control white wheat bread was used as the control to estimate the hydrolysis index (HI = 100). The predicted glycemic index (pGI) was calculated using the equation pGI = 0.549 × HI + 39.71 [41]. Each sample was analyzed in triplicate.

2.3.3. Prebiotic Activity Assay

To evaluate the prebiotic activity of AR inulin-enriched fresh pasta, samples were previously subjected to in vitro gastrointestinal digestion according to the method used by Kamiloglu and Capanoglu [42], with slight modifications. The in vitro gastrointestinal digestion was performed, comprising of a pepsin-HCl digestion for 3 h at 37 °C (to simulate gastric digestion) and pancreatin digestion with pancreatin and bile salts for 3 h at 37 °C (to simulate small intestinal digestion). As reported by Caponio et al. [43], 10 mL of each fresh pasta extract was added to α-amylase (56 mg/mL) (Sigma-Aldrich Chemistry, St. Louis, MO, USA) and to 10 mL of pepsin solution composed of NaCl 125 mM/L + KCl 7 mM/L + $NaHCO_3$ 45 mM/L + pepsin (Sigma-Aldrich Chemistry, St. Louis, MO, USA) at 3 g/L. Then, the pH was adjusted to 2 using HCl and incubated at 37 °C for 180 min in a water bath under shaking. After incubation, an aliquot of the gastric-digested extract was added in equal volume to an intestinal solution. The intestinal solution was simulated by dissolving 0.1 g/100 mL of pancreatin (Sigma-Aldrich Chemistry, St. Louis, MO, USA) and 0.15 g/100 mL bile salts (Oxoid™, Hampshire, UK). The pH was adjusted to 8 using NaOH and incubated at 37 °C for 180 min in a water bath under shaking. After incubation, an aliquot of intestinal-digested extract was ultra-filtrated with 3000 Da membrane (Vivaspin 20, Sartorius, Goettingen, Germany) to eliminate free carbohydrates. The retentate fractions were diluted in water and then filtered using 0.45 μm Whatman filter paper and further analyzed for prebiotics activities, as follows.

Twenty-two probiotic strains of probiotics and one strain of *Escherichia coli* (*E. coli*) available in the culture collection of the Department of Plant, Soil, Food Science of University of Bari, Italy were used to carry out the experiments in fecal batches. The fecal medium (FM) was constituted as previously described [44] without the addition of glucose. This was labelled as FM (absence of carbohydrates), FMPC (FM + pasta not containing inulin), FMP5 (FM + pasta with 5% of inulin), FMP10 (FM + pasta with 10% of inulin), FMP15 (FM + pasta with 15% of inulin). For those fecal samples containing pasta, this was added in a ratio of 1:5 (*w/v*) in media after cooking and digestion was simulated as previously described. Viable probiotics and *E. coli* were inoculated in fecal media at a cell density of 7 UFC/mL (log_{10}), measured through OD at 620 nm. Inoculated batches were incubated in anaerobic conditions for 36 h at 37 °C, under slight stirring (150 rpm). After the incubation, plate counts for lactic acid bacteria and *E. coli* were respectively made in De Man, Rogosa, and Sharpe agar (MRS) and Violet Red Bile Glucose agar (VRBGA). Both agar media were purchased from Oxoid Ltd. (Basingstoke, Hampshire, England, UK). Probiotic growth was also profiled in terms of ΔpH, as the difference between final (36 h) and initial (pH 7.0 ± 0.02) values of pH.

2.4. *Statistical Analysis*

The experimental data were subjected to one-way and two-way ANOVA, followed by a Tukey's HSD test. The two-way ANOVA analysis was carried out considering the rate of substitution and the physical state of pasta (raw and cooked) as factors. Significant differences among the values of all the parameters were determined at $p < 0.05$ by the Minitab 17 Statistical Software (Minitab, Inc., State College, PA, USA, 2010).

3. Results and Discussion

3.1. *Characteristic of AR Inulin Powder*

Inulin extraction using hot water is considered the conventional extraction technique and the most-used [45]. Several factors can influence the extraction yield: temperature, time of extraction, solid/liquid ratio [46]. From the data collected in our study, the extraction yield, as the mean of ten measurements, was 23.37% ± 1.55, with a purity level of 89%, estimated in comparison to commercial inulin used as standard, which had a purity of

98.5%. The degree of polymerization (DP) is a fundamental parameter to characterize inulin, being directly correlated to its technological and nutritional properties. AR inulin has shown a DP_n and DP_w equal to 45 and 60, respectively, much higher than inulin extracted from Jerusalem artichoke (5–19), agave (5–13), and dahlia (17–23) [47–49]. This result highlights the possible use of AR inulin as a functional ingredient, since higher DP is associated with the improvement of technological properties in food products [10]. Among the factors that could affect the DP are plant, period of harvesting, storage period, and extraction process [50]. Moreover, the AR inulin powder has shown a moisture content equal to 6% and water activity of 0.40 ± 0.0, values which assure a high glass transition temperature, lower cohesiveness and, consequently, higher physical and microbiological stability [46,51].

Figure 1 shows the gel permeation chromatography (GPC) profile of AR inulin. Two different peaks were observed, the first peak in the range of 15–19 mL, relative to polysaccharides elution, and a second very sharp peak centred at 20.5 mL, presumably due to the elution of very short polysaccharide chains, monomers, and impurities present in the sample. The ratio between the two peaks was 75/25.

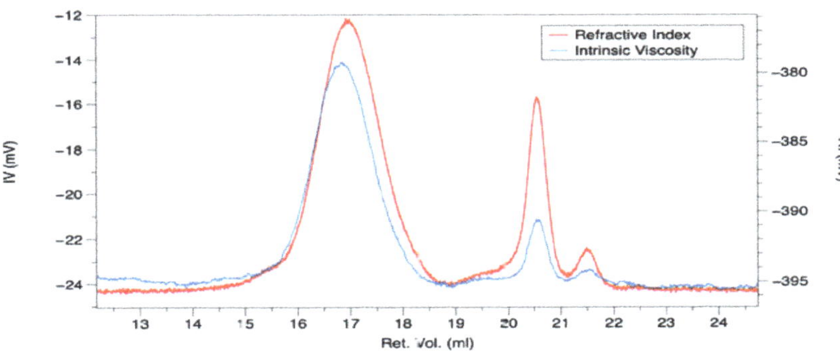

Figure 1. Gel permeation chromatography (GPC) profile of artichoke roots inulin, considering the refractive index (red line) and the intrinsic viscosity (blue line).

3.2. *Quality Characteristics of Fresh Pasta*

3.2.1. Cooking Properties of Fresh Pasta

Pasta cooking properties are of great importanceforo ensuring acceptability by consumers. These properties were evaluated by considering parameters such as optimal cooking time (OCT), swelling index (SI), water absorption index (WAI), and solid cooking loss (CL) (Table 1). OCT was set for each pasta sample, observing a slight increase in pasta enriched with 10% and 15% of inulin. The same trend was observed by Foschia et al. [31] in pasta fortified by inulin with high DP and Simonato et al. [52] after the addition of *Moringa oleifera* L. leaves powder. However, the aforementioned results disagreed with other studies [32,33,53,54], where the addition of inulin caused a significant reduction of OCT due to the disruption of gluten network, which, in turn, caused an easier water penetration into starch granules. The discordant results could be attributed to the different production processes, as well as to the added fiber and its intrinsic properties, which influence the interaction with other ingredients.

Regarding the WAI, no significant differences were found among the samples, whereas P10 and P15 showed a significant lower SI than control (PC). In accordance with our results, Naji-Tabasi et al. [55] and Attanzio et al [56] noted a significant reduction of SI in pasta samples after the addition of wheat bran, mucilaginous seeds flour and *Opuntia* cladodes extract. Moreover, Renoldi et al. [57] stated that adding Psyllium fiber led to a lower peak value of storage modulus (G′) of dough, revealing differences in starch swelling of starch granules. The explanation for these results lies in the encapsulation of starch granules into fiber reticule, whose hydroxyl groups compete with starch and protein for water absorption, thus limiting the penetration of water into the starch granules and their consequent swelling [57–59]. On the contrary, the addition of whole barley flour to pasta formulation led to an increase of SI, which was related to the presence of β-glucans and their ability to absorb water [60]. Hence, different fibers could have a different effect on starch hydration and swelling.

Table 1. Cooking properties of fresh pasta.

Sample	OCT (min)	WAI (g/100 g)	SI	CL (g/100 g)	IL (g/100 g)
PC	6.30	73.20 ± 0.01 a	1.56 ± 0.03 a	2.37 ± 0.05 d	-
P5	6.30	73.24 ± 0.10 a	1.53 ± 0.04 ab	2.70 ± 0.14 c	1.01 ± 0.03 c
P10	6.45	72.51 ± 0.62 a	1.48 ± 0.03 b	3.11 ± 0.07 b	1.45 ± 0.07 b
P15	6.45	73.24 ± 0.20 a	1.47 ± 0.04 b	3.62 ± 0.06 a	2.19 ± 0.12 a

PC, control pasta without inulin addition; P5, pasta with 5% of inulin added; P10, pasta with 10% of inulin added; P15, pasta with 15% of inulin added. OCT, optimal cooking time; WAI, water absorption index, SI, swelling index; CL, cooking losses; IL, inulin losses. The values represent means of triplicates ± standard deviation; different letters in the same column mean significant statistical differences ($p < 0.05$) to one-way ANOVA followed by Tukey's HSD test.

CL refers to pasta resistance during cooking, and it is strictly connected with the strength of the protein network [61]. Good quality pasta should not have a CL higher than 7–8% [55]. In our study, CL ranged between 2.37% for the control sample (PC) and 3.62% for the P15. However, considering that inulin water solubility increases with high temperature [62], we also estimated the contribution of inulin to the observed CL. Specifically, inulin-imputable CL accounted for 1.01 ± 0.03 g/100 g of pasta in P5, 1.45 ± 0.07 g/100 g of pasta in P10, and 2.19 ± 0.12 g/100 g of pasta in P15. Consequently, making a difference between the CL values of P5, P10, and P15 and the corresponding inulin content found in cooking water, we could assert that in inulin-enriched samples, the solids different from inulin that leach into cooking water are lower than PC. Therefore, the probable formation of fibrous reticulum around starch granules [38] and additional hydrogen bonds and hydrophobic interactions between inulin and glutenin protein could provide extra support to the protein network, reducing the solid loss [63]. However, opposite results were found in several studies in the literature, in which the disruption of protein matrix due to fibers, mainly insoluble, promotes and allows the leaching of starch during cooking, causing an increase in cooking loss value [64–66].

3.2.2. Color and Firmness of Fresh Pasta

Color, together with cooking and textural properties, is another key parameter for consumer acceptability: yellow color and high luminosity are associated with high-quality pasta [67]. Table 2 shows the color parameters in raw and cooked pasta samples. According to two-way ANOVA, both the cooking process and the addition of increasing percentages of AR inulin significantly affected the colorimetric parameters. Specifically, both raw and cooked pasta samples with higher rates of AR inulin substitution of durum wheat semolina caused a reduction of luminosity (L*) and yellow index (b*). Regarding cooking, the red index (a*) followed the same trend of L* and b*, whereas the opposite trend was observed for the increasing rate of substitution of durum wheat semolina with AR inulin. After the addition of insoluble fibers, Aravind et al. [68] found the same trend in terms of luminosity (L*), yellow, and red index, while Zarroug et al. [61] and Filipović et al. [21] reported an increase of L* values and a decrease of a* values for pasta enriched with commercial inulin, probably due to the added white color of the commercial inulin.

Table 2. Color and firmness of fresh pasta.

		L*	a*	b*	Firmness
Raw pasta	PC	79.65 ± 0.09 a	2.21 ± 0.07 e	33.57 ± 0.46 a	18.25 ± 0.45 b
	P5	74.57 ± 0.19 c	3.21 ± 0.02 c	30.49 ± 0.26 b	19.63 ± 0.50 a
	P10	72.02 ± 0.38 e	3.89 ± 0.10 b	28.52 ± 0.19 c	19.89 ± 0.05 a
	P15	70.72 ± 0.25 f	4.48 ± 0.08 a	27.62 ± 0.09 cd	19.77 ± 0.23 a
Cooked pasta	PC	78.89 ± 0.43 b	0.30 ± 0.02 f	27.11 ± 0.56 d	5.85 ± 0.36 d
	P5	73.69 ± 0.18 d	2.00 ± 0.11 e	24.13 ± 0.41 e	6.22 ± 0.26 d
	P10	72.24 ± 0.06 e	2.82 ± 0.11 d	24.18 ± 0.33 e	7.37 ± 0.48 c
	P15	68.12 ± 012 g	3.79 ± 0.10 b	22.29 ± 0.43 f	7.25 ± 0.37 c
p-value	P *C	<0.0001	<0.0001	<0.0001	<0.05

PC, control pasta without inulin addition; P5, pasta with 5% inulin added; P10, pasta with 10% inulin added; P15, pasta with 15%= inulin added. P, percentage of inulin addition; C, cooking process. Values are expressed as mean of ± standard deviation; different letters in the same column mean significant statistical differences ($p < 0.05$) according to two-way ANOVA.

Firmness can be evaluated as the force necessary to cut the pasta strains, and it is strictly connected with the protein matrix development during pasta production and the hydration level of starch granules [64,67]. Cooked P10 and P15 showed a significantly higher firmness (7.37 ± 0.48 and 7.25 ± 0.37 N) than PC and P5 (5.85 ± 0.36 and 6.22 ± 0.26 N), which did not show any significant differences between them (Table 2). According to Chillo et al. [69] the mechanical properties of conventional and unconventional pasta are strictly connected with cooking properties. Therefore, the increase in firmness in cooked P10 and P15 can be reasonably linked to the lower values of the swelling index, supporting the hypothesis of pasta structure rearrangement as discussed in Section 3.2.1 [54]. Moreover, beta-glucan addition by Aravind et al. [70] and Peressini et al. [30] found a firmer pasta than control; however, significantly lower values of firmness were found by Peressini et al. [30] with the addition of 15% of inulin with a high degree of polymerization (DP = 23).

3.2.3. Sensory Evaluation of Fresh Pasta

The results of the sensory evaluation are reported in Figure 2. The addition of larger amounts of AR inulin in fresh pasta formulations did not cause significant differences in sensory properties after cooking. The sensory scores for color and firmness have been consistent with instrumental evaluation, with all inulin-enriched pasta samples perceived as browner and P10 and P15 samples slightly firmer than PC. Although brightness and yellowness are linked to high-quality pasta, over the years, consumer attitude has been changing such that darker color is considered a positive trait, as it is associated with high-fiber products [25]. Significant differences were found in the taste of inulin-enriched samples than control, while only P10 odor resulted in a decreased perception rating compared to the typical odor of durum wheat pasta. In conclusion, P10 and P15 were statistically similar to P5; indeed, except for color and odor, no significant differences were highlighted for the other parameters considered. Therefore, it may be possible to add higher amounts of inulin without compromising the sensory properties of pasta.

3.2.4. Microstructure

Figure 3 reports the SEM micrographs of cross sections of raw pasta (PC, P5, P10, and P15). Starch granules were well visible in all the samples, and the protein matrix did not appear disrupted. However, in P10 and P15, starch granules seem to be immerged in a more dense and compact structure. Moreover, P10 and P15 exhibit the presence of a reticulated structure, probably formed from the interaction of inulin with the protein network, which could have strengthened the pasta structure and thus giving a higher firmness to the cooked pasta as observed by the instrumental analysis.

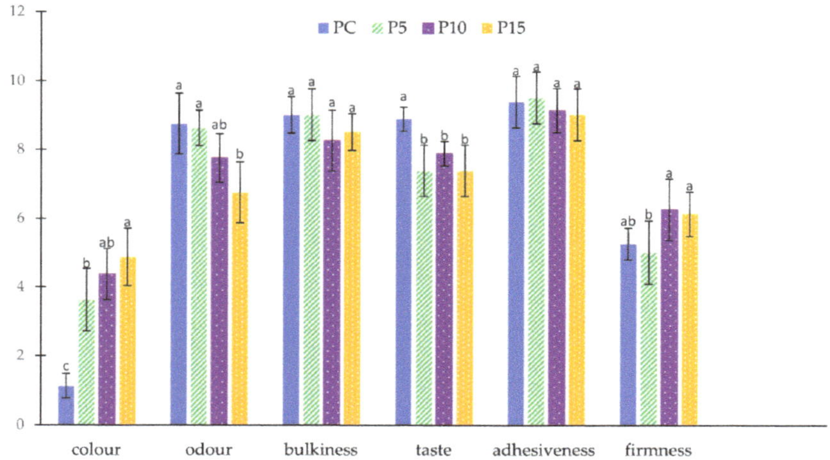

Figure 2. Sensory characteristics of control (PC) and inulin-enriched (P5, P10, P15) fresh pasta. Values are expressed as mean ± standard deviation; different letters for each parameters mean significant statistical differences ($p < 0.05$) to one-way ANOVA followed by Tukey's HSD test.

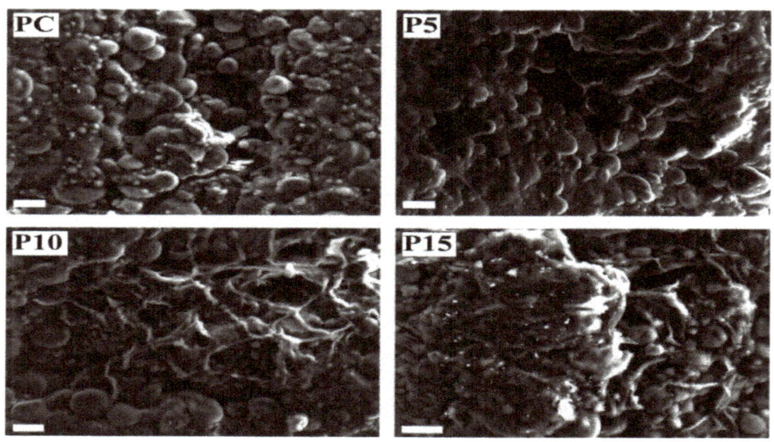

Figure 3. SE-SEM micrographs of control (PC) and inulin enriched (P5, P10, P15) fresh pasta (white bar = 20 μm).

3.3. Functional Properties of Fresh Pasta

3.3.1. Proximate Composition of Fresh Pasta

Table 3 reported the proximate composition of fresh pasta. Significant differences were found among the samples for all the parameters considered. Specifically, inulin-enriched pasta showed a decline in protein content compared to the control (PC) due to a rise in total dietary fiber, which reached values of 3.44 g/100 g in P5, 8.16 g/100 g in P10, and 12.41 g/100 g in P15, exceeding the data observed by Padalino et al. [71] in pasta fortified with tomato byproducts. Therefore, the results in terms of total dietary fiber allow for labelling P5 as a "source of fibre", while P10 and P15 could be labelled as a pasta having "high fibre content" [72], according to Reg. (EU) 1924/2006, enhancing the nutritional value of fresh pasta. Moreover, higher ash and lower lipid content were found in P10 and P15 than P5 and PC.

Table 3. Proximate composition of fresh pasta (g/100 g).

Parameters	PC	P5	P10	P15
Ash	0.41 ± 0.02 b	0.41 ± 0.01 b	0.49 ± 0.01 a	0.50 ± 0.01 a
Protein	10.75 ± 0.03 a	9.71 ± 0.44 b	9.25 ± 0.02 b	9.27 ± 0.08 b
Total dietary fiber	1.47 ± 0.04 d	3.44 ± 0.10 c	8.16 ± 0.12 b	12.69 ± 0.08 a
Lipid	0.23 ± 0.02 a	0.10 ± 0.01 b	0.06 ± 0.01 c	0.08 ± 0.01 c
Moisture	28.21 ± 0.17 a	27.00 ± 0.21 b	26.98 ± 0.10 b	28.24 ± 0.30 a
Carbohydrates	58.94 ± 0.22 a	59.34 ± 0.25 a	55.07 ± 0.02 b	49.20 ± 0.42 c

PC, control pasta without inulin addition; P5, pasta with 5% of inulin added; P10, pasta with 10% of inulin added; P15, pasta with 15% of inulin added. The values represent means of triplicates ± standard deviation; different letters indicate significant differences; different letters in the same column mean significant statistical differences ($p < 0.05$) to one way ANOVA followed by Tukey's HSD test.

3.3.2. In Vitro Starch Hydrolysis

In consideration the pivotal role of nutritional aspects for modern consumers and in recent scientific research aimed at producing foods with a lower glycemic index [73,74], cooked fresh pasta samples were analyzed for in vitro starch hydrolysis. As shown in Figure 4, the addition of AR inulin in fresh pasta promoted the decrease of HI and pGI. On average, the HI and pGI values of fresh pasta samples enriched with AR inulin (P5, P10, and P15) were statistically lower than the control (PC). Specifically, P5 reached a significantly ($p < 0.05$) lower value of pGI compared to the control (PC): 63.01 and 66.54, respectively. However, a higher concentration of AR inulin resulted in a further decrease of HI and pGI, as shown in P10 and P15. In fact, P10 has a statistically lower pGI value (58.42) compared to both the control (PC) and P5. Nevertheless, no significant differences were observed for P10 and P15 samples in terms of pGI and HI. Data demonstrate a lower pGI response following ingestion of AR inulin-enriched fresh pasta compared to the control.

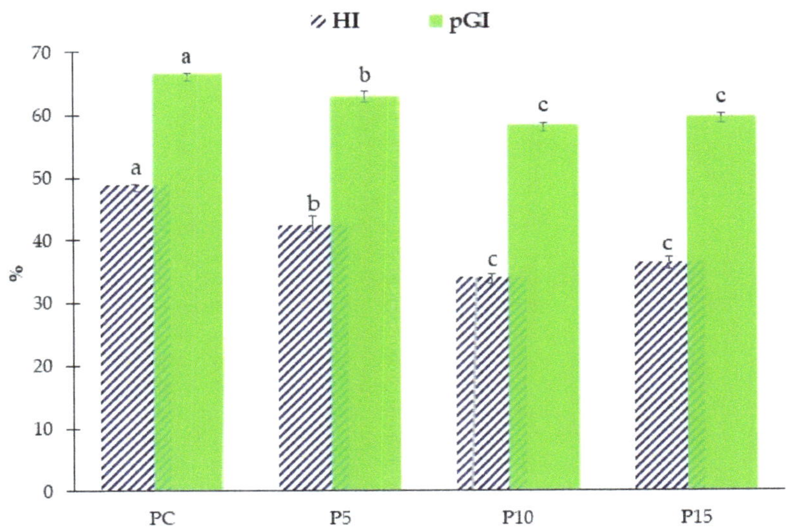

Figure 4. Results of hydrolysis index (HI) and predicted glycemic index (pGI) of control (PC) and inulin-enriched (P5, P10, P15) fresh pasta. The values represent means of triplicates ± standard deviation; different letters indicate significant differences ($p < 0.05$) according to one-way ANOVA followed by Tukey's HSD test.

Inulin, which is not digestible by humans, has interesting properties as a source of fermentable energy for some intestinal bacteria that produce short-chain fatty acids, which are essential for maintaining the intestinal homeostasis. In addition to stimulating digestion and regularity of intestinal transit, inulin may favor the presence of *Bifidobacterium* in the microbiota and, at the same time, decrease harmful bacteria [75,76]. Several studies have highlighted the beneficial role of inulin also in glycemic response [41,77–79]. Inulin—as a soluble fiber—helps keep blood sugar under control, since the fibers present in complex carbohydrates take longer to release the sugars present in the body: the slow release of glucose prevents glycemic peaks both upwards and downwards by balancing the energy intake and limiting the accumulation of fat due to the excess insulin [31,80]. The gelling effect of these fibers causes the formation of a film on the walls of the stomach and intestines with consequent lower absorption of fats and sugars [81]. In contrast to a previous study on chicory inulin-enriched pasta (from 2.5% to 10%) that did not reduce the pGI compared to the control [28], our results confirmed that AR was able to reduce the pGI and HI already at 5% of its concentration, due to the high DP of inulin used in our pasta samples.

3.3.3. Evaluation of Effects Exerted by AR Inulin, Prebiotics, and Pathogen on Probiotic Growth

Herein, the prebiotic activity of inulin-enriched pasta was assessed in vitro in terms of probiotics growth. Furthermore, the inhibition of *E. coli* when co-cultured with probiotics was also determined. Compared to batches not containing carbohydrates (FM), the addition of pasta not containing inulin (FMPC) was sufficient to increase (~0.5 $\log_{10}$ CFU/mL) the cell density of all tested probiotics (Figure 5). This reflects the prebiotic contribution of fructans and arabinoxylans, which are non-digestible oligosaccharides naturally occurring in wheats [82], since their absence in gluten-free diets seems to affect the host microbiome and metabolome [83,84]. However, the presence of 3 g/L of inulin in FMP15 was able to further increase (>0.5 cycle) the cell density by 50% compared to the used probiotics (11 out of 22), while six strains (~27%) showed an increased cell density higher than one cycle. Besides the growth of probiotics, the acidification degree in batches followed the inulin concentration in a dependent manner. Values of ΔpH were, on average, 0.07 ± 0.06, 0.14 ± 0.11, and 0.20 ± 0.17 for FMP5, FMP10, and FMP15, respectively. No significant differences were found comparing the cell densities of *E. coli* in FMP to those of FMP5 (Figure 5). Oppositely, more than 50% (12 out of 22) of used probiotics were able to significantly decrease the cell density of *E. coli* in batches containing 3 g/L of inulin (FMP15). Meanwhile, ~36% of probiotics (eight strains) significantly decreased the *E. coli* cell density in batches containing 2 g/L of inulin (FMP10).

These results are in line with those previously stated by Kareem and co-workers [85], who reported that the combination of probiotics with prebiotics in vitro exhibited a great inhibition of pathogens due to a synergistic effect. Mechanisms based on microbial cross-feeding are largely distributed within the intestinal lumen [86]. A previous study concerning β-glucans-enriched pasta determined that 3 g of daily fiber supplementation was optimal to increase the saccharolytic metabolism in terms of SCFA profiling and improving the endothelial reactivity in healthy volunteers [87]. Therefore, although some used strains did not directly decrease *E. coli* growth, evidence suggests that additional taxa belonging to the human gut microbiota (e.g., clostridial cluster IV, XIVa, and *Bifidobacterium*) can support the metabolism of inulin in SCFA in vivo, eliciting a boosted effect that contributes to the host's intestinal homeostasis [88].

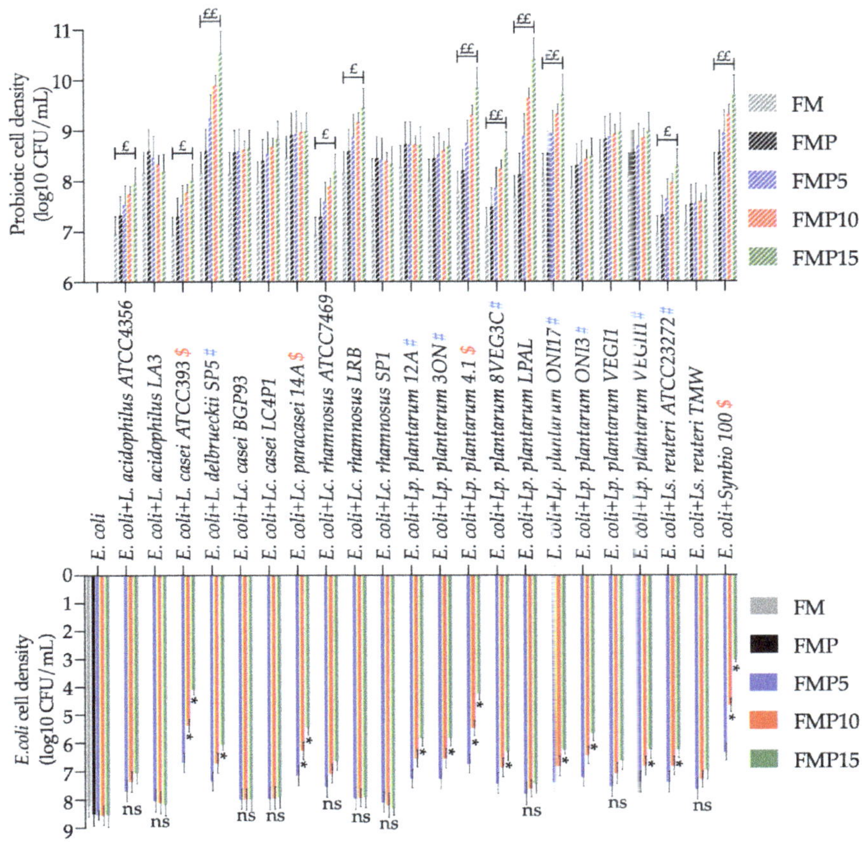

Figure 5. In vitro prebiotic assay showing the cell density of 22 probiotic lactic acid bacteria co-cultured with *E. coli* in fecal batches not containing carbohydrates (FM) or made with the addition of pasta without inulin (FMPC) and inulin-enriched pasta at 5 (FMP5), 10 (FMP10), and 15% (FMP15). "£" means >0.5 log10 cycle increased cell density (CFU/mL) of probiotic lactic acid bacte-ria in FMP15 versus FMP; "££" means >1 log10 cycle increased cell density (CFU/mL) of probiotic lactic acid bacteria in FMP15 versus FMP. "#" and "$" indicate those probiotics that had respective-ly determined a significant decrease of *E. coli* cell density in FMP15 or both FMP15 and FMP10. "*": significant decrease of *E. coli* cell density compared to FMP; "ns": no significant differences.

4. Conclusions

The obtained results show that inulin from globe artichoke represents a promising functional ingredient in terms of technological and nutritional properties. As a matter of fact, adding inulin determined changes in the structure of raw and cooked pasta, which results more compact and firmer than control. The addition of inulin resulted in increasing the optimal cooking time, reducing the swelling index and increasing cooking loss (the latter due to inulin leaching in the cooking water). In terms of color, inulin-enriched samples (P5, P10, and P15) showed lower L* and b* and higher a* than control. The sensory properties did not substantially change compared to control. From the nutritional perspective, the ability of inulin to slow down the release of glucose in the blood was assessed, showing that a significant reduction of HI and pGI occurred in inulin-enriched pasta compared to control. Moreover, the in vitro prebiotic assay demonstrated that FMP10 and FMP15, containing 2 g/L and 3 g/L of inulin, respectively, significantly increased the cell density of prebiotics able to inhibit the *E. coli* growth. Therefore, the promising results obtained highlight the possibility to upcycle, in a circular perspective, the artichoke roots from agricultural waste to a valuable food ingredient that is useful for preparing high-added-value food products that could satisfy the increasing demand of consumers for more sustainable and functional foods.

Author Contributions: Conceptualization, G.D. and A.P.; methodology, G.D., G.d.G., G.R.C., M.V., B.I. and A.P.; formal analysis, G.d.G., G.R.C., M.V., I.A. and G.d.P.; writing—original draft preparation, G.D., G.d.G., G.R.C., M.V. and B.I.; writing—review and editing, G.D., G.d.G., G.R.C. and A.P. All authors have read and agreed to the published version of the manuscript.

Funding: This research was funded by Ministero dell'Istruzione, dell'Università e della Ricerca-Programmi di Ricerca 2017 (2017JTNK78) "GOOD-BY-WASTE. Obtain GOOD products-exploit BY-products-reduce WASTE" (CUP H98D19000940006).

Institutional Review Board Statement: Ethical review and approval were waived for this study because data were completely anonymous, with no personal information being collected, and were not considered to be sensitive in nature.

Informed Consent Statement: Informed consent was obtained from all subjects involved in the study.

Data Availability Statement: The data will be available on request.

Acknowledgments: This work was supported by POR Puglia 2014/2020-Asse X-Azione 10.4 Research for Innovation- REFIN code n. E65BAEEE.

Conflicts of Interest: The authors declare no conflict of interest.

References

1. Francavilla, M.; Marone, M.; Marasco, P.; Contillo, F.; Monteleone, M. Artichoke Biorefinery: From Food to Advanced Technological Applications. *Foods* **2021**, *10*, 112. [CrossRef]
2. Domingo, C.S.; Rojas, A.M.; Fissore, E.N.; Gerschenson, L.N. Rheological behavior of soluble dietary fiber fractions isolated from artichoke residues. *Eur. Food Res. Technol.* **2019**, *245*, 1239–1249. [CrossRef]
3. Villanueva-Suárez, M.J.; Mateos-Aparicio, I.; Pérez-Cózar, M.L.; Yokoyama, W.; Redondo-Cuenca, A. Hypolipidemic effects of dietary fibre from an artichoke by-product in Syrian hamsters. *J. Funct. Foods* **2019**, *56*, 156–162. [CrossRef]
4. Zeaiter, Z.; Regonesi, M.E.; Cavini, S.; Labra, M.; Sello, G.; Di Gennaro, P. Extraction and Characterization of Inulin-Type Fructans from Artichoke Wastes and Their Effect on the Growth of Intestinal Bacteria Associated with Health. *Biomed Res. Int.* **2019**, *2019*, 1083952. [CrossRef]
5. Lattanzio, V.; Kroon, P.A.; Linsalata, V.; Cardinali, A. Globe artichoke: A functional food and source of nutraceutical ingredients. *J. Funct. Foods* **2009**, *1*, 131–144. [CrossRef]
6. Borsini, A.A.; Llavata, B.; Umaña, M.; Cárcel, J.A. Artichoke by Products as a Source of Antioxidant and Fiber: How It Can Be Affected by Drying Temperature. *Foods* **2021**, *10*, 459. [CrossRef]
7. Pasqualone, A.; Punzi, R.; Trani, A.; Summo, C.; Paradiso, V.M.; Caponio, F.; Gambacorta, G. Enrichment of fresh pasta with antioxidant extracts obtained from artichoke canning by-products by ultrasound-assisted technology and quality characterisation of the end product. *Int. J. Food Sci. Technol.* **2017**, *52*, 2078–2087. [CrossRef]
8. Holgado, F.; Campos-Monfort, G.; de las Heras, C.; Rupérez, P. In vitro fermentability of globe artichoke by-product by *Lactobacillus acidophilus* and *Bifidobacterium bifidum*. *Bioact. Carbohydr. Diet. Fibre* **2021**, *26*, 100286. [CrossRef]
9. Raccuia, S.A.; Melilli, M.G. Seasonal dynamics of biomass, inulin, and water-soluble sugars in roots of *Cynara cardunculus* L. *Field Crop. Res.* **2010**, *116*, 147–153. [CrossRef]
10. Castellino, M.; Renna, M.; Leoni, B.; Calasso, M.; Difonzo, G.; Santamaria, P.; Gambacorta, G.; Caponio, F.; De Angelis, M.; Paradiso, V.M. Conventional and unconventional recovery of inulin rich extracts for food use from the roots of globe artichoke. *Food Hydrocoll.* **2020**, *107*, 105975. [CrossRef]
11. Redondo-Cuenca, A.; Herrera-Vázquez, S.E.; Condezo-Hoyos, L.; Gómez-Ordóñez, E.; Rupérez, P. Inulin extraction from common inulin-containing plant sources. *Ind. Crops Prod.* **2021**, *170*, 113726. [CrossRef]
12. López-Molina, D.; Navarro-Martínez, M.D.; Rojas-Melgarejo, F.; Hiner, A.N.P.; Chazarra, S.; Rodríguez-López, J.N. Molecular properties and prebiotic effect of inulin obtained from artichoke (*Cynara scolymus* L.). *Phytochemistry* **2005**, *66*, 1476–1484. [CrossRef]
13. Lopes, S.M.S.; Krausová, G.; Rada, V.; Gonçalves, J.E.; Gonçalves, R.A.C.; de Oliveira, A.J.B. Isolation and characterization of inulin with a high degree of polymerization from roots of *Stevia rebaudiana* (Bert.) Bertoni. *Carbohydr. Res.* **2015**, *411*, 15–21. [CrossRef]

14. Gominho, J.; Curt, M.D.; Lourenço, A.; Fernández, J.; Pereira, H. *Cynara cardunculus* L. as a biomass and multi-purpose crop: A review of 30 years of research. *Biomass Bioenergy* **2018**, *109*, 257–275. [CrossRef]
15. Giarnetti, M.; Paradiso, V.M.; Caponio, F.; Summo, C.; Pasqualone, A. Fat replacement in shortbread cookies using an emulsion filled gel based on inulin and extra virgin olive oil. *LWT Food Sci. Technol.* **2015**, *63*, 339–345. [CrossRef]
16. Puchkova, T.S.; Pikhalo, D.M.; Karasyova, O.M. About the universal technology of processing *Jerusalem artichoke* and chicory for inulin. *Food Syst.* **2019**, *2*, 36–43. [CrossRef]
17. El-Kholy, W.M.; Aamer, R.A.; Ali, A.N.A. Utilization of inulin extracted from chicory (*Cichorium intybus* L.) roots to improve the properties of low-fat synbiotic yoghurt. *Ann. Agric. Sci.* **2020**, *65*, 59–67. [CrossRef]
18. Mitchell, C.M.; Davy, B.M.; Halliday, T.M.; Hulver, M.W.; Neilson, A.P.; Ponder, M.A.; Davy, K.P. The effect of prebiotic supplementation with inulin on cardiometabolic health: Rationale, design, and methods of a controlled feeding efficacy trial in adults at risk of type 2 diabetes. *Contemp. Clin. Trials* **2015**, *45*, 328–337. [CrossRef]
19. Caponio, G.R.; Lippolis, T.; Tutino, V.; Gigante, I.; De Nunzio, V.; Milella, R.A.; Gasparro, M.; Notarnicola, M. Nutraceuticals: Focus on Anti-Inflammatory, Anti-Cancer, Antioxidant Properties in Gastrointestinal Tract. *Antioxidants* **2022**, *11*, 1274. [CrossRef]
20. Ivkov, M.; Kosutic, M.; Filipovic, J.; Filipovic, V. Spelt pasta with addition of inulin as a functional food: Sensory evaluation and consumer attitudes. *Rom. Biotechnol. Lett.* **2018**, *23*, 13615–13624.
21. Filipovic, J.; Pezo, L.; Filipovic, V.; Ludajic, G. Spelt pasta with inulin as a functional food. *Acta Period. Technol.* **2015**, *46*, 37–44. [CrossRef]
22. Martín-Esparza, M.; Raga, A.; González-Martínez, C.; Albors, A. Micronised bran-enriched fresh egg tagliatelle: Significance of gums addition on pasta technological features. *Food Sci. Technol. Int.* **2018**, *24*, 309–320. [CrossRef]
23. Makhlouf, S.; Jones, S.; Ye, S.-H.; Sancho-Madriz, M.; Burns-Whitmore, B.; Li, Y.O. Effect of selected dietary fibre sources and addition levels on physical and cooking quality attributes of fibre-enhanced pasta. *Food Qual. Saf.* **2019**, *3*, 117–127. [CrossRef]
24. Bianchi, F.; Tolve, R.; Rainero, G.; Bordiga, M.; Brennan, C.S.; Simonato, B. Technological, nutritional and sensory properties of pasta fortified with agro-industrial by-products: A review. *Int. J. Food Sci. Technol.* **2021**, *56*, 4356–4366. [CrossRef]
25. Sobota, A.; Rzedzicki, Z.; Zarzycki, P.; Kuzawińska, E. Application of common wheat bran for the industrial production of high-fibre pasta. *Int. J. Food Sci. Technol.* **2015**, *50*, 111–119. [CrossRef]
26. Costantini, M.; Summo, C.; Faccia, M.; Caponio, F.; Pasqualone, A. Kabuli and Apulian black Chickpea Milling By-Products as Innovative Ingredients to Provide High Levels of Dietary Fibre and Bioactive Compounds in Gluten-Free Fresh Pasta. *Molecules* **2021**, *26*, 4442. [CrossRef]
27. Fradinho, P.; Oliveira, A.; Domínguez, H.; Torres, M.D.; Sousa, I.; Raymundo, A. Improving the nutritional performance of gluten-free pasta with potato peel autohydrolysis extract. *Innov. Food Sci. Emerg. Technol.* **2020**, *63*, 102374. [CrossRef]
28. Garbetta, A.; D'Antuono, I.; Melilli, M.G.; Sillitti, C.; Linsalata, V.; Scandurra, S.; Cardinali, A. Inulin enriched durum wheat spaghetti: Effect of polymerization degree on technological and nutritional characteristics. *J. Funct. Foods* **2020**, *71*, 104004. [CrossRef]
29. Krawęcka, A.; Sobota, A.; Sykut-Domańska, E. Physicochemical, Sensory, and Cooking Qualities of Pasta Enriched with Oat β-Glucans, Xanthan Gum, and Vital Gluten. *Foods* **2020**, *9*, 1412. [CrossRef]
30. Peressini, D.; Cavarape, A.; Brennan, M.A.; Gao, J.; Brennan, C.S. Viscoelastic properties of durum wheat doughs enriched with soluble dietary fibres in relation to pasta-making performance and glycaemic response of spaghetti. *Food Hydrocoll.* **2020**, *102*, 105613. [CrossRef]
31. Foschia, M.; Peressini, D.; Sensidoni, A.; Brennan, M.A.; Brennan, C.S. How combinations of dietary fibres can affect physicochemical characteristics of pasta. *LWT Food Sci. Technol.* **2015**, *61*, 41–46. [CrossRef]
32. Aravind, N.; Sissons, M.J.; Fellows, C.M.; Blazek, J.; Gilbert, E.P. Effect of inulin soluble dietary fibre addition on technological, sensory, and structural properties of durum wheat spaghetti. *Food Chem.* **2012**, *132*, 993–1002. [CrossRef]
33. Padalino, L.; Costa, C.; Conte, A.; Melilli, M.G.; Sillitti, C.; Bognanni, R.; Raccuia, S.A.; Del Nobile, M.A. The quality of functional whole-meal durum wheat spaghetti as affected by inulin polymerization degree. *Carbohydr. Polym.* **2017**, *173*, 84–90. [CrossRef]
34. Mensink, M.A.; Frijlink, H.W.; van der Voort Maarschalk, K.; Hinrichs, W.L.J. Inulin, a flexible oligosaccharide I: Review of its physicochemical characteristics. *Carbohydr. Polym.* **2015**, *130*, 405–419. [CrossRef]
35. Pasqualone, A.; Gambacorta, G.; Summo, C.; Caponio, F.; Di Miceli, G.; Flagella, Z.; Marrese, P.P.; Piro, G.; Perrotta, C.; De Bellis, L.; et al. Functional, textural and sensory properties of dry pasta supplemented with lyophilized tomato matrix or with durum wheat bran extracts produced by supercritical carbon dioxide or ultrasound. *Food Chem.* **2016**, *213*, 545–553. [CrossRef]
36. AACC, American Association of Cereal Chemists. *Approved Methods of Analysis*, 11th ed.; AACC International: St. Paul, MN, USA, 2000.
37. Bustos, M.C.; Pérez, G.T.; León, A.E. Effect of Four Types of Dietary Fiber on the Technological Quality of Pasta. *Food Sci. Technol. Int.* **2011**, *17*, 213–221. [CrossRef]
38. Pasqualone, A.; De Angelis, D.; Squeo, G.; Difonzo, G.; Caponio, F.; Summo, C. The effect of the addition of apulian black chickpea flour on the nutritional and qualitative properties of durum wheat-based bakery products. *Foods* **2019**, *8*, 504. [CrossRef]
39. AOAC International, Association of Official Analytical Chemists. *Official Method of Analyisis*, 17th ed.; AOAC International: Gaithersburg, MD, USA, 2006.
40. Liljeberg, H.; Åkerberg, A.; Björck, I. Resistant starch formation in bread as influenced by choice of ingredients or baking conditions. *Food Chem.* **1996**, *56*, 389–394. [CrossRef]

41. Capriles, V.D.; Arêas, J.A.G. Effects of prebiotic inulin-type fructans on structure, quality, sensory acceptance and glycemic response of gluten-free breads. *Food Funct.* **2013**, *4*, 104–110. [CrossRef]
42. Kamiloglu, S.; Capanoglu, E. In vitro gastrointestinal digestion of polyphenols from different molasses (pekmez) and leather (pestil) varieties. *Int. J. Food Sci. Technol.* **2014**, *49*, 1027–1039. [CrossRef]
43. Caponio, G.R.; Noviello, M.; Calabrese, F.M.; Gambacorta, G.; Giannelli, G.; De Angelis, M. Effects of Grape Pomace Polyphenols and In Vitro Gastrointestinal Digestion on Antimicrobial Activity: Recovery of Bioactive Compounds. *Antioxidants* **2022**, *11*, 567. [CrossRef]
44. Vacca, M.; Celano, G.; Lenucci, M.S.; Fontana, S.; la Forgia, F.M.; Minervini, F.; Scarano, A.; Santino, A.; Dalfino, G.; Gesualdo, L.; et al. In Vitro Selection of Probiotics, Prebiotics, and Antioxidants to Develop an Innovative Synbiotic (NatuREN G) and Testing Its Effect in Reducing Uremic Toxins in Fecal Batches from CKD Patients. *Microorganisms* **2021**, *9*, 1316. [CrossRef] [PubMed]
45. Maumela, P.; van Rensburg, E.; Chimphango, A.F.A.; Görgens, J.F. Sequential extraction of protein and inulin from the tubers of *Jerusalem artichoke* (*Helianthus tuberosus* L.). *J. Food Sci. Technol.* **2020**, *57*, 775–786. [CrossRef] [PubMed]
46. Rubel, I.A.; Iraporda, C.; Novosad, R.; Cabrera, F.A.; Genovese, D.B.; Manrique, G.D. Inulin rich carbohydrates extraction from *Jerusalem artichoke* (*Helianthus tuberosus* L.) tubers and application of different drying methods. *Food Res. Int.* **2018**, *103*, 226–233. [CrossRef] [PubMed]
47. Apolinário, A.C.; de Carvalho, E.M.; de Lima Damasceno, B.P.G.; da Silva, P.C.D.; Converti, A.; Pessoa, A.; da Silva, J.A. Extraction, isolation and characterization of inulin from *Agave sisalana* boles. *Ind. Crops Prod.* **2017**, *108*, 355–362. [CrossRef]
48. Li, W.; Zhang, J.; Yu, C.; Li, Q.; Dong, F.; Wang, G.; Gu, G.; Guo, Z. Extraction, degree of polymerization determination and prebiotic effect evaluation of inulin from *Jerusalem artichoke*. *Carbohydr. Polym.* **2015**, *121*, 315–319. [CrossRef]
49. Petkova, N.T.; Sherova, G.; Denev, P.P. Characterization of inulin from dahlia tubers isolated by microwave and ultrasound-assisted extractions. *Int. Food Res. J.* **2018**, *25*, 1876–1884.
50. Xu, H.; Gunenc, A.; Hosseinian, F. Ultrasound affects physical and chemical properties of Jerusalem artichoke and chicory inulin. *J. Food Biochem.* **2022**, *46*, 1–10. [CrossRef]
51. Jirayucharoensak, R.; Khuenpet, K.; Jittanit, W.; Sirisansaneeyakul, S. Physical and chemical properties of powder produced from spray drying of inulin component extracted from *Jerusalem artichoke* tuber powder. *Dry. Technol.* **2019**, *37*, 1215–1227. [CrossRef]
52. Simonato, B.; Tolve, R.; Rainero, G.; Rizzi, C.; Sega, D.; Rocchetti, G.; Lucini, L.; Giuberti, G. Technological, nutritional, and sensory properties of durum wheat fresh pasta fortified with *Moringa oleifera* L. leaf powder. *J. Sci. Food Agric.* **2021**, *101*, 1920–1925. [CrossRef]
53. Neylon, E.; Arendt, E.K.; Zannini, E.; Sahin, A.W. Fundamental study of the application of brewers spent grain and fermented brewers spent grain on the quality of pasta. *Food Struct.* **2021**, *30*, 100225. [CrossRef]
54. Simonato, B.; Trevisan, S.; Tolve, R.; Favati, F.; Pasini, G. Pasta fortification with olive pomace: Effects on the technological characteristics and nutritional properties. *LWT* **2019**, *114*, 108368. [CrossRef]
55. Naji-Tabasi, S.; Niazmand, R.; Modiri-Dovom, A. Application of mucilaginous seeds (*Alyssum homolocarpum* and *Salvia macrosiphon Boiss*) and wheat bran in improving technological and nutritional properties of pasta. *J. Food Sci.* **2021**, *86*, 2288–2299. [CrossRef] [PubMed]
56. Attanzio, A.; Diana, P.; Barraja, P.; Carbone, A.; Spanò, V.; Parrino, B.; Cascioferro, S.M.; Allegra, M.; Cirrincione, G.; Tesoriere, L.; et al. Quality, functional and sensory evaluation of pasta fortified with extracts from Opuntia ficus-indica cladodes. *J. Sci. Food Agric.* **2019**, *99*, 4242–4247. [CrossRef] [PubMed]
57. Renoldi, N.; Brennan, C.S.; Lagazio, C.; Peressini, D. Evaluation of technological properties, microstructure and predictive glycaemic response of durum wheat pasta enriched with psyllium seed husk. *LWT* **2021**, *151*, 112203. [CrossRef]
58. Brennan, C.S.; Tudorica, C.M. Fresh Pasta Quality as affected by enrichment of nonstarch polysaccharides. *J. Food Sci.* **2007**, *72*, S659–S665. [CrossRef]
59. Xu, F.; Li, X.; Li, J.; Chen, J. The interaction between inulin and wheat starch and effects of inulin on frozen storage quality of noodles. *Int. J. Food Sci. Technol.* **2021**, *56*, 2423–2431. [CrossRef]
60. Zarroug, Y.; Djebali, K.; Sfayhi, D.; Khemakhem, M.; Boulares, M.; El Felah, M.; Mnasser, H.; Kharrat, M. Optimization of barley flour and inulin addition for pasta formulation using mixture design approach. *J. Food Sci.* **2022**, *87*, 68–79. [CrossRef]
61. Lucisano, M.; Cappa, C.; Fongaro, L.; Mariotti, M. Characterisation of gluten-free pasta through conventional and innovative methods: Evaluation of the cooking behaviour. *J. Cereal Sci.* **2012**, *56*, 667–675. [CrossRef]
62. Li, Y.; Ma, X.; Liu, X. Physicochemical and rheological properties of cross-linked inulin with different degree of polymerization. *Food Hydrocoll.* **2019**, *95*, 318–325. [CrossRef]
63. Liu, J.; Luo, D.; Li, X.; Xu, B.; Zhang, X.; Liu, J. Effects of inulin on the structure and emulsifying properties of protein components in dough. *Food Chem.* **2016**, *210*, 235–241. [CrossRef] [PubMed]
64. Wang, J.; Brennan, M.A.; Brennan, C.S.; Serventi, L. Effect of Vegetable Juice, Puree, and Pomace on Chemical and Technological Quality of Fresh Pasta. *Foods* **2021**, *10*, 1931. [CrossRef] [PubMed]
65. Cárdenas-Hernández, A.; Beta, T.; Loarca-Piña, G.; Castaño-Tostado, E.; Nieto-Barrera, J.O.; Mendoza, S. Improved functional properties of pasta: Enrichment with amaranth seed flour and dried amaranth leaves. *J. Cereal Sci.* **2016**, *72*, 84–90. [CrossRef]
66. Minarovičová, L.; Lauková, M.; Kohajdová, Z.; Karovičová, J.; Dobrovická, D.; Kuchtová, V. Qualitative properties of pasta enriched with celery root and sugar beet by-products. *Czech J. Food Sci.* **2018**, *36*, 66–72. [CrossRef]

67. Bustos, M.C.; Perez, G.T.; Leon, A.E. Structure and quality of pasta enriched with functional ingredients. *RSC Adv.* **2015**, *5*, 30780–30792. [CrossRef]
68. Aravind, N.; Sissons, M.; Egan, N.; Fellows, C. Effect of insoluble dietary fibre addition on technological, sensory, and structural properties of durum wheat spaghetti. *Food Chem.* **2012**, *130*, 299–309. [CrossRef]
69. Chillo, S.; Civica, V.; Iannetti, M.; Suriano, N.; Mastromatteo, M.; Del Nobile, M.A. Properties of quinoa and oat spaghetti loaded with carboxymethylcellulose sodium salt and pregelatinized starch as structuring agents. *Carbohydr. Polym.* **2009**, *78*, 932–937. [CrossRef]
70. Aravind, N.; Sissons, M.; Egan, N.; Fellows, C.M.; Blazek, J.; Gilbert, E.P. Effect of β-Glucan on Technological, Sensory, and Structural Properties of Durum Wheat Pasta. *Cereal Chem. J.* **2012**, *89*, 84–93. [CrossRef]
71. Padalino, L.; Amalia, C.; Lucia, L.; Desislava, L.; Vincenzo, S.; Teresa Maria, P.; Marco, P.; Matteo Alessandro Del, N. Functional pasta with tomato by-product as a source of antioxidant compounds and dietary fibre. *Czech J. Food Sci.* **2017**, *35*, 48–56. [CrossRef]
72. European Parliament. Regulation (EC) No. 1924/2006 of the European Parliament and of the Council of 20 December 2006 on Nutrition and Health Claims Made on Foods. *Off. J. Eur. Union* **2006**, *404*, 9. Available online: https://eur-lex.europa.eu/legal-content/en/ALL/?uri=CELEX%3A32006R1924 (accessed on 30 July 2022).
73. Alongi, M.; Melchior, S.; Anese, M. Reducing the glycemic index of short dough biscuits by using apple pomace as a functional ingredient. *LWT* **2019**, *100*, 300–305. [CrossRef]
74. Caponio, G.R.; Difonzo, G.; de Gennaro, G.; Calasso, M.; De Angelis, M.; Pasqualone, A. Nutritional Improvement of Gluten-Free Breadsticks by Olive Cake Addition and Sourdough Fermentation: How Texture, Sensory, and Aromatic Profile Were Affected? *Front. Nutr.* **2022**, *9*, 830932. [CrossRef] [PubMed]
75. Ramirez-Farias, C.; Slezak, K.; Fuller, Z.; Duncan, A.; Holtrop, G.; Louis, P. Effect of inulin on the human gut microbiota: Stimulation of *Bifidobacterium adolescentis* and *Faecalibacterium prausnitzii*. *Br. J. Nutr.* **2008**, *101*, 541–550. [CrossRef] [PubMed]
76. Le Bastard, Q.; Chapelet, G.; Javaudin, F.; Lepelletier, D.; Batard, E.; Montassier, E. The effects of inulin on gut microbial composition: A systematic review of evidence from human studies. *Eur. J. Clin. Microbiol. Infect. Dis.* **2020**, *39*, 403–413. [CrossRef] [PubMed]
77. Brennan, C.S.; Kuri, V.; Tudorica, C.M. Inulin-enriched pasta: Effects on textural properties and starch degradation. *Food Chem.* **2004**, *86*, 189–193. [CrossRef]
78. Shoaib, M.; Shehzad, A.; Omar, M.; Rakha, A.; Raza, H.; Sharif, H.R.; Shakeel, A.; Ansari, A.; Niazi, S. Inulin: Properties, health benefits and food applications. *Carbohydr. Polym.* **2016**, *147*, 444–454. [CrossRef]
79. Romão, B.; Falcomer, A.L.; Palos, G.; Cavalcante, S.; Botelho, R.B.A.; Nakano, E.Y.; Raposo, A.; Shakeel, F.; Alshehri, S.; Mahdi, W.A.; et al. Glycemic Index of Gluten-Free Bread and Their Main Ingredients: A Systematic Review and Meta-Analysis. *Foods* **2021**, *10*, 506. [CrossRef]
80. Brennan, C.S.; Tudorica, C.M. Evaluation of potential mechanisms by which dietary fibre additions reduce the predicted glycaemic index of fresh pastas. *Int. J. Food Sci. Technol.* **2008**, *43*, 2151–2162. [CrossRef]
81. Kumar, M.; Nagpal, R.; Kumar, R.; Hemalatha, R.; Verma, V.; Kumar, A.; Chakraborty, C.; Singh, B.; Marotta, F.; Jain, S.; et al. Cholesterol-Lowering Probiotics as Potential Biotherapeutics for Metabolic Diseases. *Exp. Diabetes Res.* **2012**, *2012*, 902917. [CrossRef]
82. Tako, E.; Glahn, R.P.; Knez, M.; Stangoulis, J.C. The effect of wheat prebiotics on the gut bacterial population and iron status of iron deficient broiler chickens. *Nutr. J.* **2014**, *13*, 58. [CrossRef]
83. Caio, G.; Lungaro, L.; Segata, N.; Guarino, M.; Zoli, G.; Volta, U.; De Giorgio, R. Effect of Gluten-Free Diet on Gut Microbiota Composition in Patients with Celiac Disease and Non-Celiac Gluten/Wheat Sensitivity. *Nutrients* **2020**, *12*, 1832. [CrossRef] [PubMed]
84. Vacca, M.; Porrelli, A.; Calabrese, F.M.; Lippolis, T.; Iacobellis, I.; Celano, G.; Pinto, D.; Russo, F.; Giannelli, G.; De Angelis, M. How Metabolomics Provides Novel Insights on Celiac Disease and Gluten-Free Diet: A Narrative Review. *Front. Microbiol.* **2022**, *13*, 859467. [CrossRef] [PubMed]
85. Kareem, K.Y.; Hooi Ling, F.; Teck Chwen, L.; May Foong, O.; Anjas Asmara, S. Inhibitory activity of postbiotic produced by strains of *Lactobacillus plantarum* using reconstituted media supplemented with inulin. *Gut Pathog.* **2014**, *6*, 23. [CrossRef]
86. Baxter, N.T.; Schmidt, A.W.; Venkataraman, A.; Kim, K.S.; Waldron, C.; Schmidt, T.M. Dynamics of Human Gut Microbiota and Short-Chain Fatty Acids in Response to Dietary Interventions with Three Fermentable Fibers. *MBio* **2019**, *10*, e02566-18. [CrossRef]
87. Cosola, C.; De Angelis, M.; Rocchetti, M.T.; Montemurno, E.; Maranzano, V.; Dalfino, G.; Manno, C.; Zito, A.; Gesualdo, M.; Ciccone, M.M.; et al. Beta-Glucans Supplementation Associates with Reduction in P-Cresyl Sulfate Levels and Improved Endothelial Vascular Reactivity in Healthy Individuals. *PLoS ONE* **2017**, *12*, e0169635. [CrossRef]
88. Saa, P.; Urrutia, A.; Silva-Andrade, C.; Martín, A.J.; Garrido, D. Modeling approaches for probing cross-feeding interactions in the human gut microbiome. *Comput. Struct. Biotechnol. J.* **2022**, *20*, 79–89. [CrossRef]

foods

MDPI

Article

Effect of the Application of a Green Preservative Strategy on Minced Meat Products: Antimicrobial Efficacy of Olive Mill Wastewater Polyphenolic Extract in Improving Beef Burger Shelf-Life

Rossana Roila [1], Beatrice Sordini [2], Sonia Esposto [2,*], David Ranucci [1,*], Sara Primavilla [3], Andrea Valiani [3], Agnese Taticchi [2], Raffaella Branciari [1,†] and Maurizio Servili [2,†]

1 Department of Veterinary Medicine, University of Perugia, Via San Costanzo 4, 06126 Perugia, Italy
2 Department of Agricultural, Food and Environmental Sciences, University of Perugia, 06126 Perugia, Italy
3 Istituto Zooprofilattico Sperimentale dell'Umbria e delle Marche "Togo Rosati", Via G. Salvemini 1, 06126 Perugia, Italy
* Correspondence: sonia.esposto@unipg.it (S.E.); david.ranucci@unipg.it (D.R.); Tel.: +39-075-585-7952 (S.E.); +39-075-585-7931 (D.R.)
† These authors contributed equally to this work.

Abstract: The mincing process of raw meat favors microbial spoilage as well as chemical and enzymatic oxidation processes. In order to limit this degradative process, preservatives are routinely added to minced meat products. The role of olive mill wastewater polyphenolic extract as a replacement for synthetic preservatives in beef burger was assessed. The antioxidant capacity of the extract experimentally added to beef burger was evaluated using the oxygen radical absorbance capacity method ($ORAC_{FL}$) to assess the shelf-life, while the lipid oxidation was measured by thiobarbituric reactive substance (TBAR) determination. The antimicrobial activity was assayed by means of classical methods and predictive microbiology. The experimental addition of polyphenolic extract led to 62% lower lipid oxidation and 58% higher antioxidant capacity; it also successfully modulated spoilage microbial populations with an average growth reduction of 15% on day 7. Results indicate that olive mill wastewater polyphenolic extracts could be added to raw ground beef meat to act as natural antioxidants and to modulate microbial growth.

Keywords: microbial spoilage; lipidic oxidation; antioxidant; predictive microbiology; food preservation; food safety; sustainable strategy; by-product reuse; kinetic parameters; *Olea europaea*

Citation: Roila, R.; Sordini, B.; Esposto, S.; Ranucci, D.; Primavilla, S.; Valiani, A.; Taticchi, A.; Branciari, R.; Servili, M. Effect of the Application of a Green Preservative Strategy on Minced Meat Products: Antimicrobial Efficacy of Olive Mill Wastewater Polyphenolic Extract in Improving Beef Burger Shelf-Life. *Foods* **2022**, *11*, 2447. https://doi.org/10.3390/foods11162447

Academic Editors: Marco Poiana, Francesco Caponio and Antonio Piga

Received: 19 July 2022
Accepted: 11 August 2022
Published: 14 August 2022

Publisher's Note: MDPI stays neutral with regard to jurisdictional claims in published maps and institutional affiliations.

Copyright: © 2022 by the authors. Licensee MDPI, Basel, Switzerland. This article is an open access article distributed under the terms and conditions of the Creative Commons Attribution (CC BY) license (https://creativecommons.org/licenses/by/4.0/).

1. Introduction

Burger is one of the most appreciated meat products worldwide for its ease of preparation and versatility of consumption, which is considered a time-saving strategy in the modern lifestyle [1–3]. Nevertheless, the grinding process for raw meat, resulting in the disruption of muscle structure, leads to a less stable food matrix favoring microbial spoilage as well as chemical and enzymatic oxidation processes with possible repercussions on safety and health [4,5]. Several strategies, such as peculiar production processes, packaging and food additives, have been studied during the last few decades in order to reduce the above-mentioned phenomena and enhance the shelf-life of these meat products [6–10]. Studies in the literature demonstrate that antioxidant molecules protect the grinded meat from oxidation and delay the microbial growth [11]. Consequently, in fresh ground meat preparations, additives with antioxidant properties are usually employed. Although food additives are strictly regulated (Regulation (EC) No. 1333/2008, s.m.i.) [12], chemical compounds intentionally added into food are considered with mistrust by consumers due to their potential long-term adverse effects linked to hypersensitivity, asthma, cancer, skin irritation, allergies and gastrointestinal problems [13,14]. As a consequence, growing interest has been demonstrated by consumers towards products with natural antioxidants, encouraging food industries to continuously research for the newest natural food additives to replace synthetic ones [15].

Several spices, essential oils, extracts, powders and other plant by-products have been studied in recent decades in order to assess their activity and their effects on meat products as food supplementation [16–19]. Among these, olive oil by-products can be considered a source of bioactive molecules that are potentially applicable for processed meat preservation [20]. It is known that olive oil by-products are characterized by a high number of hydrophilic phenols, mainly secoiridoids, found exclusively in the *Oleaceae* family, that have been proven to inhibit or delay the rate of growth of a wide range of Gram-positive and Gram-negative bacteria as well as to have high antioxidant properties [21,22]. In particular, the olive mill wastewater generated in olive oil production has a high generation rate (49% of total mass), and the possible exploitation of this agro-industrial waste through the recovery of high-value bioactive compounds could positively affect the economic and environmental sustainability of agro-industry [23].

Although the chemical composition and the antioxidant capacity of olive oil by-products as well as their application in foodstuff have been previously studied by several authors [20,24,25], information on the effect of the olive mill wastewater polyphenolic extract on the microbial population of minced meat products to improve their shelf-life is still limited.

In this sense, the role of this natural extract in the replacement of synthetic preservatives in meat products is postulated. With this aim, the antioxidant activity and antimicrobial capacity of olive mill wastewater polyphenolic extract during beef burger shelf-life was evaluated. In order to estimate the potential activity of this polyphenolic extract in comparison with synthetic additives, different formulations were tested and compared to a control formulation and a control formulation with synthetic preservatives (sodium ascorbate) during a period of 7 days of cold storage.

2. Materials and Methods

2.1. Olive Mill Waste Water Extract and Compositon

The crude phenolic extract (PE) used in the beef burger formulation was obtained through a three-step membrane filtration process using fresh olive mill wastewaters from processing olives of the Umbrian Moraiolo cultivar, as reported by Ianni et al. [26]. To obtain a stable powder formulation, the extracts were spray-dried after their combination with maltodextrin (1:1 dw), which functioned as a carrier, and is largely used in spray drying in the food industry [27]. The phenolic composition of the spray-dried extracts was assessed by means of high-performance liquid chromatography (HPLC) [28] and is reported in Table 1.

Table 1. Composition of spray-dried crude phenolic extract (PE).

Crude PE (mg/g)	
3,4-DHPEA *	9.2 ± 0.2
p-HPEA	4.3 ± 0.0
Verbascoside	5.9 ± 0.2
3,4-DHPEA-EDA	8.1 ± 0.2
Sum of phenols	*27.6 ± 0.3*
Purity	*2.7%*

* Results are the mean of two independent analytical determinations ± standard deviation. 3,4-DHPEA = hydroxytyrosol, *p*-HPEA = tyrosol, 3,4-DHPEA-EDA = oleacein.

2.2. Beef Burger Formulation

Beef burgers were produced in an EU-approved meat processing plant located in Umbria, Central Italy. The meat for the preparation of beef burger was obtained from cuts (beef rump and shoulder muscle) of 18-month-old female Chianina cattle reared and slaughtered in Italy in accordance with European Union Regulation (Regulation (EC) No. 853/2004 s.m.i.) [29]. After 12 days of carcass aging, 40 kg of meat cuts was ground twice in a professional trimmer equipped with a 4 mm-hole plate. The grinded meat was divided into four different formulations, each 5 kg in weight, and aseptically hand-mixed with the following ingredients for 2 min (basic recipe): 10 g/kg NaCl, 60 g/kg grated parmesan, 80 g/kg breadcrumbs and 2 eggs/kg. The four different formulations were elaborated as follows: C, basic recipe with no addition; A, basic recipe plus 10 g/kg commercial antioxidant mix (CM) (CondiHamb, MEC Import, Perugia, Italy); AP, basic recipe plus 5 g/kg CM plus 350 mg/kg PE crude extract; P, basic recipe plus 700 mg/kg PE crude extract. The chemical composition of the burger was characterized by an average of 22% protein and 8 % fat. Each batch was further mixed for 2 min, and the burgers were then molded (about 100 g each) and placed in a display refrigerator at 4 ± 2 °C for 7 days, under alternating exposure to fluorescent light (12 h light/12 h darkness) to simulate retail storage conditions.

2.3. Antioxidant Capacity of PE and Mix Extracts and Beef Burger

The antioxidant capacity was evaluated for the synthetic and natural additives and burgers over the course of their shelf-life using the oxygen radical absorbance capacity method ($ORAC_{FL}$). One gram each of CM, PE and burger samples was separately mixed with a buffer, 75 mM, pH 7.2, containing 13.19 g of K_2HPO_4 and 10.26 g of KH_2PO_4 in 900 mL of deionized water, homogenized with an Ultra-Turrax homogenizer (Ultra Turrax T25 Basic, IKA Labortechnik Janke & Kunkel GmbH, Stavfen, Germany) for 1 min, and then vortexed for 2 min. The homogenates were centrifuged at 6000 rpm at 4 °C for 20 min, and the supernatant was used for the determination of the antioxidant capacity using the oxygen radical absorbance capacity method ($ORAC_{FL}$) based on the fluorescence decay rate of a probe in the presence of a radical oxygen species (ROO) and compared with that of a reference standard, Trolox (6-hydroxy-2,5,7,8-tetramethylchroman-2-carboxylic acid, Sigma-Aldrich, Steinheim, Germany). The $ORAC_{FL}$ assays were carried out on a FLUO-star OPTIMA microplate fluorescence reader (BMGLABTECH, Offenburg, Germany) at an excitation wavelength of 485 nm and an emission wavelength of 520 nm. The results are expressed as µg of Trolox equivalents (TE) per 100 g sample.

2.4. Lipid Oxidation of Beef Burger

Lipid oxidation over the course of shelf-life was measured by thiobarbituric reactive substance (TBAR) determination, performed according to Tarladgis et al. [30] and measured by an Ultrospec 2100 pro UV–visible spectrometer (Amersham Pharmacia Biotech, Amersham, UK) at 532 nm. Quantification was performed using a standard calibration curve and with a concentration range of 1E6 to 1E5 M (y = 2E + 07x + 0.0046, R^2 = 0.9999), corresponding to a range of 0.4–4 mg of malonaldehyde (MDA)/kg meat. The MDA recovery was determined by spiking the samples with a known volume of 0.2 mM TMP. The TBAR concentration was expressed as mg MDA/kg meat.

2.5. Antimicrobial Activity of Commercial Mix and PE Extract

The evaluation of the antimicrobial activity of CM and PE extract was performed through the agar well diffusion technique [31,32] on some micro-organisms relevant for the food industry, such as *Staphylococcus aureus* (WDCM 00034), *Escherichia coli* (WDCM 00013) and *Pseudomonas fluorescens* (WDCM 00115). The selected reference strains were revitalized in Brain Heart Infusion (BHI) broth and incubated at 37 °C for 24 h with the exception of *P. fluorescens*, which was incubated at 25 °C for 24–48 h. An initial suspension of 0.5 McFarland in 0.9% sterile saline solution was prepared for each micro-organism, and 100 µL was then distributed on Mueller–Hinton Agar (MHA, Thermo Fisher Scientific, Milan, Italy) plates with a swab, making four 90° rotations. At the time of use, the extracts were suspended with sterile demineralized water to obtain a concentration of 750 mg/mL, and then two dilutions were performed at 375 and 187 mg/mL. In each inoculated MHA plate, 7 mm-diameter holes were produced with a sterilized cork borer and then filled with 50 μL of extract suspension at different concentrations [31,32]. The plates were incubated according to the most suitable growth conditions, as reported above. For each bacterial strain, CM and PE extract were tested, and a negative control was set up with sterile demineralized water. At the end of the incubation period, the diameter of the inhibition halo was measured by a gauge and expressed in mm.

2.6. Microbial Analysis of Beef Burger

After 0, 2, 5 and 7 days of storage (T0, T1, T2 and T3, respectively), 10 g of each sample was aseptically removed and placed in a sterile stomacher bag with 90 mL of Buffered Peptone Water (Oxoid Ltd., Basingstoke, UK). After homogenization (Stomacher 400 circulator, Seward Ltd., Norfolk, UK), decimal serial dilutions were performed, and the below-reported microbiological determination was carried out in duplicate. Total viable count (TVC) was performed on Plate Count Agar (PCA, Oxoid Ltd.) incubated at 30 °C for 72 h according to ISO 4833-1 [33]. *Enterobacteriaceae* were enumerated according to a validated alternative method of ISO 21528-2 [34] (AFNOR AES 10/07-01/08) on Rebecca™ EB (bioMérieux, Mercy Etoile, France) incubated at 37 °C for 24 h. The lactic acid bacteria (LAB) count was performed on de Man, Rogosa and Sharpe agar (MRS, Oxoid Ltd.) incubated at 30 °C for 72 h, while the *Pseudomonas* spp. count was performed on Pseudomonas Agar Base with CFC selective agar supplement (Oxoid Ltd.) and incubated for 48 h at 25 °C. Coagulase-positive staphylococci were enumerated on Baird Parker Agar with the addition of RPF supplement (Biolife, Milano, Italy) and incubated at 37 °C for 48 h. Results were recorded as colony-forming units (CFUs) and converted into log10 values to obtain Log CFU/g of meat prior to statistical analysis according to Gill and Jones [35]. In order to assess the microbiological safety over the course of beef burgers' shelf-life, the *Salmonella* spp. detection was performed according to ISO 6579–1: 2017 [36] at T0, T1, T2 and T3, in compliance with Regulation (EC) No. 2073/2005 [37].

Corresponding with the end of the manufacturing process (T0), the enumeration of *E. coli* was performed according to ISO 16649 [38] on all experimental groups as a process hygiene criterion of Regulation (EC) No. 2073/2005 [37].

2.7. Statistical Analysis

Data were analyzed using the GLM procedure of SAS [39]. An ANOVA model was used with sample (C, A, AP, P) and time (T0, T1, T2, T3) as the fixed factors. The replicate effect was found not significant and removed from the model. The differences in the means were detected using the Tukey's test and considered significant when $p < 0.05$. The effects of formulation on the growth of the target micro-organisms were evaluated with the DMFit tool of the free predictive microbiology software Combase (https://www.combase.cc/index.php/en/DMFit, accessed on 20 June 2022), allowing for the definition of growth parameters such as lag phase duration (λ, 1/h) and maximum growth rate (μmax, 1/h) by means of the Baranyi and Roberts model [40]. The fitted results were analyzed by one-way ANOVA (with the sample as a fixed variable) and Tukey's test ($p < 0.05$).

3. Results and Discussion

A higher antioxidant activity was found in the PE in comparison with the CM containing ascorbic acid (554.42 ± 3.38 and 1149.02 ± 13.69 μg TE/100 g in CM and PE, respectively). The high antioxidant activity recorded in the PE is in agreement with a previous studies that reported the powerful antioxidant activity of olive phenolic compounds [22,41–44]. The results for the lipid oxidation and antioxidant activity of minced beef meat indicated by TBARs and $ORAC_{FL}$ values are reported in Table 2.

Table 2. Lipid oxidation ($TBAR_S$) and antioxidant capacity ($ORAC_{FL}$) in beef burger during storage.

		Days of Storage				SEM	p-Value		
		0	2	5	7		T	S	TXS
TBARS (mg MDA/kg)	C	0.13 ^{a}	0.34 ^{bW}	0.47 ^{cW}	0.63 ^{dW}	0.012	<0.001	<0.001	<0.001
	A	0.14 ^{a}	0.20 ^{bX}	0.25 ^{cX}	0.30 ^{dX}				
	AP	0.14 ^{a}	0.17 ^{abXZ}	0.19 ^{bZ}	0.29 ^{cX}				
	P	0.15 ^{a}	0.16 ^{aZ}	0.18 ^{aZ}	0.24 ^{bZ}				
$ORAC_{FL}$ (µg TE /100 g)	C	24.44 ^{W}	24.43 ^{W}	24.44 ^{W}	24.41 ^{W}	0.475	<0.001	<0.001	<0.001
	A	34.19 ^{aX}	32.11 ^{bX}	27.08 ^{cX}	25.59 ^{dW}				
	AP	38.87 ^{aY}	38.18 ^{aY}	36.17 ^{bY}	35.89 ^{bX}				
	P	44.45 ^{aZ}	43.89 ^{aZ}	40.01 ^{cZ}	38.73 ^{bY}				

C = control group; A = basic recipe with addition of 10 g/kg commercial antioxidant; AP = basic recipe with addition of 5 g/kg commercial antioxidant and 350 mg/kg phenolic extract; P = basic recipe with addition of 700 mg/kg phenolic extract. Different letters in the same row (a, b, c, d) indicate differences between mean values during sampling times ($p \leq 0.001$); different letters in the same column (W, X, Y, Z) indicate differences between mean values for different experimental groups ($p \leq 0.001$). SEM, standard error of the mean. T = time; S = sample.

Concerning lipid oxidation, increased $TBAR_S$ values were detected during storage for all experimental formulations, and, among groups, differences were reported starting from 2 days with higher values for C samples. The addition of 700 mg/kg PE in meat (P group) guaranteed the lowest level of $TBAR_S$ in the second part of storage (days 5–7) with a reduction of 62% of lipidic oxidation on day 7 for P compared to C samples. Similar results, albeit with a lower magnitude, have been reported by Martínez-Zamora et al. [43], demonstrating how the incorporation of synthetic hydroxytyrosol reduced the oxidation of lamb patties by 35% with respect to the control sample at the end of shelf-life. The hamburger formulation and storage time significantly affected the lipid oxidation, and the interaction between these two factors was also significant (Table 2). A PE-concentration-dependent effect preserving lipid oxidation was also observed by other studies in the literature both in raw and grilled beef burger [20].

The $ORAC_{FL}$ assay revealed differences in the antioxidant activity of four experimental groups, with the highest mean values recorded in the P group (Table 2). A reduction in the antioxidant activity during storage was recorded in A, AP and P. This reduction, as found in a previous study, is due mainly to the oxidative degradation phenomena of phenols or antioxidant molecule that occur during storage [20,44]. Despite the degradation of phenols, the higher level of integration in beef meat (P group) ensures considerable antioxidant activity until the end of storage time. It is well-known that phenols can act as hydrogen donors and compounds linked with a o-dihydroxyl functionality possess a high antioxidant activity, due to the formation of intramolecular hydrogen bonds observed during the reaction with free radicals [22]. Among these, the highest antioxidant activity was attributed to 3,4-DHPEA (hydroxytyrosol) and secoiridoid derivatives such as 3,4-DHPEA-EDA (oleacein) [22]. In particular, hydroxytyrosol's strong antioxidant potential is strictly related to its chemical structure: a phenol ring formed by a catechol group and three hydroxyl groups [43]. The combination of these functional groups could represent the main explanation for its preservative action in products of animal origin, as previously demonstrated in the literature [45,46].

The preliminary evaluation of the antimicrobial activity of the CM and PE performed through the agar well diffusion technique revealed that the PE possessed greater in vitro antimicrobial activity compared to the CM. Indeed, after incubation, the inhibition halos were measured for each strain, and the PE showed halos of 16, 13 and 11 mm for *P. fluorescens*, 14, 11 and 9 mm for *S. aureus*, and 9, 7 and 0 mm for *E. coli* for 750, 375 and 187 mgPE/mL, respectively. The assay's results suggest that the CM had no effect on microbial growth as the inhibition halos were absent for all the concentrations and micro-organisms tested.

Concerning the microbial analysis of beef burger, the *Salmonella* spp. detection showed that the pathogen was absent during the entire duration of products' shelf-life in all the experimental groups, complying with the food safety criterion of EU Regulation [37]. Similarly, the process hygiene criterion of *E. coli* in meat preparations was fully respected as the microbial count at T0 was below 2 Log CFU/g for all experimental groups [37]. This evidence confirms both the satisfactory safety and the hygiene levels of the beef burger production process.

The results of microflora evolution during storage for refrigerated beef burgers are depicted in Table 3. As shown, a significant ($p < 0.001$) increase was observed for all microbial populations and for all experimental groups studied as storage time elapsed. Immediately after production (T0), higher microbial populations were recorded for the TVC and LAB count followed by *Pseudomonas* spp. The initial (T0) TVC in all studied groups was approximately 4.4 Log CFU/g, which can be considered a characteristic value for minced meat products after manufacturing [47]. Indeed, this result is in agreement with the levels reported in the available literature for similar minced beef meat products [48,49], albeit other studies have found higher values [50] or lower ones [51]. It has been reported that a possible explanation for this relatively high initial TVC contamination in beef burgers may be attributed to the mincing process, which contributes to the total viable counts, likely as a consequence of the disruption of muscle structure, making nutrients easily available to micro-organisms [49]. However, the low initial value of *Enterobacteriaceae* counts (average value 1.33 Log CFU/g) confirms the optimal initial microbiological quality attributable to the good physiological status of the animal at slaughter and to proper postmortem meat acidification as well as to the good hygienic conditions during slaughter, handling and production processes [47,52].

Table 3. Microbial quality (Log CFU/g) of the four formulations of beef burger stored at 4 °C under aerobic conditions for 7 days.

		Days of Storage				SEM	*p*-Value		
		0	2	5	7		T	S	TXS
TVC	C	4.29 ^{a}	5.50 ^{b}	6.85 ^{cW}	7.41 ^{dW}	0.124	<0.001	<0.001	0.002
	A	4.37 ^{a}	5.48 ^{b}	6.82 ^{cW}	7.38 ^{dW}				
	AP	4.61 ^{a}	5.31 ^{b}	6.36 ^{cX}	6.98 ^{dWX}				
	P	4.48 ^{a}	5.11 ^{b}	6.24 ^{cX}	6.52 ^{cX}				
Staphylococcus spp.	C	1.46 ^{a}	2.07 ^{b}	2.78 ^{cW}	2.89 ^{c}	0.159	<0.001	0.005	0.671
	A	1.28 ^{a}	1.90 ^{b}	2.67 ^{cW}	2.87 ^{c}				
	AP	1.32 ^{a}	1.78 ^{a}	2.47 ^{bWX}	2.67 ^{b}				
	P	1.38 ^{a}	1.67 ^{ab}	2.05 ^{bcX}	2.49 ^{c}				
Pseudomonas spp.	C	4.02 ^{a}	5.18 ^{b}	6.67 ^{cW}	7.05 ^{dW}	0.199	<0.001	<0.001	0.163
	A	4.11 ^{a}	5.23 ^{b}	6.75 ^{cW}	7.20 ^{dW}				
	AP	4.09 ^{a}	4.74 ^{b}	5.96 ^{cX}	6.91 ^{dWX}				
	P	3.98 ^{a}	4.47 ^{a}	5.63 ^{bX}	6.11 ^{bX}				
LAB	C	4.15 ^{a}	4.46 ^{ab}	4.83 ^{b}	4.96 ^{b}	0.142	<0.001	0.508	0.999
	A	4.20 ^{a}	4.48 ^{ab}	4.79 ^{b}	4.88 ^{b}				
	AP	4.04 ^{a}	4.39 ^{ab}	4.82 ^{b}	4.94 ^{c}				
	P	3.98 ^{a}	4.29 ^{ab}	4.75 ^{b}	4.83 ^{c}				
Enterobacteriaceae	C	1.38 ^{a}	1.80 ^{ab}	2.58 ^{c}	3.60 ^{dW}	0.182	<0.001	<0.001	0.102
	A	1.30 ^{a}	1.66 ^{a}	2.53 ^{b}	3.58 ^{cW}				
	AP	1.24 ^{a}	1.46 ^{a}	2.27 ^{b}	3.17 ^{cW}				
	P	1.28 ^{a}	1.36 ^{a}	2.10 ^{b}	2.36 ^{bX}				

C = control group; A = basic recipe with addition of 10 g/kg commercial antioxidant; AP = basic recipe with addition of 5 g/kg commercial antioxidant and 350 mg/kg phenolic extract; P = basic recipe with addition of 700 mg/kg phenolic extract. Different letters in the same row (a, b, c, d) indicate differences between mean values during sampling times ($p \leq 0.001$); different letters in the same column (W, X, Y, Z) indicate differences between mean values for different experimental groups ($p \leq 0.001$). SEM, standard error of the mean. T = time; S = sample.

As above mentioned, following TVC, LAB and *Pseudomonas* spp. where the two microbial population with the highest initial value (T0), with average levels among experimental groups of 4.09 and 4.05 Log CFU/g, respectively. Similar counts were recorded for analogous products by Zamuz et al. [47] and Andres et al [53], while slightly higher values were recorded by Parafati et al. [54] and Marrone et al. [55] for *Pseudomonas* in minced beef meat products.

At the end of the storage period (day 7), significant ($p < 0.001$) differences between experimental groups were observed for TVC, *Pseudomonas* and *Enterobatteriaceae* (Table 3). Specifically, the lowest value for this microbial population corresponded to those burgers formulated with the highest amount of PE (P group), suggesting that the bioactive molecules contained in the extract affected this microbial population by limiting its growth, as preliminary suggested by an in vitro assay.

The parameters characterizing the growth curves of targeted microbial populations in the four experimental groups were obtained by modeling growth data by means of the Baranyi equation [40] and are summarized in Table 4.

Regarding the TVC, the addition of polyphenols in beef burger resulted in an extended lag phase (λ) in P samples in comparison with C and A and a reduction in the maximum growth rate (μmax) in AP compared with all other groups and the final value for P. For *Staphylococcus* spp., the experimental treatment affected the microbial growth by extending the λ in P samples in comparison with C and A and by reducing the μmax in P compared with all other groups and in AP compared to C. The final value was also affected by a significant reduction in the two experimentally manufactured burger groups (AP, P). Considering *Pseudomonas* spp., the P group recorded a longer λ compared to C and A samples while AP was not statistically different from the other experimental groups; concerning μmax, both AP and P showed lower values compared to C and A. For this microbial population, the final value was significantly reduced in the P group.

Neither λ nor the final value of the LAB population were different among groups, while the μmax was slightly higher in AP and P compared to C and A, agreeing with previous results from Servili et al. [56], who reported that olive by-product polyphenols did not affect LAB growth in fortified foodstuff. For *Enterobacteriaceae*, data show a reduction in the final value in the AP and P, while no significant differences were highlighted for λ and μmax; however, an increasing trend was recorded for λ and a decreasing one was noted for μmax ($p = 0.06$ and $p = 0.056$, respectively, data not shown).

In agreement with what is reported above, Mexis et al. [57] noticed a reduced growth rate in TVC, LAB and pseudomonads in ground chicken meat with the addition of Citrus spp. extracts. In Mortadella meat products with citrus fiber, thyme and rosemary, essential oil lowered the growth rate of the TVC during storage [58]. Roila et al. [25] reported that the addition of olive oil by-product polyphenols in Fior di Latte cheese brine resulted in an extended λ for *P. fluorescens* and *Enterobacteriaceae* and in a reduced μmax for *P. fluorescens*.

As shown, the result of growth data modeling appears to be related to the influence exerted by the experimental addition of olive-mill-wastewater-derived polyphenols on extending the λ and reducing the μmax and the final value. Analogous conclusions have been previously reported for similar compounds and for other preservation methods, albeit a systematic comparison is difficult as the characteristics of microbial growth curves can be affected by differences in food matrices, storage condition and duration [25,59].

Concerning the antimicrobial activity, it has been demonstrated that the dialdehydic structure of olive phenols exerts an antimicrobial effect by strongly interacting with amino acids, proteins and membrane molecules, promoting membrane permeabilization and bacterial cell lysis [60]. Indeed, studies have shown that tyrosol inhibits the activity of cyclooxygenase enzymes and hydroxytyrosol has a protein-denaturing ability [60]. Other studies report that many polyphenolic compounds are potent iron scavengers, and the lack of iron affects the growth of certain pathogenic bacteria by a reduction in the ribonucleotide precursor of DNA [61]. Besides the molecular mechanisms, the microbial growth inhibition exerted by this phenolic compound is strongly related to its chemical structure; as a consequence, the key factor determining the antibacterial activity of the phenolic extracts is their phenolic profile [62].

Table 4. Output parameters estimated by the DMFit program for each microbial population in the four formulations of beef burgers.

Micro-Organism and Parameters	C	A	AP	P
TVC				
λ	8.97 ± 3.45 [a]	6.12 ± 3.45 [a]	15.84 ± 8.20 [ab]	23.27 ± 11.74 [c]
μmax	0.0286 ± 0.0014 [b]	0.0286 ± 0.0013 [b]	0.0218 ± 0.0024 [a]	0.0248 ± 0.0048 [b]
Final value	7.44 ± 0.03 [b]	7.44 ± 0.03 [b]	7.11 ± 0.07 [b]	6.62 ± 0.10 [a]
R^2	*0.999*	*0.999*	*0.995*	*0.98*
SE of Fit	*0.041*	*0.0397*	*0.080*	*0.139*
Staphylococcus spp.				
λ	10.78 ± 5.35 [a]	11.86 ± 3.64 [a]	17.95 ± 3.82 [ab]	20.84 ± 6.03 [b]
μmax	0.0163 ± 0.0015 [bc]	0.0171 ± 0.0010 [c]	0.0151 ± 0.0009 [b]	0.0092 ± 0.0001 [a]
Final value	2.92 ± 0.02 [b]	2.89 ± 0.02 [b]	2.69 ± 0.02 [a]	2.66 ± 0.00 [a]
R^2	*0.997*	*0.999*	*0.998*	*0.998*
SE of Fit	*0.0346*	*0.0262*	*0.0263*	*0.065*
Pseudomonas spp.				
λ	12.33 ± 9.84 [a]	14.33 ± 8.32 [a]	20.38 ± 9.25 [ab]	29.73 ± 2.07 [b]
μmax	0.0324 ± 0.004 [b]	0.033 ± 0.004 [b]	0.024 ± 0.003 [a]	0.025 ± 0.000 [a]
Final value	7.16 ± 0.10 [b]	7.30 ± 0.09 [b]	7.07 ± 0.11 [b]	6.10 ± 0.02 [a]
R^2	*0.99*	*0.993*	*0.998*	*0.999*
SE of Fit	*0.136*	*0.12*	*0.11*	*0.025*
LAB				
λ	10.66 ± 8.58	9.86 ± 4.44	11.33 ± 3.19	17.03 ± 4.67
μmax	0.0083 ± 0.0010 [a]	0.0071 ± 0.0000 [a]	0.0095 ± 0.0040 [b]	0.0102 ± 0.0002 [b]
Final value	4.94 ± 0.02	4.89 ± 0.09	4.95 ± 0.08	4.83 ± 0.12
R^2	*0.993*	*0.998*	*0.999*	*0.997*
SE of Fit	*0.0294*	*0.013*	*0.0128*	*0.0209*
Enterobacteriaceae				
λ	34.38 ± 10.82	38.59 ± 6.21	44.40 ± 2.55	49.036 ± 10.74
μmax	0.0205 ± 0.0026	0.022 ± 0.0015	0.0201 ± 0.0006	0.0176 ± 0.0035
Final value	3.80 ± 0.12 [c]	3.88 ± 0.07 [c]	3.38 ± 0.025 [b]	2.41 ± 0.05 [a]
R^2	*0.989*	*0.996*	*0.999*	*0.987*
SE of Fit	*0.112*	*0.0672*	*0.0238*	*0.063*

λ = lag phase (h); μmax = maximum growth rate (Log/CFU/g/h); final value (Log/CFU/g); SE = standard error of fitting; R^2 = adjusted R-square statistics of the fitting. C = control group; A= basic recipe with addition of 10 g/kg commercial antioxidant; AP = basic recipe with addition of 5 g/kg commercial antioxidant and 350 mg/kg phenolic extract; P = basic recipe with addition of 700 mg/kg phenolic extract. Different letters in the same row (a, b, c) indicate differences between mean values for different experimental groups ($p \leq 0.001$).

4. Conclusions

Consumers' interest in meat products formulated with natural preservatives has motivated researchers to evaluate the effectiveness and applicability of naturally occurring compounds with antioxidant and antimicrobial purposes. This study has proved that olive mill wastewater extracts are characterized by a phenolic profile and are able to significantly improve the oxidative and microbial stability of beef burger during cold storage.

Therefore, it was concluded that olive mill wastewater extracts could be successfully added to raw ground beef meat to act as natural antioxidants and antimicrobials with added health, environmental and economic benefits as well as increased consumer appeal.

Author Contributions: Conceptualization, R.B., S.E. and R.R.; methodology, S.E. and R.B.; validation, S.E.; formal analysis, R.R., S.P. and A.T.; investigation, R.R., A.V. and B.S.; resources, A.V. and M.S.; data curation, R.R. and R.B.; writing—original draft preparation, R.R.; writing—review and editing, D.R., R.B. and S.P.; visualization, B.S.; supervision, R.B., S.E. and M.S.; project administration, M.S. All authors have read and agreed to the published version of the manuscript.

Funding: This paper is part of the research project (PSR GO-Nuovi alimenti—PSR "Dalla tradizione Umbra nuovi alimenti di origine animale ricchi in molecole bioattive per migliori caratteristiche di sicurezza e qualità salutistica" Programma di Sviluppo Rurale per l'Umbria 2014–2020—Misura 16—Sottomisura 16.1) funded by the European Commision, through the Umbrian Region (Italy), to develop rural agriculture and animal production in the member states.

Institutional Review Board Statement: Not applicable.

Informed Consent Statement: Not applicable.

Data Availability Statement: The datasets generated for this study are available on request from the corresponding author.

Acknowledgments: The authors would like to thank Fattoria Luchetti and Clarita Cavallucci for their collaboration in the research activities. The study has been developed in the context of the project: Study of "green" strategies to ensure the hygiene and safety of prepared and processed foods of animal origin.

Conflicts of Interest: The authors declare no conflict of interest.

References

1. Moghtadaei, M.; Soltanizadeh, N.; Goli, S.A.H. Production of sesame oil oleogels based on beeswax and application as partial substitutes of animal fat in beef burger. *Int. Food Res. J.* **2018**, *108*, 368–377. [CrossRef] [PubMed]
2. Rabadán, A.; Álvarez-Ortí, M.; Martínez, E.; Pardo-Giménez, A.; Zied, D.C.; Pardo, J.E. Effect of replacing traditional ingredients for oils and flours from nuts and seeds on the characteristics and consumer preferences of lamb meat burgers. *LWT* **2021**, *136*, 110307. [CrossRef]
3. Branciari, R.; Ranucci, D.; Urbani, E.; Valiani, A.; Trabalza-Marinucci, M.; Dal Bosco, A.; Franceschini, R. Freshwater fish burgers made from four different fish species as a valuable strategy appreciated by consumers for introducing EPA and DHA into a human diet. *J. Aquat. Food Prod. Technol.* **2017**, *26*, 686–694. [CrossRef]
4. Mancini, S.; Paci, G.; Fratini, F.; Torracca, B.; Nuvoloni, R.; Dal Bosco, A.; Preziuso, G. Improving pork burgers quality using Zingiber officinale Roscoe powder (ginger). *Meat Sci.* **2017**, *129*, 161–168. [CrossRef]
5. Mancini, R.; Hunt, M. Current research in meat color. *Meat Sci.* **2005**, *71*, 100–121. [CrossRef]
6. Hygreeva, D.; Pandey, M.C. Novel approaches in improving the quality and safety aspects of processed meat products through high pressure processing technology-A review. *Trends Food Sci. Technol.* **2016**, *54*, 175–185. [CrossRef]
7. Jiang, J.; Xiong, Y.L. Natural antioxidants as food and feed additives to promote health benefits and quality of meat products: A review. *Meat Sci.* **2016**, *120*, 107–117. [CrossRef] [PubMed]
8. Overholt, M.F.; Mancini, S.; Galloway, H.O.; Preziuso, G.; Dilger, A.C.; Boler, D.D. Effects of salt purity on lipid oxidation, sensory characteristics, and textural properties of fresh, ground pork patties. *LWT* **2016**, *65*, 890–896. [CrossRef]
9. Shahidi, F.; Ambigaipalan, P. Phenolics and polyphenolics in foods, beverages and spices: Antioxidant activity and health effects–A review. *J. Funct. Foods* **2015**, *18*, 820–897. [CrossRef]
10. Yang, H.J.; Lee, J.H.; Won, M.; Song, K.B. Antioxidant activities of distiller dried grains with solubles as protein films containing tea extracts and their application in the packaging of pork meat. *Food Chem.* **2016**, *196*, 174–179. [CrossRef] [PubMed]
11. Falowo, A.B.; Fayemi, P.O.; Muchenje, V. Natural antioxidants against lipid–protein oxidative deterioration in meat and meat products: A review. *Food Res. Int.* **2014**, *64*, 171–181. [CrossRef]
12. EC (European Community). Commission regulation (EC) No. 1333/2008: On food additives. *Off. J.* **2008**, *354*, 16.
13. Silva, M.M.; Lidon, F.C. An overview on applications and side effects of antioxidant food additives. *Emir. J. Food Agric.* **2016**, *28*, 823–832. [CrossRef]
14. Etienne, J.; Chirico, S.; McEntaggart, K.; Papoutsis, S.; Millstone, E. EU Insights–Consumer perceptions of emerging risks in the food chain. *EFSA Support. Publ.* **2018**, *15*, 1394. [CrossRef]
15. Aminzare, M.; Hashemi, M.; Ansarian, E.; Bimakr, M.; Hassanzad Azar, H.; Mehrasbi, M.R.; Afshari, A. Using natural antioxidants in meat and meat products as preservatives: A review. *Adv. Anim. Vet.* **2019**, *7*, 417–426. [CrossRef]
16. Ranucci, D.; Miraglia, D.; Branciari, R.; Morganti, G.; Roila, R.; Zhou, K.; Haiyang, J.; Braconi, P. Frankfurters made with pork meat, emmer wheat (*Triticum dicoccum Schübler*) and almonds nut (*Prunus dulcis Mill.*): Evaluation during storage of a novel food from an ancient recipe. *Meat Sci.* **2018**, *145*, 440–446. [CrossRef] [PubMed]
17. Bellés, M.; Alonso, V.; Roncalés, P.; Beltrán, J.A. Sulfite-free lamb burger meat: Antimicrobial and antioxidant properties of green tea and carvacrol. *J. Sci. Food Agric.* **2019**, *99*, 464–472. [CrossRef] [PubMed]

18. Mancini, S.; Preziuso, G.; Dal Bosco, A.; Roscini, V.; Szendrő, Z.; Fratini, F.; Paci, G. Effect of turmeric powder (*Curcuma longa* L.) and ascorbic acid on physical characteristics and oxidative status of fresh and stored rabbit burgers. *Meat Sci.* **2015**, *110*, 93–100. [CrossRef]
19. Beya, M.M.; Netzel, M.E.; Sultanbawa, Y.; Smyth, H.; Hoffman, L.C. Plant-Based Phenolic Molecules as Natural Preservatives in Comminuted Meats: A Review. *Antioxidants* **2021**, *10*, 263. [CrossRef]
20. Barbieri, S.; Mercatante, D.; Balzan, S.; Esposto, S.; Cardenia, V.; Servili, M.; Rodriguez-Estrada, M.T. Improved Oxidative Stability and Sensory Quality of Beef Hamburgers Enriched with a Phenolic Extract from Olive Vegetation Water. *Antioxidants* **2021**, *10*, 1969. [CrossRef]
21. Fasolato, L.; Cardazzo, B.; Balzan, S.; Carraro, L.; Taticchi, A.; Montemurro, F.; Novelli, E. Minimum bactericidal concentration of phenols extracted from oil vegetation water on spoilers, starters and food-borne bacteria. *Ital. J. Food Saf.* **2015**, *4*, 4519. [CrossRef] [PubMed]
22. Servili, M.; Sordini, B.; Esposto, S.; Urbani, S.; Veneziani, G.; Maio, I.D.; Selvaggini, R.; Taticchi, A. Biological activities of phenolic compounds of extra virgin olive oil. *Antioxidants* **2013**, *3*, 1–23. [CrossRef] [PubMed]
23. Kowalska, H.; Czajkowska, K.; Cichowska, J.; Lenart, A. What's new in biopotential of fruit and vegetable by-products applied in the food processing industry. *Trends Food Sci. Tech.* **2017**, *67*, 150–159. [CrossRef]
24. Miraglia, D.; Esposto, S.; Branciari, R.; Urbani, S.; Servili, M.; Perucci, S.; Ranucci, D. Effect of a Phenolic extract from olive vegetation water on fresh salmon steak quality during storage. *Ital. J. Food Saf.* **2016**, *5*, 6167. [CrossRef]
25. Roila, R.; Valiani, A.; Ranucci, D.; Ortenzi, R.; Servili, M.; Veneziani, G.; Branciari, R. Antimicrobial efficacy of a polyphenolic extract from olive oil by-product against "Fior di latte" cheese spoilage bacteria. *Int. J. Food Microbiol.* **2019**, *295*, 49–53. [CrossRef] [PubMed]
26. Ianni, F.; Gagliardi, A.; Taticchi, A.; Servili, M.; Pinna, N.; Schoubben, A.; Sardella, R.; Bruscoli, S. Exploiting food-grade mesoporous silica to preserve the antioxidant properties of fresh olive mill wastewaters phenolic extracts. *Antioxidants* **2021**, *10*, 1361. [CrossRef] [PubMed]
27. Zbicinski, I.; Delag, A.; Strumillo, C.; Adamiec, J. Advanced experimental analysis of drying kinetics in spray drying. *Chem. Eng. J.* **2002**, *86*, 207–216. [CrossRef]
28. Selvaggini, R.; Esposto, S.; Taticchi, A.; Urbani, S.; Veneziani, G.; Di Maio, I.; Sordini, B.; Servili, M. Optimization of the temperature and oxygen concentration conditions in the malaxation during the oil mechanical extraction process of four Italian olive cultivars. *J. Agric. Food Chem.* **2014**, *62*, 3813–3822. [CrossRef] [PubMed]
29. EC (European Community). *Commission regulation (EC) No. 853/2004*; Laying down Specific Hygiene Rules for on the Hygiene of Foodstuffs. *Off. J.* **2004**, *L139*, 55. Available online: https://eur-lex.europa.eu/LexUriServ/LexUriServ.do?uri=OJ:L:2004:139:0055:0205:EN:PDF (accessed on 23 May 2022).
30. Tarladgis, B.G.; Watts, B.M.; Younathan, M.T.; Dugan, L., Jr. A distillation method for the quantitative determination of malonaldehyde in rancid foods. *JAOCS J. Am. Oil Chem. Soc.* **1960**, *37*, 44–48. [CrossRef]
31. Balouiri, M.; Sadiki, M.; Ibnsouda, S.K. Methods for in vitro evaluating antimicrobial activity: A review. *J. Pharm. Anal.* **2016**, *6*, 71–79. [CrossRef] [PubMed]
32. Di Michele, A.; Pagano, C.; Allegrini, A.; Blasi, F.; Cossignani, L.; Raimo, E.D.; Faieta, M.; Oliva, E.; Pittia, P.; Primavilla, S.; et al. Hazelnut shells as source of active ingredients: Extracts preparation and characterization. *Molecules* **2021**, *26*, 6607. [CrossRef] [PubMed]
33. *ISO 4833–1: 2013*; 44–Microbiology of the Food Chain-Horizontal Method for the Enumeration of Microorganisms-Part 1: Colony Count at 30 °C by the Pour Plate Technique. International Organization for Standardization: Geneva, Switzerland, 2013.
34. *ISO 21528-2: 2017*; Microbiology of the Food Chain–Horizontal Method for the Detection and Enumeration of Enterobacteriaceae–Part 2: Colony-Count Technique. International Organization for Standardization (ISO): Geneva, Switzerland, 2017.
35. Gill, C.; Jones, T. Microbiological sampling of carcasses by excision or swabbing. *J. Food Prot.* **2000**, *63*, 167–173. [CrossRef] [PubMed]
36. *ISO 6579-1: 2017*; Microbiology of the Food Chain–Horizontal Method for the Detection, Enumeration and Serotyping of Salmonella–Part 1: Horizontal Method for the Detection of *Salmonella* spp. International Organization for Standardization (ISO): Geneva, Switzerland, 2017.
37. EC (European Community). *Commission regulation (EC) No. 2073/2005*; Microbiological Criteria for Foodstuffs. *Off. J.* **2005**, *338*, 1.
38. *ISO 16649*; Microbiology of Food and Animal FeedingStuffs—Horizontal Method for the Enumeration of b-Glucuronidase-Positive *Escherichia coli*. International Organization for Standardization: Geneva, Switzerland, 2001.
39. SAS Institute. *JMP®9 Basic Analysis and Graphing*; SAS Institute Inc.: Cary, NC, USA, 2010.
40. Baranyi, J.; Roberts, T.A. A dynamic approach to predicting bacterial growth in food. *Int. J. Food Microbiol.* **1994**, *23*, 277–294. [CrossRef]
41. Ranucci, D.; Branciari, R.; Cobellis, G.; Acuti, G.; Miraglia, D.; Olivieri, O.; Roila, R.; Trabalza-Marinucci, M. Dietary essential oil mix improves oxidative stability and hygienic characteristic of lamb meat. *Small Rum. Res.* **2019**, *175*, 104–109. [CrossRef]
42. Roila, R.; Valiani, A.; Miraglia, D.; Ranucci, D.; Forte, C.; Trabalza-Marinucci, M.; Servili, M.; Codini, M.; Branciari, R. Olive mill wastewater phenolic concentrate as natural antioxidant against lipid-protein oxidative deterioration in chicken meat during storage. *Ital. J. Food Saf.* **2018**, *7*, 7342. [CrossRef] [PubMed]

43. Martínez-Zamora, L.; Ros, G.; Nieto, G. Synthetic vs. Natural Hydroxytyrosol for Clean Label Lamb Burgers. *Antioxidants* **2020**, *9*, 851. [CrossRef] [PubMed]
44. Di Maio, I.; Esposto, S.; Taticchi, A.; Selvaggini, R.; Veneziani, G.; Urbani, S.; Servili, M. HPLC–ESI-MS investigation of tyrosol and hydroxytyrosol oxidation products in virgin olive oil. *Food Chem.* **2011**, *125*, 21–28. [CrossRef]
45. Nieto, G.; Martínez-Zamora, L.; Castillo, J.; Ros, G. Hydroxytyrosol extracts, olive oil and walnuts as functional components in chicken sausages. *J. Sci. Food Agric.* **2017**, *97*, 3761–3771. [CrossRef] [PubMed]
46. Shah, M.A.; Bosco, S.J.D.; Mir, S.A. Plant extracts as natural antioxidants in meat and meat products. *Meat Sci.* **2014**, *98*, 21–33. [CrossRef]
47. Zamuz, S.; López-Pedrouso, M.; Barba, F.J.; Lorenzo, J.M.; Domínguez, H.; Franco, D. Application of hull, bur and leaf chestnut extracts on the shelf-life of beef patties stored under MAP: Evaluation of their impact on physicochemical properties, lipid oxidation, antioxidant, and antimicrobial potential. *Food Res. Int.* **2018**, *112*, 263–273. [CrossRef] [PubMed]
48. Fernandes, R.D.P.P.; Trindade, M.A.; Lorenzo, J.M.; Munekata, P.E.S.; De Melo, M.P. Effects of oregano extract on oxidative, microbiological and sensory stability of sheep burgers packed in modified atmosphere. *Food Control* **2016**, *63*, 65–75. [CrossRef]
49. Hayes, J.E.; Stepanyan, V.; Allen, P.; O'grady, M.N.; Kerry, J.P. Effect of lutein, sesamol, ellagic acid and olive leaf extract on the quality and shelf-life stability of packaged raw minced beef patties. *Meat Sci.* **2010**, *84*, 613–620. [CrossRef] [PubMed]
50. Comi, G.; Tirloni, E.; Andyanto, D.; Manzano, M.; Iacumin, L. Use of bio-protective cultures to improve the shelf-life and the sensorial characteristics of commercial hamburgers. *LWT-Food Sci. Technol.* **2015**, *62*, 1198–1202. [CrossRef]
51. Sun, Y.; Zhang, M.; Bhandari, B.; Bai, B. Nanoemulsion-based edible coatings loaded with fennel essential oil/cinnamaldehyde: Characterization, antimicrobial property and advantages in pork meat patties application. *Food Control* **2021**, *127*, 108151. [CrossRef]
52. Branciari, R.; Galarini, R.; Trabalza-Marinucci, M.; Miraglia, D.; Roila, R.; Acuti, G.; Giuseppone, D.; Dal Bosco, A.; Ranucci, D. Effects of olive mill vegetation water phenol metabolites transferred to muscle through animal diet on rabbit meat microbial quality. *Sustainability* **2021**, *13*, 4522. [CrossRef]
53. Andrés, A.I.; Petrón, M.J.; Adámez, J.D.; López, M.; Timón, M.L. Food by-products as potential antioxidant and antimicrobial additives in chill stored raw lamb patties. *Meat Sci.* **2017**, *129*, 62–70. [CrossRef] [PubMed]
54. Parafati, L.; Palmeri, R.; Trippa, D.; Restuccia, C.; Fallico, B. Quality maintenance of beef burger patties by direct addiction or encapsulation of a prickly pear fruit extract. *Front. Microbiol.* **2019**, *10*, 1760. [CrossRef] [PubMed]
55. Marrone, R.; Smaldone, G.; Ambrosio, R.L.; Festa, R.; Ceruso, M.; Chianese, A.; Anastasio, A. Effect of beetroot (Beta vulgaris) extract on Black Angus burgers shelf life. *Ital. J. Food Saf.* **2021**, *10*, 9031. [CrossRef] [PubMed]
56. Servili, M.; Rizzello, C.G.; Taticchi, A.; Esposto, S.; Urbani, S.; Mazzacane, F.; Di Maio, I.; Selvaggini, M.; Gobbetti, R.; Di Cagno, R. Functional milk beverage fortified with phenolic compounds extracted from olive vegetation water, and fermented with functional lactic acid bacteria. *Int. J. Food Microbiol.* **2011**, *147*, 45–52. [CrossRef] [PubMed]
57. Mexis, S.F.; Chouliara, E.; Kontominas, M.G. Shelf life extension of ground chicken meat using an oxygen absorber and a citrus extract. *LWT* **2012**, *49*, 21–27. [CrossRef]
58. Manuel, V.M.; Yolanda, R.N.; Juana, F.L.; José Angel, P.A. Effect of packaging conditions on shelf-life of Mortadella made with citrus fibre washing water and thyme or rosemary essential oil. *Food Nutr. Sci.* **2011**, *2*, 1–10. [CrossRef]
59. Altieri, C.; Scrocco, C.; Sinigaglia, M.; Del Nobile, M.A. Use of chitosan to prolong mozzarella cheese shelf life. *J. Dairy Sci.* **2005**, *88*, 2683–2688. [CrossRef]
60. Medina, E.; Brenes, M.; Garcia, A.; Romero, C.; de Castro, A. Bactericidal activity of glutaraldehyde-like compounds from olive products. *J. Food Prot.* **2009**, *72*, 2611–2614. [CrossRef]
61. Lim, C.K.; Penesyan, A.; Hassan, K.A.; Loper, J.E.; Paulsen, I.T. Effect of tannic acid on the transcriptome of the soil bacterium Pseudomonas protegens Pf-5. *Appl. Environ. Microbiol.* **2013**, *79*, 3141–3145. [CrossRef]
62. Belaqziz, M.; Tan, S.P.; El-Abbassi, A.; Kiai, H.; Hafidi, A.; O'Donovan, O.; McLoughlin, P. Assessment of the antioxidant and antibacterial activities of different olive processing wastewaters. *PLoS ONE* **2017**, *12*, e0182622. [CrossRef]

foods

Article

Sustainable Design of Innovative Kiwi Byproducts-Based Ingredients Containing Probiotics

Gheorghe-Ionuț Ilie [1], Ștefania-Adelina Milea [1], Gabriela Râpeanu [1], Adrian Cîrciumaru [2] and Nicoleta Stănciuc [1,*]

1 Faculty of Food Science and Engineering, Dunarea de Jos University of Galati, 800201 Galați, Romania
2 Cross-Border Faculty, Dunarea de Jos University of Galati, 800201 Galați, Romania
* Correspondence: nicoleta.stanciuc@ugal.ro

Abstract: Industrial processing of kiwifruits generates a large quantity of byproducts, estimated to be one million tons per year. The resulting byproducts are rich sources of bioactive components that may be used as additives, hence minimizing economic and environmental issues. In this study, kiwifruit byproducts were used to develop added-value food-grade ingredients containing probiotics. The byproducts were divided into peels and pomace. Both residues were inoculated with a selected strain of probiotic (*Lacticaseibacillus casei* 431®), and two variants were additionally enhanced with prebiotic sources (buckwheat and black rice flours). The inoculated powders were obtained by freeze-drying, and the final ingredients were coded as KP (freeze-dried kiwi peels), KBR (freeze-dried kiwi pomace and black rice flour), KPB (freeze-dried kiwi pomace and buckwheat flour), and KPO (freeze-dried kiwi pomace). The phytochemical profile was assessed using different spectrophotometric methods, such as the determination of polyphenols, flavonoids, and carotenoids. The kiwi byproduct-based formulations showed a polyphenolic content varying from 10.56 ± 0.30 mg AGE/g DW to 13.16 ± 0.33 mg AGE/g, and the survival rate of lactic acid bacteria after freeze-drying ranged from 73% to 88%. The results showed an increase in total flavonoid content from the oral to gastric environment and controlled release in the intestinal environment, whereas a maximum survival rate of probiotics at the intestinal end stage was 48%. The results of SEM and droplet size measurements revealed vesicular and polyhedral structures on curved surfaces linked by ridge sections. The CIEL*a*b* color data were strongly associated with the particular pigment in kiwi pulp, as well as the color of the additional flour. Finally, the ingredients were tested in protein bars and enhanced the value of the final food product regarding its phytochemical and probiotic content.

Keywords: kiwi byproducts; probiotic; prebiotic; antioxidant activity; *Lactobacillus casei*; ingredients; functional foods

Citation: Ilie, G.-I.; Milea, Ș.-A.; Râpeanu, G.; Cîrciumaru, A.; Stănciuc, N. Sustainable Design of Innovative Kiwi Byproducts-Based Ingredients Containing Probiotics. *Foods* **2022**, *11*, 2334. https://doi.org/10.3390/foods11152334

Academic Editors: Marco Poiana, Francesco Caponio and Antonio Piga

Received: 13 July 2022
Accepted: 2 August 2022
Published: 5 August 2022

Publisher's Note: MDPI stays neutral with regard to jurisdictional claims in published maps and institutional affiliations.

Copyright: © 2022 by the authors. Licensee MDPI, Basel, Switzerland. This article is an open access article distributed under the terms and conditions of the Creative Commons Attribution (CC BY) license (https://creativecommons.org/licenses/by/4.0/).

1. Introduction

The industrial processing of fruits and vegetables leads to large quantities of byproducts, such as peels, seeds, and residual pulp, which are usually used as animal feed or/and discarded in the environment. Nowadays, enormous efforts, from both scientific and application points of view, are made to explore the functional potential of these byproducts as sources of important chemical compounds with significant added value for human and animal health. In general, these compounds are represented by polyphenols, carotenoids, and triterpenes [1], which display several biological activities, namely, antioxidant [2], anti-inflammatory [3], antimicrobial [4], and antidiabetic effects [5], among many others [6].

The genus *Actinidia* contains about 60 species, with kiwi probably being the most consumed fruit in natura due to its well-described health benefits [7]. Kiwifruits can be preserved in the fresh state for a prolonged time (for many months) without any decrease in quality, requiring controlled temperature (0 °C). At the industrial level, innovative kiwi products are processed and commercialized, such as juice, frozen juice, sweets, and ice creams, among other products [8]. Industrial processing of kiwifruits causes significant amounts of peels and pomace, which are still under-explored and have aroused great interest due to their high contents of bioactive molecules, such as phenolic compounds [9]. The development of kiwi-derived products is based on their nutritional and biological properties due to their rich contents of dietary fiber and bioactive compounds, such as vitamins (C, E, and A), phenolic compounds, and minerals [6].

Significant data from the literature have shown that one of the most recommended ways to prevent and treat gastrointestinal alterations is to alter the commensal microbiota [10–12]. It is well known that microorganisms coexist with eukaryotic cells at the mucosal surfaces of vertebrates in a complex and harmonious symbiosis [11]. However, the administration of antibiotics at a large scale causes various side effects, such as disrupting normal microflora in the human body by destroying normal gut and genital tract bacteria [13]. These imbalances can be adjusted by probiotic supplementation in the diet, providing a load of live bacterial cells sufficient for beneficial effects on human health at both metabolic and immune levels [11]. The engineering of probiotic foods is challenging due to their low acid-bile tolerance and requirements for their non-pathogenicity, nontoxicity, ability to survive and metabolize in the gut environment, resistance to low pH and organic acid, and potential to remain viable for a long period under storage and other conditions [14]. An optimal population of probiotic bacteria is critical for the preservation and proper functioning of the digestive system [15].

Prebiotics are defined as non-fermentable components that are transferred into the colon to be selectively used by host microorganisms [16]. Multiple benefits have been suggested for prebiotics, such as the mediation of host health (such as improved intestinal function), the regulation of glucose and lipid metabolism, immune response, bone health, and the regulation of satiety [17]. It has been suggested that prebiotics can be classified into three categories: oligosaccharides (fructooligosaccharide, xylooligosaccharide, galactooligosaccharide, isomaltose, inulin, etc.), fiber (β-glucan, pectin, cellulose, dextrin, etc.), and polyols (xylitol, mannitol, lactulose, etc.) [17]. Other compounds have been proposed as candidates for prebiotics, such as linoleic acid, polyunsaturated fatty acids, phenols, and polyphenols, such as anthocyanins [18].

Given the need to use prebiotics in the engineering of probiotic foods, different complex food matrices should be considered, including fibers, polyphenolics, and oligosaccharides. For example, black rice (*Oryza sativa* L.) flour is a valuable source of protein, fat, carbohydrates, phenols, flavonoids, and anthocyanins [19] and could be a potent candidate for complex prebiotic activity. Additionally, buckwheat is a type of underutilized pseudocereal belonging to the genus *Fagopyrum*, and it could be regarded as a potential source for food and nutritional applications [20]. According to FoodData Central [21], the proximate composition of buckwheat includes starch as the major component (~70%), followed by protein (~12%), dietary fiber (~10%), lipids (~3%), and ash (2.5%); minor components with biological significance, such as polyphenols, D-chiro-inositol, and vitamins, were also reported [22].

Therefore, our study aims to contribute to the innovative development of kiwi byproduct-based products containing probiotics. Although the kiwi phytochemical profile is already advanced in the scientific literature, based on our knowledge, no studies have explored the potential of transforming kiwi byproducts into foods and ingredients containing probiotics, consequently reducing food waste. Kiwi pomace was enhanced with two prebiotic sources, namely, black rice and buckwheat flour, and inoculated with *Lacticaseibacillus casei* (*L. casei*) 431®, whereas kiwi peels were inoculated to the same extent with *L. casei*. Four powders containing probiotics were obtained by freeze-drying and characterized for phytochemicals (anthocyanins, polyphenols, and carotenoids), antioxidant activity, color, cell viability, and the bioaccessibility of polyphenols and probiotics. The powders were tested for food applications by introducing them into a protein bar formulation. The foods were tested for phytochemicals, antioxidant activity, and cell viability to assess their potential as functional foods.

2. Materials and Methods

2.1. Chemicals

Analytical-grade n-hexane, acetone, 2,2-diphenyl-1-picrylhydrazyl (DPPH), 6-hydroxy-2,5,7,8-tetramethylchromane-2-carboxylic acid (Trolox), acetic acid, hydrochloric acid, aluminum chloride, ethanol, Folin–Ciocâlteu reagent, gallic acid, pectin, and methanol were purchased from Sigma Aldrich Steinheim (Darmstadt, Germany). Other reagents such as sodium bicarbonate were purchased from Honeywell, Fluka (Seelze, Germany), and *Lacticaseibacillus casei (L. casei)* 431® strain was purchased from Chr. Hansen (Hoersholm, Denmark). The probiotic character of *L. casei* 431® strain has been described in several studies [23,24] and has been associated with health benefits in several areas of health, including immune health, respiratory health, and bowel function. De Man, Rogosa, and Sharpe agar (MRS agar) was purchased from Merck (Darmstadt, Germany). All reagents and solvents were of analytical and HPLC grade.

2.2. Fruit Processing

Kiwifruits (*Actinidia deliciosa* cv. "Hayward") at full maturity were purchased in November 2021 from a local supermarket in Galati County, Romania. The fruits were sorted and processed immediately. Once they arrived at the laboratory, the fruits were washed with distilled water, and peels were separated from the fruits. The fruits were further cut in half and squeezed using a juice extractor (Stainless Steel Fruit Vegetable Juice Extractor Juicer Squeezer, Guangdong, China). The fresh material, specifically peels and pomace, was used for initial phytochemical content and antioxidant activity analyses.

2.3. Extraction of Phytochemicals from Fresh Kiwi Samples

In order to characterize the freeze-dried powders of the fresh material, a subsequent extraction was performed using a solid–liquid ultrasound-assisted method with two different solvents. The design of extraction procedures was based on a comprehensive comparison of both lipophilic and hydrophilic profiles by using two different combinations of solvents: a mixture of ethanol and water (ratio of 70:30, v/v) and n-hexane–acetone (ratio 1:3, v/v). The fresh material was extracted with the corresponding solvent solutions in a solid–liquid ratio of 1:10 by mixing and ultrasound treatment at 35 °C for 30 min. Each extraction was repeated twice to obtain enriched extracts.

2.4. Customized Design to Develop Kiwi-Based Ingredients Enriched with Probiotics

About 50 g of fresh material (kiwi peels and pomace) was added to ultrapure water in a ratio of 1:2 and mixed by homogenization (ProBlend Crush Blender, Global Headquarters Netherlands, Eindhoven). Prior to inoculation, the mixtures were enhanced with different flours, i.e., black rice and buckwheat (20%), and allowed to hydrate, followed by pH adjustment to 5.0 with 1.0 N NaOH. The designed samples were coded as follows: KP (freeze-dried kiwi peels), KBR (freeze-dried kiwi pomace and black rice flour), KPB (freeze-dried kiwi pomace and buckwheat flour), and KPO (freeze-dried kiwi pomace). All samples were sterilized using a UV lamp and inoculated with 2% *L. casei* 431® and freeze-dried (CHRIST Alpha 1–4 LD plus, Osterode am Harz Germany) at −42 °C under a pressure of 10 Pa for 48 h. Afterwards, the powders were collected and packed in glass containers and kept at 4 °C until further analysis.

2.5. Total Polyphenol (TPC) and Total Flavonoid (TFC) Analysis

Spectrophotometric methods were used for the TPC and TFC evaluation by using the Folin–Ciocâlteu reagent and aluminum chloride methods, respectively, as described by Milea et al. [25]. For TP content, from each sample extract, a volume of 0.2 mL was diluted with 15.8 mL of distilled water, followed by the addition of 1 mL of Folin–Ciocâlteu reagent and 3 mL of 20% Na_2CO_3. The mixtures were left to stand for one hour in the dark to react, followed by reading the absorbance at λ = 765 nm (Jenway Scientific Instruments, Essex, UK). TP was expressed as mg gallic acid equivalents/g dry weight (mg GAE/g

DW) using a calibration curve. For TF evaluation, a volume of 0.25 mL from each extract was sequentially mixed with 0.075 mL of sodium nitrite (5%), 0.15 mL of 10% $AlCl_3$, and 0.5 mL of 1 M NaOH. The absorbance of the mixtures was immediately measured at 510 nm against the suitable blank. TF was expressed in mg catechin equivalents (CE) per g of dry powder (mg CE/g DW).

2.6. Total Monomeric Anthocyanin Content

Due to the special formulation of the KBR sample, which contained freeze-dried kiwi pomace and black rice flour, the total monomeric anthocyanin content (TAC) was assessed using the official AOAC method [26] and expressed as mg cianidin-3-O-glucoside equivalents/g DW (mg C3G/g DW).

2.7. Carotenoid Content Evaluation

Carotenoid content, specifically β-carotene and lycopene, was determined by the spectrophotometric method by selecting different wavelengths: 470 nm (β-carotene) and 503 nm (lycopene). The amount of carotenoids was calculated according to the following equation [27]:

$$Carotenoids\left(\frac{mg}{g}\right) = \frac{A \times M_w \times D_f}{M_a \times L} \quad (1)$$

where A is the absorbance of the extracts at the corresponding wavelength; M_w is molecular weight, D_f is the sample dilution rate, M_a is molar absorptivity (2500 L mol^{-1} cm^{-1} and 3450 L mol^{-1} cm^{-1}, respectively), and L is the cell diameter of the spectrophotometer (1 cm).

2.8. Radical Scavenging Activity

In order to evaluate the radical scavenging activity of the extracts, two methods were comparatively used, namely, the scavenging percentages of DPPH and ABTS radicals [28]. In order to evaluate the DPPH radical scavenging activity, a volume of 0.1 mL from the hydrophilic extract was mixed with 3.9 mL of DPPH solution (1 mM in methanol), mixed, and allowed to react for 1 h at room temperature in the dark. For ABTS radical scavenging activity, a volume of 0.15 mL from the lipophilic extract was mixed with a volume of 2.9 mL ABTS solution (7 mM ABTS in 2.45 mM $K_2S_2O_8$) and allowed to react for 2 h at room temperature in the dark. The radical scavenging activity was expressed as mMol Trolox/g DW using a calibration curve.

2.9. Viable Counts of *L. casei* 431®

In order to evaluate the viable counts of *L. casei* 431®, 10-fold serial dilutions of the freeze-dried powders were performed using sterile physiological serum (0.9 NaCl %, *w*/*v*) by using the pour plate technique, as described by Vasile et al. [29]. The viable cell number was determined by estimating the number of colony-forming units (CFU) by cultivation on MRS agar plates (medium at pH 5.7) after 72 h of aerobic incubation at 37 °C. The counts were expressed as CFU/g DW.

2.10. In Vitro Digestion

In order to evaluate the bioaccessibility of the bioactives and probiotics from powders, the KPB sample was selected in order to evaluate the flavonoids and probiotic viability in a simulated in vitro digestion model. The samples were selected based on the phytochemical profile and viable counts of *L. casei* 431®. The simulated model was prepared using a modified method described by Kim et al. [30]. In brief, the method involved successive steps of dissolving the powder in 10 mL of simulated saliva (Tris-HCl buffer, pH 7.7) containing 1 mg/mL α-amylase, followed by gastric digestion for 2 h and then intestinal digestion for another 2 h. An amount of 0.5 g of powder was dissolved in 10 mL of simulated saliva, which was then added to 20 mL of simulated gastric juice (SGS) composed of pepsin (1 mg/mL in 0.1 N HCl) at pH 2.0 and 1N HCl.

After 2 h of gastric digestion at 37 °C, with continuous stirring at 150 rpm, a volume of 15 mL of the digested sample was transferred to simulated intestinal juice (SIS) consisting of pancreatin with sodium bicarbonate at pH 7.0, followed by incubating the samples for 2 h at 37 °C with continuous stirring at 150 rpm on an SI-300R orbital shaker (Medline Scientific, UK). The TFC of the sample was measured in the initial and final phases of each digestion stage (after the oral phase, before and after 2 h of gastric digestion, and before and after 2 h of intestinal digestion).

The bioavailability of flavonoids refers to the amounts of compounds released from the powder after gastrointestinal digestion that could become available for absorption into the systemic circulation [31]. Bioaccessibility was calculated using Equation (2):

$$Bioaccesibility\ (\%) = \frac{TFC_1}{TFC_0} \times 100 \tag{2}$$

where TFC_1 is the total flavonoid content (mg/g DW) in samples digested after gastrointestinal digestion, and TFC_0 is the total flavonoid content (mg/g DW) in powder before gastrointestinal digestion.

For the survival rate of probiotics, the inoculated powder was dissolved in 10 mL of simulated saliva, and the aforementioned protocol describing the in vitro digestion of flavonoids was applied. In the initial step and after 2 h of gastric digestion or at 2 h of intestinal digestion, samples were analyzed in terms of the viability of probiotic cells. The survival rate of probiotic cells was calculated using Equation (3):

$$Survival\ rate\ (\%) = \frac{Log\ (N_1)}{Log\ (N_0)} \times 100 \tag{3}$$

where N_1 is the total number of viable cells after each stage of simulated digestion, and N_0 is the initial total number of viable cells before exposure at each stage of digestion.

2.11. CIEL*a*b* Analysis of Freeze-Dried Powders

For the color analysis of powders, a CR 410 Chroma Meter (Konica Minolta, Tokyo, Japan) colorimeter was used to determine the color coordinates: L* (illumination, brightness, 0 black, 100 white), a* (positive value red, negative value green), and b* (positive value yellow, negative value blue).

2.12. Powder Structure and Morphology

The powder morphology was observed by scanning electron microscopy (SEM). In order to assess the structural and morphological characteristics, each powder was placed on a sample holder and fixed with double-sided tape. Then, it was sputter-coated with gold and observed in a Quanta FEG 250 SEM (FEI, United States of America).

2.13. Food Formulation

Based on the phytochemical profile and cell viability, two powders were selected to formulate protein bars (KP and KPB). Then, the selected powders were added in a ratio of 6% to a formulation containing 25.0 g of nuts, 25.0 g of dates, 50.0 g of dried prunes, 25.0 g of rice flour, and 10 g of wheat germ. Prior to powder addition, the abovementioned ingredients were crushed and mixed well for 5 min, followed by powder addition and mixing to the extent that ensured the uniform distribution of the powders in the mixtures. The samples were molded into spherical shapes and stored at 4 °C for 24 h. Three variants were obtained, coded as V1 (with 6% addition of KPB), V2 (with 6% addition of KP), and control (C) with no powder added. The corresponding variants of protein bars were analyzed for total polyphenols, total flavonoids, carotenoid content, antioxidant activity, and *L. casei* 431® cell counts.

2.14. Statistical Analyses

Analyses were performed in triplicate, and the results were expressed as mean and standard deviation (SD). Experimental data were subjected to one-way analysis of variance (ANOVA) after checking the normality and homoscedasticity to find significant differences. For post hoc analysis, the Tukey technique with a 95% confidence interval was used; $p < 0.05$ was deemed statistically significant. Minitab 18 software was used to perform the statistical analysis.

3. Results and Discussion

3.1. The Phytochemical Profiles of the Kiwi Extracts

In order to have a comprehensive view of kiwi byproducts' phytochemical content, selected solvents were used in order to obtain extracts enriched in hydrophilic and lipophilic compounds, such as polyphenols, flavonoids, and carotenoids. The phytochemical profile and antioxidant activity are shown in Table 1.

Table 1. Phytochemical characterization and antioxidant activity of kiwifruit pomace and peels.

Phytochemicals	Pomace	Peels
Hydrophilic profile		
Total flavonoid content, mg CE/g DW	1.97 ± 0.04 [b]	5.34 ± 0.39 [a]
Total polyphenolic content, mg GAE/g DW	3.79 ± 0.15 [b]	9.71 ± 0.28 [a]
Antioxidant activity, μM Trolox/g DW	9.68 ± 0.43 [b]	51.09 ± 2.57 [a]
Hydrophobic profile		
β-Carotene, mg/g DW	1.19 ± 0.065 [b]	4.23 ± 0.027 [a]
Lycopene, mg g DW	0.55 ± 0.042 [b]	1.06 ± 0.027 [a]
Total carotenoids, mg/g DW	1.55 ± 0.093 [b]	5.70 ± 0.033 [a]
Antioxidant activity, μM Trolox/g DW	5.92 ± 0.13 [b]	8.62 ± 0.09 [a]

Means in the same row that do not share a letter (a,b) are significantly different based on Tukey's method with 95% confidence.

As expected, the kiwi peels showed higher contents of polyphenols, flavonoids, and carotenoids, leading to significantly higher antioxidant activity. Our results support the hypothesis of the accumulation of bioactives in the outer layer of fruits. Wang et al. [32] suggested that most phenolic compounds, such as protocatechuic acid, chlorogenic acid, caffeic acid, rutin, p-hydroxybenzoic acid, and quercitin, are found in different parts of kiwifruit, with higher contents in the peels. These authors reported higher concentrations of flavonoids and polyphenols in kiwi peels, up to 8.13 mg gallic acid/g, compared with values ranging from 2.89 to 6.91 mg gallic acid/g in kiwi pulp.

As expected, the carotenoid content had the same trend, with β-carotene and lycopene contents of 4.23 ± 0.027 mg/g DW and 1.06 ± 0.027 mg/g DW, respectively, in kiwi peels. Our results are consistent with the hypothesis that the carotenoid content in kiwifruits is due to the cumulative content of β-carotene and chlorophylls [33]. As regards antioxidant activity, the kiwi peel extracts showed higher values for both DDPH and ABTS radical scavenging activities, which were correlated with the higher contents of polyphenols and carotenoids. Antioxidant activity is often associated with the prevention or inhibition of processes that lead to cellular degradation, caused mainly by the effects of free radicals. Using multivariate correlation analysis, Zhang et al. [34] established that different groups of polyphenolic compounds are mainly responsible for antioxidant activity.

3.2. The Phytochemical Profiles and Cell Viability of Inoculated Freeze-Dried Kiwi Powders

In order to valorize the kiwi pomace and peels as added-value ingredients, the fresh materials were enhanced with different flours in order to improve additional properties, such as color, prebiotic activity, enhanced antioxidant activity, etc. Therefore, four variants of inoculated kiwi-based powders containing *L. casei* 431® were obtained. The phytochemical profiles of the four variants are given in Table 2.

Table 2. Phytochemical characterization of freeze-dried samples.

Variants	KPO	KPB	KBR	KP
		Hydrophilic profile		
Total polyphenolic content, mg GAE/g DW	12.46 ± 0.34 a	10.56 ± 0.30 c	11.51 ± 0.16 b	13.16 ± 0.33 a
Total flavonoid content, mg CE/g DW	1.49 ± 0.10 b	2.03 ± 0.29 a	2.17 ± 0.07 a	1.89 ± 0.03 a,b
Antioxidant activity, μM Trolox/g DW	57.67 ± 0.14 a	56.32 ± 1.04 a	56.93 ± 0.22 a	56.91 ± 0.29 a
		Hydrophobic profile		
β-Carotene, mg/g DW	0.90 ± 0.07 b	0.84 ± 0.05 b	0.81 ± 0.03 b	1.47 ± 0.05 a
Lycopene, mg/g DW	0.57 ± 0.05 a	0.55 ± 0.01 a	0.56 ± 0.07 a	0.57 ± 0.04 a
Total carotenoids, mg/g DW	1.04 ± 0.02 b	1.04 ± 0.06 b	1.04 ± 0.03 b	1.89 ± 0.05 a
AA, μM Trolox/g DW	3.90 ± 0.08 a	3.99 ± 0.08 a	4.07 ± 0.04 a	4.06 ± 0.11 a
L. casei 431®, log CFU/g DW	7.89 c	9.27 b	8.86 c	9.46 a

KPO—kiwi pomace with *L. casei* 431®; KPB—kiwi pomace with buckwheat flour and *L. casei* 431®; KBR—kiwi pomace with black rice flour and *L. casei* 431®; KP—kiwi peels and *L. casei* 431®. Means in the same row that do not share a letter (a–d) are significantly different based on Tukey's method with 95% confidence.

It is well known that fruits and vegetables are rich in polysaccharides. Kiwifruits contain large amounts of pectin and dietary fiber, which have been found to improve the immune system and chronic diseases associated with constipation. Therefore, it is fair to consider that kiwifruits confer prebiotic effects on the intestinal microbiota [35]. These authors suggested that kiwifruit can act as a prebiotic in the selective intensification of the growth of lactic acid bacteria (*Lactobacillus* and *Bifidobacterium*) while inhibiting *Clostridium* and *Bacteriodes* ssp.

Table 2 gives the number of colony-forming units in each sample before and after freeze-drying. The variant based on kiwi pomace (KPO) showed an initial cell concentration of 10.63 log CFU/g DW, while after freeze-drying, a decrease to 7.89 log CFU/g DW was observed. In the case of the variant based on kiwi peels (KP), the initial probiotic load (10.64 log CFU/g DW) slightly decreased to 9.46 log CFU/g DW, whereas for the variants with black rice and buckwheat flour addition, a prebiotic effect was observed, with a higher value for the KPB variant. Although the freeze-drying process is considered a friendly preservation method with a low rate of biologically active compound degradation, a decrease in the viable counts of *L. casei* 431® was observed to different extents in all samples. It can be inferred that the higher survival rate observed in samples based on peels with flour addition is due to the prebiotic effects of the fiber, vitamins, and bioactives. For example, Xie et al. [36] suggested a link between polyphenolic compounds in kiwi byproducts and the viable counts of lactic acid bacteria cells. The survival rate of lactic acid bacteria ranged from 73% to 88%. Based on these results, it is fair to attribute a probiotic character to all of the kiwi-based variants, given the suggested minimum concentration of viable cells of log 6 CFU/g.

The polyphenolic content of the samples was highly influenced by the particular formulation. The kiwi byproduct-based formulations showed a polyphenolic content varying from 10.56 ± 0.30 mg AGE/g DW to 13.16 ± 0.33 mg AGE/g, whereas the flavonoid content was influenced by the type of flour added (Table 2). No significant differences were found in antioxidant activity values. Table 2 shows the carotenoid contents of the samples, highlighting the higher contents of β-carotene and lycopene in the KPO sample (0.90 ± 0.07 mg/g DW and 0.57 ± 0.05 mg/g DW, respectively).

3.3. Color Parameters

The color of the kiwi-based powders was evaluated by CIEL*a*b* color parameters, including brightness L* (0 black/100 white), a* (red/green), and b* (yellow/blue) (Table 3).

The L* values differed significantly among freeze-dried variants and were affected by flour addition. Therefore, the brightest sample was the kiwi pomace-based variant (KPO), followed by the sample with buckwheat flour addition (KPB). The variants based on kiwi peels and kiwi pomace with black rice addition showed the darkest shade. The a* parameters had positive values in all samples, with the KPR variant closest to red, correlated with anthocyanin content. The kiwi peel-based variant showed an a* value close to 0, whereas the b* parameter had a shade close to yellow-green. These values of the color parameters are linked to the predominant bioactive in each powder. For example, the predominance of chlorophylls and carotenoids in the KPO variant, anthocyanin in KPR, and flavonoids in KP and KPB can be observed visually.

Table 3. Color parameters of freeze-dried powders.

Parameter	KPO	KPB	KBR	KP
L*	39.48 ± 0.30 [a]	37.76 ± 0.30 [b]	13.63 ± 0.01 [d]	33.78 ± 0.31 [c]
a*	7.22 ± 0.10 [b]	6.055 ± 0.40 [c]	13.67 ± 0.09 [a]	0.605 ± 0.04 [d]
b*	22.32 ± 0.20 [b]	13.67 ± 0.09 [c]	0.91 ± 0.01 [d]	25.43 ± 0.21 [a]

L*—brightness, a*—redness/greenness, b*—yellowness/ blueness, KPO—kiwi pomace with *L. casei*; KPB—kiwi pomace with buckwheat flour and *L. casei*; KBR—kiwi pomace with black rice flour and *L. casei*; KP—kiwi peels and *L. casei* 431®. Means in the same row that do not share a letter (a–d) are significantly different based on Tukey's method with 95% confidence.

3.4. Survival of L. Casei 431® during Exposure to Simulated Gastrointestinal Conditions

Probiotics are microorganisms that, when administered in appropriate amounts, confer a health benefit to the host [37]. It is well known that probiotic survival in the gastrointestinal tract highly depends on the host, the presence of antibiotics, and interactions with different food constituents [38]. Probiotics are sensitive to temperature, pH variation, oxygen, bile acids, and salts [39]. Therefore, in order to assess the survival rate of *L. casei* 431® in simulated gastrointestinal conditions, the powder with the highest microbial load was tested for the cell survival rate in simulated gastrointestinal conditions, that is, kiwi pomace with buckwheat addition. During oral digestion, a decrease in viable counts from 9.27 log CFU/g DW to 8.46 log CFU/g DW was found. After 2 h of gastric digestion followed by 2 h of intestinal digestion, the survival rate of *L. casei* 431® was 80% and 48%, respectively.

3.5. In Vitro Bioaccessibility of Flavonoids

Flavonoids are a subgroup of polyphenols known for their high antioxidant activity, but they have low stability due to their molecular structure characteristics. Flavonoids are sensitive to factors such as temperature, light, pH, oxygen, enzymatic hydrolysis, etc. In order to evaluate the bioaccessibility of flavonoids from the KPB sample, simulated in vitro digestion was performed. The results showed an increase in total flavonoid content from the oral (1.70 ± 0.19 mg CE/g DW) to gastric environment (2.84 ± 0.14 CE/g DW), suggesting the release of flavonoids from the freeze-dried matrices. The controlled release of the flavonoids continued in the intestinal environment, reaching up to 4.61 ± 0.07 mg CE/g DW. The complex composition of the selected variant, rich in flavonoids and fiber, caused favorable conditions for controlled release during digestion. Xie et al. [36] reported a decrease in polyphenols and flavonoids from kiwi pomace during digestion. These authors reported a total polyphenol content ranging from 1.10 ± 0.17 mg AGE/100 g plant material in the oral cavity to 0.17 ± 0.02 mg AGE/100 g plant material in the intestine. The flavonoid content decreased from 4.68 ± 0.23 mg CE/100 g plant material in the oral environment to 0.75 ± 0.06 mg CE/100 g plant material in the intestine.

3.6. Microscopic Structure of the Powders

The microstructure of the inoculated powders highlights potentially important physical features and potential controlled release of the bacterial load and bioactives. Figure 1 shows the scanning electron microscopy (SEM) images used to visualize the network architectures formed in the structurally complex freeze-dried matrices. In the case of freeze-dried samples based on kiwi pomace and *L. casei* (Figure 1a) (KP), thin sheets with apparent flat faces showing corrugated surfaces can be observed. The structure shows some vesicular formations with regular shapes. Similar views can be observed in the case of the KPB variant (Figure 1b), with a more complex structure involving the presence of numerous vesicular formations with sizes of few micrometers. The variant with black rice addition (Figure 1c) shows a more complex aspect on the surfaces, with more polyhedral formations. The KPO variant has a non-uniform appearance, similar to KP, with smaller areas containing both vesicular and polyhedral formations distributed on curved surfaces connected by ridge areas (Figure 1d).

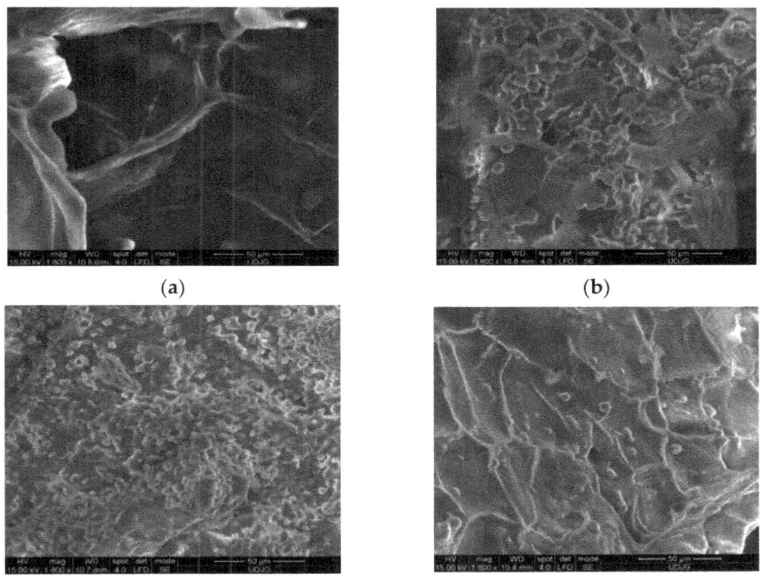

Figure 1. Structural and morphological characteristics of freeze-dried kiwi byproduct-based powders inoculated with *Lacticaseibacillus casei* 431® (freeze-dried samples based on: (**a**) kiwi pomace; (**b**)—kiwi pomace and buckwheat flour; (**c**) kiwi pomace and black rice flour and (**d**)—kiwi peels.

3.7. Food Formulations

The selected powders were added to a formula of protein bars at a ratio of 6%. Three product variants were obtained by adding KPB and KP powders to the protein bars and using a control sample without any powder added. The protein bars were tested for phytochemicals, antioxidant activity, and cell viability. The addition of kiwi peel powder led to a statistically different phytochemical profile of the protein bar (Table 4), with a higher flavonoid content (1.48 ± 0.01 mg CE/g DW), polyphenol content (4.76 ± 0.07 mg AGE/g DW), and antioxidant activity (23.75 ± 0.16 mM Trolox/g DW) when compared with the control sample.

The carotenoid content increased compared to the control sample, with values for β-carotene of 0.72 ± 0.02 and 0.43 ± 0.06 mg/g DW, respectively. The phenolic compound behavior is strongly dependent on food composition. Kiwifruit is a good source of various bioactive compounds and therefore has a high level of antioxidant activity; therefore, it can be introduced into foods to increase their functional value. Tylewicz et al. [40] developed snack formulas with the addition of kiwifruit and proved that kiwi sticks had higher contents of total phenolic compounds, vitamin C, and flavonoids. The viable *L. casei* 431® counts reached 9.36 log CFU/g DW in the sample with KPB addition and 9.95 log CFU/g DW in the sample with KP addition, highlighting the high functionality of both added-value foods.

Table 4. Phytochemical profile and antioxidant activity of protein bars with freeze-dried powders addition.

Variants	C	V1	V2
Total polyphenolic content, mg GAE/g DW	3.95 ± 0.12 [b]	4.18 ± 0.17 [b]	4.76 ± 0.07 [a]
Total flavonoid content, mg CE/g DW	1.36 ± 0.01 [b]	1.37 ± 0.03 [b]	1.48 ± 0.01 [a]
Antioxidant activity, μM Trolox/g DW	22.51 ± 0.18 [b]	22.67 ± 0.25 [b]	23.75 ± 0.16 [a]
β-caroten, mg/g DW	0.37 ± 0.02 [b]	0.43 ± 0.02 [b]	0.72 ± 0.06 [a]
Lycopen, mg/g DW	0.17 ± 0.01 [b]	0.19 ± 0.01 [b]	0.37 ± 0.03 [a]
Total carotenoids, mg/g DW	0.37 ± 0.02 [c]	0.52 ± 0.02 [b]	0.84± 0.06 [a]

C—control protein bar with no powder added, V1—protein bars with 6% addition of freeze-dried powder based on kiwi-pomace, buckwheat and *Lacticaseibacillus casei* 431®, V2—protein bars with 6% addition of freeze-dried powder based on kiwi peels *Lacticaseibacillus casei* 431®. Means that on the same row do not share a letter (a,b,c) are significantly different, based on Tukey method and 95% confidence.

L. casei 431® is one of the most widely used probiotic cultures in the entire industry. It has also been found to produce many bioactive metabolites that can confer benefits to the host when consumed [41]. There is great potential in the field of novel functional foods and pharmacobiotics derived from the genus *Lacticaseibacillus* [42]. As the mechanisms behind their health-promoting capabilities are revealed, possible applications for these strains are being developed in the food, biotechnology, and medical fields [43].

4. Conclusions

Nowadays, consumers are more interested in sustainable food production due to the growing awareness of environmental pollution and the large amount of waste generated during conventional food processing. Moreover, consumers are more oriented toward the healthy aspects of food. Given that the food industry is facing the handling of thousands of tons of kiwi byproducts discarded each year, it is necessary to consider kiwi byproducts as good sources of functional ingredients. The present study is based on the formulation of multifunctional bio-ingredients based on kiwifruit pomace and peels, various flours, and cultures of lactic acid bacteria. These powders were obtained by freeze-drying and characterized in terms of phytochemical content, bioaccessibility of flavonoids after digestion, color, morphology, and structure, as well as the viability of *Lacticaseibacillus casei* 431® cells. In order to test the potential for food functionalization, the powder was added to protein bars, and the products were subsequently characterized in terms of added value. The results confirm that kiwi pomace has the potential to be used as a functional food ingredient with a beneficial effect on consumer health and a prebiotic effect on lactic acid bacteria. The results are also promising for the utilization of natural antioxidants in combination with lactic acid bacteria in order to develop multifunctional ingredients for food, pharmaceutical, and cosmetic applications.

Author Contributions: Conceptualization, N.S.; methodology, N.S. and Ș.-A.M.; software, Ș.-A.M.; validation, Ș.-A.M., A.C., G.R. and N.S.; formal analysis, G.-I.I.; investigation, G.-I.I.; resources, G.R. and N.S.; writing—original draft preparation, G.-I.I.; writing—review and editing, Ș.-A.M., G.R., A.C. and N.S.; visualization, N.S.; supervision, Ș.-A.M. and N.S.; project administration, N.S.; funding acquisition, G.R. All authors have read and agreed to the published version of the manuscript.

Funding: This research was financed by the Internal Grant financed by Dunarea de Jos University of Galati, Romania, contract no. 3637/30.09.2021.

Institutional Review Board Statement: Not applicable.

Informed Consent Statement: Not applicable.

Data Availability Statement: Data is contained within the article.

Acknowledgments: The Integrated Center for Research, Expertise and Technological Transfer in Food Industry is acknowledged for providing technical support.

Conflicts of Interest: The authors declare no conflict of interest.

References

1. Santana-Méridas, O.; González-Coloma, A.; Sánchez-Vioque, R. Agricultural residues as a source of bioactive natural products. *Phytochem. Rev.* **2012**, *11*, 447–466. [CrossRef]
2. Talukder, P.; Talapatra, S.; Ghoshal, N.; Sen Raychaudhuri, S. Antioxidant activity and high-performance liquid chromatographic analysis of phenolic compounds during in vitro callus culture of *Plantago ovata* Forsk. and effect of exogenous additives on accumulation of phenolic compounds. *J. Sci. Food Agric.* **2016**, *96*, 232–244. [CrossRef] [PubMed]
3. Li, W.; Sun, Y.N.; Yan, X.T.; Yang, S.Y.; Kim, S.; Chae, D.; Hyun, J.W.; Kang, H.K.; Koh, Y.-S.; Kim, Y.H. Anti-inflammatory and antioxidant activities of phenolic compounds from *Desmodium caudatum* leaves and stems. *Arch. Pharm. Res.* **2014**, *37*, 721–727. [CrossRef] [PubMed]
4. Soquetta, M.B.; Stefanello, F.S.; Huerta, K.M.; Monteiro, S.S.; da Rosa, C.S.; Terra, N.N. Characterization of physiochemical and microbiological properties, and bioactive compounds, of flour made from the skin and bagasse of kiwi fruit (*Actinidia deliciosa*). *Food Chem.* **2016**, *199*, 471–478. [CrossRef]
5. Asgar, M.A. Anti-diabetic potential of phenolic compounds: A review. *Int. J. Food Prop.* **2013**, *16*, 91–103. [CrossRef]
6. Latocha, P.; Krupa, T.; Wołosiak, R.; Worobiej, E.; Wilczak, J. Antioxidant activity and chemical difference in fruit of different *Actinidia* sp. *Int. J. Food Sci. Nutr.* **2010**, *61*, 381–394. [CrossRef]
7. Chamorro, F.; Carpena, M.; Fraga-Corral, M.; Echave, J.; Riaz Rajoka, M.M.; Barba, F.J.; Cao, H.; Xiao, J.; Prieto, M.A.; Simal-Gandara, J. Valorization of kiwi agricultural waste and industry by-products by recovering bioactive compounds and applications as food additives: A circular economy model. *Food Chem.* **2022**, *370*, 131315. [CrossRef]
8. Zhu, C.H.; Gong, Q.; Li, J.X.; Zhang, Y.; Yue, J.Q.; Gao, J.Y. Research progresses of the comprehensive processing and utilization of kiwifruit. *Storage Proc.* **2013**, *13*, 57–62.
9. Dias, M.; Caleja, C.; Pereira, C.; Calhelha, R.C.; Kostic, M.; Sokovic, M.; Tavares, D.; Baraldi, I.J.; Barros, L.; Ferreira, I.C.F.R. Chemical composition and bioactive properties of byproducts from two different kiwi varieties. *Food Res. Int.* **2020**, *127*, 108753. [CrossRef]
10. Kim, S.; Covington, A.; Pamer, E.G. The intestinal microbiota: Antibiotics, colonization resistance, and enteric pathogens. *Immunol. Rev.* **2017**, *279*, 90–105. [CrossRef]
11. Yadav, A.; Chandra, H.; Maurya, V.K. Probiotics: Recent advances and future prospects. *J. Plant Dev. Sci.* **2017**, *9*, 967–975.
12. Song, H.Y.; Zhou, L.; Liu, D.Y.; Yao, X.J.; Li, Y. What roles do probiotics play in the eradication of Helicobacter pylori? Current knowledge and ongoing research. *Gastroenterol. Res. Pract.* **2018**, *2018*, 9379480. [CrossRef]
13. Tegegne, B.A.; Kebede, B. Probiotics, their prophylactic and therapeutic applications in human health development: A review of the literature. *Heliyon* **2022**, *8*, e0972. [CrossRef] [PubMed]
14. Anosike, F.C.; Onyemah, C.O.; Ossai, C.U.; Ofoegbu, J.N.G.; Okpaga, F.O.; Ikpeama, C.C.; Nkwegu, F.M.; Nwankwo, S.C.; Onyeji, G.N.; Inyang, P.; et al. Probiotic potential and viability of bacteria in fermented African oil bean seed (*Pentaclethra macropyhlla*): A mini review. *Appl. Food Res.* **2022**, *2*, 1000082. [CrossRef]
15. Likotrafiti, E.; Rhoades, J. *Probiotics, Prebiotics, Synbiotics, and Foodborne Illness, in Probiotics, Prebiotics, and Synbiotics. Bioactive Foods in Health Promotion*; Watson, R.R., Preedy, V.R., Eds.; Academic Press: Cambridge, MA, USA, 2016; pp. 469–476.
16. Wang, M.; Zhang, Z.; Sun, H.; He, S.; Liu, S.; Zhang, T.; Wang, L.; Ma, G. Research progress of anthocyanin prebiotic activity: A review. *Phytomedicine* **2022**, *102*, 154145. [CrossRef] [PubMed]
17. Gibson, G.R.; Hutkins, R.; Sanders, M.E.; Prescott, S.L.; Reimer, R.A.; Salminen, S.J.; Scott, K.; Stanton, C.; Swanson, K.S.; Cani, P.D.; et al. Expert consensus document: The International Scientific Association for Probiotics and Prebiotics (ISAPP) consensus statement on the definition and scope of prebiotics. *Nat. Rev. Gastroenterol Hepatol.* **2017**, *14*, 491–502. [CrossRef]
18. Alves-Santos, A.M.; Araújo Sugizaki, C.S.; Lima, G.C.; Veloso Naves, M.M. Prebiotic effect of dietary polyphenols: A systematic review. *J. Funct. Foods.* **2020**, *74*, 104169. [CrossRef]
19. Shen, Y.; Jin, L.; Xiao, P.; Lu, Y.; Bao, J.S. Total phenolics, flavonoids, antioxidant capacity in rice grain and their relations to grain color, size and weight. *J. Cereal Sci.* **2009**, *49*, 106–111. [CrossRef]
20. Zhu, F. Buckwheat proteins and peptides: Biological functions and food applications. *Trends Food Sci. Technol.* **2021**, *110*, 155–167. [CrossRef]
21. FAOSTAT. The Food and Agriculture Organization Corporate Statistical Database. 2020. Available online: http://www.fao.org/faostat/en/#data/QC (accessed on 29 July 2022).
22. Zhu, F. Buckwheat starch: Structures, properties, and applications. *Trends Food Sci. Technol.* **2016**, *49*, 121–135. [CrossRef]
23. Jespersen, L.; Tarnow, I.; Eskesen, D.; Melsaether Morberg, C.; Michelsen, B.; Bügel, S.; Dragsted, L.O.; Rijkers, G.T.; Calder, F.C. Effect of Lactobacillus paracasei subsp. paracasei, L. casei 431 on immune response to influenza vaccination and upper respiratory tract infections in healthy adult volunteers: A randomized, double-blind, placebo-controlled, parallel-group study. *Am. J. Clin. Nutr.* **2015**, *101*, 1188–1196. [CrossRef] [PubMed]
24. Trachootham, D.; Chupeerach, C.; Tuntipopipat, S.; Pathomyok, L.; Boonnak, K.; Praengam, K.; Promkam, C.; Santivarangkn, C. Drinking fermented milk containing Lactobacillus paracasei 431 (IMULUS™) improves immune response against H1N1 and cross-reactive H3N2 viruses after influenza vaccination: A pilot randomized triple-blinded placebo controlled trial. *J. Funct. Foods* **2017**, *33*, 1–10. [CrossRef]

25. Milea, Ș.A.; Vasile, M.A.; Crăciunescu, O.; Prelipcean, A.-M.; Bahrim, G.E.; Râpeanu, G.; Stănciuc, N. Co-Microencapsulation of flavonoids from yellow onion skins and lactic acid bacteria lead to multifunctional ingredient for nutraceutical and pharmaceutics applications. *Pharmaceutics* **2020**, *12*, 1053. [CrossRef]
26. *AOAC Official Method 2005.02 Total Monomeric Anthocyanin Pigment Content of Fruit Juices, Beverages, Natural Colorants, and Wines. pH Differential Method*; AOAC: Gaithersburg, MD, USA, 2005.
27. Souza, A.L.R.; Hidalgo-Chávez, D.W.; Pontes, S.M.; Gomes, F.S.; Cabral, L.M.C.; Tonon, R.V. Microencapsulation by spray drying of a lycopene-rich tomato concentrate: Characterization and stability. *LWT* **2018**, *91*, 286–292. [CrossRef]
28. Fathollahi, M.; Aminzare, M.; Mohseni, M.; Hassanzadazar, H. Antioxidant capacity, antimicrobial activities and chemical composition of Pistacia atlantica subsp. kurdica essential oil. *Vet. Res. Forum.* **2019**, *10*, 299–305. [PubMed]
29. Vasile, M.A.; Milea, Ș.A.; Enachi, E.; Barbu, V.; Cîrciumaru, A.; Bahrim, G.E.; Râpeanu, G.; Stănciuc, N. Functional enhancement of bioactives from black beans and lactic acid bacteria into an innovative food ingredient by co-microencapsulation. *Food Bioproc. Technol.* **2020**, *13*, 978–987. [CrossRef]
30. Kim, I.; Moon, J.K.; Hur, S.J.; Lee, J. Structural changes in mulberry (*Morus Microphylla.* Buckl) and chokeberry (*Aronia melanocarpa*) anthocyanins during simulated in vitro human digestion. *Food Chem.* **2020**, *318*, 126449. [CrossRef]
31. Paucar-Menacho, M.; Peñas, E.; Hernandez-Ledesma, B.; Frias, J.; Martínez-Villaluenga, C. A comparative study on the phenolic bioaccessibility, antioxidant and inhibitory effects on carbohydrate-digesting enzymes of maca and mashua powders. *LWT* **2020**, *131*, 109798. [CrossRef]
32. Wang, Y.; Li, L.; Liu, H.; Zhao, T.; Meng, C.; Liu, Z.; Liu, X. Bioactive compounds and in vitro antioxidant activities of peel, flesh and seed powder of kiwi fruit. *Int. J. Food Sci. Technol.* **2018**, *53*, 2239–2245. [CrossRef]
33. Liu, Y.; Zhou, B.; Qi, Y.; Chen, X.; Liu, C.; Liu, Z.; Ren, X. Expression differences of pigment structural genes and transcription factors explain flesh coloration in three contrasting kiwifruit cultivars. *Front. Plant Sci.* **2017**, *8*, 1507. [CrossRef]
34. Zhang, J.; Gao, N.; Shu, C.; Cheng, S.; Sun, X.; Liu, C.; Xin, G.; Li, B.; Tian, J. Phenolics profile and antioxidant activity analysis of kiwi berry (*Actinidia arguta*) flesh and peel extracts from four regions in China. *Front. Plant Sci.* **2021**, *12*, 689038. [CrossRef] [PubMed]
35. Lee, Y.K.; Low, K.Y.; Siah, K.; Drummond, L.M.; Gwee, K.-A. Kiwifruit (*Actinidia deliciosa*) changes intestinal microbial profile. *Microb. Ecol. Health Dis.* **2012**, *23*, 18572. [CrossRef] [PubMed]
36. Xie, X.; Chen, C.; Huang, Q.; Fu, X. Digestibility, bioactivity and prebiotic potential of phenolics released from whole gold kiwifruit and pomace by in vitro gastrointestinal digestion and colonic fermentation. *Food Funct.* **2020**, *11*, 9613–9623. [CrossRef] [PubMed]
37. Hill, C.; Guarner, F.; Reid, G.; Gibson, G.R.; Merenstein, D.J.; Pot, B.; Morelli, L.; Canani, R.B.; Flint, H.J.; Salminen, S.; et al. Expert consensus document. The International Scientific Association for Probiotics and Prebiotics consensus statement on the scope and appropriate use of the term probiotic: Expert consensus document. *Nat. Rev. Gastroenterol. Hepatol.* **2014**, *11*, 506–514. [CrossRef] [PubMed]
38. Nakkarach, A.; Withayagiat, U. Comparison of synbiotic beverages produced from riceberry malt extract using selected free and encapsulated probiotic lactic acid bacteria. *Agric. Nat. Resour.* **2018**, *52*, 467–476. [CrossRef]
39. Soccol, C.R.; Vandenberghe, L.P.S.; Spier, M.R.; Medeiros, A.B.P.; Yamaguishi, C.T.; Lindner, J.D.; Pandey, A. The potential of probiotics: A review. *Food Technol. Biotechnol.* **2010**, *48*, 413–434.
40. Tylewicz, U.; Nowacka, M.; Rybak, K.; Drozdzal, K.; Dalla Rosa, M.; Mozzon, M. Design of Healthy Snack Based on Kiwifruit. *Molecules* **2020**, *25*, 3309. [CrossRef]
41. Dietrich, C.G.; Kottmann, T.; Alavi, M. Commercially available probiotic drinks containing Lactobacillus casei Dn-114001 reduce antibiotic-associated diarrhea. *World J. Gastroenterol.* **2014**, *20*, 15837–15844. [CrossRef]
42. Hill, D.; Sugrue, I.; Tobin, C.; Hill, C.; Stanton, C.; Ross, R.P. The *Lactobacillus casei* Group: History and Health Related Applications. *Front Microbiol.* **2018**, *109*, 2107. [CrossRef]
43. Sanders, M.E.; Shane, A.L.; Merenstein, D.J. Advancing probiotic research in humans in the United States: Challenges and strategies. *Gut Microbes* **2016**, *7*, 97–100. [CrossRef]

foods

MDPI

Article

Technological Properties and Composition of Enzymatically Modified Cranberry Pomace

Jolita Jagelaviciute [1], Loreta Basinskiene [1,*], Dalia Cizeikiene [1] and Michail Syrpas [1,2]

[1] Department of Food Science and Technology, Kaunas University of Technology, Radvilenu Rd. 19, LT-50254 Kaunas, Lithuania; jolita.jagelaviciute@ktu.lt (J.J.); dalia.cizeikiene@ktu.lt (D.C.); michail.syrpas@ktu.lt (M.S.)

[2] Bioprocess Research Centre, Kaunas University of Technology, Radvilenu Rd. 19, LT-50254 Kaunas, Lithuania

* Correspondence: loreta.basinskiene@ktu.lt

Abstract: Cranberry pomace obtained after juice production is a good source of dietary fiber and other bioactive compounds. In this study, cranberry pomace was hydrolyzed with Viscozyme® L, Pectinex® Ultra Tropical, Pectinex® Yieldmash Plus, and Celluclast® 1.5L (Novozyme A/S, Denmark). The soluble and insoluble dietary fiber was determined using the Megazyme kit, while the changes in mono-, disaccharide and oligosaccharides' contents were determined using HPLC-RI; the total phenolic contents were determined by Folin–Ciocalteu's Assay. Prebiotic activity, using two probiotic strains *Lactobacillus acidophilus* DSM 20079 and *Bifidobacterium animalis* DSM 20105, was investigated. The technological properties, such as hydration and oil retention capacity, were evaluated. The enzymatic treatment increased the yield of short-chain soluble saccharides. The highest oligosaccharide content was obtained using Viscozyme® L and Pectinex® Ultra Tropical. All of the tested extracts of cranberry pomace showed the ability to promote growth of selected probiotic bacteria. The insoluble dietary fiber content decreased in all of the samples, while the soluble dietary fiber increased just in samples hydrolyzed with Celluclast® 1.5L. The highest content of total phenolic compounds was obtained using Viscozyme® L and Pectinex® Ultra Tropical (10.9% and 13.1% higher than control, respectively). The enzymatically treated cranberry pomace exhibited lower oil and water retention capacities in most cases. In contrast, water swelling capacity increased by 23% and 70% in samples treated with Viscozyme® L and Celluclast® 1.5L, respectively. Enzymatically treated cranberry pomace has a different composition and technological properties depending on the enzyme used for hydrolysis and can be used in various novel food products.

Keywords: cranberry pomace; enzymatic hydrolysis; dietary fiber; technological properties

Citation: Jagelaviciute, J.; Basinskiene, L.; Cizeikiene, D.; Syrpas, M. Technological Properties and Composition of Enzymatically Modified Cranberry Pomace. *Foods* **2022**, *11*, 2321. https://doi.org/10.3390/foods11152321

Academic Editors: Marco Poiana, Francesco Caponio and Antonio Piga

Received: 14 July 2022
Accepted: 1 August 2022
Published: 3 August 2022

Publisher's Note: MDPI stays neutral with regard to jurisdictional claims in published maps and institutional affiliations.

Copyright: © 2022 by the authors. Licensee MDPI, Basel, Switzerland. This article is an open access article distributed under the terms and conditions of the Creative Commons Attribution (CC BY) license (https://creativecommons.org/licenses/by/4.0/).

1. Introduction

Agricultural by-products form a large part of waste, with approximately 88 metric tons per year generated in Europe alone [1]. However, these by-products are considered to be potential sources of various high added-value, bioactive compounds (dietary fibers, pigments, essential minerals, fatty acids, antioxidant polyphenolic compounds, etc.) [2]. Berry pomace (or press cake) obtained after juice pressing usually contains skin, stems, and seed parts [1]. Various fruit and berry products are substantially composed of structural polysaccharides and their related oligosaccharides and these compounds are commonly described as insoluble (IDF) and soluble (SDF) dietary fibers [3]. The dietary fiber in berry pomace is usually composed of pectin, lignin, cellulose, hemicellulose, and inulin [1]. These types of compounds have a potentially significant impact on human health [3], such as a decrease in the risk of hypertension, obesity, diabetes, coronary heart disease, stroke, and certain gastrointestinal disorders [1]. The Food and Drug Administration recommendation for the use of dietary fiber in the diet is approximately 25–35 g per day, and 6 g of it should be SDF. Furthermore, some studies show that 10 g of extra dietary fiber in a diet could help decrease the risk of death caused by coronary heart disease by up to 35% [1,4].

Polyphenols and cell-wall polysaccharides can also positively modulate the profile of the gut microbiota, which may have implications in the prevention against metabolic diseases [5]. Other studies indicate that SDF can be used as a prebiotic [1]. Islam et al. [6] reported that the incorporation of cranberry pomace in the food not only improved the short-term iron and cholesterol levels in the blood serum of broilers, but also increased the levels of beneficial bacterial genera and decreased the undesirable ones. Other studies indicated that the fractions of cranberry oligosaccharide and the xyloglucan and purified cranberry xyloglucan compounds have shown a distinct lack of antimicrobial properties or cytotoxic effects in numerous bacterial and human cell lines [3]. The oligosaccharides extracted from cranberries even shown the ability to modify the biofilm formation of some *E. coli* strains [7]. Liu et al. [8] reported that cranberry oligosaccharides can be used as a carbon source for *Lactobacillus*, however the oligosaccharide metabolization is strain specific. The supplementation of fermented meat with cranberry pomace significantly inactivates *Salmonella* and increases the growth of lactic acid bacteria [9].

An increase in the level of dietary fiber in food products usually negatively influences the products' texture and color, depending on the properties and level of fiber. However, some studies indicated that enzymatic treatment is practical and a potential modification method to obtain functional food materials with better technological properties from agriculture by-products [10,11]. For example, the enzymatically treated tomato peels have higher contents of lycopene and dietary fiber [10]. The enzymatic treatment can effectively increase the release of antioxidants and phenolic compounds and the fiber composition. However, this is not the case for all of the fruit and berry matrixes; in some cases, enzymatic treatment can reduce the fiber content and increase the free monosaccharides [11]. The treatment of fruit fiber with enzymes of different activities has resulted in structural changes that have altered the technological properties, such as the water-holding and swelling capacities [12].

In this study, cranberry pomace was enzymatically treated with several commercially available enzymes to modify their technological properties and the composition of the dietary fiber. The technological properties were determined, such as water-holding, oil-holding, swelling, emulsion capacities, dietary fiber composition, and prebiotic potential. Most previously published studies indicated cranberry by-products as a good source for isolating bioactive compounds. However, there is a lack of information regarding the effect of enzymatic treatments on cranberry pomace properties. This study provides experimental support for the whole cranberry pomace as a source of dietary fiber and for developing functional and innovative food products.

2. Materials and Methods

2.1. Cranberry Pomace

The cranberry pomace was kindly donated by the company "Įvairios sultys" (Kėdainiai, Lithuania). The moisture content was 3.96%. The pomace was ground to a particle size of 0.5 mm and stored at 4 °C.

2.2. Proximate Composition Analysis of Cranberry Pomace

The moisture content was determined by drying 1 g of cranberry pomace in an oven at 105 °C to a constant weight, according to the Association of Official Analytical Collaboration (AOAC) Method 925.10-1925. After drying, the final weight of the pomace was subtracted from the initial weight divided by the weight of the sample multiplied by 100 and expressed as g/100 g. The proteins were determined on 1 g of pomace by the Kjeldahl method (N × 6.25), according to the AOAC Method 978.04. The results were expressed as g/100 g of the dry weight sample. The lipids were determined on 1 g of pomace by drying the sample and subsequently the Soxhlet extraction with hexane was performed for 5 h, according to the AOAC Method 948.22. After the extraction, the residues were dried in an oven at 105 °C to a constant weight. The lipid content was calculated from the initial dry weight of pomace by subtracting the final weight of the residues divided by the weight of the dry sample multiplied by 100 and expressed as g/100 g of the dry weight sample. The ash content was determined by charring 2 g of pomace for 30 min, followed by incineration in a muffle furnace at 525 °C for 4 h, according to the AOAC Method 930.05. The residues obtained after incineration were recorded and divided by the weight of the sample weight and multiplied by 100 and expressed as g/100 g of the dry weight sample. The total carbohydrate content was calculated by subtracting the value of the protein, ash, and lipid contents from 100% of the dry weight. The soluble and insoluble dietary fiber were quantified using a Megazyme kit, based on the American Association of Cereal Chemists (AACC) Method 32-07.01 and the AOAC Method 991.43 [13]. One g of pomace (duplicate) was mixed with 40 mL MES-Tris buffer (0.05 M pH 8.2) and hydrolyzed with 50 µL of α-amylase (Megazyme E-BLAAM) at 95 °C for 30 min under constant stirring (120 rpm). After hydrolysis, the temperature was cooled to 60 °C and the sample was hydrolyzed with 100 µL of protease (Megazyme E-BSPRT) at 60 °C for 30 min under constant stirring (120 rpm). Then, the pH was adjusted to 4.1–4.8 with 0.561 N HCl and further hydrolysis was performed with 200 µL of amyloglucosidase (Megazyme E-AMGDF) at 60 °C for 30 min under constant stirring (120 rpm). The suspension was vacuum-filtered in the Fibertec 1023 E equipment (Foss System, Hilleroed, Denmark) through a celite (Megazyme cat. No. G-CELITE) in a bed crucible. The residues were washed with hot distilled water (70 °C), 95% ethanol and acetone, dried and recorded as IDF. The obtained supernatant was mixed with four volumes of 95% ethanol (60 °C) and left to stand at room temperature for 1 h; the SDF was recorded by filtration of the ethanolic suspension in the same conditions as the IDF and washed with 78% ethanol, 95% ethanol, and acetone. The IDF and SDF were calculated by subtracting the protein and ash contents determined in the obtained residues and expressed as g/100 g of the dry weight sample.

2.3. Enzymatic Hydrolysis

The enzymatic hydrolysis of the cranberry pomace was produced using commercially available Novozyme A/S enzymes (Novozyme, Denmark), such as Viscozyme® L (β-glucanases, pectinases, hemicellulases, and xylanases), Pectinex® Ultra Tropical (pectinases, cellulases, hemicellulases, and β-glucanases), Pectinex® Yieldmash Plus (pectinases), and Celluclast® 1.5L (cellulases). The cranberry pomace was mixed with distilled water in a 1:10 ration (2.5 g of pomace with 25 mL of distilled water) and the enzyme was added at various concentration combinations in a range from 0.02 to 0.1 mL/g (control without enzymes). The mixture was incubated at 50 °C with shaking (200 rpm), and various time combinations from 1 to 7 h were used. The independent variable concentration of enzymes and duration of hydrolysis were chosen on the manufacturer's recommendations, and literature data [14,15]. After the hydrolysis, the enzyme activity was terminated by heating the sample in a 95 °C water bath for 20 min. For the water-soluble fraction yield calculation, the prebiotic activity and the mono-, disaccharides and oligosaccharides analysis, the mixture was cooled at room temperature (20 °C) and centrifugated (8000 rpm, 20 min). The water-soluble part (the resulting supernatant) was collected and freeze-dried (Harvest Right, North Salt Lake, UT, USA). The yield of the water-soluble fraction was determined gravimetrically after freeze-drying. The hydrolyzed cranberry pomace samples used for the determination of the functional properties were collected whole and freeze-dried. All of the freeze-dried samples were kept in a dark place at room temperature.

2.4. Mono-, Disaccharides' and Oligosaccharides' Analysis by High-Pressure Liquid Chromatography with Refractive Index Detector (HPLC-RI)

For the saccharide analysis, 10 mg of the freeze-dried water-soluble fraction, was dissolved in 1 mL of Millipore water (10 mg/mL). The saccharide analysis was performed as previously described by Syrpas et al. [16]. The total oligosaccharides were recorded as the sum of all of the detected and quantified fractions of oligosaccharides with a degree of polymerization (DP) DP 7–10, DP 5–6, DP4, DP3, while the total mono- and disaccharides were recorded as a sum of the sucrose, glucose, fructose, sugar alcohols, and galacturonic acids. According to the obtained results, the samples enzymatically hydrolyzed for 1 h were selected for further analysis.

2.5. In Vitro Assessment of the Prebiotic Activity

The in vitro prebiotic activity of the water-soluble fraction was assessed by evaluating the probiotic bacteria growth stimulation, as described by Moreno-Vilet et al. [17] with some modifications. Two probiotic strains, *Lactobacillus acidophilus* DSM 20079 and *Bifidobacterium animalis* DSM 20105, were used. The probiotic strains were grown in De Man Rogosa and Sharpe (MRS) medium (Biolife, Milan, Italy), supplemented with 0.05% L-cysteine (*w*/*v*) under anaerobic conditions at 37 °C for 18 h. The bacterial inoculants were prepared by centrifugation (5000 g, 10 min, 4 °C) and washing twice with saline solution. After that, the bacterial cell pellets were collected and re-suspended in saline solution, using McFarland standard No. 0.5 (Liofilchem, Waltham, MA, USA) to standardize the bacterial concentrations. The growth of probiotics was evaluated using carbohydrate-free culture media with the same composition as the MRS supplemented with 1% (*w*/*v*) from a different carbohydrate source. The water-soluble fraction obtained after 1 h hydrolysis was used as the carbohydrate source. Glucose was used as the negative control; commercial inulin was used as the positive control; and carbohydrate-free media was used as the blank control. Each tube was inoculated with 1% of the probiotic suspension and incubated for 48 h at 37 °C under anaerobic conditions. The probiotic bacteria counts were determined after 24 and 48 h. The probiotic growth was performed in quadruplet and measured by serial dilutions using the standard plate count technique following the calculation of the number of cells on the agar plate after 72 h incubation at optimal temperature and was expressed as log10 value of CFU/g. The proliferative index (PI) was calculated by the following equation:

$$\text{PI} = \text{average LogA} - \text{average LogB} \quad (1)$$

where B is the bacterial count at 0 h (CFU/g) and A is the bacterial count at 24 or 48 h (CFU/g).

2.6. Total Phenolic Content (TPC) by Folin–Ciocalteu's Assay

The TPC of the enzymatically modified cranberry pomace was determined by Folin–Ciocalteu's Assay according to Singleton et al. [18] with some modifications. The extracts were obtained by mixing 1 g of enzymatically hydrolyzed cranberry pomace with 10 mL of methanol solution (0.1% HCl in methanol (*v*/*v*)) and left overnight in the dark at 4 °C. One ml of the extract solution in methanol was mixed with 5 mL of Folin–Ciocalteu's reagent (1:9 *v*/*v*) and 4 mL sodium carbonate solution (7.5%). The samples were kept in the dark for 30 min and absorption was measured at 765 nm. The TPC was expressed in mg of gallic acid equivalents (GAE)/g of dry weight of cranberry pomace.

2.7. Water Retention Capacity (WRC)

The water retention capacity was performed according to Yu et al. [19] with some modifications. The cranberry pomace (0.2 g) was mixed with distilled water (6 mL) at room temperature and left for 18 h. After that, the sample was centrifugated at 5000 rpm for 20 min. The resulting residues were weighted before and after drying at 105 °C. The WRC was calculated by the following equation:

$$\text{WRC (g/g)} = (\text{Mb} - \text{Ma})/\text{Ma} \quad (2)$$

where Mb is the weight of residues before drying (g) and Ma is the weight of residues after drying (g).

2.8. Water Swelling Capacity (WSC)

The water-swelling capacity was determined according to Yu et al. [19] with some modifications. The cranberry pomace (0.2 g) was weighted in graduated test tubes and hydrated in 6 mL of distilled water at room temperature for 18 h. The volume of the pomace was recorded before and after hydration; the WSC was calculated by the following equation:

$$\text{WSC (mL/g)} = (\text{V1} - \text{V0})/\text{W} \tag{3}$$

where V1 is the volume of pomace after hydration (mL); V0 is the volume of pomace before the hydration; W is the weight of the dry pomace prior to hydration (g).

2.9. Oil Retention Capacity (ORC)

The ORC was performed according to Yu et al. [19] with some modifications. The dried pomace (0.2 g) was mixed with 2 g of sunflower oil and left for 1 h at room temperature. After that, the mixture was centrifugated at 300 rpm for 10 min and the supernatant was carefully decanted, and the pellet recovered was weighed. ORC was calculated by the following equation:

$$\text{ORC (g/g)} = (\text{W1} - \text{W0})/\text{W0} \tag{4}$$

where W1 is the weight of the pellet (g) and W0 is the weight of the dry sample (g).

2.10. Solubility

The yield of soluble material in the pomace was determined, using the supernatant from WRC analysis. The supernatant, after decantation, was dried at 105 °C to a constant weight. The solubility was calculated by the following equation:

$$\text{Solubility (\%)} = (\text{M/M0}) \times 100 \tag{5}$$

where M is the weight of dried soluble material (g); M0 is the weight of the dry pomace used for analysis (g).

2.11. Emulsion Stability

The emulsion was prepared by mixing the pomace (0.16 g) with sodium phosphate (0.02 M) and citric acid (0.1 M) buffer pH 4, 6, 8 or distilled water (8 g), then the sunflower oil was added (8 g) and emulsified using a rotor–stator homogenizer (IKA® T-25 digital, Ultra-Turrax, Staufen, Germany) for 5 min at 10,000 rpm. The emulsions were transferred into graduated tubes and stored for 3 weeks at 4 °C. The emulsion stability was recorded in terms of phase separation and expressed by the following equation:

$$\text{Emulsion stability} = (\text{VE/VT}) \times 100 \tag{6}$$

where VE is the volume of emulsion; VT is the total volume.

To evaluate the thermal stability of emulsions, the tubes were heated in an 80 °C water bath for 30 min. Then, the tubes were cooled to room temperature and stored for 3 weeks at 4 °C. The thermal stability of the emulsion was recorded and expressed as previously described.

2.12. Statistical Analyses

The mean values and standard deviations were calculated using MS Excel 2019 (Microsoft Corp, Albuquerque, NM, USA). The statistical analysis was performed using the Statgraphics Centurion 19 statistical package. One-way analysis of variation (ANOVA), followed by Tukey's honest significant difference (HSD) test were performed to determine significant differences ($p < 0.05$).

3. Results and Discussion

3.1. Chemical Composition

The chemical composition of the cranberry pomace is shown in Table 1. The major component of the pomace is carbohydrates, in which dietary fibers represent 72.67 g/100 g of dry weight (IDF—59.93 g/100 g and SDF—12.74 g/100 g). The dietary fibers obtained from the various berries mainly include: lignin, pectin, cellulose, and hemicellulose [1]. Islam et al. [6] reported a similar amount of SDF (11–12.3 g/100 g) in cranberry pomace, while the protein content was reported as lower (5.75 g/100 g). White et al. [20] reported a similar carbohydrate content in cranberry pomace, but a lower SDF and protein content and a higher IDF and fat content. The chemical composition of the cranberry pomaces varies and is influenced by several conditions, such as the cultivar, ripeness, or processing conditions.

Table 1. Chemical composition of cranberry pomace.

Parameter	Content (g/100 g)
Moisture	5.57 ± 0.11
Total fat [a]	9.83 ± 0.46
Ash [a]	0.96 ± 0.04
Protein [a]	7.40 ± 0.06
Carbohydrate [b]	81.81
Insoluble dietary fiber [a]	59.93 ± 1.46
Soluble dietary fiber [a]	12.74 ± 0.09

[a] Expressed as g/100 g of dry weight; [b] Carbohydrate content was determined by subtracting the value of protein, ash, and lipid content from 100% of dry weight. Data values are expressed as means with the standard deviation ($n = 3$).

3.2. Yield of Water-Soluble Fraction

The enzymatic hydrolysis of the cranberry pomace was performed using four different commercial enzymes. The different times for hydrolysis and different concentrations of enzymes were evaluated (rationale). The yield of the water-soluble fraction obtained after enzymatic hydrolysis is presented in Table 2. As could be expected, the enzymatic hydrolysis increased the water-soluble material content in all of the samples. The extraction yield depended on the composition and concentration of enzymes, and the duration of the hydrolysis. The highest yield of the water-soluble fraction was obtained using Viscozyme® L and Pectinex® Ultra Tropical enzymes. Compared with the yield of the control sample after 1 h of incubation, the lowest increase in the water-soluble fraction was observed with Celluclast® 1.5L (increased by 19.68%), while the highest increase was in the sample hydrolyzed with Pectinex® Ultra Tropical (increased by 93.34%). Previous studies have reported that enzymatic treatment increases the water-soluble fraction yield [10]. Yoon et al. [21] reported an increase in the alcohol-insoluble dietary fiber by prolonging the time of enzyme hydrolysis, while the alcohol-soluble dietary fiber yield did not change significantly after 24 h of enzymatic hydrolysis. However, increasing the enzyme level can reduce hydrolysis time (Table 2). In most cases, there were no significant differences ($p < 0.05$) between the yields of water-soluble fraction obtained using the same enzyme by increasing the enzyme level and decreasing enzymatic hydrolysis time.

Table 2. Water-soluble fraction yield and saccharide composition of cranberry pomace after enzymatic hydrolysis.

Time (h)	E/S Ratio (mL/g)	Enzyme	Water-Soluble Fraction Yield, %	Total Oligosaccharides, mg/g DW	Total Mono- and Disaccharides, mg/g DW
1	0.04	Viscozyme® L	25.56 ± 1.33 [e]	71.6 ± 4.4 [b]	178.5 ± 8.9 [abcd]
5.5	0.1	Viscozyme® L	27.98 ± 0.20 [e]	61.3 ± 2.1 [b]	222.9 ± 4.3 [d]
1	0.1	Pectinex® Yieldmash Plus	21.50 ± 0.78 [d]	21.2 ± 0.6 [a]	197.0 ± 0.3 [bcd]
7	0.02	Pectinex® Yieldmash Plus	19.95 ± 0.23 [cd]	16.3 ± 3.9 [a]	173.7 ± 11.4 [abcd]
1	0.08	Pectinex® Ultra Tropical	27.34 ± 1.56 [e]	59.0 ± 6.0 [b]	211.9 ± 24.3 [cd]
4	0.06	Pectinex® Ultra Tropical	27.41 ± 0.71 [e]	56.4 ± 4.3 [b]	216.3 ± 12.7 [cd]
1	0.02	Celluclast® 1.5L	18.50 ± 1.35 [bc]	32.2 ± 3.3 [a]	123.3 ± 14.8 [ab]
7	0.1	Celluclast® 1.5L	21.74 ± 0.71 [d]	56.5 ± 3.1 [b]	150.2 ± 9.0 [abcd]
1	-	-	14.12 ± 0.34 [a]	19.9 ± 2.2 [a]	108.9 ± 2.5 [a]

Table 2. *Cont.*

Time (h)	E/S Ratio (mL/g)	Enzyme	Water-Soluble Fraction Yield, %	Total Oligosaccharides, mg/g DW	Total Mono- and Disaccharides, mg/g DW
4	-	-	16.14 ± 0.14 [ab]	18.9 ± 2.5 [a]	118.9 ± 7.5 [a]
5.5	-	-	16.56 ± 0.68 [abc]	20.9 ± 2.9 [a]	140.4 ± 20.9 [abc]
7	-	-	17.08 ± 0.17 [abc]	22.0 ± 2.9 [a]	143.3 ± 21.9 [abc]

E/S—enzyme/substrate (pomace) ratio. - without enzyme (control samples). Data values are expressed as means with the standard deviation ($n = 3$). Values in one column followed by the same letter are not significantly different ($p < 0.05$).

3.3. Saccharide Analysis

The total oligosaccharides, mono- and disaccharides increased in all of the enzymatically hydrolyzed samples (Table 2). The quantity of oligosaccharides and mono- and disaccharides depended on the type and concentration of the enzyme as well as the duration of hydrolysis. The highest increase in total oligosaccharides after 1 h hydrolysis was obtained in the sample hydrolyzed with Viscozyme® L (3.6 times higher than control), while the lowest increase was determined in the sample hydrolyzed with Pectinex® Yieldmash Plus. Viscozyme® L is composed of β-glucanases, pectinases, hemicellulases, and xylanases. The enzymatic hydrolysis of β-glucan-type hemicellulose polymers in cranberry pomace increases the level of oligosaccharides [3].

Spadoni Andreani et al. [22] reported cranberry extracts after enzymatic hydrolysis with a significantly increased glucose level. The highest mono- and disaccharide content was observed in the sample hydrolyzed with Pectinex® Ultra Tropical. The lowest mono- and disaccharide content was determined in the sample hydrolyzed with Celluclast® 1.5L (1.7 times lower than in the pomaces hydrolyzed with Pectinex® Ultra Tropical) compared with other enzymatically hydrolyzed pomaces. A longer enzymatic hydrolysis increased the mono- and disaccharide content, except for the pomace treated with Pectinex® Yieldmash Plus. However, a lower enzyme concentration was used for the longer hydrolysis. The greater increase in the saccharide content was determined using enzymes, which were mainly pectinases. Viscozyme® L used in this study (0.04 mL/g for 1 h hydrolysis) showed the highest level of total oligosaccharides, while the longer duration of hydrolysis and the higher enzyme concentration decreased the level of the oligosaccharides and increased the level of the mono- and disaccharides.

For these reasons, the determination of the further functional properties was evaluated in the enzymatically hydrolyzed samples for 1 h.

3.4. In Vitro Prebiotic Activity

The growth of the probiotic bacteria in a medium supplemented with different carbohydrate sources is shown in Table 3. Both of the tested probiotic bacteria showed an ability to utilize the carbohydrates from cranberry water-soluble fractions as their carbon sources. The growth of the probiotic bacteria in the medium supplemented with carbohydrates was significantly higher than in the carbohydrate-free media. *L. acidophilus* DSM 20079 did not survive in the carbohydrate-free media after 48 h, while *B. animalis* DSM 20105 showed an ability to grow in this media. The cranberry water-soluble fraction after 24 h promoted the growth of *L. acidophilus* DSM 20079 better than glucose or inulin. However, with *L. acidophilus* DSM 20079 viability decreased in all of the samples after 48 h, except for the media supplemented with glucose when viability was compared after 24 h, whereas with *B. animalis* DSM 20105, growth increased in all of the samples after 48 h (Table 3). The highest increase was observed in the media supplemented with glucose and inulin. The cranberry water-soluble fraction obtained after treatment with Pectinex® Ultra Tropical promoted the growth of *L. acidophilus* DSM 20079 better than the other water-soluble fractions after 24 h. Furthermore, after 48 h, the *L. acidophilus* DSM 20079 in this medium showed a better cell viability than in the other media supplemented with cranberry extracts. However in most of the cases, no statistically significant differences were observed ($p < 0.05$). It may be due to a higher concentration of mono- and disaccharides (Table 2) in this extract compared with the other cranberry extracts. The same proliferation effect of prebiotics with a higher sugar content was reported in other studies [23,24]. *L. acidophilus* could utilize a variety of carbohydrates, such as mono-, di-, and polysaccharides [25]. The *L. acidophilus* could use oligosaccharides as a carbon source and grow without sugar [26]. All of the cranberry water-soluble fractions had oligosaccharides which may have an impact on the faster proliferation of *L. acidophilus* DSM 20079, compared with glucose and inulin. However, the faster utilization of carbohydrates may lead to a decreased viability of the probiotic cells after 48 h.

Table 3. Growth of probiotics in medium supplemented with different carbohydrate source.

Prebiotics	*Lactobacillus acidophilus* DSM 20079					*Bifidobacterium animalis* DSM 20105				
	0 h	24 h		48 h		0 h	24 h		48 h	
	Log CFU/g		PI	Log CFU/g	PI	Log CFU/g		PI	Log CFU/g	PI
Carbohydrate-free MRS	5.03 ± 0.07 [a]	4.27 ± 0.13 [a]	−0.66	- [a]	−5.03	5.03 ± 0.05 [a]	6.64 ± 0.09 [a]	1.61	6.45 ± 0.03 [a]	1.42
Glucose	4.78 ± 0.20 [a]	7.31 ± 0.22 [bc]	2.53	7.62 ± 0.08 [e]	2.84	5.28 ± 0.05 [b]	7.99 ± 0.22 [b]	2.71	9.00 ± 0.41 [cd]	3.72
Inulin	4.81 ± 0.23 [a]	7.11 ± 0.25 [b]	2.29	5.52 ± 0.57 [b]	0.71	5.28 ± 0.08 [b]	8.36 ± 0.25 [bc]	3.08	9.07 ± 0.12 [d]	3.78
Control	4.77 ± 0.06 [a]	7.72 ± 0.10 [de]	2.95	5.86 ± 0.06 [bc]	1.09	5.52 ± 0.09 [c]	8.03 ± 0.52 [b]	2.51	8.83 ± 0.12 [cd]	3.30
Viscozyme® L	4.79 ± 0.09 [a]	7.69 ± 0.11 [cde]	2.90	6.21 ± 0.04 [cd]	1.41	5.60 ± 0.06 [c]	8.37 ± 0.13 [bc]	2.77	8.37 ± 0.05 [b]	2.77
Pectinex® Ultra Tropical	4.72 ± 0.07 [a]	7.84 ± 0.23 [e]	3.13	6.45 ± 0.14 [d]	1.73	5.58 ± 0.05 [c]	8.40 ± 0.02 [bc]	2.81	8.63 ± 0.05 [bc]	3.04
Pectinex® Yieldmash Plus	5.00 ± 0.01 [a]	7.53 ± 0.14 [cde]	2.53	5.81 ± 0.19 [bc]	0.82	5.52 ± 0.09 [ab]	8.03 ± 0.52 [bc]	2.51	8.77 ± 0.19 [bcd]	3.24
Celluclast® 1.5L	4.74 ± 0.06 [a]	7.44 ± 0.15 [cd]	2.70	5.69 ± 0.09 [b]	0.95	5.62 ± 0.09 [c]	8.66 ± 0.03 [c]	3.04	8.93 ± 0.07 [cd]	3.31

- No growth; data values are expressed as means with the standard deviation (n = 4). Values in one column followed by the same letter are not significantly different ($p < 0.05$).

The extract obtained after treatment with Celluclast® 1.5L promoted the growth of *B. animalis* DSM 20105 after 24 h better than the other extracts, however, in most cases, no statistically significant differences were observed ($p < 0.05$). The *Bifidobacterium* shows a specific preference for prebiotic substrates within the genus, although most of the bacteria could utilize a range of different carbohydrates [27]. The molecular weight, degree of polymerization, and the type of linkage between the comprising units are known to influence the prebiotic activity of carbohydrates [23,28–30].

3.5. Dietary Fiber and TPC Content after Enzymatic Hydrolysis

The dietary fiber content after enzymatic hydrolysis was evaluated in whole freeze-dried samples hydrolyzed for 1 h (Table 4). The IDF significantly decreased ($p < 0.05$) in all of the samples. The lowest content of IDF was observed in a sample hydrolyzed with Pectinex® Ultra Tropical (13.3% lower than the control). The pectinolytic enzymes increased the plant cell-wall breakdown of the pomace [31]. The content of the SDF decreased significantly ($p < 0.05$) in all of the samples, except for the sample hydrolyzed with Celluclast® 1.5L. The lowest SDF content was observed in the sample treated with Pectinex® Ultra Tropical (84.41% lower than the control) and the highest content was determined in the sample treated with Celluclast® 1.5 L, however, no significant differences were observed ($p < 0.05$) compared to the control. The results indicate that the SDF can be hydrolyzed to smaller fragments and does not precipitate with ethanol. Mrabet et al. [32] reported a similar decrease in the dietary fiber content treated with Viscozyme® L, while Yoon et al. [21] reported an increase in the alcohol-soluble dietary fiber after the enzymatic hydrolysis of carrot pomace. Spadoni Andreani et al. [33] reported an increased yield of oligosaccharides in cranberry carbohydrate extracts after enzymatic treatment with Viscozyme® L and Pectinex® Ultra SPL. The obtained results suggest that Celluclast® 1.5L could be used for increasing the SDF, while Viscozyme® L and Pectinex® Ultra Tropical showed promising results for the production of oligosaccharides.

Table 4. Dietary fiber, TPC content and technological properties of enzymatically treated cranberry pomace.

Sample	Time (h)	E/S Ratio (mL/g)	IDF, g/100 g	SDF, g/100 g	ORC, g/g	WSC mL/g	WRC, g/g	Solubility, %	TPC, mg of GAE/g
Control	1	-	62.48 ± 0.46 [d]	11.23 ± 0.84 [c]	6.50 ± 0.08 [a]	1.80 ± 0.3 [a]	10.59 ± 0.20 [a]	15.4 ± 0.9 [a]	7.04 ± 0.41 [a]
Viscozyme® L	1	0.04	56.46 ± 0.15 [ab]	2.64 ± 0.26 [a]	4.37 ± 0.25 [b]	2.22 ± 0.01 [b]	8.58 ± 0.24 [b]	27.5 ± 1.3 [b]	7.81 ± 0.36 [b]
Pectinex® Yieldmash Plus	1	0.1	58.48 ± 0.44 [bc]	5.44 ± 0.30 [b]	3.99 ± 0.13 [b]	1.65 ± 0.01 [a]	7.25 ± 0.22 [c]	24.0 ± 0.6 [c]	7.42 ± 0.24 [ab]
Pectinex® Ultra Tropical	1	0.08	54.17 ± 1.38 [a]	1.75 ± 0.29 [a]	4.11 ± 0.16 [b]	1.57 ± 0.11 [a]	8.10 ± 0.43 [b]	29.6 ± 1.2 [d]	7.96 ± 0.29 [b]
Celluclast® 1.5 L	1	0.02	60.52 ± 0.17 [cd]	11.90 ± 0.53 [c]	5.80 ± 0.37 [c]	3.06 ± 0.16 [c]	11.02 ± 0.24 [a]	20.3 ± 1.0 [e]	7.06 ± 0.28 [a]

Data values are expressed as means with the standard deviation of three replicates for SDF, IDF, ORC, WRC, WSC, solubility and of six replicates for TPC. Values in one column followed by the same letter are not significantly different ($p < 0.05$).

The TPC of the enzymatically treated cranberry pomace was evaluated (Table 4). The enzymatic hydrolysis increased the TPC in all of the samples, and varied in range from 7.04 to 7.96 mg GAE/g, however, no significant difference ($p < 0.05$) was determined between the control and the pomaces hydrolyzed with Celluclast® 1.5L and Pectinex® Yieldmash Plus. The highest amount of TPC was determined after hydrolysis with Pectinex® Ultra Tropical and Viscozyme® L (13.07 and 10.94% higher than the control, respectively). The results showed that decreasing the total dietary fiber content increases the TPC content. The antioxidants are usually stored in natural cell compartments, so they must be released during digestion to be absorbed in the gut [34]. Gouw et al. [35] reported TPC in dried cranberry pomace of 7.58 mg GAE/g; other studies reported phenolic content of 6 mg/g [20], while Ross et al. [36] reported higher values of TPC in cranberry pomace (~13.55–15.17 mg GAE/g) depending on the drying conditions. The TPC content in the berries depends on many factors, such as cultivar peculiarities, cultivation technologies, region, weather conditions, ripeness, harvesting time, and storage conditions/time [37]. The higher content of TPC was reported in grape pomace (38.7 ± 0.36 mg GAE/g) [38] and blueberry pomace (~17.76–20.82 mg GAE/g) [36]. The phenolic compounds associated with soluble dietary fiber may present different structures, including soluble flavonoids and phenolic acids [34].

3.6. Technological Properties

Fiber-rich by-products can be included in food products as low-cost and non-caloric bulking agents to partially replace flour, fat or sugar, as water and oil retention agents, and to improve the stability of emulsions [39]. The technological properties were evaluated of samples hydrolyzed for 1 h (Table 4). The ORC after enzymatic hydrolysis significantly decreased ($p < 0.05$) in all of the samples compared to the control. The lowest ORC was observed in a sample hydrolyzed with Pectinex® Yieldmash Plus; however, a significant difference ($p < 0.05$) between the samples treated with enzyme preparations with pectinases was not observed. Zheng et al. [40] reported a decrease in the coconut cake dietary fiber ORC after enzymatic hydrolysis with cellulase. Oil retention is related mainly to the surface properties, overall charge density, and the constituents' hydrophilic nature [39]. Enzymatic hydrolysis decreased the total dietary fiber ratio and quantity, which may influence the surface properties of the enzymatically treated pomace. Despite the decreased ORC, all of the pomaces showed good ORC properties compared to the other berry pomaces. Gouw et al. [35] reported a lower ORC of cranberry pomaces of 1.97 g/g; other authors also reported lower ORC values of berry pomaces (blackcurrant, redcurrant, chokeberry, rowanberry, and gooseberry) that ranged from 1.91 to 2.27 g/g [41]. Dietary fiber with a good ability to retain oil can be used for the stabilization of emulsions and high-fat food products [39].

The WSC increased in the samples hydrolyzed with Viscozyme® L and Celluclast® 1.5L (23.33% and 70%, respectively) compared to the control, while treatment with other enzymes decreased the WSC. The enzymatic hydrolysis of cellulose and hemicellulose leads to the exposure of more hydrogen bonds, which influence a higher WSC [40]. However, a high IDF and a low SDF content (especially pectin) have an adverse effect on the WSC [42]. The particle sizes also affect the WSC, and, in most cases, the decrease in the particle size increases the WSC; however, it can also be reduced due to the destruction of the dietary fiber matrix and the links between polysaccharides [43]. Reißner et al. [41] reported higher WSC (5.50–7.09 mL/g) of other berry pomaces (blackcurrant, redcurrant, gooseberry, rowanberry, and chokeberry). Gouw et al. [35] reported a higher WSC of cranberry pomace (5.87 mL/g). A previous study showed that the WSC in most cases increased after enzymatic treatments. However, it depends not only on the kind of fiber or enzyme, but on other conditions used in the treatment [12].

The WRC decreased significantly after enzymatic treatment with most of the enzyme preparations, while enzymatic treatment with Celluclast® 1.5L increased the WRC, however, no significant differences were observed ($p < 0.05$) compared with the control. The fibers consisting mainly of primary cell walls generally have higher water hydration values than the fibers consisting mainly of secondary cell walls [44]. The properties of water hydration decreased after enzymatic hydrolysis with enzymes, the main declared activities being pectinases. Spadoni Andreani et al. [33] indicated that the cranberry pomace cell wall is rich in pectic polysaccharides. The hydration properties of dietary fibers are strongly related to the source of the dietary fiber [39,41].

The stability of the emulsion is the ability to maintain the emulsion and its rupture resistance [45]. Kalla-Bertholdt et al. [46] reported that an emulsion prepared with high amounts of SDF has a micelle-like network, leading to a faster initial fat digestion. The emulsion stability of the enzymatically hydrolyzed cranberry pomaces was determined (Figure 1). The stability of the emulsion depended not only on the enzymatically treated pomace, but also on the pH value. This also confirms other studies [47]. The stability of the emulsion decreased during storage in all of the samples, however, in most cases, the stability of the emulsion did not change significantly ($p < 0.05$) during 168 and 504 h of storage (Figure 1). The pomaces hydrolyzed with Pectinex® Yieldmash Plus showed the lowest emulsion stability compared with the other samples. The pomaces hydrolyzed with Pectinex® Ultra Tropical and Celluclact® 1.5 L showed the highest emulsion stability in water (pH 3.27) and pH 4 buffer solution, while, at pH 6, better emulsion stability was shown in the pomaces hydrolyzed with Viscozyme® L and Celluclast® 1.5L. The control sample and the sample hydrolyzed with Celluclast® 1.5L showed a higher stability of the emulsion in a pH 8 buffer compared to the other enzymatically treated pomaces. The thermal stability of the emulsions was lower (Figure 2). The lowest thermal stability in all of the cases was observed using pomaces hydrolyzed with Pectinex® Yieldmash Plus. The highest stability was obtained in most of the cases using pomace hydrolyzed with Celluclast® 1.5L. Huc-Mathis et al. [48] reported that IDF helps maintain the stability of the emulsions through the Pickering mechanism and/or network formation in the continuous phase, probably favored by the stabilization of the proteins and the pectins in the soluble fraction.

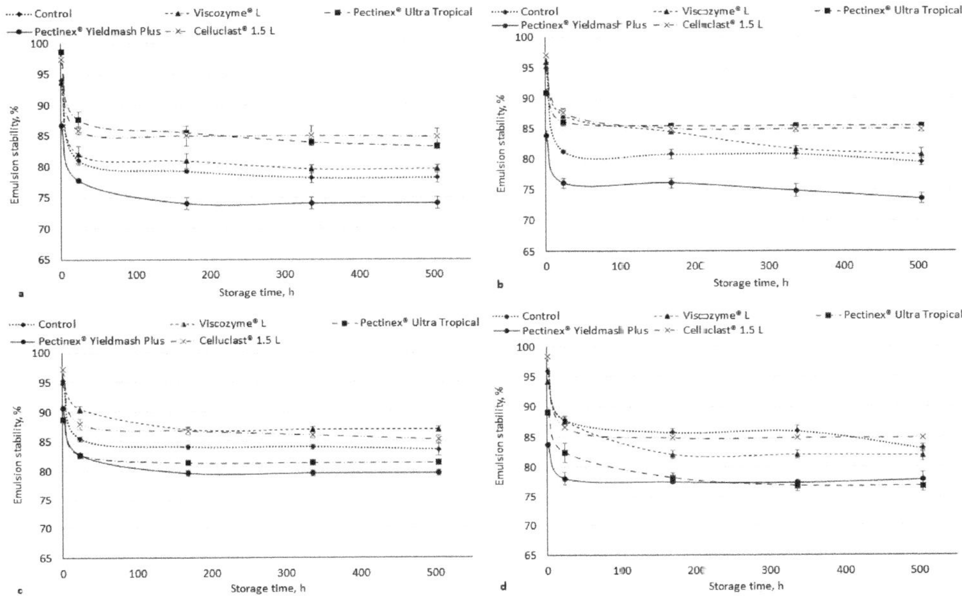

Figure 1. Emulsion stability of enzymatically hydrolyzed cranberry pomaces at pH 3.27 (water) (**a**); 4 (**b**); 6 (**c**); 8 (**d**). Data values are expressed as means with the standard deviation (n = 3).

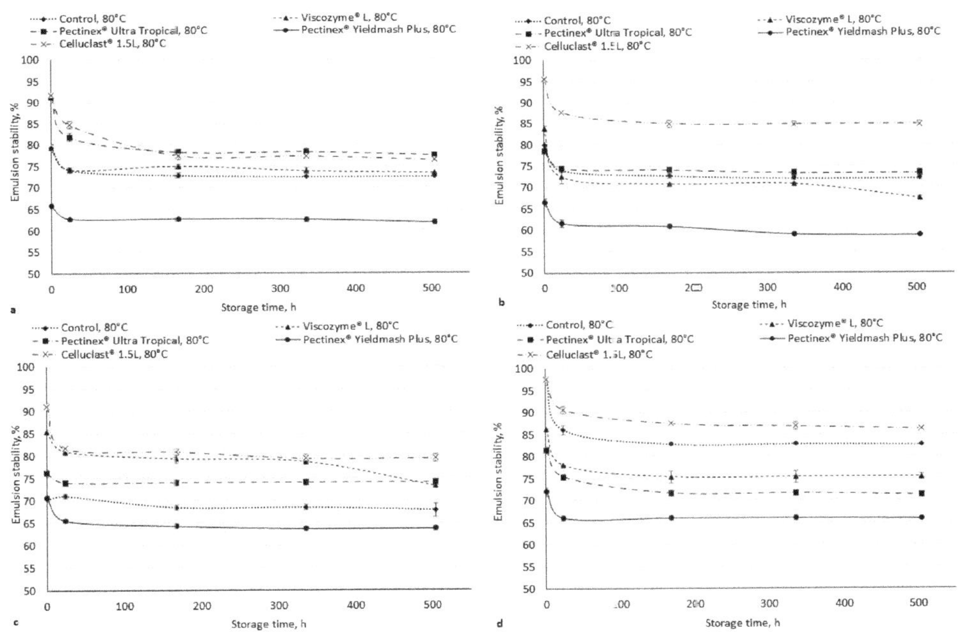

Figure 2. Thermal emulsion stability of enzymatically hydrolyzed cranberry pomaces at pH 3.27 (water) (**a**); 4 (**b**); 6 (**c**); 8 (**d**). Data values are expressed as means with the standard deviation (n = 3).

4. Conclusions

Cranberry pomace is a good source of dietary fiber, containing 59.93 g/100 g (dry weight) of IDF and 12.74 g/100 g of SDF. The enzymatic hydrolysis changed the technological properties of cranberry pomace, the ratio of SDF and IDF, and their quantities. The pomace treated with Celluclast® 1.5L resulted in the highest SDF content and an increase in the WSC and WRC. The highest amount of oligosaccharides was obtained with Viscozyme® L, while hydrolysis with Pectinex® Ultra Tropical resulted in the highest amount of mono- and disaccharides and TPC content. The pomace treated with these enzymes can be used to enhance products with oligosaccharides and phenolic compounds. All of the tested water-soluble fractions showed prebiotic activity and enhanced the growth of *Lactobacillus acidophilus* DSM 20079 and *Bifidobacterium animalis* DSM 20105 after 24 h of fermentation, while viability after 48 h of fermentation was strain and carbon source-dependent. The ORC after the enzymatic hydrolysis significantly decreased ($p < 0.05$) in all of the samples. However, all of the pomaces showed good ORC properties and could be used to stabilize high-fat foods. The highest emulsion stability was obtained in most cases using the pomace hydrolyzed with Celluclast® 1.5L. The enzymatic hydrolysis using different enzymes gave cranberry pomaces with different compositions and technological properties, which could be used in developing different novel food products. Therefore, enzymatic hydrolysis can open new market perspectives for the application of cranberry pomace as a new, cheap and valuable ingredient to improve the nutritional values and technological properties of food products.

Author Contributions: Conceptualization, L.B. and D.C.; methodology, D.C.; formal analysis, J.J.; investigation, J.J., D.C. and M.S; resources, L.B.; data curation, L.B. and D.C; writing—original draft preparation, J.J.; writing—review and editing, L.B., D.C. and M.S.; visualization, J.J.; supervision, L.B. and D.C. All authors have read and agreed to the published version of the manuscript.

Funding: This research received no external funding.

Institutional Review Board Statement: Not applicable.

Informed Consent Statement: Not applicable.

Data Availability Statement: The data presented in this study are available in the article.

Acknowledgments: The authors acknowledge Novozymes A/S, Denmark, for kindly providing the enzymes required for this research.

Conflicts of Interest: The authors declare no conflict of interest.

References

1. Hussain, S.; Jõudu, I.; Bhat, R. Dietary Fiber from Underutilized Plant Resources—A Positive Approach for Valorization of Fruit and Vegetable Wastes. *Sustainability* **2020**, *12*, 5401. [CrossRef]
2. Banerjee, J.; Singh, R.; Vijayaraghavan, R.; MacFarlane, D.; Patti, A.F.; Arora, A. Bioactives from Fruit Processing Wastes: Green Approaches to Valuable Chemicals. *Food Chem.* **2017**, *225*, 10–22. [CrossRef]
3. Coleman, C.M.; Ferreira, D. Oligosaccharides and Complex Carbohydrates: A New Paradigm for Cranberry Bioactivity. *Molecules* **2020**, *25*, 881. [CrossRef] [PubMed]
4. FDA, authors Health Claims: Fruits, Vegetables, and Grain Products That Contain Fiber, Particularly Soluble Fiber, and Risk of Coronary Heart Disease. Available online: https://www.ecfr.gov/current/title-21/chapter-I/subchapter-B/part-101/subpart-E/section-101.77 (accessed on 28 May 2022).
5. Rodríguez-Daza, M.-C.; Roquim, M.; Dudonné, S.; Pilon, G.; Levy, E.; Marette, A.; Roy, D.; Desjardins, Y. Berry Polyphenols and Fibers Modulate Distinct Microbial Metabolic Functions and Gut Microbiota Enterotype-Like Clustering in Obese Mice. *Front. Microbiol.* **2020**, *11*, 2032. [CrossRef] [PubMed]
6. Islam, M.R.; Hassan, Y.I.; Das, Q.; Lepp, D.; Hernandez, M.; Godfrey, D.V.; Orban, S.; Ross, K.; Delaquis, P.; Diarra, M.S. Dietary Organic Cranberry Pomace Influences Multiple Blood Biochemical Parameters and Cecal Microbiota in Pasture-Raised Broiler Chickens. *J. Funct. Foods* **2020**, *72*, 104053. [CrossRef]
7. Sun, J.; Marais, J.P.J.; Khoo, C.; LaPlante, K.; Vejborg, R.M.; Givskov, M.; Tolker-Nielsen, T.; Seeram, N.P.; Rowley, D.C. Cranberry (Vaccinium Macrocarpon) Oligosaccharides Decrease Biofilm Formation by Uropathogenic Escherichia Coli. *J. Funct. Foods* **2015**, *17*, 235–242. [CrossRef]

8. Liu, S.-Q.; Hotchkiss, A.T.; Renye, J.A.; White, A.K.; Nunez, A.; Guron, G.K.P.; Chau, H.; Simon, S.; Poveda, C.; Walton, G.; et al. Cranberry Arabino-Xyloglucan and Pectic Oligosaccharides Induce Lactobacillus Growth and Short-Chain Fatty Acid Production. *Microorganisms* **2022**, *10*, 1346. [CrossRef]
9. Lau, A.T.Y.; Arvaj, L.; Strange, P.; Goodwin, M.; Barbut, S.; Balamurugan, S. Effect of Cranberry Pomace on the Physicochemical Properties and Inactivation of Salmonella during the Manufacture of Dry Fermented Sausages. *Curr. Res. Food Sci.* **2021**, *4*, 636–645. [CrossRef] [PubMed]
10. Gu, M.; Fang, H.; Gao, Y.; Su, T.; Niu, Y.; Yu, L. Characterization of Enzymatic Modified Soluble Dietary Fiber from Tomato Peels with High Release of Lycopene. *Food Hydrocoll.* **2020**, *99*, 105321. [CrossRef]
11. Alberici, N.; Fiorentini, C.; House, A.; Dordoni, R.; Bassani, A.; Spigno, G. Enzymatic Pre-Treatment of Fruit Pomace for Fibre Hydrolysis and Antioxidants Release. *Chem. Eng. Trans.* **2020**, *79*, 175–180. [CrossRef]
12. Canela-Xandri, A.; Balcells, M.; Villorbina, G.; Cubero, M.Á.; Canela-Garayoa, R. Effect of Enzymatic Treatments on Dietary Fruit Fibre Properties. *Biocatal. Biotransform.* **2018**, *36*, 172–179. [CrossRef]
13. Total Dietary Fiber Assay Procedure. *Megazyme* **2017**, 1–19. Available online: https://www.megazyme.com/documents/Assay_Protocol/K-TDFR-200A_DATA.pdf (accessed on 1 May 2022).
14. Yoshida, B.Y.; Prudencio, S.H. Physical, Chemical, and Technofunctional Properties of Okara Modified by a Carbohydrase Mixture. *LWT* **2020**, *134*, 110141. [CrossRef]
15. Kitrytė, V.; Kraujalienė, V.; Šulniūtė, V.; Pukalskas, A.; Venskutonis, P.R. Chokeberry Pomace Valorization into Food Ingredients by Enzyme-Assisted Extraction: Process Optimization and Product Characterization. *Food Bioprod. Process.* **2017**, *105*, 36–50. [CrossRef]
16. Syrpas, M.; Valanciene, E.; Augustiniene, E.; Malys, N. Valorization of Bilberry (*Vaccinium myrtillus* L.) Pomace by Enzyme-Assisted Extraction: Process Optimization and Comparison with Conventional Solid-Liquid Extraction. *Antioxidants* **2021**, *10*, 773. [CrossRef]
17. Moreno-Vilet, L.; Garcia-Hernandez, M.H.; Delgado-Portales, R.E.; Corral-Fernandez, N.E.; Cortez-Espinosa, N.; Ruiz-Cabrera, M.A.; Portales-Perez, D.P. In Vitro Assessment of Agave Fructans (*Agave salmiana*) as Prebiotics and Immune System Activators. *Int. J. Biol. Macromol.* **2014**, *63*, 181–187. [CrossRef] [PubMed]
18. Singleton, V.L.; Orthofer, R.; Lamuela-Raventós, R.M. [14] Analysis of Total Phenols and Other Oxidation Substrates and Antioxidants by Means of Folin-Ciocalteu Reagent. *Methods Enzymol.* **1999**, *299*, 152–178. [CrossRef]
19. Yu, G.; Bei, J.; Zhao, J.; Li, Q.; Cheng, C. Modification of Carrot (*Daucus carota* Linn. Var. Sativa Hoffm.) Pomace Insoluble Dietary Fiber with Complex Enzyme Method, Ultrafine Comminution, and High Hydrostatic Pressure. *Food Chem.* **2018**, *257*, 333–340. [CrossRef]
20. White, B.L.; Howard, L.R.; Prior, R.L. Proximate and Polyphenolic Characterization of Cranberry Pomace. *J. Agric. Food Chem.* **2010**, *58*, 4030–4036. [CrossRef] [PubMed]
21. Yoon, K.Y.; Cha, M.; Shin, S.R.; Kim, K.S. Enzymatic Production of a Soluble-Fibre Hydrolyzate from Carrot Pomace and Its Sugar Composition. *Food Chem.* **2005**, *92*, 151–157. [CrossRef]
22. Spadoni Andreani, E.; Karboune, S. Comparison of Enzymatic and Microwave-Assisted Alkaline Extraction Approaches for the Generation of Oligosaccharides from American Cranberry (*Vaccinium macrocarpon*) Pomace. *J. Food Sci.* **2020**, *85*, 2443–2451. [CrossRef]
23. Wang, X.; Huang, M.; Yang, F.; Sun, H.; Zhou, X.; Guo, Y.; Wang, X.; Zhang, M. Rapeseed Polysaccharides as Prebiotics on Growth and Acidifying Activity of Probiotics in Vitro. *Carbohydr. Polym.* **2015**, *125*, 232–240. [CrossRef] [PubMed]
24. Huang, F.; Liu, H.; Zhang, R.; Dong, L.; Liu, L.; Ma, Y.; Jia, X.; Wang, G.; Zhang, M. Physicochemical Properties and Prebiotic Activities of Polysaccharides from Longan Pulp Based on Different Extraction Techniques. *Carbohydr. Polym.* **2019**, *206*, 344–351. [CrossRef]
25. Altermann, E.; Russell, W.M.; Azcarate-Peril, M.A.; Barrangou, R.; Buck, B.L.; McAuliffe, O.; Souther, N.; Dobson, A.; Duong, T.; Callanan, M.; et al. Complete Genome Sequence of the Probiotic Lactic Acid Bacterium *Lactobacillus acidophilus* NCFM. *Proc. Natl. Acad. Sci. USA* **2005**, *102*, 3906–3912. [CrossRef]
26. Pan, X.; Wu, T.; Zhang, L.; Cai, L.; Song, Z. Influence of Oligosaccharides on the Growth and Tolerance Capacity of Lactobacilli to Simulated Stress Environment. *Lett. Appl. Microbiol.* **2009**, *48*, 362–367. [CrossRef] [PubMed]
27. Vernazza, C.L.; Gibson, G.R.; Rastall, R.A. Carbohydrate Preference, Acid Tolerance and Bile Tolerance in Five Strains of Bifidobacterium. *J. Appl. Microbiol.* **2006**, *100*, 846–853. [CrossRef] [PubMed]
28. Li, D.; Kim, J.M.; Jin, Z.; Zhou, J. Prebiotic Effectiveness of Inulin Extracted from Edible Burdock. *Anaerobe* **2008**, *14*, 29–34. [CrossRef]
29. Manning, T.S.; Gibson, G.R. Prebiotics. *Best Pract. Res. Clin. Gastroenterol.* **2004**, *18*, 287–298. [CrossRef] [PubMed]
30. Stewart, M.L.; Timm, D.A.; Slavin, J.L. Fructooligosaccharides Exhibit More Rapid Fermentation than Long-Chain Inulin in an in Vitro Fermentation System. *Nutr. Res.* **2008**, *28*, 329–334. [CrossRef]
31. Landbo, A.K.; Meyer, A.S. Enzyme-Assisted Extraction of Antioxidative Phenols from Black Currant Juice Press Residues (*Ribes nigrum*). *J. Agric. Food Chem.* **2001**, *49*, 3169–3177. [CrossRef] [PubMed]
32. Mrabet, A.; Rodríguez-Gutiérrez, G.; Rubio-Senent, F.; Hamza, H.; Rodríguez-Arcos, R.; Guillén-Bejarano, R.; Sindic, M.; Jiménez-Araujo, A. Enzymatic Conversion of Date Fruit Fiber Concentrates into a New Product Enriched in Antioxidant Soluble Fiber. *LWT* **2017**, *75*, 727–734. [CrossRef]

33. Spadoni Andreani, E.; Karboune, S.; Liu, L. Extraction and Characterization of Cell Wall Polysaccharides from Cranberry (*Vaccinium macrocarpon* Var. Stevens) Pomace. *Carbohydr. Polym.* **2021**, *267*, 118212. [CrossRef] [PubMed]
34. Palafox-Carlos, H.; Ayala-Zavala, J.F.; González-Aguilar, G.A. The Role of Dietary Fiber in the Bioaccessibility and Bioavailability of Fruit and Vegetable Antioxidants. *J. Food Sci.* **2011**, *76*, R6–R15. [CrossRef]
35. Gouw, V.P.; Jung, J.; Zhao, Y. Functional Properties, Bioactive Compounds, and in Vitro Gastrointestinal Digestion Study of Dried Fruit Pomace Powders as Functional Food Ingredients. *LWT* **2017**, *80*, 136–144. [CrossRef]
36. Ross, K.A.; DeLury, N.; Fukumoto, L.; Diarra, M.S. Dried Berry Pomace as a Source of High Value-Added Bioproduct: Drying Kinetics and Bioactive Quality Indices. *Int. J. Food Prop.* **2020**, *23*, 2123–2143. [CrossRef]
37. Česoniene, L.; Daubaras, R. Phytochemical Composition of the Large Cranberry (*Vaccinium macrocarpon*) and the Small Cranberry (*Vaccinium oxycoccos*). In *Nutritional Composition of Fruit Cultivars*, 1st ed.; Simmonds, M.S.J., Preedy, V.R., Eds.; Elsevier: Cambridge, MA, USA, 2016; pp. 173–194. [CrossRef]
38. Milinčić, D.D.; Stanisavljević, N.S.; Kostić, A.; Soković Bajić, S.; Kojić, M.O.; Gašić, U.M.; Barać, M.B.; Stanojević, S.P.; Lj Tešić, Ž.; Pešić, M.B. Phenolic Compounds and Biopotential of Grape Pomace Extracts from Prokupac Red Grape Variety. *LWT* **2021**, *138*, 110739. [CrossRef]
39. Elleuch, M.; Bedigian, D.; Roiseux, O.; Besbes, S.; Blecker, C.; Attia, H. Dietary Fibre and Fibre-Rich by-Products of Food Processing: Characterisation, Technological Functionality and Commercial Applications: A Review. *Food Chem.* **2011**, *124*, 411–421. [CrossRef]
40. Zheng, Y.; Li, Y. Physicochemical and Functional Properties of Coconut (*Cocos nucifera* L.) Cake Dietary Fibres: Effects of Cellulase Hydrolysis, Acid Treatment and Particle Size Distribution. *Food Chem.* **2018**, *257*, 135–142. [CrossRef] [PubMed]
41. Reißner, A.M.; Al-Hamimi, S.; Quiles, A.; Schmidt, C.; Struck, S.; Hernando, I.; Turner, C.; Rohm, H. Composition and Physicochemical Properties of Dried Berry Pomace. *J. Sci. Food Agric.* **2019**, *99*, 1284–1293. [CrossRef] [PubMed]
42. Navarro-González, I.; García-Valverde, V.; García-Alonso, J.; Periago, M.J. Chemical Profile, Functional and Antioxidant Properties of Tomato Peel Fiber. *Food Res. Int.* **2011**, *44*, 1528–1535. [CrossRef]
43. Ma, M.M.; Mu, T.H. Effects of Extraction Methods and Particle Size Distribution on the Structural, Physicochemical, and Functional Properties of Dietary Fiber from Deoiled Cumin. *Food Chem.* **2016**, *194*, 237–246. [CrossRef] [PubMed]
44. Guillon, F.; Champ, M. Structural and Physical Properties of Dietary Fibres, and Consequences of Processing on Human Physiology. *Food Res. Int.* **2000**, *33*, 233–245. [CrossRef]
45. Lan, G.; Chen, H.; Chen, S.; Tian, J. Chemical Composition and Physicochemical Properties of Dietary Fiber from Polygonatum Odoratum as Affected by Different Processing Methods. *Food Res. Int.* **2012**, *49*, 406–410. [CrossRef]
46. Kalla-Bertholdt, A.M.; Nguyen, P.V.; Baier, A.K.; Rauh, C. Influence of Dietary Fiber on In-Vitro Lipid Digestion of Emulsions Prepared with High-Intensity Ultrasound. *Innov. Food Sci. Emerg. Technol.* **2021**, *73*, 102799. [CrossRef]
47. Abdolmaleki, K.; Mohammadifar, M.A.; Mohammadi, R.; Fadavi, G.; Meybodi, N.M. The Effect of PH and Salt on the Stability and Physicochemical Properties of Oil-in-Water Emulsions Prepared with Gum Tragacanth. *Carbohydr. Polym.* **2016**, *140*, 342–348. [CrossRef]
48. Huc-Mathis, D.; Journet, C.; Fayolle, N.; Bosc, V. Emulsifying Properties of Food By-Products: Valorizing Apple Pomace and Oat Bran. *Coll. Surf. A Physicochem. Eng. Asp.* **2019**, *568*, 84–91. [CrossRef]

foods

MDPI

Article

Attempts to Create Products with Increased Health-Promoting Potential Starting with Pinot Noir Pomace: Investigations on the Process and Its Methods

Stephen Lo [1,*], Lisa I. Pilkington [1], David Barker [1,2] and Bruno Fedrizzi [1,2]

1 School of Chemical Sciences, University of Auckland, 23 Symonds St, Auckland 1010, New Zealand; lisa.pilkington@auckland.ac.nz (L.I.P.); d.barker@auckland.ac.nz (D.B.); b.fedrizzi@auckland.ac.nz (B.F.)
2 Centre for Green Chemical Science, School of Chemical Sciences, University of Auckland, 23 Symonds St, Auckland 1010, New Zealand
* Correspondence: s.lo@auckland.ac.nz

Abstract: A process for using grape (Pinot noir) pomace to produce products with improved health-promoting effects was investigated. This process integrated a solid–liquid extraction (SLE) method and a method to acylate the polyphenolics in the extract. This report describes and discusses the methods used, including the rationale and considerations behind them, and the results obtained. The study begins with the work to optimize the SLE method for extracting higher quantities of (+)-catechin, (−)-epicatechin and quercetin by trialing 28 different solvent systems on small-scale samples of Pinot noir pomace. One of these systems was then selected and used for the extraction of the same flavonoids on a large-scale mass of pomace. It was found that significantly fewer quantities of flavonoids were observed. The resultant extract was then subject to a method of derivatization using three different fatty acylating agents. The antiproliferative activities of these products were measured; however, the resulting products did not display activity against the chosen cancer cells. Limitations and improvements to the methods in this process are also discussed.

Keywords: Pinot noir pomace; solid–liquid extraction; valorization; flavonoids; fatty acids; derivatization

Citation: Lo, S.; Pilkington, L.I.; Barker, D.; Fedrizzi, B. Attempts to Create Products with Increased Health-Promoting Potential Starting with Pinot Noir Pomace: Investigations on the Process and Its Methods. *Foods* **2022**, *11*, 1999. https://doi.org/10.3390/foods11141999

Academic Editors: Marco Poiana, Francesco Caponio and Antonio Piga

Received: 20 May 2022
Accepted: 3 July 2022
Published: 6 July 2022

Publisher's Note: MDPI stays neutral with regard to jurisdictional claims in published maps and institutional affiliations.

Copyright: © 2022 by the authors. Licensee MDPI, Basel, Switzerland. This article is an open access article distributed under the terms and conditions of the Creative Commons Attribution (CC BY) license (https://creativecommons.org/licenses/by/4.0/).

1. Introduction

The International Organisation of Vine and Wine (OIV) recently reported that 248–294 (an average of 266) million hectoliters of wine was produced each year, globally, in between the years of 2016 to 2020 [1]. These figures highlight the scale at which wine is produced to satisfy a large global demand. It is estimated that for every 100 L of wine produced, 18–35 kg of grape pomace (the unwanted grape stems, skins and seeds) exits out as a waste product [2]. There are concerns that this large and accumulating waste mass will become unmanageable and have negative impacts towards the environment [3–6]. Other than to discard it as landfill, other practices currently employed to manage this waste are to reuse the material either as compost or livestock feed [5,6]. However, employments of these methods to break down the pomace are not without their limitations. The high phenolic content within the material leads to germination problems, and the polymeric compounds (i.e., lignins) poses issues for animal digestion [3]. The environmental concerns have led government bodies to place stringent policies around managing this waste and, in some cases, has resulted in financial penalties on wine producers [7,8]. This has encouraged a need to explore novel strategies in utilizing grape pomace for higher value purposes. Development of these strategies has also been driven by consumer demands for both naturally sourced starting materials and sustainable practices, which include implementing the reusing and recycling of waste products in the production process to achieve a circular economy model [9,10].

Grape pomace contains a variety of valuable compounds, including lignins, aroma precursors, sugars, proteins, unsaturated fatty acids and phenolics [3,4]. These compounds have properties that can be exploited to provide benefits to consumers and are attractive for various commercial applications. Phenolics (which includes polyphenols) in grape pomace, for example, comprise of a diverse set of compounds that are further classified into groups (such as phenolic acids, flavonoids and stilbenes) based on their core chemical structures. Polyphenols are well known for their biological activity and have demonstrated their ability to exert antioxidant, antiproliferative, anti-inflammatory, antimicrobial and neuroprotective effects [3,4,11,12]. It has been proposed that the recovery of valuable compounds from grape pomace could be further processed and applied for purposes such as food fortification [13], food processing [9,14], food preservation [15], improving wine quality [16], antibacterial materials [17] and cosmetic use [18]. Additionally, given their bioactive properties, it has been suggested that polyphenolic compounds could be sourced as starting materials and converted into products that directly deliver health-promoting effects [4,19]. In the literature, there are a number of examples in which grape pomace extracts, which have higher or more concentrated phenolic contents, were found to exhibit antioxidant, antimicrobial, anti-inflammatory and antiplatelet activities [20–22], as well as promising antiproliferative or cytotoxic activities against cancer cells [23,24]. These studies were validated based on chemical, biochemical and in vitro studies and are promising indications for the health-promoting potential of the pomace. There are, however, fewer examples of in vivo studies that demonstrate the health-promoting effects of grape pomace products in animals or humans. As one example, Olivero-David et al. demonstrated that their grape by-product extract was able to the lower cholesterol levels in Wistar rats, which has implications towards hypercholesterolemia [25]. The scarcity of such studies is due to the fact that bioactive polyphenols have poor oral bioavailability, and this severely limits their effects in live models [26,27]. One major contributing factor to their low bioavailability is that these compounds are highly susceptible to metabolism by the enzymes they encounter as they pass through compartments such as the intestinal cells and the liver before reaching the systemic circulation [28,29]. These enzymatic activities convert these compounds into metabolite products that are inactive and more easily eliminated from the body. This is a large barrier that needs to be overcome. The ability to increase their bioavailability enables higher amounts of the active compounds to enter into the systemic circulation, be transported to the cells and tissues of interest and be able to exert the desired health-promoting activities. Further, if these compounds can be structurally manipulated to have increased activity (i.e., increased binding affinity to their biological targets), then this reduces the amounts of active compounds required around the active sites to exert these health effects. Therefore, the strategy for developing health-promoting products from grape pomace must thoroughly consider a process that can both effectively extract the bioactive polyphenolic compounds from the waste material and make the appropriate adjustments to these compounds that would enhance the properties needed for health promotion, including increased bioavailability and bioactivity.

The aim of the study was to develop a process whereby bioactive polyphenolics could be sourced from grape pomace, used as starting materials and modified into products with improved health-promoting properties. This work drew inspiration from previously reported studies conducted by our research group, which presented findings on the use of a simple method for extracting polyphenolics from grape pomace [30] and the structural modifications to flavonoids that led to improved bioactivity [31,32]. It was proposed that the integration of these methods could be a feasible process to achieve the study aim. Unfortunately, in following this proposed process, the expected outcome was not achieved. Despite that, the investigations of these methods did generate results that are worth presenting. This report aims to describe and discuss these methods, including the rationale and considerations underlying the decisions to use and adapt them, and the results. Following that, limitations to these methods are explored and improvements are suggested.

2. Materials and Methods

2.1. Chemicals and Reagents

Commercial standards include (+)-catechin hydrate (≥98%), (−)-epicatechin (≥90%), rutin hydrate (≥95%) and quercetin (≥95%) were purchased from Sigma-Aldrich (Auckland, New Zealand). Acetonitrile (MeCN, HPLC grade), L-(+)-tartaric acid (reagent grade), acetone (analytical grade) and ethanol (analytical grade) were purchased from ECP (Auckland, New Zealand). Glacial acetic acid (AcOH, HPLC grade) was purchased from ThermoFisher Scientific (Auckland, New Zealand). Ultrapure water (H_2O) had resistivity of 18.2 MΩ cm and was obtained from Sartorius Arium® Pro system. Pinot noir grape pomace (seeds, skins and stems) was industrially sourced from a New Zealand vineyard in 2015. Acylating agents (octanoyl chloride, lauroyl chloride and palmitoyl chloride) were purchased from Sigma-Aldrich (Auckland, New Zealand).

2.2. Preparation of Model Red Wine and Solvent Systems

Model red wine was prepared by dissolving tartaric acid (5 g) into 13.5% EtOH in ultrapure water (1 L). The pH of this solution was adjusted to 3.5 with NaOH (10 M and 1 M).

A subset of 28 systems were selected (from those used in the literature) and trialed in extraction on small-scale mass Pinot noir pomace samples (see Section 3.1) [30]. The ratios of the solvents (acetone, H_2O and EtOH) that made up the selected systems (50 mL) are found in Table 1.

Table 1. Solvent systems trialed for solvent–liquid extraction of Pinot noir pomace.

Acetone:H_2O:EtOH						
80:20:0	60:30:10	50:30:20	40:40:20	30:60:10	30:20:50	20:50:30
70:30:0	60:20:20	50:20:30	40:30:30	30:50:20	20:80:0	20:40:40
60:40:0	50:50:0	40:60:0	40:20:40	30:40:30	20:70:10	20:30:50
60:30:10	50:40:10	40:50:10	30:70:0	30:30:40	20:60:20	20:20:60

2.3. HPLC Analysis

All standards and samples were analyzed using the Agilent 1290 Infinity Liquid Chromatography (Santa Clara, CA, USA), fitted with a quaternary pump and coupled to a UV-Vis DAD. DAD wavelengths were set to 280 nm for (+)-catechin hydrate and (−)-epicatechin; 350 nm for rutin hydrate; and 365 nm for quercetin. The eluents are 100% H_2O (solvent A), 5% AcOH in H_2O (solvent B) and 100% MeCN (solvent C). The HPLC method used was reported in the literature (Table S1) [33]. The standards and samples were eluted through the Phenomenex Luna C-18 Column (250 mm × 4.6 mm, 5 µm particle size, Phenomenex, Torrance, CA, USA) at a flow rate of 0.8 mL/min at 25 °C.

Retention times (tr) for (+)-catechin hydrate, (−)-epicatechin, rutin hydrate and quercetin were 38.0 min, 54.5 min, 76.1 min and 94 6 min, respectively. External calibration standard curves for (+)-catechin hydrate, (−)-epicatechin and rutin hydrate were generated, with each curve consisting of six calibration points that were measured in triplicate. Each curve displays excellent linearity with a correlation coefficient (r^2) of >0.99. The limit of detection and limit of quantitation for each flavonoid per kg pomace are 13.2 mg and 43.9 mg for **1**; 1.9 mg and 6.6 mg for **2**; and 2.8 mg of rutin hydrate and 9.4 mg (of rutin) for **3**.

2.4. Materials and Solid–Liquid Extraction Protocol

The solid–liquid extraction (SLE) protocol used in this study was similar to that described in the literature with some adaptations (discussed in Sections 3.1 and 3.2) [30]. Pinot noir pomace was homogenized with a kitchen blender. For the small-scale extraction study, the pomace samples (10 g) were then immersed in the solvents (50 mL) and stirred at r.t. for one hour in an open vessel. The solvent was decanted and then centrifuged at 6000 rpm for 10 min. The supernatant (5 mL) was taken, and the solvent removed in vacuo.

The resulting extract was dissolved in model red wine (2 mL). This solution was passed through a Phenomenex regenerated cellulose membrane syringe filter (0.45 µm pore size, 15 mm diameter, from CA, USA) and stored at −20 °C until HPLC analysis. The quantities of each flavonoid in each of the samples were analyzed with HPLC according to the method described in Section 2.3. Extractions of the pomace samples with each solvent system were conducted in triplicate.

For the large-scale extraction, Pinot noir pomace (2 kg) was separated into batches (200 g). Each batch was immersed in 40:30:30 acetone:H_2O:EtOH solvent (1 L) and manually stirred for 1 h. After extraction, the solid–solvent mixture was poured over a kitchen sieve to filter out the solid mass and to collect only the phenolic-enriched solvent, which was then centrifuged. For quantification, the supernatant (10 mL) from each batch was taken and combined together (total of 100 mL). The solvent was removed in vacuo. Model red wine (6 mL) was then added to the extract and separated into 3 equal portions (2 mL), which were passed through the Phenomenex syringe filter (0.45 µm pore size, 15 mm diameter, from CA, USA). The quantities of each flavonoid in each of the samples were analyzed with HPLC. The rest of the phenolic-rich solvent were combined together, and the solvent was removed in vacuo to give a resulting extract that would be used for derivatization. Various stages of this process can be seen in Figure 1.

(a) (b) (c) (d)

Figure 1. Large-scale Pinot noir extraction procedure: (**a**) Pinot noir pomace batch (200 g) with solvent (1 L); (**b**) filtering phenolic-enriched solvent; (**c**) centrifugation of phenolic-enriched solvent; (**d**) removal of solvent from extract.

2.5. *Derivatization of the Flavonoid-Enriched Extract and Infrared Analysis*

The starting material extract (1 g) was dissolved in dimethyl formamide (DMF, 15 mL) solvent. Triethylamine was added to this solution and then the acylating agents (octanoyl, lauroyl and palmitoyl chlorides). The reaction was stirred for 24 h and then was quenched with $NaHCO_3$. The mixture was then filtered and washed with excess water. The resulting products were collected and dried to give red, solid products. The infrared (FT-IR) spectra of non-derivatized and derivatized products were obtained using Perkin-Elmer Spectrum 100 FT-IR spectrometer (Santa Clara, CA, USA).

2.6. *Antiproliferative Activity Procedures*

The antiproliferative studies on HCT116 and MDA-MB-231 cell lines were conducted using the [^{3}H]-thymidine incorporation assay method, as previously reported by [34]. Essentially, the method was conducted by seeding 3000 cells in each well of the 96-well plates with a solution of non-derivatized and derivatized extracts in dimethyl sulfoxide (DMSO) for 3 days. Then, [^{3}H]-thymidine is added to the cells, and they are incubated for 6 h. The cells were then counted using a Trilux/Betaplate counter, which shows the percentage of the cells with [^{3}H]-thymidine that are incorporated into the DNA helix. This was conducted in duplicate. The antiproliferative activity was determined as cell growth percentages relative to the 100% growth in the control (non-treated).

2.7. Statistical Analysis

Ternary plots were used to identify and depict regions of high and low concentrations of compounds extracted across the investigated solvent ratios. R (version 3.6.0) was used to conduct the analysis using the "ggtern" R package [35,36]. To create the plots, interpolation was carried out using a multivariate linear regression with a fitted polynomial expression that was chosen on the basis of best fit and accurate representation of the measured data.

Statistical analyses of the antiproliferative testing were conducted using one sample, a one-tail *t*-test to determine whether the mean proliferation value of each product was lower than that of the control (100%) and a one-way ANOVA to determine whether there were any differences between these values.

3. Results and Discussion

3.1. Optimization of the Solid–Liquid Extraction Method (Small-Scale Extraction Studies)

This study began with an investigation into an appropriate method for extracting polyphenolics from grape pomace. Many considerations were made to develop this part of the process. These included the variety of grape pomace that would be used, the specific polyphenolic compounds to target, the selection of an appropriate extraction technique and how the extraction protocol could be optimized.

In consideration of the pomace, it was decided to use a grape variety that was of significance to New Zealand (NZ). Sauvignon blanc has been recognized as the most significant grape to NZ, as its production mass is the highest of all grapes (with an annual production mass of around 10-fold higher than the grape variety that follows behind it) [37,38]. Our research group had already investigated the use of this grape pomace, including the extraction of valuable compounds (aroma precursors and phenolics) from it and processes to valorize them [17,30]. Alternatively, Pinot noir is another grape type that is of significance to NZ, with a production mass (average of 30,000 tonnes produced each year between the years from 2012 to 2021) often following behind Sauvignon blanc as the second most produced grape type in the country [37,38]. For this reason, Pinot noir pomace was selected for use in this study.

It was mentioned earlier that grape pomace contains a wide range of polyphenolic compounds. Among them, the flavonoid compounds are one prominent class. In line with polyphenolics, flavonoids also possess a wide range of biological activities, including antioxidant and antiproliferative activity, as well as potential for anti-inflammatory, anti-thrombotic, cardio-protective and neuro-protective effects [19,39,40]. Therefore, it was considered that the extraction strategy should be tailored to selectively target the flavonoids that are present in the pomace. Raising the profile of these compounds also serves to assist the next method in the process, which was chosen based on knowledge about the lipophilic flavonoid derivatives and their improved bioactivity (see Section 3.3). Upon review of the literature that have reported the extraction of polyphenolics (particularly focusing on SLE extraction techniques) from grape pomace, it was noted that (+)-catechin (**1**), (−)-epicatechin (**2**) and quercetin (**3**), are the most frequently cited flavonoids (Figure 2 and Table 2) [20,30,41–46]. Therefore, it was decided that the extraction method would be developed to obtain as much of these three flavonoids as possible.

In the literature, a wide variety of extraction techniques have been presented for extracting phenolics from plant-based materials. Examples include conventional SLE [30,41], SLE assisted with ultrasound [47,48], SLE assisted with microwave [49,50], and supercritical fluid extraction [51,52]. Each technique has advantages (or disadvantages) based on factors such as cost, ease of use, efficiency of extraction, selectivity of phenolics, requirement for specialized equipment and being environmentally friendly. Since this study intended to be an initial proof of concept, the primary criteria for selecting the extraction technique were simple to conduct, relatively low cost and reduced reliance on sophisticated and specialized equipment. The conventional SLE satisfies these criteria, and, therefore, this technique was chosen. In the literature, this technique has been employed on different types of grape pomace, and selected examples of these works and the quantities of flavonoids **1**, **2** and **3** extracted are presented in Table 2.

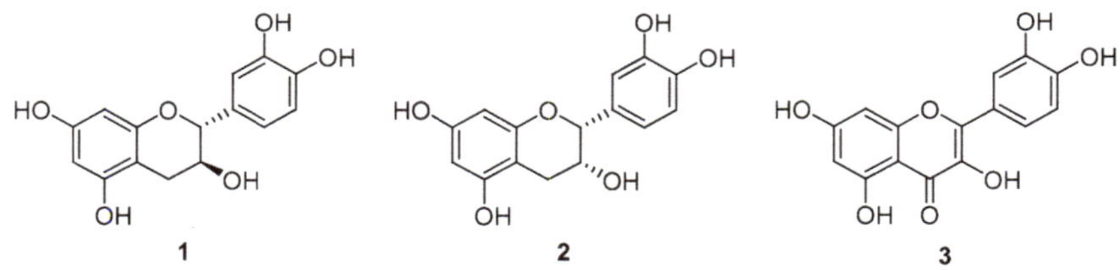

Figure 2. Chemical structures of the flavonoids (+)-catechin (**1**), (−)-epicatechin (**2**) and quercetin (**3**).

Table 2. Examples of flavonoid content extracted from the pomace of various grape types using SLE method; [dw] = dry weight.

Grape Type (Component)	Extraction Technique (Conditions)	Solvent System	Flavonoid	Average Amount (mg/kg Pomace)	Reference
Sauvignon blanc (whole pomace)	SLE (r.t., 1 h)	Acetone:H_2O:EtOH (50:50:0)	1	466.29 (±83.01)	[30]
		Acetone:H_2O:EtOH (30:50:20)	2	151.14 (±7.91)	
		Acetone:H_2O:EtOH (40:30:30)	3	31.89 (±2.77) *	
Weisser Riesling (skins)	SLE (r.t., 2 h)	MeOH:HCl (99.9:0.1 v/v)	1	226.7 (±24.6) [dw]	[45]
			2	134.6 (±12.1) [dw]	
Weisser Riesling (seeds)			1	790.2 (±11.2) [dw]	
			2	674.5 (±24.9) [dw]	
Pinot noir (seed)	SLE (25 °C, 19 h)	MeOH	1	1583 [dw]	[46]
			3	1386 [dw]	
		EtOH	1	1450 [dw]	
			3	1386 [dw]	

* measured as rutin equivalence per kg pomace.

The SLE protocol selected for this study was based on one that had previously been implemented by our research group [30]. In that study, the SLE technique was investigated for its ability to extract a wide range of aroma precursors and phenolic compounds from Sauvignon blanc pomace. The study trialed the protocol with a series of 66 different solvent systems (made up of the three different polar solvents—acetone, H_2O and EtOH) to identify the optimal systems for extracting each compound of interest. From those findings, the optimal systems proposed for flavonoids **1**, **2** and **3** were of relevance to the purpose of the present study (Table 2 and Figure 3). However, before proceeding to apply the SLE protocol with the exact same solvent systems to the Pinot noir pomace, it was considered that those findings were specific to Sauvignon blanc pomace and that differences in the pomace matrix between the two grape types could result in differences in the extraction outcomes. In essence, what is optimal for one pomace is not necessarily the same for another. Therefore, it was necessary to conduct another trial to independently ascertain which solvent systems would be optimal for extracting these three flavonoids from Pinot noir pomace. Rather than conducting the same trial with all 66 solvent systems, a more focused approach was taken by narrowing down the number of systems in the trial. Based on an analysis of the ternary diagrams for each of the three flavonoids (presented from the report by Jelley et al.), a subset of 28 solvent systems were considered as having generally good ability to extract those desired flavonoids (see Table 1 and depicted as the red triangle on Figure 3) [30]. Therefore, it was decided that these solvent systems would be selected for the trial. The trial was conducted on a small mass (10 g) of Pinot noir pomace. It is envisaged that, by determining the quantities of flavonoids extracted from each system, this would then help in the identification of the few systems that are optimal. From these, one would then be selected and taken forward for use in the large-scale pomace mass extraction study (see Section 3.2).

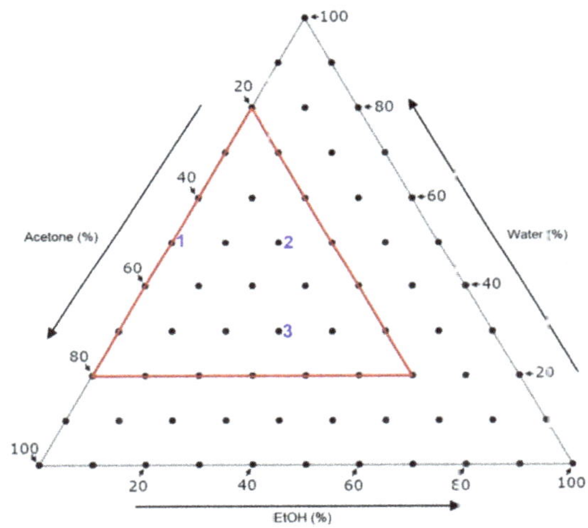

Figure 3. Ternary diagram with **1**, **2** and **3** (in blue) to indicate the optimal solvent systems for extracting the corresponding flavonoids from Sauvignon blanc pomace as previously identified in the literature [30] and the range of solvent systems (outlined in red) that were trialed in this study for extracting the three flavonoids from Pinot noir pomace.

The small-scale extraction trial was conducted according to Section 2.4. At the end of each extraction (and after the solvents had been removed in vacuo), a pomace extract resulted, all of which were viscous residues that were red in color. Prior to HPLC analysis, this extract was dissolved in a small volume of model red wine. The reason for this was the need to dissolve all of the contents in the extract in a solvent with low volatility to minimize solvent loss through evaporation. Model red wine was an ideal choice as it was low in volatility and, proposedly, its resemblance to actual red wine enabled dissolution of all the contents in the extract. The dissolved extracts were then analyzed with HPLC to determine the quantities of the three flavonoids. The quantities of these flavonoids extracted from each individual solvent system are found in Table S2 From these results, it was found that the optimal systems for extracting **1**, **2** and **3** were the 80:20:0, 40:40:20 and 40:50:10 systems, respectively (Table 3).

Table 3. Solvent systems that had extracted the highest amount of each flavonoid and the average (of triplicate extractions) quantity (mg/kg pomace) as measured in the extract.

Flavonoid	Best Solvent System (Acetone:H_2O:EtOH)	Average Quantity Extracted (mg/kg Pomace)
1	80:20:0	174.1 (±17.1)
2	40:40:20	269.6 (±34.1)
3	40:50:10	87.0 (±46.4) *

* measured as mg rutin hydrate equivalence per kg pomace.

To develop some perspective on the performance of this method, these flavonoid quantities were compared (by determining the relative and absolute differences) to those reported in the selected literature presented in Table 2. This provided an indication as to whether the SLE method used in the present study was more, less or the same in effectiveness for extracting these compounds compared to the method employed by others. It should be noted that there are significant differences between the SLE protocol used in the selected literature and that of the present study. These differences include factors such as grape type that was used, the pomace component (i.e., whole pomace, only skins, only seeds), preparation of the pomace (i.e., lyophilized, blended, grounded), solvent systems (i.e., solvent type, mono or multi, different ratios) and the conditions (i.e., temperature and length of extraction time). The limitation of comparing results that have been obtained from very different methods is the inability to deduce the extent to which each factor would have contributed to the quantitative differences. In comparison, the quantities of **1**, **2** and **3** obtained from the Pinot noir pomace were 2.7-fold lower (by 292.2 mg/kg pomace), 1.8-fold higher (118.5 mg/kg) and 2.8-fold higher (55.1 mg/kg), respectively, compared to those obtained from the Sauvignon blanc pomace [30]; the quantities of **1** and **2** were 1.3-fold lower (52.6 mg/kg) and 2.0-fold higher (135 mg/kg), respectively, compared to those obtained from the Weisser Riesling pomace skins [45]. These differences were not large and suggest that the present method was slightly less effective for **1**, slightly more effective for **2** and slightly more effective for **3**, compared to that of the two studies. In contrast, the differences in the quantities of the flavonoids were much more pronounced compared to those obtained from the pomace seeds. For example, quantities of **1** and **2** in the present study were 4.5-fold lower (616.1 mg/kg) and 2.5-fold lower (404.9 mg/kg), respectively, to those obtained from the Weisser Riesling pomace seeds [45]; the quantities of **1** and **3** were 8- to 16-fold lower (1299 to 1409 mg/kg) to those obtained from the Pinot noir seeds (extracted using both MeOH and EtOH solvent) [46]. The increased effectiveness of those methods for these flavonoids may be attributed to the fact that seeds in general carry higher amounts of these flavonoids. This notion is supported by comparing the results within the report by Kammerer et al., which showed that the contents of **1** and **2**, extracted from the pomace seeds, were much higher than those obtained from the pomace skins [45].

For a more general understanding on the solvent systems' effectiveness for extracting these flavonoids, the data (average quantities of the flavonoids extracted) were statistically analyzed and visually interpreted as model ternary diagrams (Figure 4b–d). It should be noted that the data displayed in these diagrams fall within a range of solvent systems (depicted as a green triangle in Figure 4a) that differ from the experimental range (depicted as a red triangle in Figure 4a). Therefore, the limitations of these diagrams are that the data for solvent systems with <10% EtOH are omitted and that the data generated for solvent systems with <20% water were extrapolated. From analyses of each individual diagram, it was observed that the extraction of **1** from the pomace was better achieved with systems containing higher amounts of acetone (50–80%) combined with low amounts of H_2O (10–30%) and EtOH (10–30%). The extraction of **2** was better achieved with systems containing moderate amounts of acetone (40–60%) combined with low amounts of H_2O (10–30%) and low-to-moderate amounts of EtOH (20–40%). The extraction of **3** was better achieved with systems containing moderate amounts of acetone (40–50%) combined with moderate amounts of H_2O (30–50%) and moderate amounts of EtOH (30–40%). From an analysis of all three ternary diagrams together, it was deduced that these six systems–60:30:10, 50:40:10, 50:30:20 40:50:10, 40:40:20 and 40:30:30 (depicted as a blue triangle in Figure 4a)—would be optimal for the overall extraction of all three flavonoids. This is supported by looking at the ranking of these solvent systems (out of all 28) for each compound (Table 4). Five of these systems were found to be within the top five in their ability to extract at least one of the three flavonoids, and the other system ranked just above the middle for all three flavonoids. A process was then taken to narrow down further the most suitable solvent system that would be used in the SLE method for extraction of flavonoids from a large-scale mass of pomace. This process was executed by choosing the top few systems based on the sum of ranks and then validating these systems according to the ternary diagrams. By looking at the sum of ranks, the top three were 40:40:20 (sum = 12), 50:30:20 (14) and 40:50:10 (17). These systems ranked within the top five for extracting two flavonoids and were ranked as being moderately effective for the other flavonoid. The next best system was 40:30:30 (24), which was relatively good for all three flavonoids. In comparing these systems on the ternary diagrams, it was seen that 40:50:10 was the least effective for **1** and **2** and that 50:30:10 was the most effective for **2** but least effective for **3**. The 40:30:30 and 40:40:20 systems were similar to each other in that they were both effective for all three flavonoids. It was reasoned that, out of the two, the 40:30:30 system had a better balance of all three solvent systems. For this reason, it was decided that this system would be used for the subsequent SLE of a large-scale mass (2 kg) of Pinot noir pomace.

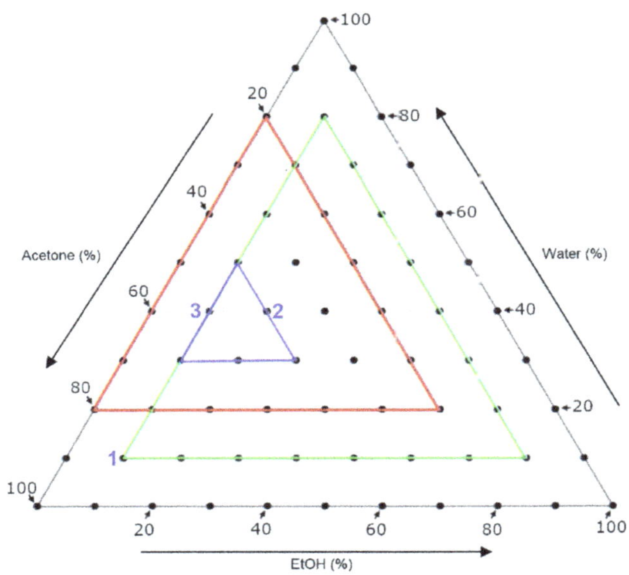

(**a**)

Figure 4. *Cont.*

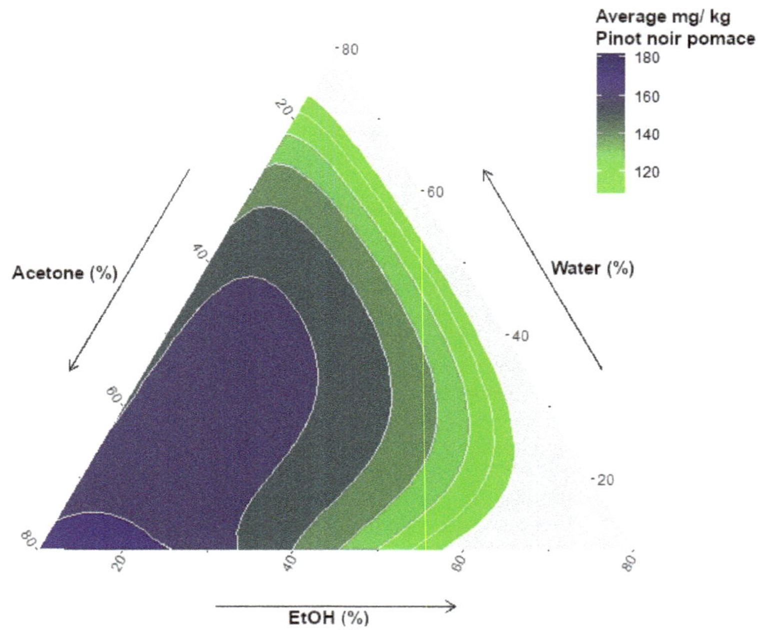

(b)

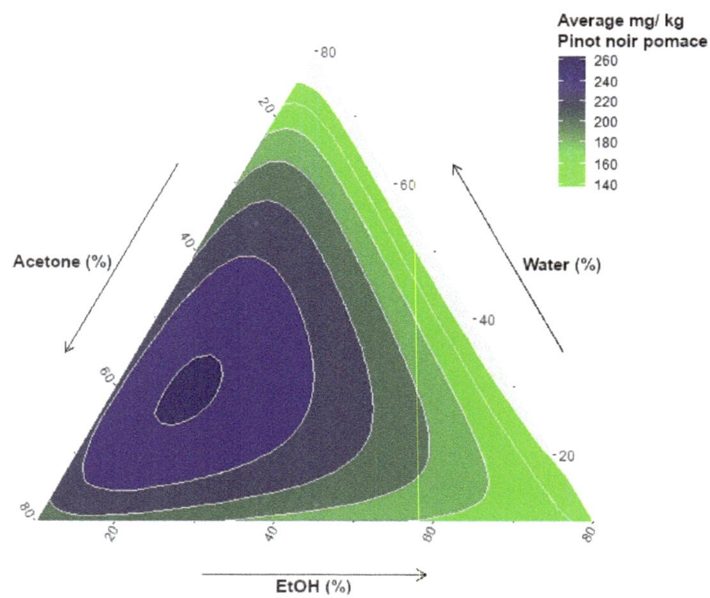

(c)

Figure 4. *Cont.*

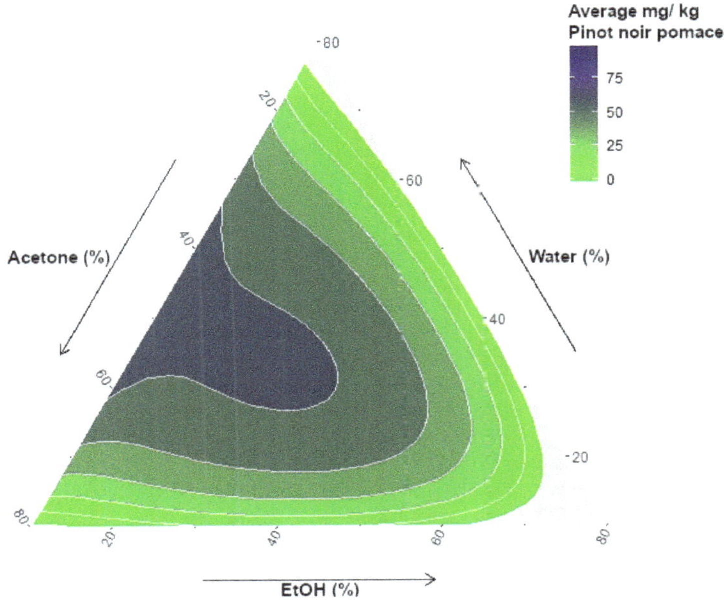

(d)

Figure 4. Ternary diagrams of: (**a**) the range of solvent systems trialed in this study (outlined in red), the optimal solvents for individual flavonoids (labelled 1, 2 and 3 in blue), the solvent boundaries to generate model diagrams of each flavonoid (outlined in green) and the optimal range of solvent systems for simultaneous extraction of all three flavonoids (outlined in blue); (**b**) solvent effectiveness for extraction of **1**; (**c**) solvent system effectiveness for extraction of **2**; and (**d**) solvent system effectiveness for extraction of **3**.

Table 4. The quantity of flavonoids extracted and their rank (out of 28) for each of the six solvent systems identified as optimal systems for the SLE for large-scale extraction.

Acetone:H_2O:EtOH	(+)-Catechin (1)		(−)-Epicatechin (2)		Quercetin (3)	
	mg/kg Pomace	Rank	mg/kg Pomace	Rank	mg/kg Pomace	Rank
60:30:10	148.4 (±25.7)	16	218.8 (±39.5)	16	49.0 (±14.3)	16
50:40:10	157.0 (±32.1)	10	231.1 (±39)	12	66.5 (±4.2)	4
50:30:20	165.9 (±16.3)	7	250.86 (±25.3)	5	72.3 (±39.1)	2
40:50:10	169.9 (±28.3)	3	227.1 (±18.8)	14	87.0 (±46.4)	1
40:40:20	173.3 (±23.9)	2	269.59 (±34.1)	1	55.0 (±1.3)	9
40:30:30	164.7 (±18.4)	8	240.1 (±8.9)	9	58.9 (±01.0)	7

3.2. Large-Scale Solid–Liquid Extraction of Pinot noir Pomace

After identifying an optimal solvent system from the small-scale trials, the next part of this study was to implement the SLE protocol on a large-scale mass of the Pinot noir pomace. It was expected that this would deliver a larger mass of the extract and, therefore, higher total amounts of the three flavonoids. In this part of the study, the pomace mass was scaled up (from 10 g) to 2 kg, which was thought of as both a mass that was manageable and also appropriate for a proof of concept study. It was intended to keep all procedures similar to those of the small-scale extraction; however, some adjustments were needed. To remain consistent with the pomace's mass-to-solvent volume ratio, the 2 kg mass would have needed to be immersed into 10 L of solvent. This was considered impractical for a few reasons. Firstly, without bespoke equipment, the handling of such a large volume of volatile solvent (acetone and MeOH) posed safety issues; secondly, this would have required a specialized stirring apparatus to consistently move a large body of pomace around the large volume of solvent for the duration of the extraction time. Thirdly, the transfer of solvent from one vessel to another would have resulted in unavoidable and non-ideal spillages and, thus, the loss of both solvent and content. To mitigate these issues, the 2 kg pomace mass was divided into 10 batches of 200 g of pomace, in which each batch was immersed into 1 L of solvent. Another adjustment was made to the way in which the pomace was moved around the solvent. Consistent motion of the pomace around the solvent was necessary to ensure maximum contact between each pomace particle and the solvent. However, in the absence of an appropriate stirring apparatus (possessing sufficient power to move the thick pomace around the liquid), the stirring was conducted manually. Stirring of some capacity is a better alternative to non-stirring; however, this manual stirring, which is further addressed in Section 3.5, was a major limitation of this method. At the end of the extraction, the resultant extracts had the same appearance (red-colored viscous residue) as that which was observed from the small-scale extraction. The extracts obtained from each batch were combined, and the quantities of the flavonoids in the combined extract were analyzed.

It was found that the quantities of the flavonoids obtained from the large-scale extraction were lower, each to varying degrees, compared to those obtained from the small-scale extraction using the same solvent system (Table 5). The amount of **1** and **2** was significantly reduced by 10 times and 240 times, respectively, and with absolute differences of 150.7 mg/kg and 239.0 mg/kg of pomace, respectively. As for the amount of **3**, this difference was less pronounced and had only reduced by half, with an absolute difference of 25.2 mg/kg of pomace. It is speculated that one of the reasons for the less pronounced difference observed between the quantities of **3**, compared to those of **1** and **2**, is the fact that the 40:30:30 solvent system comprises moderate amounts of all three solvents. As discussed in Section 3.1, the systems with moderate amounts of the three solvents were found generally to be more suitable for extracting this flavonoid. There are some limitations to this extraction method that would have contributed to the reduced amounts of all three flavonoids, which are further discussed in Section 3.5.

Table 5. Quantities of flavonoids extracted from small-scale (200 g) and large-scale (2 kg) Pinot noir pomace using 40:30:30 (acetone:H_2O:EtOH) solvent system.

Flavonoid	Extraction on Small-Scale Pomace (mg/kg Pomace)	Extraction on Large-Scale Pomace (mg/kg Pomace)
1	164.7 (±18.4)	14.0 (±5.3)
2	240.1 (±8.9)	1.1 (±0.6)
3	54.9 (±1.0) *	29.7 (±1.1) *

* measured as mg rutin hydrate equivalence per kg of pomace.

3.3. *Fatty Acyl Derivatization of the Extract*

Despite the smaller than expected amounts of flavonoids obtained from the large-scale extraction method, it was still decided to take the pomace extract forward to the next method of the process. The lack of flavonoids in the extract required the original study concept (of developing an extract enriched with derivatized flavonoids) to be reframed. It was considered that these small amounts of flavonoids were present in a mixture with other phenolic compounds. This study did not implement methods to identify and measure quantities of other bioactive compounds in the extract, but, according to literature, it is suggested that there would be other flavonoids (i.e., anthocyanins and anthocyanidins), stilbenes (i.e., resveratrol) and phenolic acids (i.e., caffeic acid, *p*-coumaric acid) available [3,4]. Thus, it was proposed that subjecting the extract to the derivatization method would lead to structural modifications in the bioactive phenolic compounds, as a collective, and that this would raise the health-promoting properties in the final product compared to those of the original starting extract.

The rationale for derivatizing these compounds is based on the reported evidence regarding structurally modified phenolic compounds that have enhanced biological activities [53,54] and properties associated with bioavailability [55–57]. Additionally, our group had also reported studies on fatty acyl modifications to flavonoid structures and their effects on bioactivity. In these studies, it was found that by selectively acylating specific hydroxy positions of both luteolin [32] and quercetin [31], these products displayed improved antiproliferative activity against both HCT116 and MDA-MB-231 cancer cells. It was also found that the radical scavenging activity did not change, which was a good indication that the modifications did not reduce their antioxidant activities. Although these studies did not investigate whether they had properties related to good bioavailability, there are examples in the literature in which an increase in the lipophilicity of phenolic compounds could improve their absorption and metabolic stability [55–57]. These findings led to the proposal that acylation modifications to the polyphenolics in the extract, especially the flavonoids, could also result in an enhanced overall health-promoting potential. The feasibility of this proposal is further supported by the success of a similar work, recently reported by Lei et al., who produced fatty acyl derivatives of polyphenolics in grape seed extract and demonstrated their radical scavenging activity and antiproliferative effects against HepG2 cells [58].

The were some considerations around the method for derivatizing the compounds in the extract. Since our reported studies identified three acylated flavonoids (which are octanoyl, lauroyl and palmitoyl derivatives of both luteolin or quercetin) as having better antiproliferative activity, it was decided that the corresponding acylating agents would be used [31,32]. The differences between previous studies and the present study that would affect the acylation approach were also considered. Where previous works were able to utilize a multi-step synthetic approach, starting with a single flavonoid and developing reaction steps to selectively acylate only one hydroxy position on the flavonoid, the same synthetic approach would not be achievable on the pomace extract. The extract comprises a mixture of many different polyphenolic compounds with many hydroxy sites, which are all susceptible to acylation. It was therefore accepted that the acylation approach in this study would be much less targeted and more random (occurring on many compounds and at multiple hydroxy sites) and would give a mixture of different fatty acyl products.

The method of derivatization was conducted by dissolving 1 g samples of the extract in DMF. This solvent was chosen for its moderate polarity (with a relative polarity of 0.386 [59]), and, therefore, it has the ability to dissolve both the polar polyphenolic content in the extract as well as the hydrophobic acylating agents. This was important as it increases the chances of these compounds contacting and colliding, thereby facilitating the reaction. The reaction was initiated by adding the organic base, triethylamine, to remove hydroxy protons from the phenolics. The acylating agents (either octanoyl chloride, lauroyl chloride or palmitoyl chloride) were then added to acylate these deprotonated hydroxy sites. At the end of the reaction, the mixture was then quenched with $NaHCO_3$, filtered and washed excessively with water to remove any remaining salt byproducts. Interestingly, the products that resulted from this procedure were red solids, which is different to those of the starting material.

The non-derivatized and the fatty-acyl-derivatized extract products were characterized by infrared (IR) spectroscopy (Figure S1). From the IR spectrum of the non-derivatized extract, there is a strong broad signal observed at 3275 cm^{-1}, representing the hydroxy groups. In this same region in the IR spectra of the derivatized extract products, the intensity of this signal significantly reduced, which indicates that some of the hydroxy groups of the polyphenolics were masked, likely due to the fatty acyl derivatization.

3.4. Antiproliferative Activity Studies of the Products

HCT116 and MDA-MB-231 were chosen for study as they are cancer cell lines representative of two very significant cancer types: colon (or colorectal) cancer and breast cancer, respectively. In 2020, colorectal cancer was the third most prevalent cancer for all sexes, and breast cancer was the most prevalent cancer in females [60]. There is a clear need to develop agents that can target the cell lines contributing to these cancer types. The production of lipophilic flavonoids and their improved activity against these two cell lines (compared to that of their parent compounds) have previously been exemplified in the literature. For example, Omonga et al. demonstrated that *O*-alkyl derivatives of chrysin increased activity against HCT116; whilst Nair et al. demonstrated fatty esters of phloridzin increased the inhibitory effects against MDA-MB-231 [61,62]. Further, our group also found that lauroyl, octanoyl and palmitoyl derivatives of quercetin and luteolin had the greatest improvements against these two cell lines. As discussed in Section 3.3, this provided the rationale for derivatizing the polyphenolics in the extracts with these same acyl groups, in hopes that this would raise the extracts' ability to inhibit the growth of these two cell lines.

The ^{3}[H] thymidine incorporation assay was used to study the antiproliferative activity of these extracts against the two cell lines and was conducted according to Section 2.6. Unfortunately, when preparing the extract samples, it was found these products lacked the important solubility in the DMSO solvent, and thus, this was a significant limitation to the results obtained. This lack of solubility would have reduced the amounts of bioactive products available in solution. This unexpected problem was a limitation to this study. Despite that, the procedure was carried out, and the proliferation rates of the cell lines treated with the products were obtained (Table 6). From statistical analyses of these values, it was found that the proliferation rates of these products (both non-derivatized and non-derivatized) were not significantly lower (p-value > 0.05, one sample one-tail t-test) compared to that of the control (100%). Therefore, there is no evidence that these products limited the growth of these cells. There was also no evidence of a statistical difference (p-value > 0.05, one-factor ANOVA) between the mean values of each of the products. Although these results were not promising, they do not necessarily imply a lack of the desired bioactivity. Improvements to this methodology are explored in the following section.

Table 6. Cell proliferation rates of cell lines treated with non-derivatized and derivatized extracts.

Crude Extract Type	% Cell Proliferation (Compared to Control)	
	HCT116	**MDA-MB-231**
Non-derivatized	95.6 (±2.2)	75.1 (±27.8)
Octanoyl	91.9 (±4.1)	82.6 (±21.0)
Lauroyl	91.4 (±4.2)	88.9 (±15.0)
Palmitoyl	88.6 (±6.6)	75.8 (±24.9)

3.5. Limitations to the Methods and Suggested Improvements

Overall, this process was not successful in achieving the study aim. However, exploring some of the limitations to the methods provides insights into improvements that could be made. This discussion remains within the scope of using SLE to extract polyphenolics (particularly the flavonoids) from Pinot noir pomace and the lipophilic derivatization of these polyphenolics. Within this scope, there are three areas of the process that are explored. These are the pomace's preparation, SLE method for large-scale Pinot noir pomace and the method for acylating polyphenolics from the extract.

One limitation was that the pomace that was used was a wet material and contained significant amounts of water, present on the surface of the pomace particles and inside the pomace particles. The implication of using wet pomace is that when it is immersed into the solvent systems, the water passes into the surrounding solvent and raises the water profile in the system. It was discussed in Section 3.1 that flavonoids are generally best extracted with low water solvent systems or that higher water systems are less effective for these compounds. Therefore, an increase in the proportion of water could reduce the ability for the flavonoids to be extracted. To mitigate this limitation, the pomace material could be lyophilized to eliminate water from the materials before extraction.

The pomace material was prepared by blending in attempt to increase homogeneity. This also reduced the pomace particle sizes and, therefore, increased the surface-area-to-volume ratio. However, the resultant pomace particle size after blending may not have been small enough. The importance of lowering this ratio as much as possible is that it enables more solvent to come into contact with more parts of the pomace, thereby maximizing the extraction ability. It is suggested that a more superior preparation process would be to mechanically grind the pomace down into a finer powder. This step would require the pomace material to be dry and, therefore, should be completed after conduct from the first suggestion.

The method used immersed the pomace in the solvent for 1 h, which may not have been long enough to enable total extraction. If the extraction time was extended to at least 6 h, and up to 12 h or 24 h, this could have significantly maximized the extraction. Either alternatively or additionally, the extraction could also be maximized by repeating the extraction, where the pomace is immersed again into freshly prepared solvent. Furthermore, it was questioned whether the conduct of extraction at room temperature was sufficient. It is proposed that raising the temperature (i.e., from 20–25 °C) could improve solubility of the compounds in solvent, and, therefore, it would be worthwhile to investigate the effects of temperature. However, with higher temperatures, the mitigation of solvent (acetone and EtOH) evaporation must be carefully considered.

The choice to manually stir the pomace in solvent in the large-scale extraction was a great limitation to the method. The motion was inconsistent, and it was reasoned that the lack of even contact between the pomace particles and the solvent was a major contributor to the lower amounts of flavonoids extracted from the large pomace mass. A suggested improvement is to implement an overhead mechanical stirrer (with sufficient power) to produce a continuous and consistent motion for moving the pomace around the solvent.

From the small-scale extraction investigation, six solvent systems were identified as ideal (see Section 3.1). However, only one system was taken forward and investigated on the large-scale pomace mass. Another trial, using all six systems on the large-scale pomace, could be conducted to validate which system is the best system for extracting the desired flavonoids.

This study placed high emphasis on extraction and analysis of the flavonoids **1**, **2** and **3**. The original concept operated under the premise that these flavonoids and their lipophilic derivatives would be able to contribute to enhancing the health-promoting properties of grape pomace extract. One significant improvement would be to expand the investigation to understand the profile of other bioactive phenolic compounds that would typically be found in grape pomace. This would further increase the overall understanding of the components in the extract that could also contribute to the desired health effects.

Returning the focus back onto flavonoids, these compounds existed in a mixture and among a number of other polar compounds in the extract. The limitation of this is that the subsequent derivatization could not be selective of only those compounds. It would be more ideal if these flavonoids (either as a collective or individually) could be isolated and then be subjected to the derivatization step. Alternatively, the process could be followed through to the end, and then the acylated flavonoid derivatives in the resulting product could be purified from the mixture. One possible way of doing this is to use preparative HPLC, which is an effective technique for purifying compounds. However, the tradeoff to this is the high costs of both the equipment and its operation.

Although the reduced intensities observed on the IR spectra are promising indications that the phenolic compounds in the extract may have been masked by acylation, the limitation to this characterization technique is that it remains indicative and does not specify which compounds are present. The use of HPLC–mass spectrometry and (proton and carbon) nuclear magnetic resonance (NMR) are suggested as further characterization techniques that could provide more detailed information on these compounds. However, implementation of these techniques does require some further preparative work. For example, HPLC–mass spectrometry would require an appropriate method that enables good separation of the lipophilic compounds, and NMR would require a degree of purity in order to resolve the signals corresponding to the (proton and carbon) atoms for each compound.

One major limitation to the final derivatized products' antiproliferative activities was their lack of solubility in the solvent that was required to perform the assay. It is hard to ascertain what exactly caused the lack of solubility. It would be worth conducting a quick assessment on the solubility of this product with a range of other solvents that are appropriate for this test. Another lengthier, but more ideal, improvement is to implement the already mentioned suggestion of further purifying the compounds within the product. This would enable individual (or fractions of only a few) compounds to be prepared separately, and, for those compounds that can be solubilized, these can be taken forward for testing. In turn, this would also help in the identification of compounds (or fractions of compounds) that display activity against the chosen cell lines.

The above suggestions were made with the aim of increasing the prospect of obtaining a final product with improved health-promoting outcomes. However, each additional step introduced to the process will unavoidably increase the cost, complexity, and time. These factors have opposing effects on the commercial viability of the strategy. Therefore, before the adoption of any improvements to the process, the tradeoffs, such as whether the additional benefits will outweigh the drawbacks, and how these drawbacks can be attenuated must be carefully considered.

4. Conclusions

This study investigated a process that uses the SLE method on Pinot noir pomace, which was optimized to extract as many of the three flavonoids (**1**, **2** and **3**) as possible and a method that derivatizes the polyphenolics in the extract, converting them into fatty acyl derivatives. It was hoped that this process would result in products that could display some improved health-promoting properties. Unfortunately, this process did not achieve the expected outcome; however, the findings from these investigations remained interesting. In the small-scale extraction study, the trial of 28 selected solvent systems, led to identification of the most effective solvent system for each individual flavonoid and the identification of 6 systems that were ideal for extracting all three flavonoids. Implementing one of these solvent systems for the SLE of large-scale mass of pomace, interestingly, led to significantly reduced quantities of the flavonoids compared to those observed in the small-scale study. Despite that, this extract was taken into the next derivatization step. The IR spectra of these products was used to indicate the success of this derivatization. The subjection of these products to antiproliferative testing did not display improved results. The exploration of the limitations to these methods and suggested improvements may serve as foundations for future work with possibilities to increase success.

Supplementary Materials: The following are available online at https://www.mdpi.com/article/10.3390/foods11141999/s1. Table S1: the HPLC method, Table S2: flavonoid quantities extracted using different solvent systems, Figure S1: IR spectra of the non-derivatized and derivatized products.

Author Contributions: Conceptualization, S.L., D.B. and B.F.; data curation, S.L.; formal analysis, S.L. and L.I.P.; funding acquisition, B.F.; investigation, S.L.; methodology, S.L., D.B. and B.F.; project administration, B.F.; resources, D.B. and B.F.; supervision, D.B. and B.F.; validation, S.L. and B.F.; visualization, S.L.; writing—original draft, S.L.; writing—review and editing, S.L., L.I.P., D.B. and B.F. All authors have read and agreed to the published version of the manuscript.

Funding: This work was financially supported by Marlborough Research Centre for the purpose of "extracting value from grape marc".

Institutional Review Board Statement: Not Applicable.

Informed Consent Statement: Not Applicable.

Data Availability Statement: Not applicable.

Acknowledgments: The authors wish to thank the University of Auckland for providing the Doctoral Scholarship and Euphemia Leung for conducting the antiproliferative studies on the derivatized products.

Conflicts of Interest: The authors declare no conflict of interest.

References

1. International Organisation of Vine and Wine State of the World Vitivinicultural Sector in 2020. Available online: http://www.oiv.int/public/medias/6782/oiv-2019-statistical-report-on-world-vitiviniculture.pdf (accessed on 3 June 2022).
2. Mendes, J.A.S.; Xavier, A.M.R.B.; Evtuguin, D.V.; Lopes, L.P.C. Integrated Utilization of Grape Skins from White Grape Pomaces. *Ind. Crops Prod.* **2013**, *49*, 286–291. [CrossRef]
3. Fontana, A.R.; Antoniolli, A.; Bottini, R. Grape Pomace as a Sustainable Source of Bioactive Compounds: Extraction, Characterization, and Biotechnological Applications of Phenolics. *J. Agric. Food Chem.* **2013**, *61*, 8987–9003. [CrossRef] [PubMed]
4. Teixeira, A.; Baenas, N.; Dominguez-Perles, R.; Barros, A.; Rosa, E.; Moreno, D.A.; Garcia-Viguera, C. Natural Bioactive Compounds from Winery By-Products as Health Promoters: A Review. *Int. J. Mol. Sci.* **2014**, *15*, 15638–15678. [CrossRef] [PubMed]
5. Bordiga, M.; Travaglia, F.; Locatelli, M. Valorisation of Grape Pomace: An Approach That Is Increasingly Reaching Its Maturity–a Review. *Int. J. Food Sci. Technol.* **2019**, *54*, 933–942. [CrossRef]
6. Maicas, S.; Mateo, J.J. Sustainability of Wine Production. *Sustainability* **2020**, *12*, 559. [CrossRef]
7. Muhlack, R.A.; Potumarthi, R.; Jeffery, D.W. Sustainable Wineries through Waste Valorisation: A Review of Grape Marc Utilisation for Value-Added Products. *Waste Manag.* **2018**, *72*, 99–118. [CrossRef]
8. Devesa-Rey, R.; Vecino, X.; Varela-Alende, J.L.; Barral, M.T.; Cruz, J.M.; Moldes, A.B. Valorization of Winery Waste vs. the Costs of Not Recycling. *Waste Manag.* **2011**, *31*, 2327–2335. [CrossRef]
9. Dwyer, K.; Hossenian, F.; Pod, M. The Market Potential of Grape Waste Alternatives. *J. Food Res.* **2014**, *3*, 91–106. [CrossRef]
10. Ferri, M.; Vannini, M.; Ehrnell, M.; Eliasson, L.; Xanthakis, E.; Monari, S.; Sisti, L.; Marchese, P.; Celli, A.; Tassoni, A. From Winery Waste to Bioactive Compounds and New Polymeric Biocomposites: A Contribution to the Circular Economy Concept. *J. Adv. Res.* **2020**, *24*, 1–11. [CrossRef]
11. Xia, E.-Q.; Deng, G.-F.; Guo, Y.-J.; Li, H.-B. Biological Activities of Polyphenols from Grapes. *Int. J. Mol. Sci.* **2010**, *11*, 622–646. [CrossRef]
12. Silva, R.F.M.; Pogačnik, L. Polyphenols from Food and Natural Products: Neuroprotection and Safety. *Antioxidants* **2020**, *9*, 61. [CrossRef] [PubMed]
13. Chouchouli, V.; Kalogeropoulos, N.; Konteles, S.J.; Karvela, E.; Makris, D.P.; Karathanos, V.T. Fortification of Yoghurts with Grape (*Vitis vinifera*) Seed Extracts. *LWT-Food Sci. Technol.* **2013**, *53*, 522–529. [CrossRef]
14. Rivera, O.M.P.; Moldes, A.B.; Torrado, A.M.; Domínguez, J.M. Lactic Acid and Biosurfactants Production from Hydrolyzed Distilled Grape Marc. *Process Biochem.* **2007**, *42*, 1010–1020. [CrossRef]
15. Sánchez-Alonso, I.; Jiménez-Escrig, A.; Saura-Calixto, F.; Borderías, A.J. Effect of Grape Antioxidant Dietary Fibre on the Prevention of Lipid Oxidation in Minced Fish: Evaluation by Different Methodologies. *Food Chem.* **2007**, *101*, 372–378. [CrossRef]
16. Jelley, R.E.; Deed, R.C.; Barker, D.; Parish-Virtue, K.; Fedrizzi, B. Fermentation of Sauvignon Blanc Grape Marc Extract Yields Important Wine Aroma 3-Sulfanylhexan-1-Ol (3SH). *LWT* **2020**, *131*, 109653. [CrossRef]
17. Jelley, R.E.; Lee, A.J.; Zujovic, Z.; Villas-Boas, S.G.; Barker, D.; Fedrizzi, B. First Use of Grape Waste-Derived Building Blocks to Yield Antimicrobial Materials. *Food Chem.* **2022**, *370*, 131025. [CrossRef]
18. Nunes, M.A.; Rodrigues, F.; Oliveira, M.B.P.P. 11-Grape Processing By-Products as Active Ingredients for Cosmetic Proposes. In *Handbook of Grape Processing By-Products*; Galanakis, C.M., Ed.; Academic Press: Cambridge, MA, USA, 2017; pp. 267–292. ISBN 978-0-12-809870-7.
19. Georgiev, V.; Ananga, A.; Tsolova, V. Recent Advances and Uses of Grape Flavonoids as Nutraceuticals. *Nutrients* **2014**, *6*, 391–415. [CrossRef]
20. Apostolou, A.; Stagos, D.; Galitsiou, E.; Spyrou, A.; Haroutounian, S.; Portesis, N.; Trizoglou, I.; Wallace Hayes, A.; Tsatsakis, A.M.; Kouretas, D. Assessment of Polyphenolic Content, Antioxidant Activity, Protection against ROS-Induced DNA Damage and Anticancer Activity of Vitis Vinifera Stem Extracts. *Food Chem. Toxicol.* **2013**, *61*, 60–68. [CrossRef]
21. Choleva, M.; Tsota, M.; Boulougouri, V.; Panara, A.; Thomaidis, N.S.; Antonopoulou, S.; Fragopoulou, E. Anti-Platelet and Anti-Inflammatory Properties of an Ethanol-Water Red Grape Pomace Extract. *Proc. Nutr. Soc.* **2020**, *79*, E370. [CrossRef]
22. Gerardi, C.; Pinto, L.; Baruzzi, F.; Giovinazzo, G. Comparison of Antibacterial and Antioxidant Properties of Red (cv. Negramaro) and White (cv. Fiano) Skin Pomace Extracts. *Molecules* **2021**, *26*, 5918. [CrossRef]
23. Pérez-Ortiz, J.M.; Alguacil, L.F.; Salas, E.; Hermosín-Gutiérrez, I.; Gómez-Alonso, S.; González-Martín, C. Antiproliferative and Cytotoxic Effects of Grape Pomace and Grape Seed Extracts on Colorectal Cancer Cell Lines. *Food Sci. Nutr.* **2019**, *7*, 2948–2957. [CrossRef] [PubMed]

24. Milinčić, D.D.; Stanisavljević, N.S.; Kostić, A.Ž.; Soković Bajić, S.; Kojić, M.O.; Gašić, U.M.; Barać, M.B.; Stanojević, S.P.; Lj Tešić, Ž.; Pešić, M.B. Phenolic Compounds and Biopotential of Grape Pomace Extracts from Prokupac Red Grape Variety. *LWT* **2021**, *138*, 110739. [CrossRef]
25. Olivero-David, R.; Ruiz-Roso, M.B.; Caporaso, N.; Perez-Olleros, L.; De las Heras, N.; Lahera, V.; Ruiz-Roso, B. In Vivo Bioavailability of Polyphenols from Grape By-Product Extracts, and Effect on Lipemia of Normocholesterolemic Wistar Rats. *J. Sci. Food Agric.* **2018**, *98*, 5581–5590. [CrossRef] [PubMed]
26. Hu, M. Commentary: Bioavailability of Flavonoids and Polyphenols: Call to Arms. *Mol. Pharm.* **2007**, *4*, 803–806. [CrossRef] [PubMed]
27. Di Lorenzo, C.; Colombo, F.; Biella, S.; Stockley, C.; Restani, P. Polyphenols and Human Health: The Role of Bioavailability. *Nutrients* **2021**, *13*, 273. [CrossRef] [PubMed]
28. Manach, C.; Scalbert, A.; Morand, C.; Rémésy, C.; Jiménez, L. Polyphenols: Food Sources and Bioavailability. *Am. J. Clin. Nutr.* **2004**, *79*, 727–747. [CrossRef] [PubMed]
29. D'Archivio, M.; Filesi, C.; Varì, R.; Scazzocchio, B.; Masella, R. Bioavailability of the Polyphenols: Status and Controversies. *Int. J. Mol. Sci.* **2010**, *11*, 1321–1342. [CrossRef]
30. Jelley, R.E.; Herbst-Johnstone, M.; Klaere, S.; Pilkington, L.I.; Grose, C.; Martin, D.; Barker, D.; Fedrizzi, B. Optimization of Ecofriendly Extraction of Bioactive Monomeric Phenolics and Useful Flavor Precursors from Grape Waste. *ACS Sustain. Chem. Eng.* **2016**, *4*, 5060–5067. [CrossRef]
31. Lo, S.; Leung, E.; Fedrizzi, B.; Barker, D. Synthesis, Antiproliferative Activity and Radical Scavenging Ability of 5-O-Acyl Derivatives of Quercetin. *Molecules* **2021**, *26*, 1608. [CrossRef]
32. Lo, S.; Leung, E.; Fedrizzi, B.; Barker, D. Syntheses of Mono-Acylated Luteolin Derivatives, Evaluation of Their Antiproliferative and Radical Scavenging Activities and Implications on Their Oral Bioavailability. *Sci. Rep.* **2021**, *11*, 12595. [CrossRef]
33. Kilmartin, P.A.; Zou, H.; Waterhouse, A.L. Correlation of Wine Phenolic Composition versus Cyclic Voltammetry Response. *Am. J. Enol. Vitic.* **2002**, *53*, 294–302.
34. Leung, E.Y.; Askarian-Amiri, M.E.; Singleton, D.C.; Ferraro-Peyret, C.; Joseph, W.R.; Finlay, G.J.; Broom, R.J.; Kakadia, P.M.; Bohlander, S.K.; Marshall, E.; et al. Derivation of Breast Cancer Cell Lines Under Physiological (5%) Oxygen Concentrations. *Front. Oncol.* **2018**, *8*, 425. [CrossRef] [PubMed]
35. R Core Team. *R: A Language and Environment for Statistical Computing. R Foundation for Statistical Computing*; R Foundation for Statistical Computing: Vienna, Austria, 2019; Available online: https://www.r-project.org/ (accessed on 1 July 2021).
36. Hamilton, N. Ggtern: An Extension to 'ggplot2' for the Creation of Ternary Diagrams. R Package Version 2.2.2. 2018. Available online: https://cran.r-project.org/web/packages/ggtern/index.html (accessed on 1 July 2021).
37. New Zealand Winegrowers. *Annual Report 2020*; New Zealand Winegrowers: Auckland, New Zealand, 2020; Available online: https://www.nzwine.com/en/media/statistics/annual-report (accessed on 6 June 2022).
38. New Zealand Winegrowers. *Annual Report 2021*; New Zealand Winegrowers: New Zealand, Auckland, 2021; Available online: https://www.nzwine.com/en/media/statistics/annual-report (accessed on 6 June 2022).
39. Boots, A.W.; Haenen, G.R.M.M.; Bast, A. Health Effects of Quercetin: From Antioxidant to Nutraceutical. *Eur. J. Pharmacol.* **2008**, *585*, 325–337. [CrossRef] [PubMed]
40. Liu, H.L.; Jiang, W.B.; Xie, M.X. Flavonoids: Recent Advances as Anticancer Drugs. *Recent Pat. Anti-Cancer Drug Discov.* **2010**, *5*, 152–164. [CrossRef]
41. Agustin-Salazar, S.; Medina-Juárez, L.A.; Soto-Valdez, H.; Manzanares-López, F.; Gámez-Meza, N. Influence of the Solvent System on the Composition of Phenolic Substances and Antioxidant Capacity of Extracts of Grape (*Vitis vinifera* L.) Marc. *Aust. J. Grape Wine Res.* **2014**, *20*, 208–213. [CrossRef]
42. De Sá, M.; Justino, V.; Spranger, M.I.; Zhao, Y.Q.; Han, L.; Sun, B.S. Extraction Yields and Anti-Oxidant Activity of Proanthocyanidins from Different Parts of Grape Pomace: Effect of Mechanical Treatments. *Phytochem. Anal.* **2014**, *25*, 134–140. [CrossRef]
43. Amico, V.; Chillemi, R.; Mangiafico, S.; Spatafora, C.; Tringali, C. Polyphenol-Enriched Fractions from Sicilian Grape Pomace: HPLC–DAD Analysis and Antioxidant Activity. *Bioresour. Technol.* **2008**, *99*, 5960–5966. [CrossRef]
44. Chafer, A.; Pascual-Martí, M.C.; Salvador, A.; Berna, A. Supercritical Fluid Extraction and HPLC Determination of Relevant Polyphenolic Compounds in Grape Skin. *J. Sep. Sci.* **2005**, *28*, 2050–2056. [CrossRef]
45. Kammerer, D.; Claus, A.; Carle, R.; Schieber, A. Polyphenol Screening of Pomace from Red and White Grape Varieties (*Vitis vinifera* L.) by HPLC-DAD-MS/MS. *J. Agric. Food Chem.* **2004**, *52*, 4360–4367. [CrossRef]
46. Casazza, A.A.; Aliakbarian, B.; Mantegna, S.; Cravotto, G.; Perego, P. Extraction of Phenolics from *Vitis Vinifera* Wastes Using Non-Conventional Techniques. *J. Food Eng.* **2010**, *100*, 50–55. [CrossRef]
47. Altemimi, A.; Watson, D.G.; Choudhary, R.; Dasari, M.R.; Lightfoot, D.A. Ultrasound Assisted Extraction of Phenolic Compounds from Peaches and Pumpkins. *PLoS ONE* **2016**, *11*, e0148758. [CrossRef] [PubMed]
48. Carullo, G.; Ramunno, A.; Sommella, E.M.; De Luca, M.; Belsito, E.L.; Frattaruolo, L.; Brindisi, M.; Campiglia, P.; Cappello, A.R.; Aiello, F. Ultrasound-Assisted Extraction, Chemical Characterization, and Impact on Cell Viability of Food Wastes Derived from Southern Italy Autochthonous Citrus Fruits. *Antioxidants* **2022**, *11*, 285. [CrossRef] [PubMed]
49. Pedroza, M.A.; Amendola, D.; Maggi, L.; Zalacain, A.; De, F.D.M.; Spigno, G. Microwave-Assisted Extraction of Phenolic Compounds from Dried Waste Grape Skins. *Int. J. Food Eng.* **2015**, *11*, 359–370. [CrossRef]

50. Zhang, L.; Wang, Y.; Wu, D.; Xu, M.; Chen, J. Microwave-Assisted Extraction of Polyphenols from Camellia Oleifera Fruit Hull. *Molecules* **2011**, *16*, 4428–4437. [CrossRef]
51. Da Porto, C.; Natolino, A. Supercritical Fluid Extraction of Polyphenols from Grape Seed (*Vitis Vinifera*): Study on Process Variables and Kinetics. *J. Supercrit. Fluids* **2017**, *130*, 239–245. [CrossRef]
52. Khaw, K.-Y.; Parat, M.-O.; Shaw, P.N.; Falconer, J.R. Solvent Supercritical Fluid Technologies to Extract Bioactive Compounds from Natural Sources: A Review. *Molecules* **2017**, *22*, 1186. [CrossRef]
53. Vergara, D.; Domenico, S.D.; Tinelli, A.; Stanca, E.; Mercato, L.L.D.; Giudetti, A.M.; Simeone, P.; Guazzelli, N.; Lessi, M.; Manzini, C.; et al. Anticancer Effects of Novel Resveratrol Analogues on Human Ovarian Cancer Cells. *Mol. BioSyst.* **2017**, *13*, 1131–1141. [CrossRef]
54. Zhu, S.; Yang, C.; Zhang, L.; Wang, S.; Ma, M.; Zhao, J.; Song, Z.; Wang, F.; Qu, X.; Li, F.; et al. Development of M10, Myricetin-3-O-β-d-Lactose Sodium Salt, a Derivative of Myricetin as a Potent Agent of Anti-Chronic Colonic Inflammation. *Eur. J. Med. Chem.* **2019**, *174*, 9–15. [CrossRef]
55. Wen, X.; Walle, T. Methylated Flavonoids Have Greatly Improved Intestinal Absorption and Metabolic Stability. *Drug Metab. Dispos.* **2006**, *34*, 1786–1792. [CrossRef]
56. Lambert, J.D.; Sang, S.; Hong, J.; Kwon, S.-J.; Lee, M.-J.; Ho, C.-T.; Yang, C.S. Peracetylation as a Means of Enhancing in Vitro Bioactivity and Bioavailability of Epigallocatechin-3-Gallate. *Drug Metab. Dispos.* **2006**, *34*, 2111–2116. [CrossRef]
57. Walle, T.; Wen, X.; Walle, U.K. Improving Metabolic Stability of Cancer Chemoprotective Polyphenols. *Expert Opin. Drug Metab. Toxicol.* **2007**, *3*, 379–388. [CrossRef] [PubMed]
58. Lei, C.; Tang, X.; Li, H.; Chen, H.; Yu, S. Molecular Hybridization of Grape Seed Extract: Synthesis, Structural Characterization and Anti-Proliferative Activity in Vitro. *Food Res. Int.* **2020**, *131*, 109005. [CrossRef] [PubMed]
59. Reichardt, C.; Welton, T. Empirical Parameters of Solvent Polarity. In *Solvents and Solvent Effects in Organic Chemistry*, 4th ed.; Wiley-VCH Verlag GmbH & Co. KGaA: Weinheim, Germany, 2010; pp. 425–508. ISBN 978-3-527-32473-6.
60. World Cancer Research Fund International. Cancer Data | World Cancer Research Fund International. WCRF International. Available online: https://www.wcrf.org/cancer-trends/worldwide-cancer-data (accessed on 10 June 2022).
61. Omonga, N.; Zia, Z.; Ghanbour, H.; Ragazzon-Smith, A.; Foster, H.; Hadfield, J.; Ragazzon, P. Facile Synthesis and Biological Evaluation of Chrysin Derivatives. *J. Chem. Res.* **2021**, *45*, 1083–1092. [CrossRef]
62. Nair, S.V.G.; Ziaullah; Rupasinghe, H.P.V. Fatty Acid Esters of Phloridzin Induce Apoptosis of Human Liver Cancer Cells through Altered Gene Expression. *PLoS ONE* **2014**, *9*, e107149. [CrossRef] [PubMed]

foods

Article

Recovery and Concentration of Polyphenols from Roasted Hazelnut Skin Extract Using Macroporous Resins

Negin Seif Zadeh and Giuseppe Zeppa *

Department of Agriculture, Forest and Food Sciences (DISAFA), University of Turin, 10095 Grugliasco, Italy; negin.seifzadeh@unito.it
* Correspondence: giuseppe.zeppa@unito.it; Tel.: +39-011-670-8705

Abstract: Hazelnut skin is a rich source of polyphenols but is generally discarded during the roasting process of hazelnuts. Previous studies reported the extraction and identification of these compounds using different solvents and procedures; however, there are few reports on their enrichment and purification. In this study, three types of Amberlite macroporous resins (XAD 16, XAD 4, and XAD 7) were compared to evaluate the enrichment of polyphenols via adsorption and desorption mechanisms. The operating condition parameters for polyphenol adsorption/desorption of each resin were determined, the kinetics of adsorption were examined, and a method for polyphenol recovery was developed using static and dynamic adsorption/desorption. Antioxidant activity and high-performance liquid chromatography-diode array detection were used to confirm the increase in polyphenols obtained using the adsorption/desorption technique. XAD16 showed the highest adsorption capacity, with a recovery of 87.7%, and the adsorption kinetics fit well with a pseudo-second-order model. The highest poly-phenol desorption ratio was observed using an ethanol/water solution (70% v/v) at a flow rate of 1.5 bed volume/h.

Keywords: Amberlite resin; hazelnut skin; polyphenols; by-product

Citation: Seif Zadeh, N.; Zeppa, G. Recovery and Concentration of Polyphenols from Roasted Hazelnut Skin Extract Using Macroporous Resins. *Foods* **2022**, *11*, 1969. https://doi.org/10.3390/foods11131969

Academic Editors: Marco Poiana, Francesco Caponio and Antonio Piga

Received: 8 June 2022
Accepted: 30 June 2022
Published: 2 July 2022

Publisher's Note: MDPI stays neutral with regard to jurisdictional claims in published maps and institutional affiliations.

Copyright: © 2022 by the authors. Licensee MDPI, Basel, Switzerland. This article is an open access article distributed under the terms and conditions of the Creative Commons Attribution (CC BY) license (https://creativecommons.org/licenses/by/4.0/).

1. Introduction

Hazelnut is one of the most widely consumed nut crops. Its worldwide production in 2020 was reported as 528,070 tons, with Turkey, Italy, and Spain as the major producers [1]. Approximately 50–60% of the nut is discarded as by-products, such as shell, skin, and damaged nuts, during the dehulling, roasting, and sorting processes [2]. Many parts of these by-products are rich in bioactive compounds, particularly polyphenols, which can be extracted and used in the food, feed, cosmetic, and pharmaceutical industries [3]. The skin is considered as one of the most useful components of hazelnut by-products, accounting for 2.5% of the total weight of the nut which is separated from the kernel during the roasting process [4]. The antioxidant capacity of hazelnut skin is significantly higher than that of the hazelnut [5], and the roasted skin is even richer in total phenols and has higher antioxidant activity compared to in the natural skin [6–8]. Recovering bioactive compounds from hazelnut skin would increase the availability of large amounts of molecules of natural origin and positively impact disposal management, providing considerable economic advantages, such as minimizing the challenges of waste management occurring due to the lack of proper disposal sites to avoid the spread of insects and unwanted wildlife, and production of value-added products from low-cost material. Polyphenols can be extracted from plants using solvents or supercritical fluids. Although the use of solvents is less costly and simpler, it is not selective and results in diluted extracts with low polyphenol concentrations. Macroporous resins are physiochemically stable polymers with polar, non-polar, or slightly polar characteristics and high adsorption capacities for organic compounds [9]. They can be used to purify and concentrate active compounds from complex extracts [10]. The target molecules in aqueous and non-aqueous solutions can be adsorbed by macroporous resins via electrostatic forces, hydrogen bonding interactions, complexation, and size sieving [3]. Resins with large surface areas and pore sizes provide numerous active sites that can interact with the target molecules, and thus, are useful for extracting active compounds. The extraction and regeneration processes are simple and inexpensive [11]. In this study, three poly aromatic Amberlite resins (XAD 4, XAD 7, and XAD 16) were used to concentrate and purify polyphenols in the ethanol extracts of roasted hazelnut skins. Amberlite XAD 16 and XAD 4 is hydrophobic, whereas Amberlite XAD 7 is moderately hydrophilic. Strongly polar resins contain sulfur or nitrogen oxide groups and are not suitable for purifying polyphenols. Slightly polar macroporous resins such as XAD 7 are generally composed of polyacrylate polymers with multifunctional methacrylate crosslinking agents, whereas non-polar macroporous resins such as XAD 16 and XAD 4 consist of styrene and divinylbenzene polymers and are suitable for separating weakly polar compounds [10]. The objective of this research is to propose an efficient method of recovering and purifying polyphenols from hazelnut skin extract, comparing three Amberlite resins and optimizing the operating condition parameters of polyphenol adsorption/desorption. For this purpose, the static experiments were performed to select the best resin and solvent and the kinetics of adsorption were studied. The optimum flow rates were selected through dynamic adsorption/desorption and validation of method for polyphenols recovery from hazelnut skin extract has been measured comparing the phenolic content and antioxidant activity of the crude and concentrated extracts.

2. Materials and Methods

2.1. Materials

Roasted hazelnut skin of Tonda Gentile delle Langhe PGI was provided by La Gentile srl (Cortemilia, Italy) and obtained by roasting for 7 min at 190 °C. All chemicals used were of analytical grade. Folin–Ciocalteu phenol reagent (2 M), 2,2-diphenyl-1-picrylhydrazyl (DPPH), 6-hydroxy-2,5,7,8-tetramethylchroman-2-carboxylic acid (97%; Trolox), methanol (99.9%), formic acid (98–100%), and ethanol (99.9%) were provided by Sigma-Aldrich (St. Louis, MO, USA). High-performance liquid chromatography (HPLC)-grade standards (gallic acid, protocatechuic acid, catechin, epicatechin, and quercetin) were purchased from Fluka (Buchs, Switzerland). Ultrapure water was prepared using a Milli-Q filter system (Millipore, Billerica, MA, USA). Amberlite macroporous resins (XAD 4, XAD 7, and XAD 16) were supplied by Sigma-Aldrich (Table 1). Before use, the resins (50 g) were soaked in pure ethanol (100 mL) for 3 h at 22 °C with constant rotatory agitation on a VDRL 711 orbital shaker (Asal S.r.l, Milan, Italy) at 60 rpm. The ethanol was removed, and the resins were rinsed with excess ultra-pure water.

Table 1. Physicochemical characteristics of Amberlite resins used for the recovery of polyphenolic compounds by a roasted hazelnut skin extract.

	XAD 4	XAD 7	XAD 16
Polarity	non polar	moderately polar	non polar
Chemical structure	Hydrophobic polyaromatic	Acrylic ester	Hydrophobic polyaromatic
Dry density (g/mL)	1.08–1.02	1.24–1.05	1.08–1.02
Surf. Area (m^2/g)	725	450	900
Pore diameter (nm)	5	9	10
Pore mesh size	20–60	20–60	20–60
Pore volum (mL/g)	0.98	1.14	0.82
Particle size (mm)	0.3–1.2	0.3–1.2	0.3–1.2

2.2. Polyphenol Extraction

Polyphenols were extracted from hazelnut skin as described by Locatelli et al. [6]. Prior to extraction, the hazelnut skin was defatted with *n*-hexane using the method described by Özdemir et al. [12]. Defatted hazelnut skins (2 g) were extracted with 50 mL of pure ethanol at 22 °C for 1 h in the dark using a VDRL 711 orbital shaker under constant rotatory agitation at 60 rpm. The extract was centrifugated at 2800× *g* for 10 min at 4 °C. The solid residue was re-extracted for 30 min with 25 mL of pure ethanol and then centrifuged. The two supernatants were mixed, filtered through a 0.45 μm nylon membrane, and stored in amber vials at −18 °C until analysis. All extractions were performed in triplicate.

2.3. Total Polyphenolic Content

The total phenolic content (TPC) of the extracts was determined spectrophotometrically as described by Barbosa-Pereira et al. [13] using a BioTek Synergy HT spectrophotometric multi-detection 96-well microplate reader (BioTek Instruments, Winooski, VT, USA). The absorbance was measured in triplicate at 740 nm. A standard curve of gallic acid (100–600 μM; R^2 = 0.9994) was used to quantify the phenolic content, which was expressed as milligrams of gallic acid equivalents per milliliter of the fresh extract (mg GAE/mL).

2.4. Antioxidant Capacity

The antioxidant capacity of the extracts was determined using the DPPH$^\bullet$ radical scavenging method with a BioTek Synergy HT spectrophotometric multidetection 96-well microplate reader [13]. The decrease in DPPH absorbance was measured at 515 nm. The antioxidant capacity was expressed as the inhibition percentage (IP) of DPPH radicals and was calculated using the following equation:

$$\% \text{ Inhibition} = ((A_0 - A_e)/A_0) \times 100 \quad (1)$$

where A_0 is the absorbance of the blank and A_e is the absorbance at 30 min. A standard curve of Trolox (12.5–350 μM; R^2 = 0.9982) was used to determine the radical-scavenging activity, and the results are expressed as micromoles of Trolox equivalent per mL of extract. Three analyses were evaluated for each sample.

2.5. ABTS$^+$ Assay

The antioxidant capacity of the extracts was determined using the ABTS$^{\bullet+}$ assay as described by Re et al. [14] with some modifications. An aliquot of ABTS solution (7 mM) was reacted with 2.45 mM potassium persulfate to produce an ABTS radical cation (ABTS$^+$). The solution was incubated in the dark for 12 h at 22 °C to ensure its stability. Before use, the ABTS$^+$ stock solution was diluted with ethanol to an absorbance of 0.70 ± 0.02 at 734 nm and 30 °C. The diluted ABTS$^+$ solution (3 mL) was mixed with 30 μL of sample, and the absorbance was measured after 6 min. The ABTS$^-$ scavenging activity was calculated using the following equation:

$$\% \text{ Inhibition} = ((A_0 - A_e)/A_0) \times 100 \quad (2)$$

where A_0 is the absorbance of the blank and A_e is the absorbance at 6 min. The results were expressed as μM Trolox equivalent (TE)/mL of extract, using a dose–response curve for Trolox (0–350 μM; R^2 = 0.996) as the standard. Each sample was evaluated in triplicate.

2.6. Reversed Phase-HPLC-Diode Array Detector Analysis

A reversed-phase HPLC coupled with a Thermo-Finnigan Spectra System diode array detector (Thermo-Finnigan, Waltham, MA, USA) was used to quantify the phenolic components in the extracts. The instrument was equipped with an SCM 1000 degasser, an AS 3000 automatic injector, a P2000 binary gradient pump, and a Finnigan Surveyor PDA Plus detector. ChromQuest software (version 5.0; Thermo-Finnigan, Waltham, MA, USA) was used for data acquisition. Separation was performed using a reverse-phase Kinetex Phenyl-Hexyl C18 column (150 × 4.6 mm id and 5 μm particle size; Phenomenex, Torrance, CA, USA) at a flow rate of 1 mL/min. The mobile phase was composed of formic acid at 0.1% *v*/*v* (A) and acetonitrile (B), and the sample injection volume was 10 μL. The following gradient elution was utilized: 5% B for 0–7 min; a linear gradient from 5% to 50% B for 7–35 min; a linear gradient from 50% to 80% B for 35–37 min; a linear gradient from 80% to 90% B for 37–38 min; and a linear gradient until 90% A and 10% B were reached for 38–41 min. Gallic acid, protocatechuic acid, catechin and epicatechin, and quercetin were quantified by measuring the absorbance at 271, 293, 279, and 366 nm, respectively. Quantification was performed using the external standard linear calibration curves obtained under the same conditions.

2.7. Static Adsorption/Desorption Evaluation

To define the adsorption capacity of the resins for polyphenolic compounds, five amounts of activated macroporous resin (1, 2, 3, 4, and 5 g) were placed in 100 mL flasks and 50 mL of hazelnut skin extract was added. The sealed flasks were shaken at 22 °C for 24 h in the dark using a VDRL 711 orbital shaker under constant rotatory agitation at 120 rpm. TPC was evaluated before and after adsorption, and the adsorption capacity and adsorption ratio were calculated using the following equations:

Adsorption capacity:

$$q_a = ((C_0 - C_e) \times V_i)/M \tag{3}$$

Adsorption ratio:

$$A\ (\%) = ((C_0 - C_e)/C_0) \times 100 \tag{4}$$

where q_a is the adsorption capacity (mg/g dry resin), C_0 is the TPC value of the extract before the adsorption phase (mg GAE/mL), C_e is the TPC value of the extract after the adsorption phase (mg GAE/mL), V_i is the volume of the extract (mL), and M is the weight of the resin (g).

After adsorption, the resins were placed in a 100 mL flask and treated with 50 mL of three ethanol solutions (50%, 70%, and 99.9% *v*/*v*). The sealed flasks were shaken at 22 °C for 24 h in the dark using a VDRL 711 orbital shaker under constant rotatory agitation at 120 rpm. The desorption ratio (D%) was calculated using the following equation:

$$D\ (\%) = (C_d \times V_d)/((C_0 - C_e) \times V_i) \times 100 \tag{5}$$

where C_d is the TPC value of the ethanol solution after desorption (mg GAE/mL), V_d is the volume of the ethanol solution used for the desorption phase (mL), V_i is the volume of the extract (mL), C_0 is the TPC value of the extract before the adsorption phase (mg GAE/mL), and C_e is the TPC value of the extract after the adsorption phase (mg GAE/mL).

To evaluate the adsorption kinetics, 1 g of each resin was mixed in a flask containing 50 mL of extract with a TPC value of 5 mg GAE/mL and shaken at 22 °C in the dark on a VDRL 711 orbital shaker under constant rotatory agitation at 120 rpm. The TPC of the solution was analyzed every 15 min for the first 2 h and then every 30 min for 6 h. To evaluate the adsorption kinetics, the obtained results were fitted using two widely used kinetic models, the pseudo-first-order [6] and pseudo-second-order models [7]:

$$\mathrm{Ln}\,(q_e - q_t) = -k_1\,t + \ln q_e \tag{6}$$

$$t/q_t = 1/(k_2\,q^2{}_e) + t/q_e \tag{7}$$

where q_e (mg/g) and q_t (mg/g) are the adsorption capacities at equilibrium and at time t (min), respectively, and k_1 (min^{-1}) and k_2 (g/mg min) are the rate constants of the pseudo-first and pseudo-second order models, respectively [15].

The fit of each model to the experimental data was estimated using the linear regression correlation coefficient (R^2).

2.8. Dynamic Adsorption/Desorption

Dynamic adsorption was performed by loading the hazelnut skin extracts (5 mg GAE/mL) into a stainless-steel column (300 × 78 mm) packed with the amount of activated macroporous resin that showed the best adsorption/desorption values in the static experiments; the column was connected to a P2000 Thermo-Finnigan pump to obtain different flow rates (1.5, 3, and 5 bed volumes (BV)/h). Dynamic desorption was performed after dynamic adsorption, and the column was eluted with three different ethanol solutions (50%, 70%, and 99.9% v/v) at varying flow rates (1.5, 3, 5 BV/h) to recover the adsorbate. The inlet volume for both adsorption and desorption was 10 BV.

2.9. Statistical Analysis

The results are expressed as the mean ± standard deviation. Significant differences between means were identified using one-way analysis of variance (ANOVA) followed by Duncan's post-hoc or two-tailed Student t-test. Statistical significance was set at $p < 0.05$. The effect of factor interaction on static desorption was examined using the generalized linear model. The data were analyzed using SPSS version 28.0 software (SPSS Inc., Chicago, IL, USA).

3. Results

3.1. Static Adsorption/Desorption

The adsorption capacity and desorption ratio are typically regarded as the two main benchmarks for selecting resins. We first selected the best resin type and amount for adsorbing polyphenolic compounds from hazelnut skin extracts. The adsorption ratios and capacities of the resins according to their amounts are shown in Tables 2 and 3, respectively. The absorption ratio increased with increasing amounts of resin, and the maximum value was obtained using 5 g of resin. XAD 16 and XAD 4 showed better adsorption ratios and capacities compared to that of XAD 7 for all amounts of resin evaluated. Using 5 g of XAD 16, 65.06 ± 0.14% of polyphenols in the extract were adsorbed, and this resin showed the highest adsorption capacity (40.05 ± 0.55 mg GAE/g dry resin). The adsorption capabilities of resins are related to both the target compound and absorbent properties, such as polarity, particle size, surface area, pore diameter, and chemical structure. Particularly, polyphenol compounds can be absorbed by macroporous resins via physical mechanisms, such as van der Waals forces, hydrogen bonds, and π-π conjugation between phenolics and the benzene rings of resins [16]. Polyphenols contain hydrogen groups and benzene rings and, depending on their structure, exhibit different polarities. Although XAD 16 and XAD 4 have similar polarities, XAD 16 provides a higher surface area and pore volume size and absorbs more polyphenols.

Table 2. Adsorption ratios (A%; mean ± standard deviation) of polyphenols from an extract of roasted hazelnut skin by different resin types and amount and results of ANOVA with Duncan's test.

	Resin Amount (g)					
	1	2	3	4	5	Significance
XAD 4	13.87 ± 0.74 ^{Be}	24.82 ± 1.20 ^{Bd}	34.82 ± 0.58 ^{Bc}	51.89 ± 0.79 ^{Bb}	58.73 ± 0.75 ^{Ba}	***
XAD 7	10.95 ± 1.15 ^{Ce}	20.65 ± 0.55 ^{Cd}	24.25 ± 1.03 ^{Cc}	38.23 ± 1.09 ^{Cb}	42.66 ± 0.89 ^{Ca}	***
XAD 16	15.58 ± 0.47 ^{Ae}	29.03 ± 1.87 ^{Ad}	38.93 ± 1.95 ^{Ac}	58.81 ± 1.17 ^{Ab}	65.06 ± 0.14 ^{Aa}	***
Significance	***	***	***	***	***	

Means in each column with the same uppercase letter are not significantly different according to Duncan's test ($p < 0.05$); means in each row with the same lowercase letter are not significantly different according to Duncan's test ($p < 0.05$); *** $p < 0.001$.

Table 3. Adsorption capacity (q_a; mean ± standard deviation) of polyphenols from an extract of roasted hazelnut skin by different resin types and amount and results of ANOVA with Duncan's test.

	Resin Amount (g)					
	1	2	3	4	5	
XAD 4	17.70 ± 0.95 ^{Be}	20.82 ± 0.36 ^{Ad}	23.38 ± 0.39 ^{Bc}	26.61 ± 0.40 ^{Bb}	36.14 ± 0.46 ^{Ba}	***
XAD 7	13.97 ± 1.48 ^{Cd}	15.25 ± 0.40 ^{Bd}	16.28 ± 0.69 ^{Cc}	19.60 ± 0.56 ^{Cb}	26.25 ± 0.54 ^{Ca}	***
XAD 16	19.88 ± 0.61 ^{Ae}	22.39 ± 1.44 ^{Ad}	26.14 ± 1.30 ^{Ac}	30.16 ± 0.60 ^{Ab}	40.05 ± 0.55 ^{Aa}	***
Significance	***	***	***	***	***	

Means in each column with the same uppercase letter are not significantly different according to Duncan's test ($p < 0.05$); means in each row with the same lowercase letter are not significantly different according to Duncan's test ($p < 0.05$); *** $p < 0.001$.

The desorption ratios of absorbed polyphenols from resin using different concentrations of ethanol (50%, 70%, and 99.9% *v/v*) are shown in Table 4. A significant two-way interaction was observed ($p < 0.001$), confirming that the change in the number of resins or concentration of solvent affected the amount of polyphenol desorption for each resin type. Because a significant difference was found between each of the two variables and the main effects were significant, one-way ANOVA was performed to compare the differences within each group. During the static adsorption stage, XAD 4 and XAD 16 exhibited the maximum desorption ratio when 70% *v/v* ethanol was used, whereas XAD 7 showed the highest desorption when 50% *v/v* of ethanol was used as the solvent. Non-polar resins showed better adsorption and desorption of polyphenols compared to the slightly polar resin. Using a 70% *v/v* ethanol solution, 76.64% and 81.17% of the polyphenols were desorbed that had been absorbed by 5 g of XAD 4 and XAD 16, respectively, whereas the lowest efficiency was observed when 99.9% ethanol was used to recover the polyphenols from 1 g of XAD 7. Similarly, Wang et al. [10] compared different concentrations of ethanol solution (10–100%) to recover polyphenols adsorbed by HPD-300 (non-polar) resin. They observed that the highest content of polyphenol was recovered using 60% aqueous ethanol, which was eight-fold that of the crude extract. In addition, approximately 95% of the polyphenol was present in the 60% and 80% ethanol fractions. As explained by Xi et al. [17], pure ethanol increases the desorption of some impurities, but polyphenols are not completely dissolved at lower ethanol concentrations. Leyton et al. [11] obtained similar results in a comparison of different Amberlite XAD resins for purification of phlorotannins from *Macrocystis pyrifera*, and XAD 16 N showed good results with a desorption ratio of 38.2%.

Table 4. Static desorption ratios (mean ± standard deviation) of XAD 16, XAD 4, and XAD 7 Amberlite resins using different ethanol solutions, and ANOVA results with Duncan's test.

	Ethanol Concentration (%)	Resin Amount (g)					Significance
		1	2	3	4	5	
XAD 4	99.99	35.27 ± 0.48 ^{Be}	44.70 ± 1.41 ^{Bd}	58.46 ± 0.98 ^{Bc}	65.85 ± 2.90 ^{Bb}	72.53 ± 0.78 ^{Ba}	***
	70	40.70 ± 2.04 ^{Ad}	51.07 ± 1.94 ^{Ac}	64.26 ± 1.66 ^{Ab}	73.94 ± 1.45 ^{Aa}	76.64 ± 0.93 ^{Aa}	***
	50	32.40 ± 1.18 ^{Ce}	40.28 ± 1.21 ^{Cd}	50.37 ± 1.70 ^{Cc}	58.17 ± 2.28 ^{Cb}	65.10 ± 1.93 ^{Ca}	***
	Significance	***	***	***	***	***	
XAD 7	99.99	14.73 ± 0.46 ^{Ce}	19.57 ± 0.37 ^{Cd}	30.26 ± 0.77 ^{Cc}	38.34 ± 3.13 ^{Cb}	43.58 ± 1.43 ^{Ca}	***
	70	17.76 ± 0.91 ^{Be}	29.92 ± 0.77 ^{Bd}	37.49 ± 0.79 ^{Bc}	44.46 ± 0.92 ^{Bb}	48.75 ± 2.13 ^{Ba}	***
	50	19.89 ± 0.59 ^{Ae}	33.45 ± 1.08 ^{Ad}	40.57 ± 0.75 ^{Ac}	48.64 ± 0.83 ^{Ab}	54.37 ± 1.65 ^{Aa}	***
	Significance	***	***	***	***	***	
XAD 16	99.99	39.46 ± 1.50 ^{Be}	47.49 ± 1.19 ^{Bd}	58.97 ± 1.23 ^{Bc}	71.38 ± 2.28 ^{Bb}	75.80 ± 2.35 ^{Ba}	***
	70	45.06 ± 1.47 ^{Ae}	53.05 ± 2.31 ^{Ad}	65.65 ± 1.22 ^{Ab}	76.79 ± 2.41 ^{Ab}	81.17 ± 1.19 ^{Aa}	***
	50	35.66 ± 0.70 ^{Ce}	43.89 ± 0.66 ^{Cd}	53.28 ± 0.92 ^{Cb}	61.99 ± 2.60 ^{Cb}	67.01 ± 2.46 ^{Ca}	***
	Significance	***	***	***	***	***	

Means with same uppercase letter are not significantly different between ethanol concentration for each resin type, according to the Duncan's test ($p < 0.05$); means in each row with the same lowercase letter are not significantly different according to Duncan's test ($p < 0.05$); *** $p < 0.001$.

The adsorption process consists of different stages and does not remain stable during the adsorption phase. Generally, target molecules are absorbed through their mass transfer from the boundary layer, diffusion into the pores of the adsorbent, and/or adsorption at the surface-active sites of the adsorbent [18]. The adsorption rate is directly correlated with the duration, solvent, and adsorbent material used.

The adsorption kinetics of polyphenols from the roasted hazelnut skin extracts obtained from the three Amberlite macroporous resins are shown in Figure 1. The adsorption quantity increased over time and adsorption was faster in the initial stages. The TPC value decreased by approximately 50% during the first 30 min by XAD 16 and XAD 4, and 1 h by XAD 7. After 1 h, the rate of adsorption gradually decreased; after 120 min of contact, only a minor change was observed because the surface binding sites of the macroporous resin were mostly saturated. The system may have reached equilibrium after 120 min. This trend is similar to that reported by Park and Lee [19]. Le et al. also observed that the adsorption equilibrium of polyphenols (sinapine) from rapeseed meal protein isolate by-products was obtained after 120 min on Amberlite XAD 16 resin [19]. Hou and Zhang reported that polyphenol adsorption equilibrium was reach after 4 h using a highly polar resin (NKA–II) [9]. The surface area of this resin is very close to that of XAD 16 and XAD 4; hence, use of nonpolar or slightly polar resin accelerates the adsorption process. As shown in Table 5, the adsorption kinetics are not best-described by a pseudo-first-order model because the absorbance capacity values were inconsistent with the values predicted using the first-order model [19].

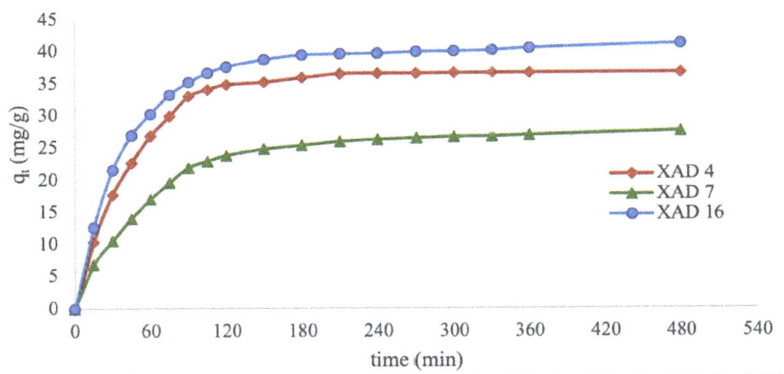

Figure 1. Adsorption kinetic curve of polyphenols from extracts of roasted hazelnut skin with Amberlite microporous resins.

Table 5. Kinetic parameters of static adsorption phase evaluated using two model equations.

		Pseudo-First Order			Pseudo-Second Order		
	q_e exp.	k_1	q_e	r^2	k_2	q_e	r^2
XAD 4	36.68	0.0210	33.08	0.9853	0.00065	38.17	0.9947
XAD 7	27.49	0.0105	18.43	0.9560	0.00065	29.07	0.9972
XAD 16	41.15	0.0105	20.12	0.9014	0.00061	41.32	0.9950

In contrast, there was good agreement between the experimental and calculated absorbance capacities predicted by the pseudo-second-order model, and the correlation coefficients were close to unity ($R^2 > 0.99$) for all types of resins. This suggests that the pseudo-second-order model can be applied to predict the kinetics of polyphenol adsorption from hazelnut skin extract using macroporous resins.

In addition, previous studies reported that the pseudo-second-order model is suitable for the adsorption of polyphenols from extracts [20–22]. Wang et al. reported that the adsorption patterns of polyphenols from *Eucommia ulmoides* Oliv. leaves on HPD-300, HPD-600, D-3250, X-5, D-140, NKA-9, D-101, and AB-8 resins fit well to pseudo-second-order kinetics [10]. Soto et al. also demonstrated that the pseudo-second-order model fits better than the pseudo-first-order model with the adsorption process of phenols from wine vinasses by SP700 and XAD16 HP resins. They suggested that the process fits pseudo-second-order models better when the sorption system is controlled by chemisorption mechanisms [23].

3.2. Dynamic Adsorption/Desorption

The effect of the feed flow rate on the adsorption of polyphenols by XAD 16 resin is shown as a breakthrough curve in Figure 2A. The breakthrough point (BP) was obtained when the ratio of the TPC value of outlet extract (C_o) was 5% of the TPC value of inlet extract (C_i). Generally, BP is considered as the completion time of adsorption in industrial applications, as the adsorption capacity of the resin decreases and the absorbent cannot hold all target molecules, and thus, the solute begins to leak [24]. The best dynamic adsorption performance of the resins was obtained using the lowest flow rate (1.5 BV/h), where the BP was achieved after 120 min. At higher flow rates of 3 and 5 BV/h, the BP was reached more quickly after 75 and 45 min than at slower flow rates. These results indicate that increasing the flow rate negatively affects the dynamic adsorption of polyphenols on XAD 16 resin because as the eluent passes faster through the column, target molecules have less time to interact with active sites on the resin surface. A slow flow rate would positively impact the adsorption capacity of resins and prolong the breakthrough time [21,25,26]. Xi et al. reported the same trend for the adsorption of polyphenols from sweet potato leaves using AB-8 resin. This resin is slightly polar, with an average diameter similar to that of XAD 16. At higher flow rates, some polyphenols leaked out without being adsorbed by the resin because of the high flow speed [17]. Soto et al. showed that by increasing the flow rate from 1 to 2.5 and 5 mL/min, breakthrough decreased, and thus, the efficiency of polyphenol sorption from wine vinasses was reduced for both XAD16 HP and SP700 polymeric resins [23].

Dynamic desorption was performed after the adsorption stage of hazelnut skin extract when the BP was obtained and using ethanol solution (70% v/v) at three flow rates (1.5, 3, 5 BV/h). The desorption curves are shown in Figure 2B. Higher polyphenol recovery was observed at a desorption flow rate of 1.5 BV/h. By increasing the flow rate, the time required to recover a higher quantity of polyphenols was reduced. Similarly, Li et al. [27] and Park and Lee [21] indicated that a higher flow rate can shorten the time required to reach maximum recovery. Based on the results, using 5 g of XAD 16 resin, 87.7% of polyphenols was recovered from hazelnut skin extract with a desorption ratio of 92.36% by eluting 10 BV of ethanol solution (70% v/v) as solvent at adsorption and desorption flow rates of 1.5 BV/h. Hou and Zhang recovered 85.74% of total phenol in *Vernonia patula* extract using NKA-II resin, which is 2.48-fold higher than the total phenol of the crude extract [9]. Vavouraki showed that FPX66 resin absorbed 60% of phenolic compounds from olive mill wastewater and, using a solvent mixture of ethanol/isopropanol (1:1), recovered 70% of polyphenols [28].

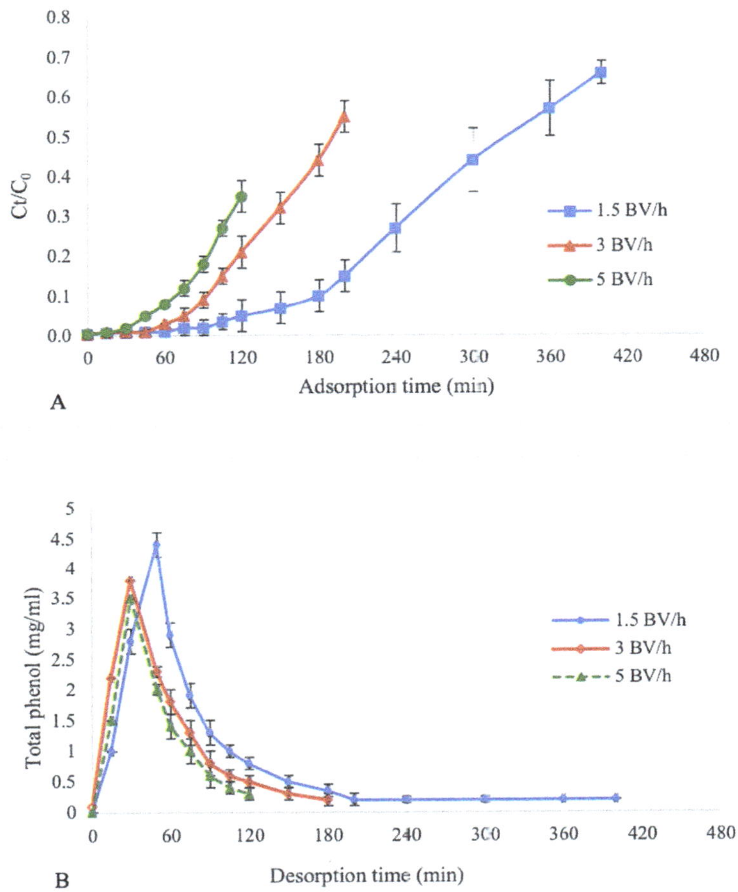

Figure 2. Effect of different flow rates on the (**A**) breakthrough adsorption curve and (**B**) desorption curve of polyphenols using XAD 16 resin.

3.3. Phenolic Content and Antioxidant Activity

A comparison between the phenolic content and antioxidant activity of the hazelnut skin extract before and after purification with the dynamic adsorption/desorption phases is shown in Table 6. Among the polyphenolic compounds identified in hazelnut skin, gallic acid and protocatechuic acid belong to subclasses of phenolic acids, (+)-catechin and (−)-epicatechin belong to the subclass of flavan-3-ols, and quercetin belongs to flavonols. These compounds were selected, identified, and quantified in both the initial and purified extracts.

Table 6. Concentration (mean ± standard deviation) of polyphenolic compounds and values (mean ± standard deviation) of antioxidant activities for the extract obtained from roasted hazelnut skin before and after dynamic absorption/desorption performed with 5 g of Amberlite XAD 16 resin.

	Raw Extract	Purified Extract	Increment% *
Gallic acid (μg/mL)	9.16 ± 0.74	30.85 ± 2.01	237
Protocatechuic acid (μg/mL)	2.80 ± 0.22	8.18± 0.12	192
Catechin (μg/mL)	4.39 ± 0.34	22.06 ± 0.44	402
Epicatechin (μg/mL)	2.72 ± 0.31	14.83 ± 1.05	445
Quercitin (μg/mL)	9.34 ± 0.76	47.23 ± 2.25	406
DPPH (mM TE/mL)	13.44 ± 0.85	83.51 ± 1.25	521
ABTS (mM TE/mL)	8.71 ± 1.23	51.83 ± 1.45	495

* (amount in purified extract-amount in raw extract) * 100/amount in raw extract.

The levels of all phenols were increased in the purified extract. The concentrations of gallic acid and protocatechuic acid in the purified extract increased by approximately three-fold, whereas catechin, epicatechin, and quercetin were increased by over five-fold compared to those in the crude extract. Hou and Zhang reported that the contents of chlorogenic acid and caffeic acid in *V. patula* extracts were increased 2.5-fold after column desorption from NKA-II resin using 5.5 BV of 60% ethanol at a flow rate of 3 BV/h [9]. Johnson and Mitchell showed that Amberlite resins assist in the debittering of olives during normal brine storage by adsorbing bitter phenols. They reported higher adsorption of oleuropein, ligstroside, and oleacein on FPX66 and XAD 16 N resins than on XAD 7 HP and XAD 4 resins [29]. Zheng and Wang utilized AB-8 resin to purify anthocyanins from *Aronia melanocarpa* fruits. The anthocyanin purity increased by 11.5-fold in the final product, using 80% ethanol as desorbing solvent at an elution flow rate of 2.0 BV/h [30].

The results of antioxidant activity were in accordance with the levels of phenolic compounds, with DPPH and ABTS values significantly higher after dynamic adsorption/desorption processes than before these processes, confirming that phenolic compounds were present in the purified extract.

4. Conclusions

The adsorption/desorption conditions of the three Amberlite resins were evaluated to optimize the extraction and purification of polyphenols from the ethanol extract of roasted hazelnut skin. Static adsorption and desorption tests showed that 5 g of Amberlite XAD16 resin had the highest adsorption capacity (40.06 ± 0.55 mg GAE/g) and adsorption ratio. The adsorption kinetics were well-fitted by a pseudo-second order model. Among the tested concentrations of desorbing solvent, 70% *v/v* ethanol solution showed the highest desorption ratio (81.17 ± 1.19%). In the dynamic adsorption/desorption processes performed using 5 g of XAD 16, the breakthrough point increased with decreasing adsorption flow rates, whereas the higher flow rate of solvent in dynamic desorption shortened the desorption time, but polyphenol recovery (87.7%) was observed at the lowest flow rate (1.5 BV/h). The purified extract showed higher phenolic compound levels and antioxidant activity than the crude extract and may be useful as a natural source of bioactive compounds for producing functional foods, as well as cosmetic and pharmaceutical preparations.

Author Contributions: Conceptualization, N.S.Z.; methodology, N.S.Z.; formal analysis, N.S.Z.; resources, G.Z.; writing—original draft preparation, N.S.Z. and G.Z.; writing—review and editing, N.S.Z. and G.Z. All authors have read and agreed to the published version of the manuscript.

Funding: Research supported by VALTIFRU 4.0-Valorizzazione delle filiere di frutta a guscio e fresca trasformata ad alto valore aggiunto-ARS01_01060.

Institutional Review Board Statement: Not applicable.

Informed Consent Statement: Not applicable.

Data Availability Statement: Not applicable.

Conflicts of Interest: The authors declare no conflict of interest.

References

1. Spagnuolo, L.; della Posta, S.; Fanali, C.; Dugo, L.; de Gara, L. Antioxidant and Antiglycation Effects of Polyphenol Compounds Extracted from Hazelnut Skin on Advanced Glycation End-products (Ages) Formation. *Antioxidants* **2021**, *10*, 424. [CrossRef] [PubMed]
2. Bertolino, M.; Belviso, S.; Dal Bello, B.; Ghirardello, D.; Giordano, M.; Rolle, L.; Gerbi, V.; Zeppa, G. Influence of the Addition of Different Hazelnut Skins on the Physicochemical, Antioxidant, Polyphenol and Sensory Properties Ofyogurt. *LWT Food Sci. Technol.* **2015**, *63*, 1145–1154. [CrossRef]
3. Pérez-Larrán, P.; Díaz-Reinoso, B.; Moure, A.; Alonso, J.L.; Domínguez, H. Adsorption Technologies to Recover and Concentrate Food Polyphenols. *Curr. Opin. Food Sci.* **2018**, *23*, 165–172. [CrossRef]
4. Zeppa, G.; Belviso, S.; Bertolino, M.; Cavallero, M.C.; Dal Bello, B.; Ghirardello, D.; Giordano, M.; Giorgis, M.; Grosso, A.; Rolle, L.; et al. The Effect of Hazelnut Roasted Skin from Different Cultivars on the Quality Attributes, Polyphenol Content and Texture of Fresh Egg Pasta. *J. Sci. Food Agric.* **2015**, *95*, 1678–1688. [CrossRef] [PubMed]
5. Taş, N.G.; Gökmen, V. Bioactive Compounds in Different Hazelnut Varieties and Their Skins. *J. Food Compos. Anal.* **2015**, *43*, 203–208. [CrossRef]
6. Locatelli, M.; Travaglia, F.; Coïsson, J.D.; Martelli, A.; Stévigny, C.; Arlorio, M. Total Antioxidant Activity of Hazelnut Skin (Nocciola Piemonte PGI): Impact of Different Roasting Conditions. *Food Chem.* **2010**, *119*, 1647–1655. [CrossRef]
7. Pelvan, E.; Olgun, E.Ö.; Karadağ, A.; Alasalvar, C. Phenolic Profiles and Antioxidant Activity of Turkish Tombul Hazelnut Samples (Natural, Roasted, and Roasted Hazelnut Skin). *Food Chem.* **2018**, *244*, 102–108. [CrossRef]
8. del Rio, D.; Calani, L.; Dall'Asta, M.; Brighenti, F. Polyphenolic Composition of Hazelnut Skin. *J. Agric. Food Chem.* **2011**, *59*, 9935–9941. [CrossRef]
9. Hou, M.; Zhang, L. Adsorption/Desorption Characteristics and Chromatographic Purification of Polyphenols from Vernonia Patula (Dryand.) Merr. Using Macroporous Adsorption Resin. *Ind. Crops Prod.* **2021**, *170*, 113729. [CrossRef]
10. Wang, Z.; Peng, S.; Peng, M.; She, Z.; Yang, Q.; Huang, T. Adsorption and Desorption Characteristics of Polyphenols from Eucommia Ulmoides Oliv. Leaves with Macroporous Resin and Its Inhibitory Effect on α-Amylase and α-Glucosidase. *Ann. Transl. Med.* **2020**, *8*, 1004. [CrossRef]
11. Leyton, A.; Vergara-Salinas, J.R.; Pérez-Correa, J.R.; Lienqueo, M.E. Purification of Phlorotannins from Macrocystis Pyrifera Using Macroporous Resins. *Food Chem.* **2017**, *237*, 312–319. [CrossRef] [PubMed]
12. Özdemir, K.S.; Yilmaz, C.; Durmaz, G.; Gokmen, V. Hazelnut Skin Powder: A New Brown Colored Functional Ingredient. *Food Res. Int.* **2014**, *65*, 291–297. [CrossRef]
13. Barbosa-Pereira, L.; Guglielmetti, A.; Zeppa, G. Pulsed Electric Field Assisted Extraction of Bioactive Compounds from Cocoa Bean Shell and Coffee Silverskin. *Food Bioprocess Technol.* **2018**, *11*, 818–835. [CrossRef]
14. Re, R.; Pellegrini, N.; Proteggente, A.; Pannala, A.; Yang, M.; Rice-Evans, C. Antioxidant Activity Applying an Improved ABTS Radical Cation Decolorization Assay. *Free Radic Biol. Med.* **1999**, *26*, 1231–1237. [CrossRef]
15. Yu, Q.; Fan, L.; Li, J. A Novel Process for Asparagus Polyphenols Utilization by Ultrasound Assisted Adsorption and Desorption Using Resins. *Ultrason. Sonochem.* **2020**, *63*, 104920. [CrossRef] [PubMed]
16. Lv, C.; Yang, J.; Liu, R.; Lu, Q.; Ding, Y.; Zhang, J.; Deng, J. A Comparative Study on the Adsorption and Desorption Characteristics of Flavonoids from Honey by Six Resins. *Food Chem.* **2018**, *268*, 424–430. [CrossRef] [PubMed]
17. Xi, L.; Mu, T.; Sun, H. Preparative Purification of Polyphenols from Sweet Potato (Ipomoea batatas L.) Leaves by AB-8 Macroporous Resins. *Food Chem.* **2015**, *172*, 166–174. [CrossRef]
18. Wang, J.; Guo, X. Adsorption Kinetic Models: Physical Meanings, Applications, and Solving Methods. *J. Hazard. Mater.* **2020**, *390*, 121156. [CrossRef]
19. Park, J.J.; Lee, W.Y. Adsorption and Desorption Characteristics of a Phenolic Compound from Ecklonia Cava on Macroporous Resin. *Food Chem.* **2021**, *338*, 128150. [CrossRef]
20. Le, T.T.; Framboisier, X.; Aymes, A.; Ropars, A.; Frippiat, J.P.; Kapel, R. Identification and Capture of Phenolic Compounds from a Rapeseed Meal Protein Isolate Production Process By-Product by Macroporous Resin and Valorization Their Antioxidant Properties. *Molecules* **2021**, *26*, 5853. [CrossRef]
21. Buran, T.J.; Sandhu, A.K.; Li, Z.; Rock, C.R.; Yang, W.W.; Gu, L. Adsorption/Desorption Characteristics and Separation of Anthocyanins and Polyphenols from Blueberries Using Macroporous Adsorbent Resins. *J. Food Eng.* **2014**, *128*, 167–173. [CrossRef]
22. Wang, X.; Su, J.; Chu, X.; Zhang, X.; Kan, Q.; Liu, R.; Fu, X. Adsorption and Desorption Characteristics of Total Flavonoids from Acanthopanax Senticosus on Macroporous Adsorption Resins. *Molecules* **2021**, *26*, 4162. [CrossRef] [PubMed]
23. Soto, M.L.; Moure, A.; Domínguez, H.; Parajó, J.C. Batch and Fixed Bed Column Studies on Phenolic Adsorption from Wine Vinasses by Polymeric Resins. *J. Food Eng.* **2017**, *209*, 52–60. [CrossRef]
24. Sandhu, A.K.; Gu, L. Adsorption/Desorption Characteristics and Separation of Anthocyanins from Muscadine (Vitis Rotundifolia) Juice Pomace by Use of Macroporous Adsorbent Resins. *J. Agric. Food Chem.* **2013**, *61*, 1441–1448. [CrossRef]
25. Iheanacho, O.C.; Nwabanne, J.T.; Obi, C.C.; Onu, C.E. Packed Bed Column Adsorption of Phenol onto Corn Cob Activated Carbon: Linear and Nonlinear Kinetics Modeling. *S. Afr. J. Chem. Eng.* **2021**, *36*, 80–93. [CrossRef]

26. Ma, C.; Tao, G.; Tang, J.; Lou, Z.; Wang, H.; Gu, X.; Hu, L.; Yin, M. Preparative Separation and Purification of Rosavin in Rhodiola Rosea by Macroporous Adsorption Resins. *Sep. Purif. Technol.* **2009**, *69*, 22–28. [CrossRef]
27. Li, B.; Wang, C.; Chen, X.; Lyu, J.; Guo, S. Highly Specific Separation for Antitumor Spiropreussione A from Endophytic Fungal [Preussia Sp.] Fermentation Broth by One-Step Macroporous Resins AB-8 Treatment. *J. Chromatogr. B Anal. Technol. Biomed. Life Sci.* **2013**, *938*, 1–7. [CrossRef]
28. Vavouraki, A. Removal of Polyphenols from Olive Mill Wastewater by FPX 66 Resin: Part II. Adsorption Kinetics and Equilibrium Studies. *Int. J. Waste Resour.* **2020**, *10*, 374. [CrossRef]
29. Johnson, R.; Mitchell, A.E. Use of Amberlite Macroporous Resins to Reduce Bitterness in Whole Olives for Improved Processing Sustainability. *J. Agric. Food Chem.* **2019**, *67*, 1546–1553. [CrossRef]
30. Zheng, Y.; Wang, P. Extraction and Purification Conditions of Anthocyanin from the Fruit of Aronia Melanocarpa. *Beijing Linye Daxue Xuebao J. Beijing For. Univ.* **2016**, *38*, 118–124. [CrossRef]

foods

MDPI

Article

Radical Scavenging Activity and Physicochemical Properties of Aquafaba-Based Mayonnaises and Their Functional Ingredients

Katarzyna Włodarczyk [1], Agnieszka Zienkiewicz [2] and Aleksandra Szydłowska-Czerniak [1,*]

1 Department of Analytical Chemistry and Applied Spectroscopy, Faculty of Chemistry, Nicolaus Copernicus University in Toruń, Gagarina 7, 87-100 Toruń, Poland; kwlodarczyk@doktorant.umk.pl
2 Centre for Modern Interdisciplinary Technologies, Nicolaus Copernicus University in Toruń, Wileńska 4, 87-100 Toruń, Poland; agazet@umk.pl
* Correspondence: olasz@umk.pl

Abstract: A plant-based diet has become more popular as a pathway to transition to more sustainable diets and personal health improvement in recent years. Hence, vegan mayonnaise can be proposed as an egg-free, allergy friendly vegan substitute for full-fat conventional mayonnaise. This study intends to evaluate the effect of aquafaba from chickpeas and blends of refined rapeseed oil (RRO) with different cold-pressed oils (10% of rapeseed oil—CPRO, sunflower oil—CPSO, linseed oil—CPLO or camelina oil—CPCO) on the radical scavenging, structural, emulsifying, and optical properties of novel vegan mayonnaise samples. Moreover, the functional properties and radical scavenging activity (RSA) of mayonnaise ingredients were evaluated. Aquafaba-based emulsions had a higher RSA than commercial vegan mayonnaise, determined by QUick, Easy, Novel, CHEap and Reproducible procedures using 2,2-diphenyl-1-picrylhydrazyl (QUENCHER-DPPH) and 2,2′-azino-bis(3-ethylbenzothiazoline-6-sulfonic acid) (QUENCHER-ABTS). Oxidative parameters such as peroxide values (PV), anisidine values (AnV), total oxidation (TOTOX) indexes and acid values (AV) of the proposed vegan mayonnaises were similar to those for commercial mayonnaises. Moreover, aquafaba-based samples had smaller oil droplet sizes than commercial vegan mayonnaise, which was observed using confocal laser scanning microscopy. The novel formulas developed in this study are promising alternatives to commercial vegan emulsions.

Keywords: aquafaba; cold-pressed oils; confocal laser scanning microscopy; egg replacement; physicochemical properties; radical scavenging activity; vegan mayonnaise

Citation: Włodarczyk, K.; Zienkiewicz, A.; Szydłowska-Czerniak, A. Radical Scavenging Activity and Physicochemical Properties of Aquafaba-Based Mayonnaises and Their Functional Ingredients. *Foods* **2022**, *11*, 1129. https://doi.org/10.3390/foods11081129

Academic Editors: Marco Poiana, Francesco Caponio and Antonio Piga

Received: 15 March 2022
Accepted: 12 April 2022
Published: 14 April 2022

Publisher's Note: MDPI stays neutral with regard to jurisdictional claims in published maps and institutional affiliations.

Copyright: © 2022 by the authors. Licensee MDPI, Basel, Switzerland. This article is an open access article distributed under the terms and conditions of the Creative Commons Attribution (CC BY) license (https://creativecommons.org/licenses/by/4.0/).

1. Introduction

Mayonnaise is semi-solid oil-in-water emulsion. Traditional mayonnaise is made from vegetable oil, egg yolk, vinegar, and spices during gentle mixing. Therefore, traditional mayonnaise as a high oil-containing product is susceptible to deterioration by fast, destructive oxidation of the unsaturated fats in its oil fraction. Oxidation processes reduce the nutritional value of fat-based products due to the loss of polyunsaturated lipids and vitamins that are beneficial to human health. However, the elimination of intrinsic (prooxidants content) and extrinsic factors (high temperature, light) and the addition of high-quality antioxidant ingredients can inhibit the oxidative reactions and increase the shelf life of mayonnaise [1]. On the other hand, mayonnaise ingredients, especially types of edible oils, fat replacers, and emulsifiers, strongly influence the physicochemical properties of the final products. Although the most common oils used to make mayonnaise are refined oils, such as rapeseed oil, sunflower oil, and soybean oil [2], the effect of various nut oils on the structure and physicochemical properties of gel-like emulsions was also investigated [3]. The gel-like emulsion with sunflower oil showed better stability than the gel-like emulsions of nuts oils due to the smaller particle size and higher viscosity. Moreover, mayonnaise made with linseed oil with a high level of linolenic acid (>50%) was more susceptible to lipid oxidation than mayonnaise with saturated medium-chain triglyceride oil [4]. However, Eidhin and O'Beirne [5] did not notice any changes in the oxidative stability of salad dressing and mayonnaises after the replacement of sunflower seed oil with refined camelina oil. In particular, many studies have reported that various cold-pressed oils (linseed, camelina, soybean, sunflower, rapeseed, corn, grapeseed, hemp, rice bran, pumpkin, walnut, rosehip, milk thistle, and black cumin) contain a higher level of natural antioxidants such as tocopherols (165.4–1036.0 mg/kg), phenols (5.1–1151.2 mg/kg), flavonoids (4.5–64.2 mg/kg), carotenoids (18.3–198.0 mg/kg), and squalene (0–1324.3 mg/kg) than refined vegetable oils [6–10]. Therefore, emulsions containing cold-pressed black cumin oil and cold-pressed rice bran oil had a better oxidative status (lower amounts of primary and secondary oxidation products) and antioxidant properties (higher total phenolics content and antioxidant potential determined by 2,2-diphenyl-1-picrylhydrazyl (DPPH) assay) than control emulsions without cold-pressed oils [10,11].

Consequently, lipid oxidation in emulsions depends also on the type of emulsifier. It is well known that egg yolk is the most popular emulsifier used in traditional mayonnaise formulation. However, eggs consumption increases the risk of cardiovascular disease, particularly among diabetic patients [12–14], while today's consumers express greater concern about the quality and health benefits of eaten products. Lately, consumption of this product has decreased to promote environmental care and animal welfare. The mean daily water consumption footprint per capita in the European Union countries for eggs and egg products was approximately 1.5 times higher than this result for legumes, nuts, oilseeds and spices [15]. One other reason to reduce the consumption of eggs is an allergy caused by an allergic reaction to their proteins. Unfortunately, raw eggs present in mayonnaise are more likely to cause an allergic reaction than small amounts of cooked eggs. Egg allergy is one of the most common food allergies in infancy and small children, with a prevalence of up to 3.2% in the USA and Western Europe [16]. However, an allergy to eggs is rarely diagnosed in adults. Therefore, egg yolk in mayonnaise samples has often been replaced with plant protein isolates [17], canola protein [18], milk protein [19], soy milk [20], peanut and sesame meal milk [21], and Arabic gum [22]. Moreover, the effect of wastewater after cooking chickpeas (aquafaba) on the texture and physicochemical properties of plant-based mayonnaise was investigated [23]. Recently, the application of one of the most popular pulses cooking water (PCW)—aquafaba as a food ingredient has successfully broadened the list of new egg replacers [24]. Damian et al. [25] reported that PCW was a stable emulsifier and a rich source of phenolic compounds (0.3–0.7 mg/mL) and saponins (8–12 mg/mL). The amount of leached proteins and saponins impacted the foaming and gelling abilities of PCW. Other applications of aquafaba are in bakery products, dressing, dip, ice cream, legume-based cheese, and legume-based dairy substitutes [26]. Interestingly, chickpea ranks third among total legume production worldwide, after beans and peas and accounting for 10.1 million tons annually [27].

In recent years, the functional properties of aquafaba such as its emulsifying capacity, foaming ability, gelling attributed to its composition of proteins, carbohydrates, polysaccharide-protein complexes, coacervates, saponins, and phenolic compounds have been used in various formulations for vegans [28–30].

On the other hand, antioxidant compounds added to fat-based foods counteract lipid oxidation by acting as reducing agents, free radical scavengers and inactivators of prooxidants. Therefore, the antioxidant potential of emulsions enriched with new functional ingredients with antioxidant properties determines their quality and storage stability.

In this study, understanding the relationship between emulsion properties (mainly oxidative stability) and radical scavenging activity (RSA) is essential to improving the quality of mayonnaise production. To the best of our knowledge, there was no reference to the determination of the antioxidant properties of eggless mayonnaise containing aquafaba as a functional replacer. Therefore, the present work focused on the evaluation of radical scavenging characteristics, oxidation stability, microstructures, and optical properties of vegan mayonnaises containing aquafaba from chickpeas and blends of refined rapeseed oil (RRO) with different cold-pressed oils as key ingredients. The radical scavenging and physicochemical properties of the prepared emulsions were characterized and compared with commercial samples (egg yolk-based and plant-based mayonnaises). The modified QUick, Easy, New, CHEap, and Reproducible (QUENCHER) procedure was applied for an evaluation of the RSA of mayonnaises, whereas the radical scavenging characteristics of ingredients were analyzed by two conventional 2,2-diphenyl-1-picrylhydrazyl (DPPH) and 2,2′-azino-bis(3-ethylbenzothiazoline-6-sulfonic acid) (ABTS) methods. Moreover, the emulsifying properties of aquafaba and fatty acid composition (FAC) of cold-pressed oils were estimated and discussed. Finally, a principal component analysis (PCA) was applied to check the differences and similarities among all of the studied mayonnaise samples.

2. Materials and Methods

2.1. Reagents and Materials

Chickpeas (*Cicer arietinum* L.), mustard, salt, vinegar, sugar, nutritional yeast, and eggs were purchased in a local market. Refined rapeseed oil (RRO), cold-pressed rapeseed oil from *Brassica napus* L. (CPRO), cold-pressed sunflower oil from *Helianthus* L. (CPSO), cold-pressed linseed oil from *Linum usitatissimum* L. (CPLO) and cold-pressed camelina oil from *Camelina sativa* L. (CPCO) were kindly provided by a local vegetable oil factory in the original packing (1 L of RRO in polyethylene terephthalate (PET) bottle, 250 mL of cold-pressed oils in amber-colored glass bottles, marasca type).

Additionally, three mayonnaises (MT1, MT2—traditional recipe mayonnaises (full-fat, 76 and 68%, respectively, and MV—vegan mayonnaise containing 35% fat) packed in colorless glass jars were supplied directly after production by two different manufacturers representing the top-selling brands in the Polish market.

All oil and mayonnaise samples were within their stated shelf lives and stored in a refrigerator at 4 °C until analysis (no longer than 4 days after opening the original packaging).

All chemicals of analytical grade were purchased from Sigma-Aldrich (Poznań, Poland). Redistilled water was used for the preparation of solutions.

2.2. Preparation of Mayonnaise Samples

Aquafaba, the liquid from the chickpea jars, was separated using a stainless-steel mesh kitchen strainer. A representative sample of aquafaba was taken for analysis and mayonnaise preparation. The content of each ingredient for the preparation of mayonnaise samples was selected based on preliminary tests. Four mayonnaise batches (MRO, MSO, MLO and MCO) of 200 g were prepared according to the procedure described by Raikos et al. [23] with some modifications using the ingredients listed in Table 1. Aquafaba was mixed with mustard, vinegar, nutritional yeast, salt, and sugar using a high mixing bowl and a stick blender (BOSH® MSM67160, Robert Bosch GmbH, Gerlingen, Germany). Then, blends of RRO with each cold-pressed oil were gradually and slowly added to the aqueous mixture during the blending procedure to achieve the proper consistency. Emulsions were homogenized for 5 min utilizing a homogenizer (DT basic, Yellow Line, IKA®-Werke GmbH & Co., KG, Staufen, Germany) at 8000 rpm. The appearance of the prepared aquafaba-based emulsions is shown in Figure 1. Mayonnaises were packaged into colorless glass jars and stored at 4 °C in a refrigerator until analyses (no longer than 4 days).

Table 1. Ingredients of different formulations of vegan mayonnaise samples.

Ingredient	MRO (g/100 g)	MSO (g/100 g)	MLO (g/100 g)	MCO (g/100 g)
Aquafaba	23.70	23.70	23.70	23.70
Mustard	1.25	1.25	1.25	1.25
Nutritional yeast	0.10	0.10	0.10	0.10
Vinegar	0.65	0.65	0.65	0.65
Sugar	0.15	0.15	0.15	0.15
Salt	0.15	0.15	0.15	0.15
RRO	67.00	67.00	67.00	67.00
CPRO	7.00	–	–	–
CPSO	–	7.00	–	–
CPLO	–	–	7.00	–
CPCO	–	–	–	7.00

MRO—mayonnaise with a blend of refined rapeseed oil (RRO) and cold-pressed rapeseed oil (CPRO); MSO—mayonnaise with a blend of refined rapeseed oil (RRO) and cold-pressed sunflower oil (CPSO); MLO—mayonnaise with a blend of refined rapeseed oil (RRO) and cold-pressed linseed oil (CPLO); MCO—mayonnaise with a blend of refined rapeseed oil (RRO) and cold-pressed camelina oil (CPCO).

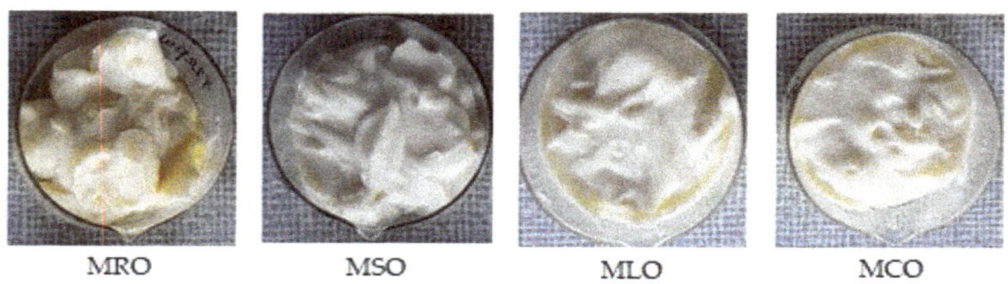

Figure 1. The appearance of the prepared aquafaba-based mayonnaise samples with blends of refined rapeseed oil and cold-pressed rapeseed oil (MRO), cold-pressed sunflower oil (MSO), cold-pressed linseed oil (MLO), and cold-pressed camelina oil (MCO).

2.3. *Characterization of Mayonnaise Ingredients*

2.3.1. Determination of Emulsifying Properties

The emulsifying activity index (EAI) and the emulsion stability index (ESI) were determined according to the procedure described by Cheung et al. [31] but developed initially by Pearce and Kinsella [32]. In brief, 5.0 g of aquafaba or egg yolk was homogenized with 5.0 g of RRO using a homogenizer at a speed of 8000 rpm for 5 min. Then, a 50 µL aliquot of the emulsion was diluted to 7.5 mL of 0.1% sodium dodecyl sulphate (SDS) and vortexed using a classic vortex mixer (Velp Scientifica Srl, Usmate (MB), Italy) for 10 s. The absorbance of the diluted emulsion samples was measured at λ = 500 nm by a Hitachi U-2900 spectrophotometer (Tokyo, Japan). The EAI and ESI were calculated using the following equations:

$$\mathrm{EAI}\left(\frac{\mathrm{m}^2}{\mathrm{g}}\right) = \frac{2 \cdot 2.303 \cdot A_0 \cdot N}{c \cdot \varphi \cdot \phi \cdot 10000} \tag{1}$$

$$\mathrm{ESI}\ (\mathrm{min}) = \frac{A_0}{A_0 - A_{10}} \cdot t \tag{2}$$

where, A_0 and A_{10} are the absorbance values measured at an initial time, and after 10 min, respectively, t is the time interval (10 min), N is the dilution factor, c is the protein concentration (g/mL), φ is the oil volume fraction of the emulsion and ϕ is an optical path (1 cm).

2.3.2. Determination of Protein Concentration

The protein concentration was determined with the Kjeldahl method, according to the official Polish Standard Method PN-75/A-04018 [33]. The protein concentration was estimated as the nitrogen content multiplied by a conversion factor of 6.25.

2.3.3. Determination of Fatty Acid Composition

Fatty acid profiles for RRO and all cold-pressed vegetable oils (CPRO, CPSO, CPLO and CPCO) were determined in accordance with the official method ISO 5508 [34]. Fatty acid methyl esters (FAMEs) were prepared by the transesterification of oil samples carried out in methanol using potassium hydroxide as a base and the derivatization of fatty acids was conducted following the procedure described by ISO 5509 [35].

The quantification of fatty acids was performed by applying a gas chromatograph (HP 5890 GC) equipped with a flame-ionization detector (FID) (Hewlett-Packard, Avondale, PA, USA) and a high polar capillary column BPX 70 (60 m × 0.25 mm, 0.25 µm). The temperatures of the injector and detector were adjusted to 250 °C, while the oven temperature program was as follows: heating from 150 to 210 °C at 1.3 °C/min and holding at 210 °C for 5 min. The carrier gas was helium at a flow rate of 0.6 mL/min.

The identification of fatty acids was accomplished using external FAME standards, and the results are presented as weight percentages of total fatty acids.

2.3.4. Determination of Radical Scavenging Activity

The RSA values of aquafaba, oils (RRO, CPRO, CPSO, CPLO and CPCO), mustard and nutritional yeast were analyzed spectrophotometrically by DPPH and ABTS methods according to the modified procedures previously described in detail [36].

The liquid aquafaba sample was dissolved in methanol and used directly for RSA measurements.

The test tubes with oils (3.00 g), mustard (2.00 g) and nutritional yeast (0.50 g) with methanol (5 mL) were shaken for 30 min using a shaker SK-0 330-PRO (CHEMLAND, Stargard Szczeciński, Poland) at room temperature. The extracts were separated from oils in a freezer (−20 °C, 30 min) and transferred quantitatively into glass bottles. Each studied sample was extracted in triplicate, and extracts were stored in a refrigerator prior to RSA analysis.

In the case of the DPPH test, 0.3 mL of 0.1% methanolic solution of aquafaba (v/v) or 0.2–0.5 mL of the methanolic extracts of oils, mustard and nutritional yeast were added to 1.8–1.5 methanol and 0.5 mL of DPPH methanolic solution (304 µmol/L). The mixtures were shaken vigorously and left in darkness for 15 min. The absorbance of each sample was measured at 517 nm against a reagent blank (2 mL of methanol and 0.5 mL of DPPH methanolic solution) using a Hitachi U-2900 spectrophotometer in a 1 cm glass cell.

For the ABTS assay, 0.1 mL of 0.1% methanolic solution of aquafaba (v/v) or 0.1–0.3 mL of the methanolic extracts of oils, mustard and nutritional yeast were added to 2.4–2.2 mL of $ABTS^{\bullet+}$ reagent (diluted with ethanol to an absorbance of 0.70 ± 0.02 at 734 nm). The obtained mixtures were incubated at 30 °C for 1 min and the absorbance of each sample was measured at 734 nm against a reagent blank (2.5 mL of $ABTS^{\bullet+}$ solution).

The RSA values were expressed as micromoles of Trolox equivalents (TE) per 100 g of the studied sample.

2.4. Characterization of Mayonnaise Samples

2.4.1. Determination of Radical Scavenging Activity

The QUENCHER-DPPH and QUENCHER-ABTS extraction-free procedures were applied for a direct evaluation of the RSA of the proposed vegan and commercial mayonnaises.

For the QUENCHER-DPPH assay, 0.0300–0.0500 g of the mayonnaise samples was transferred into centrifuge tubes. The reaction was started by adding 6 mL of DPPH solution (60.8 µmol/L). The mixture was vortexed for 5 min and left in darkness for 15 min. The samples were centrifuged at 3120× g (centrifuge MPW-54, MPW MED. INSTRUMENTS, Warsaw, Poland) for 3 min and the absorbance of optically clear supernatant was measured at 517 nm.

For the QUENCHER-ABTS assay, 0.0200–0.0300 g of the mayonnaise samples was weighed to centrifuge tubes and the reaction was started by adding 6 mL of $ABTS^{\bullet+}$ reagent (diluted with ethanol to an absorbance of 0.70 ± 0.02 at 734 nm). The mixtures were vortexed for 5 min and centrifuged at 3120× g for 3 min. The absorbance of clear supernatants was measured at 734 nm against a reagent blank (2.5 mL of $ABTS^{\bullet+}$ solution).

2.4.2. Determination of Oxidative Stability

The lipid phase of each mayonnaise sample was separated according to the procedure [37] for an analysis of oxidative stability. Mayonnaises were frozen at −18 °C for 12 h and then thawed for 12 h at 4 °C to break the emulsion. The mixture was centrifuged for 5 min. Each lipid phase, separated from the emulsion residue, was stored in a closed glass flask in the refrigerator prior to further analysis.

The peroxide value (PV), anisidine value (AnV) and acid value (AV) of the lipid phases were determined to estimate the formation of primary and secondary oxidation products as well as free fatty acids, respectively, that affect the rancidity and the mayonnaise stability.

The PV was measured by iodometric titration according to the official procedure ISO 3960:2017 [38] and was expressed as milliequivalents of active oxygen per kilogram of lipid phase (mEq O_2/kg).

The AnV was analyzed according to the ISO 6885:2016 method [39].

The oxidation state of the lipid phase given by the TOTOX index was calculated according to the formula: (TOTOX = 2PV + AnV).

However, the AV was analyzed according to the ISO 660 (1996) method [40].

2.4.3. Microstructure

Commercial mayonnaises (MT1, MT2, MV) and freshly made emulsions (MRO, MSO, MLO, MCO) were observed under an Olympus FluoView 3000 confocal laser scanning microscope (Olympus, Tokyo, Japan). All samples were prepared by adding 0.01% Nile red and analyzed using a diode 488 nm laser (excitation). The area of oil droplets (ODs) from each sample was quantified from 10 to 15 confocal images acquired by FluoView software (Olympus), corresponding to ∼1000 ODs by using ImageJ software (www.imagej.nih.gov) (accessed on 6 November 2020). Data point-box plots were generated with Origin Pro software (OriginLab Corporation, Northampton, MA, USA).

2.4.4. Color Parameters

Before an analysis of the color parameters, all mayonnaise samples were thoroughly mixed. The color was measured using a MICRO-COLOR II LCM 6 spectrophotometer (Dr. Bruno Lange GmbH & Co. KG, Berlin, Germany) based on three color coordinates, namely L*, a*, b*. The color values were expressed as L* (whiteness or brightness/darkness), a* (redness/greenness) and b* (yellowness/blueness).

2.5. Statistical Analysis

The measurements of the oxidative parameters of lipid phase, emulsifying properties, and protein content in emulsifiers were performed in triplicate on the same day to verify the repeatability of the obtained results. The RSA for ingredients and emulsions, FAC of vegetable oils, and color parameters of mayonnaises were measured in five replications. All results were presented as mean (c) ± standard deviation (SD). A one-way analysis of variance (ANOVA), which was followed by the Tukey's post hoc test, was performed to analyze the significant differences between data ($p < 0.05$). Moreover, PCA was employed to study the clustering and differentiation of seven mayonnaise samples based on QUENCHER-DPPH, QUENCHER-ABTS, PV, AnV, TOTOX, AV, area of ODs, L*, a* and b* results. The scores and loadings of the data analyzed by PCA were displayed as a biplot. Statistical analyses of data were carried out using the Statistica 8.0 software (StatSoft, Tulsa, OK, USA).

3. Results and Discussion

3.1. Characterization of Mayonnaise Ingredients

3.1.1. Emulsifying Properties and Protein Content

Two indexes, namely the emulsifying activity (EAI) and emulsifying stability (ESI) were used to characterize the emulsifying properties of aquafaba. The obtained EAI and ESI results and protein content in aquafaba were compared with those values for an egg yolk (Table 2).

Table 2. Emulsifying activity index (EAI), emulsion stability index (ESI), and protein content in aquafaba and egg yolk.

Emulsifier Type	EAI ± SD (m^2/g)	ESI ± SD (min)	Protein ± SD (%)
Aquafaba	13.75 ± 0.69 [b]	20.92 ± 0.61 [a]	1.26 ± 0.05 [a]
Egg yolk	1.78 ± 0.07 [a]	2385 ± 103 [b]	16.12 ± 0.47 [b]

n = 3; SD—standard deviation; Different letters (a, b) within the same column indicate significant differences between emulsifying parameters (EAI—emulsifying activity index, ESI—emulsifying stability index) and protein content in emulsifiers (Tukey's post hoc test, $p < 0.05$).

The EAI is an oil/water interface area stabilized per unit weight of protein. As can be seen, the EAI of aquafaba (13.75 m^2/g) is almost 8-fold higher than the EAI of egg yolk (1.78 m^2/g). This calculated EAI result for egg yolk was significantly lower than previously published values ranging between 24.5 and 30.5 m^2/g [41,42]. Moreover, Meurer et al. [29] observed a high stability of egg yolk-based emulsion during 4 days (EAI = 100%). In general, differences between the EAI results can be caused by various types and concentrations of proteins, their hydrophobicity, unfolding ability, the treatment and storage conditions of egg yolks, as well as procedures and equipment used in emulsions production. It is probable that the studied egg yolk contained a lower amount of proteins with lower solubility and structural unfolding than the egg yolk proteins investigated by other authors [41,42]. A higher EAI of aquafaba proteins could be attributed to a combination of the less compact structure and higher solubility, which enhanced the ability to form interfacial membranes around the oil droplets.

However, the ESI represents a decrease in turbidity of a diluted emulsion over time. This parameter depends on the resistance of proteins to coalescence, sedimentation, flocculation and creaming over a certain period [43]. It is noteworthy that the ESI of egg yolk-based emulsion was over 110-fold higher than the ESI of vegetable emulsion produced by aquafaba (Table 2). The EAI and ESI results obtained for aquafaba (EAI = 13.75 m^2/g and ESI = 20.92 min) were comparable with those reported by other authors (EAI = 12–38.6 m^2/g and ESI = 15–25 min) [25,28]. The emulsifying properties of aquafaba were highly dependent on chickpea cultivar, canning process conditions and additives [24,44]. Additionally, amounts of proteins, carbohydrates, saponins, and phenolic compounds, as well as thermal processes, affected the functional properties of aquafaba [45]. It is known that during emulsion generation, proteins reduced the interfacial tension at the oil–water interface due to the presence of hydrophobic and hydrophilic groups. Although the studied aquafaba had a significantly lower protein content (1.26%) than egg yolk (16.12%), aquafaba proteins aggregated at the water–oil interface and formed an intermolecular cohesive film with enough elasticity to stabilize emulsions.

At the same time, polysaccharides can stabilize the emulsion and prevent flocculation and coalescence [46]. Furthermore, proteins and polysaccharides present in aquafaba can be bound by bioactive substances such as saponins and phenolic compounds, causing changes in the emulsion capacity of this natural emulsifier [25]. Therefore, a low ESI value for aquafaba can be related to its high RSA determined by DPPH and ABTS methods and discussed in Section 3.1.3.

For comparison, Günal-Köroğlu et al. [47] observed that the EAI (13.9–41.6 m^2/g) and ESI (26.1–98.6 min) for emulsions of lentil protein isolate–phenolic solutions were inversely proportional to the concentrations of gallic acid (0.05–0.25 mg/mL) and phenolic extracts from yellow and red onion skin (0.10–0.50 mg/mL). On the other hand, the amphiphilic structure of saponins generated smaller oil droplets (ODs) during homogenization by lowering interfacial tension [48].

3.1.2. Fatty Acid Compositions of Vegetable Oils

The fatty acid compositions of oils used to prepare new vegan mayonnaise samples are presented in Table 3.

Table 3. Fatty acid compositions of oils used in making vegan mayonnaises.

Oil Type	C16:0 ± SD (%)	C18:0 ± SD (%)	C18:1 ± SD (%)	C18:2 ± SD (%)	C18:3 ± SD (%)	C20:0 ± SD (%)	C20:1 ± SD (%)	ΣSAFA (%)	ΣMUFA (%)	ΣPUFA (%)
RRO	4.4 ± 0.1 [a]	1.8 ± 0.1 [a]	62.2 ± 0.2 [d]	19.3 ± 0.1 [b,c]	9.2 ± 0.1 [c]	0.7 ± 0.0 [c]	1.4 ± 0.0 [c]	6.9	63.6	28.5
CPRO	4.8 ± 0.0 [b]	1.6 ± 0.0 [a]	63.4 ± 0.1 [e]	19.4 ± 0.0 [c]	8.0 ± 0.1 [b]	0.6 ± 0.0 [c]	1.2 ± 0.0 [b]	7.0	64.6	27.4
CPSO	6.6 ± 0.1 [e]	3.7 ± 0.1[c]	29.0 ± 0.0 [c]	58.9 ± 0.1 [d]	0.1 ± 0.0 [a]	0.3 ± 0.0 [b]	0.1 ± 0.0 [a]	10.6	29.1	59.0
CPLO	5.3 ± 0.0 [c]	2.5 ± 0.0 [b]	18.8 ± 0.1 [b]	19.2 ± 0.0 [b]	30.5 ± 0.1 [d]	1.7 ± 0.1 [d]	15.0 ± 0.1 [d]	9.5	33.8	49.7
CPCO	5.5 ± 0.0 [d]	4.1 ± 0.1 [d]	15.6 ± 0.0 [a]	16.3 ± 0.0 [a]	57.5 ± 0.1 [e]	0.1 ± 0.0 [a]	0.1 ± 0.0 [a]	9.7	15.7	73.8

$n = 5$; SD—standard deviation; Different letters (a–e) within the same column indicate significant differences between the percentages of fatty acids of vegetable oils: RRO—refined rapeseed oil, CPRO—cold-pressed rapeseed oil, CPSO—cold-pressed sunflower oil, CPLO—cold-pressed linseed oil, CPCO—cold-pressed camelina oil (Tukey's post hoc test, $p < 0.05$); SAFA—saturated fatty acids; MUFA—monounsaturated fatty acids; PUFA—polyunsaturated fatty acid.

Tukey's post hoc test indicated significant differences in the fatty acid percentages of the studied cold-pressed vegetable oils and RRO. Each investigated vegetable oil has a specific fatty acid profile depending on the plant sources. It is noteworthy that all vegetable oils contained a low amount of saturated fatty acids (SAFA = 6.9–10.6%) and a high level of monounsaturated fatty acids (MUFA = 15.7–64.6%) and polyunsaturated fatty acids (PUFA = 27.4–73.8%).

The major SAFA were palmitic acid (C16:0) and stearic acid (C18:0) which were determined in the highest concentrations in CPSO (6.6%) and CPCO (4.1%) samples, respectively. Rapeseed oils (RRO and CPRO) were the richest source of oleic acid (18:1n-9 = 62.2–63.4%), followed by linoleic acid (18:2n-6 = 19.3–19.4%).

On the contrary, linoleic acid was present at the highest level (58.9%) in CPSO, while a moderate content of oleic acid (29%) was found in this oil (Table 3). However, linolenic acid (C18:3n-3 = 30.5–57.5%) was the principal PUFA for two of the cold-pressed oils, CPLO and CPCO. Nevertheless, CPCO had an approximately two times higher C18:3 percentage than CPLO.

As can be seen, CPSO, CPLO and CPCO revealed the highest PUFA content (49.7–73.8%), whereas MUFA accounted for more than 60% of total fatty acids in CPRO and RRO.

It is known that the consumption of vegetable oils with high amounts of unsaturated fatty acids exert a substantial impact on human health, mainly in the prevention of cardiovascular and cancer diseases. On the other hand, vegetable oils rich in PUFA are sensitive to oxidative damage during storage and processing [36].

Our previous studies reported similar results of FAC for rapeseed oils (SAFA = 6.91–7.58%, MUFA = 64.14–66.14%, PUFA = 27.22–30.17%) [36]. However, Symoniuk et al. [6] obtained a higher level of PUFA (64.30–77.42%) in linseed oils.

For comparison, Ratusz et al. [7] studied the quality characteristics of commercial cold-pressed camelina oils and reported somewhat different concentrations of 7.4–10.1% for SAFA, 31.3–36.3% for MUFA, and 55.2–58.4% for PUFA. Specifically, the fatty acid profiles depend on both genetic and environmental factors.

3.1.3. Radical Scavenging Activity of Mayonnaise Ingredients

The radical scavenging properties of mayonnaise ingredients were analyzed by the DPPH and ABTS assays, and the obtained RSA results are presented in Table 4.

Table 4. Radical scavenging activity of mayonnaise ingredients determined by DPPH and ABTS assays.

Ingredient	DPPH ± SD (µmol TE/100 g)	ABTS ± SD (µmol TE/100 g)
Aquafaba	437 ± 22 [b]	2097 ± 22 [b]
Nutritional yeast	1687 ± 60 [d]	9192 ± 238 [d]
Mustard	1129 ± 16 [c]	4985 ± 107 [c]
RRO	387 ± 8.3 [b]	669 ± 29 [a]
CPRO	391 ± 19 [b]	738 ± 25 [a]
CPSO	290 ± 13 [a]	588 ± 15 [a]
CPLO	272 ± 4.5 [a]	602 ± 27 [a]
CPCO	387 ± 4.5 [b]	706 ± 34 [a]

$n = 5$; SD—Standard Deviation; Different letters (a–d) within the same column indicate significant differences between radical scavenging activity of samples: RRO—refined rapeseed oil, CPRO—cold-pressed rapeseed oil, CPSO—cold-pressed sunflower oil, CPLO—cold-pressed linseed oil, CPCO—cold-pressed camelina oil (Tukey's post hoc test, $p < 0.05$).

Among the investigated ingredients, nutritional yeast and mustard had the highest DPPH (1687 and 1129 µmol TE/100 g) and ABTS (9192 and 4985 µmol TE/100 g) values. For comparison, Bors et al. [49] reported that the RSA of mustard paste determined using the DPPH method ranged between 23.4–23.7%.

The radical scavenging properties of aquafaba (DPPH = 437 µmol TE/100 g and ABTS = 2097 µmol TE/100 g) were higher than reported by previously published data (0.15–0.38 µmol TE/g) [50]. This variability in total antioxidant content in chickpeas and aquafaba can be explained by the prolonged time of contact of chickpeas with aquafaba in the jar as well as the influence of other factors such as genetic, agronomic, environmental and technological factors. It is known that the ABTS assay is applicable to both hydrophilic and lipophilic antioxidant systems, whereas the DPPH assay is more suited to hydrophobic systems. It can be noted that the ABTS value of aquafaba (2097 µmol TE/100 g) was five times higher than the DPPH value (437 µmol TE/100 g), probably due to water-soluble antioxidants being predominant in the aquafaba containing 95% water (Table 4).

Moreover, the ABTS results (588–738 µmol TE/100 g) of oils used to make vegan mayonnaise samples were higher than DPPH values (272–391 µmol TE/100 g). Insignificant differences for ABTS values of all the studied cold-pressed vegetable oils were observed, whereas RRO, CPRO and CPCO had significantly higher DPPH results than CPSO and CPLO (Tukey's post hoc test, Table 4). Interestingly, rapeseed and camelina cold-pressed oils were found to be richer sources of antioxidants than cold-pressed sunflower and linseed oils. A similar decrease in DPPH values for cold-pressed oils (rapeseed > sunflower > flax and camelina > flaxseed) was reported by Siger et al. [8] and Grajzer et al. [9].

3.2. *Characterization of Mayonnaise Samples*

3.2.1. Radical Scavenging Activity

The QUENCHER-DPPH and QUENCHER-ABTS methods were proposed for the first time to directly measure the RSA of the developed vegan egg-free mayonnaises containing blends of RRO with different cold-pressed oils and to compare their radical scavenging properties with that of commercial mayonnaises.

These analytical methodologies rely on the surface reaction phenomenon between mayonnaise samples with bound and free radical scavengers with organic DPPH radical and ABTS cation radical solutions. The QUENCHER method, as a cost-saving and extraction-free procedure, is highly relevant and recommended for the fat industrial laboratory to quickly control the antioxidant properties of emulsions.

The QUENCHER-DPPH and QUENCHER-ABTS results for all of the studied mayonnaise samples are presented in Figure 2.

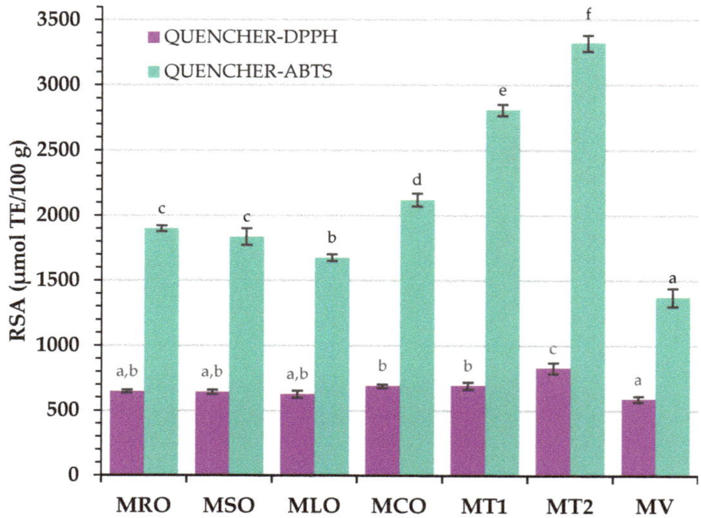

Figure 2. Radical scavenging activity of aquafaba-based mayonnaises with blends of refined rapeseed oil and cold-pressed rapeseed oil (MRO), cold-pressed sunflower oil (MSO), cold-pressed linseed oil (MLO), and cold-pressed camelina oil (MCO); MT1—commercial egg yolk mayonnaise from producer 1; MT2—commercial egg yolk mayonnaise from producer 2; MV—commercial vegan mayonnaise; QUENCHER-DPPH—QUick, Easy, Novel, CHEap and Reproducible-2,2-diphenyl-1-picrylhydrazyl method; QUENCHER-ABTS—QUick, Easy, Novel, CHEap and Reproducible-2,2′-azino-bis(3-ethylbenzothiazoline-6-sulfonic acid) method. Bars with different letters (a–f) indicate significant differences between radical scavenging activity of mayonnaises (Tukey's post hoc test, $p < 0.05$).

It is noteworthy that the richest sources of hydrophilic and lipophilic antioxidants were traditional full fat mayonnaises (QUENCHER-ABTS = 3320 μmol TE/100 g and 2808 μmol TE/100 g for MT2 and MT1, respectively), while MT2 had a higher content of lipophilic antioxidants, as determined by the QUENCHER-DPPH test (828 μmol TE/100 g), than MT1 (DPPH = 692 μmol TE/100 g). In contrast, the store-bought plant-based mayonnaise (MV) revealed the lowest QUENCHER-DPPH (589 μmol TE/100 g) and QUENCHER-ABTS (1371 μmol TE/100 g) values (Figure 2). However, among aquafaba-based mayonnaises, samples containing blends of RRO with cold-pressed oils had similar DPPH results, whereas only the ABTS values for MRO and MSO did not differ significantly (Tukey's post hoc test, Figure 2). The lowest effectiveness of MLO to radical scavenging can be explained by the lowest DPPH (270 μmol TE/100 g) for CPLO among all cold-pressed oils (Table 4). It is worth noting that the addition of CPCO into the MCO sample caused a significant increase in QUENCHER-DPPH and QUENCHER-ABTS results. Moreover, nutritional yeast and mustard had the highest radical scavenging properties (Table 4) which probably increased the RSA of aquafaba-based mayonnaise compared to commercially available MV mayonnaise.

For comparison, the methanolic extract of traditional mayonnaise had an RSA (DPPH = 267 μmol TE/100 g and ABTS = 856 μmol TE/100 g [51]) that was approximately three-fold lower than commercial samples measured by the QUENCHER procedure proposed in this study. Moreover, DPPH (78–134 μmol TE/100 g) and ABTS results (463–613 μmol TE/100 g) for methanol:water (70:30) extracts of vegan mayonnaises based on soy milk [52] were lower than our QUENCHER-DPPH values (589–828 μmol TE/100 g) and QUENCHER-ABTS values (1371–3320 μmol TE/100 g). The higher radical scavenging properties of mayonnaises determined by the QUENCHER procedure can be explained as the proposed analytical method detects both the extractable and non-extractable compounds present in the analyzed samples.

3.2.2. Oxidative Status

The oxidative status of the oil phase of each mayonnaise sample was evaluated by characteristic values such as PV, AnV, TOTOX and AV (Table 5).

Table 5. Oxidation parameters of mayonnaises.

Mayonnaise Type	PV ± SD (mEq O_2/kg)	AnV ± SD	TOTOX Index	AV ± SD (mg KOH/g)
MRO	1.43 ± 0.06 [c]	1.39 ± 0.08 [c]	4.25	0.14 ± 0.01 [a,b]
MSO	1.90 ± 0.06 [d]	1.12 ± 0.06 [b]	4.92	0.26 ± 0.01 [c]
MLO	2.27 ± 0.10 [f]	2.95 ± 0.06 [d]	7.49	0.11 ± 0.01 [a]
MCO	1.19 ± 0.01 [b]	1.55 ± 0.09 [c]	3.93	0.17 ± 0.01 [b]
MT1	2.09 ± 0.07 [e]	1.19 ± 0.08 [b,c]	5.37	0.49 ± 0.01 [d]
MT2	0.82 ± 0.01 [a]	0.76 ± 0.05 [a]	2.40	0.49 ± 0.02 [d]
MV	4.61 ± 0.08 [g]	3.07 ± 0.12 [d]	12.29	0.52 ± 0.03 [d]

n = 3; SD—Standard Deviation; Different letters (a–g) within the same column indicate significant differences between oxidation parameters (PV—peroxide value; AnV—anisidine value; TOTOX—total oxidation index; AV—acid value) of aquafaba-based mayonnaises with blends of refined rapeseed oil and cold-pressed rapeseed oil (MRO), cold-pressed sunflower oil (MSO), cold-pressed linseed oil (MLO), and cold-pressed camelina oil (MCO); MT1—commercial egg yolk mayonnaise from producer 1; MT2—commercial egg yolk mayonnaise from producer 2; MV—commercial vegan mayonnaise (Tukey's post hoc test, $p < 0.05$).

The PV shows the degree of peroxidation and measures the amount of total peroxides, whereas the content of aldehyde carbonyl bonds formed during secondary lipid oxidation is evaluated as AnV. PV and AnV measurements are commonly used together to determine the total extent of oxidation by the TOTOX index. Moreover, AV is an important variable for the quality of the fat-based product because it reveals free fatty acid content affecting its oxidative ageing.

As can be seen, PV results were significantly different in different mayonnaise samples (Tukey's post hoc test, Table 5). The lowest amount of primary oxidation products was in MT2 (PV = 0.82 mEq O_2/kg), while vegan mayonnaise (MV) had the highest PV = 4.61 mEq O_2/kg. This can be explained by the highest amount of total antioxidants in the MT2 sample (QUENCHER-DPPH = 828 µmol TE/100 g and QUENCHER-ABTS = 3320 µmol TE/100 g, Figure 2), which most effectively inhibited harmful oxidation reactions. However, the highest PV for MV can be attributed to the lowest antioxidant potential (QUENCHER-DPPH = 589 µmol TE/100 g and QUENCHER-ABTS = 1371 µmol TE/100 g), preventing the peroxidation in polyunsaturated fatty acids of the oil phase. Moreover, the addition of CPLO with a high level of unsaturated fatty acids (MUFA = 33.8% and PUFA = 49.7%, Table 3) sensitive to peroxidation raised the susceptibility of mayonnaise to oxidation. Thus, MLO had a high level of PV (2.27 mEq O_2/kg) and AnV (2.95) in its fat phase (Table 5). Meanwhile, the presence of resistant SAFA (7.0–10.6%) to degradation processes and a high percentage of MUFA (29.1–64.6%) in CPRO, and CPSO improved the oxidative stability of MRO and MSO samples (PV and AnV ranged between 1.43–1.90 mEq O_2/kg and 1.12–1.39, respectively).

Unexpectedly, low amounts of primary (PV = 1.19 mEq O_2/kg) and secondary (AnV = 1.55) oxidative products in the oil phase of MCO with the highest content of PUFA (73.8%, Table 3) can be explained by the high radical scavenging properties of CPCO (DPPH = 387 µmol TE/100 g and ABTS = 706 µmol TE/100 g, Table 4) and MCO (QUENCHER-DPPH = 690 µmol TE/100 g and QUENCHER-ABTS = 2121 µmol TE/100 g, Figure 2). The antioxidant compounds in cold-pressed oils may act as antioxidants and prevent or delay the lipid oxidation of mayonnaises.

For comparison, the effect of cold-pressed black cumin oil (CPBCO) on the oxidative stability of traditional mayonnaise was evaluated. At the end of storage for 4 weeks at

20 °C, mayonnaise with 20% of CPBCO had a lower content of primary oxidation products (PV = 17.66 mEq O_2/kg) than the control sample (36.07 mEq O_2/kg) [10]. Moreover, substituting egg yolk with milk protein and adding high levels of fish oil to the mayonnaise samples raised the PV results to approximately 75 and 25 mEq O_2/kg at 20 °C and 2 °C of storage, respectively [19].

It is noteworthy that the best stability revealed the MT2 sample with the lowest TOTOX index (2.40), which combines the amount of primary with secondary oxidation products. In contrast, the MV sample containing the lowest total antioxidants level (QUENCHER-DPPH = 589 μmol TE/ 100 g and QUENCHER-ABTS = 1371 μmol TE/100 g, Figure 2) was characterized with the highest TOTOX value = 12.29 (Table 5).

Additionally, aquafaba-based mayonnaises had a significantly lower AV (0.11–0.26 mg KOH/g) than commercial mayonnaise samples (AV = 0.49–0.52 mg KOH/g).

Similarly, commercial mayonnaise revealed higher AV (5.39 mg KOH/g) than samples made with lemon juice and a mix of lemon juice and vinegar (AV = 1.68–3.37 mg KOH/g), which can be explained by non-specific reactions during AV measurements [53].

3.2.3. Microstructure

Previous studies clearly demonstrated that the size of oil droplets (ODs) is one of the most essential parameters influencing mayonnaise stability, texture, and taste [54]. Mayonnaises characterized by small ODs usually possess significantly increased viscosity and stability. The size of ODs depends on the homogenization approach, including its duration and intensity, mayonnaise composition, and ingredient viscosity [55].

In our studies, the CLMS images revealed differences in the size of ODs between analyzed emulsions (Figure 3A–G). The largest ODs were observed for MV and smallest ODs were found for MT1 and MT2. Recently, it was reported that aquafaba-emulsifying properties caused the presence of larger ODs compared to the egg-yolk-based mayonnaise [23,30]. The size of ODs likely depends on the size of emulsifier molecules which in the egg yolk are much smaller compared to aquafaba [30]. He et al. [30] demonstrated that the average of ODs size (d_{43}) of mayonnaise analogue prepared using freeze-dried or spray-dried aquafaba was significantly larger, at 222 ± 9 μm^2 and 97.3 ± 16 μm^2, respectively, compared to egg yolk based mayonnaise (9.02 ± 1 μm^2). Our results seem to confirm that the addition of plant-based proteins, such as pea protein, results in larger average areas of ODs (101.40 μm^2) in cases of MV when compared to egg yolk-based mayonnaise MT1 and MT2, where the average areas of ODs were 4.47 μm^2 and 4.44 μm^2, respectively. We also cannot exclude that the larger size of ODs in MV might be the result of low oil content (35%) in comparison to MT1 and MT2 (around 70%).

The use of equal amounts of aquafaba—23.70% and different oil composition was accompanied by the presence of significantly larger ODs in MCO (45.28 μm^2) and in MLO (56.75 μm^2). In turn, ODs in MRO and in MSO showed an average area of 15.96 μm^2 and 14.29 μm^2, respectively (Figure 3H). The observed variations in ODs size between the analyzed mayonnaises could reflect the differences in their lipidic composition.

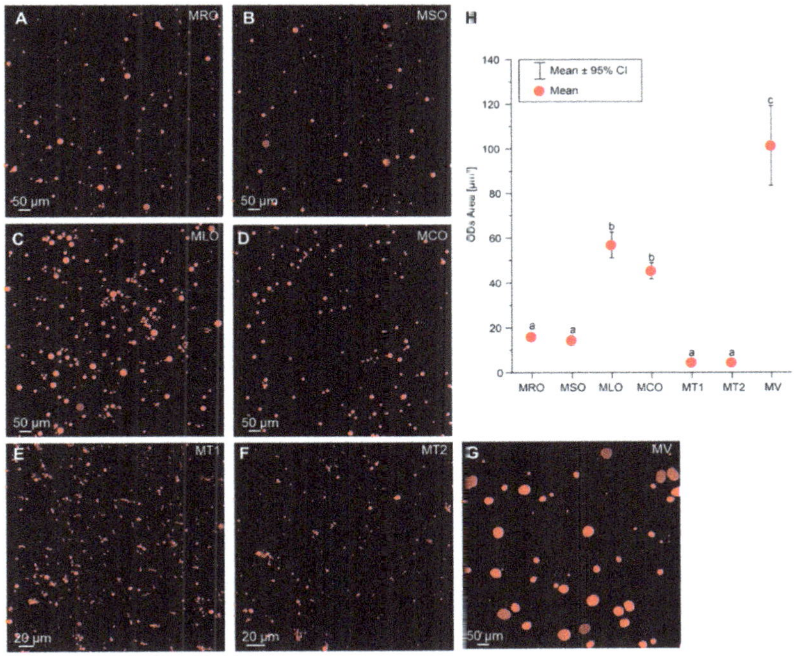

Figure 3. (**A–G**) Representative images of oil droplets (ODs) stained with Nile red. (**H**) The data point-box plot showing a mean area of ODs quantified with imaging software (ImageJ). A statistical analysis was performed by a one-way analysis of variance (ANOVA) with Tukey's post hoc test ($p < 0.05$). Different letters (a–c) indicate significant differences with $p < 0.05$ between aquafaba-based mayonnaises with blends of refined rapeseed oil and cold-pressed rapeseed oil (MRO), cold-pressed sunflower oil (MSO), cold-pressed linseed oil (MLO), and cold-pressed camelina oil (MCO); MT1—commercial egg yolk mayonnaise from producer 1; MT2—commercial egg yolk mayonnaise from producer 2; MV—commercial vegan mayonnaise.

3.2.4. Color

It is crucial to control the color of mayonnaise samples to obtain the desired color that most affects the consumer's willingness to purchase or taste a new product. Therefore, values of lightness (L*), redness (a*) and yellowness (b*) of the prepared vegan mayonnaise samples are listed in Table 6.

Table 6. Color of mayonnaises.

Mayonnaise Type	L* ± SD	a* ± SD	b* ± SD
MRO	48.7 ± 0.2^{e}	0.9 ± 0.2^{c}	10.9 ± 0.4^{c}
MSO	44.5 ± 0.1^{b}	-0.3 ± 0.1^{a}	7.1 ± 0.3^{a}
MLO	41.5 ± 0.3^{a}	0.2 ± 0.1^{b}	9.2 ± 0.2^{b}
MCO	47.1 ± 0.5^{c}	0.5 ± 0.1^{b}	6.7 ± 0.1^{a}
MT1	47.9 ± 0.1^{d}	3.4 ± 0.1^{f}	11.2 ± 0.1^{c}
MT2	47.7 ± 0.2^{d}	2.2 ± 0.2^{d}	9.0 ± 0.4^{b}
MV	41.4 ± 0.2^{a}	2.6 ± 0.2^{e}	11.3 ± 0.4^{c}

$n = 5$; SD—Standard Deviation; Different letters (a–f) within the same column indicate significant differences between color parameters (L*—lightness; a*—redness; b*—yellowness) of aquafaba-based mayonnaise samples with blends of refined rapeseed oil and cold-pressed rapeseed oil (MRO), cold-pressed sunflower oil (MSO), cold-pressed linseed oil (MLO), and cold-pressed camelina oil (MCO); MT1—commercial egg yolk mayonnaise from producer 1; MT2—commercial egg yolk mayonnaise from producer 2; MV—commercial vegan mayonnaise (Tukey's post hoc test, $p < 0.05$).

The Tukey's post hoc test indicated that the color parameters for mayonnaises with blends of RRO and various cold-pressed oils significantly differ from each other. The highest lightness value (L* = 48.7–47.7) had MRO, MT1 and MT2, respectively. It can be caused by the smaller OD sizes in these samples (Figure 3). Due to light scattering, mayonnaises with smaller OD sizes are whiter, and also the intensity of the yellow color is lower [56]. It is noteworthy that store-bought samples had higher redness values (a* = 2.2–3.4) than the prepared aquafaba-based samples (a* = −0.3–0.9). High yellowness values (b* = 9.0–11.2) for commercial samples MT1 and MT2 containing a high egg yolk content can be preferable by consumers. The highest b* value (11.3) for the MV sample was probably caused by the addition of carotenoids. High b* values (6.7–10.9) and low a* values (−0.3–0.9) for MRO, MSO, MLO and MCO samples with cold-pressed oils can be explained by the presence of natural pigments, mainly carotenoids and chlorophyll, in added press-cold oils.

For comparison, the a* and b* values of mayonnaises with different oil-to-aquafaba (O/A) ratios increased significantly for higher oil contents in mayonnaises (a* = 3.5–4.1 and b* = 36.5–39.9 for O/A = 80/15, while a* = 2.6–2.9 and b* = 30.5–31.8 for O/A = 70/25). Moreover, mayonnaise samples with smaller OD sizes had greater brightness values [23].

3.3. Principal Component Analysis

PCA was used to observe any possible groups within the commercial and prepared mayonnaises as well as to reveal the interrelationships among variables (QUENCHER-DPPH, QUENCHER-ABTS, PV, AnV, TOTOX, AV, area of ODs, L*, a* and b*) that mainly influence the similarities and differences of the analyzed samples. The first two principal components took into account 85.86% (PC1 = 57.80% and PC2 = 28.06%, respectively) of the total variation.

The entire data set was visualized by a bi-plot (combined scores and loadings plot for two components) and presented in Figure 4.

As can be seen, the QUENCHER-DPPH (0.8486), QUENCHER-ABTS (0.8497), AV (0.0010), L* (0.8625), and a* (0.0146) of mayonnaise samples had positive loadings on the PC1, whereas the PV (−0.9169), AnV (−0.9388), TOTOX indexes (−0.9656), ODs size (−0.9123), and b* (−0.3271) were the variables with negative loadings on PC1. However, all variables (except AnV) revealed loadings on the negative dimension of PC2 (−0.0045–−0.9819). It can be noted that QUENCHER-DPPH, QUENCHER-ABTS, AV, L* and a* were the variables with positive loadings on PC1 and negative loadings on PC2. On the contrary, AnV revealed loadings on the negative dimension of PC1 and positive on PC2. However, b*, PV, TOTOX and ODs were the features with negative loadings on PC1 and PC2.

The PCA graph depicted that the two vegan mayonnaises (the commercial MV and MLO sample made from a blend of RRO and CPLO) with the lowest RSA (QUENCHER-DPPH = 589–626 μmol TE/100 g, QUENCHER-ABTS = 1371–1674 μmol TE/100 g) and L* (41.4–41.5) values and the highest oxidative parameters (PV = 2.27–4.61 mEq O_2/kg, AnV = 2.95–3.07, TOTOX = 7.49–12.29) and area of ODs (56.76–101.40 μm^2) were located to the left in the score bi-plot and had negative values for PC1. Three prepared vegan emulsions (MSO, MCO and MRO) and two commercial egg yolk mayonnaises (MT1 and MT2) with high radical scavenging properties (QUENCHER-DPPH = 643–828 μmol TE/100 g, QUENCHER-ABTS = 1836–3320 μmol TE/100 g) and more lightness (L* = 44.5–48.7) were situated at the right in the score bi-plot and had positive values for PC1. Consequently, the studied samples could be divided into three groups based on their distribution on the PCA graph. The yellowest commercial plant-based mayonnaise (MV) with the longest distance from other samples had the highest of all the oxidative parameters (PV, AnV, TOTOX, AV), ODs area and the lowest radical scavenging properties determined by two analytical methods (QUENCHER-DPPH and QUENCHER-ABTS). Two traditional mayonnaises containing egg yolk (MT1 and MT2) characterized by the highest RSA (QUENCHER-DPPH = 692–828 μmol TE/100 g, QUENCHER-ABTS = 2808–3320 μmol TE/100 g), the lowest ODs (4.44–4.47 μm^2)

and with the same AV results (0.49 mg KOH/g) created a distinct cluster. Additionally, four egg-free mayonnaises made from blends of RRO and cold-pressed vegetable oils with moderate radical scavenging properties (QUENCHER-DPPH = 626–690 µmol TE/100 g, QUENCHER-ABTS = 1674–2121 µmol TE/100 g), oxidative status (PV = 1.19–2.27 mEq O_2/kg, AnV = 1.12–2.95, TOTOX = 3.93–7.49), and ODs areas (14.29–56.75 µm^2) as well as the lowest AV (0.11–0.26 mg KOH/g) and a* (−0.3–0.9) values were separated from the other studied samples and located at the upper the A1 axis.

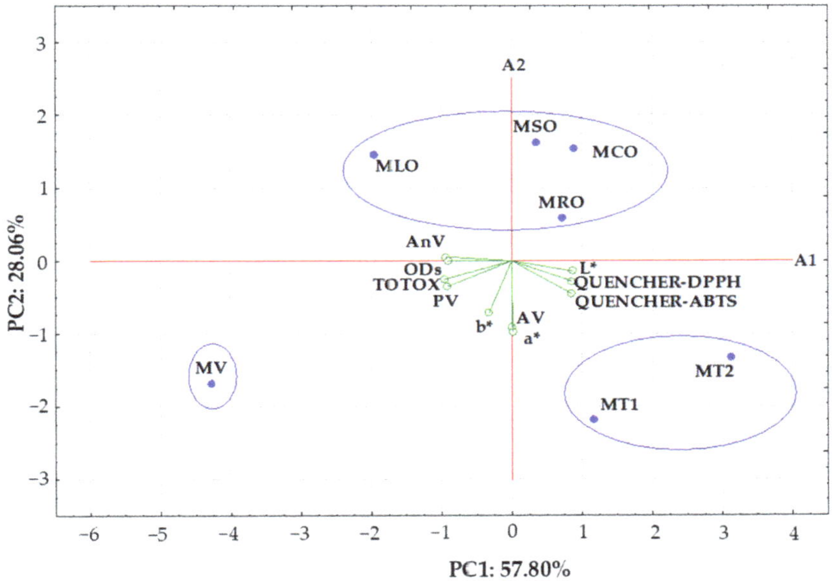

Figure 4. Biplot of scores and loadings of data obtained from radical scavenging properties (QUENCHER-DPPH, QUENCHER-ABTS), oxidation stability (PV—peroxide value; AnV—anisidine value; TOTOX—total oxidation index; AV—acid value), areas of oil droplets (ODs) and color parameters (L*—lightness; a*—redness; b*—yellowness) of aquafaba-based mayonnaise samples with blends of refined rapeseed oil and cold-pressed rapeseed oil (MRO), cold-pressed sunflower oil (MSO), cold-pressed linseed oil (MLO), and cold-pressed camelina oil (MCO); MT1—commercial egg yolk mayonnaise from producer 1; MT2—commercial egg yolk mayonnaise from producer 2; MV—commercial vegan mayonnaise.

It is noteworthy that there was a high correlation between the QUENCHER-DPPH and QUENCHER-ABTS results for all studied mayonnaises (r = 0.9427). This suggests that antioxidants present in the studied samples have the ability to scavenge both stable DPPH radical and ABTS cation radical.

Moreover, RSA values of the discussed mayonnaises were positively correlated with their AV data (r = 0.3177 and 0.4319) as well as color parameters, such as lightness (L*, r = 0.6238 and 0.6894) and redness (a*, r = 0.2615 and 0.4544), but negatively associated with oxidative parameters (PV, AnV and TOTOX, r = −0.6453–−0.7549) and ODs sizes (r = −0.6432–−0.7315). These high negative correlations between oxidative parameters and radical scavenging properties of the studied emulsions indicated the significant contribution of the antioxidant potential of samples to enhance their oxidation status. The lower PV values can be explained by the hindered oxidation of the lipid fraction in the mayonnaise samples, and thus the delay in their deterioration, due to the presence antioxidants and higher RSA values.

As expected, the PV results for seven mayonnaises contributed significantly and positively with the AnV (0.7800) and TOTOX indexes (0.9843). Additionally, a high correlation coefficient (r = 0.8781) was found between AnV and TOTOX data for these samples.

Interestingly, there were high positive correlations between the amounts of primary and secondary oxidation products in the analyzed samples and ODs (r = 0.8162–0.9117). These positive correlations indicate that emulsions with smaller areas of ODs had a better oxidative stability. Nevertheless, the area of ODs for mayonnaises was negatively associated with their lightness (L*, r = −0.7916). However, negative correlation coefficients (r = −0.7470–−0.8583) for the relationships between PV, AnV, TOTOX and L* were observed.

In contrast, there were positive correlations between the free fatty acids content and color parameters of the investigated samples (AV–a*, r = 0.8540 and a*–b*, r = 0.7187).

The results obtained by PCA indicated the differences in radical scavenging and physicochemical properties of two yolk egg mayonnaises, one commercial vegan mayonnaise, and four aquafaba-based samples made from blends of RRO and cold-pressed vegetable oils. This grouping of the samples suggests that the used ingredients in mayonnaise samples and technological conditions were responsible for their quality.

4. Conclusions

Aquafaba-based vegan mayonnaises with blends of RRO and different cold-pressed oils were made. This research paper has presented that aquafaba from chickpeas is a suitable alternative emulsifier with high radical scavenging properties for the production of egg-free vegan emulsions. Moreover, the other ingredients of the proposed mayonnaise formulations, mainly nutritional yeast and mustard, had high radical scavenging characteristics. Additionally, cold-pressed vegetable oils used as functional ingredients in developing vegan egg-free mayonnaises were rich sources of unsaturated fatty acids and natural bioactive compounds with scavenging radical properties.

The new aquafaba-based emulsions containing blends of RRO with different cold-pressed oils had higher radical scavenging properties and oxidative stability, similar color parameters, and a smaller size of ODs than commercial vegan mayonnaise (MV). A significant contribution of the radical scavenging properties of the prepared vegan mayonnaises to enhance their oxidation status was observed. Additionally, emulsions with smaller areas of ODs had better oxidative stability.

The results of PCA indicated that the radical scavenging and physicochemical properties of aquafaba-based emulsions, commercial vegan mayonnaise (MV) and traditional emulsions containing egg yolk (MT1 and MT2) as the main emulsifier differ significantly.

Novel plant-based emulsions could be a promising alternative to egg-based mayonnaises and salad dressing. Further investigations should be performed to determine the effect of aquafaba on the textural and sensory properties and consumer acceptability of mayonnaise.

Author Contributions: Conceptualization, K.W., A.S.-C. and A.Z.; methodology, K.W., A.Z. and A.S.-C.; software, A.S.-C. and A.Z.; validation, A.S.-C., A.Z. and K.W.; formal analysis, K.W. and A.Z.; investigation, K.W., A.S.-C. and A.Z.; resources, K.W., A.S.-C. and A.Z.; data curation, A.S.-C., A.Z. and K.W.; writing—original draft preparation, K.W. and A.Z.; writing—review and editing, A.S.-C. and A.Z.; visualization, K.W., A.S.-C. and A.Z.; supervision, A.S.-C. and A.Z.; project administration, A.S.-C., K.W. and A.Z.; funding acquisition, A.S.-C. and A.Z. All authors have read and agreed to the published version of the manuscript.

Funding: This research received no external funding.

Institutional Review Board Statement: Not applicable.

Informed Consent Statement: Not applicable.

Data Availability Statement: The data presented in this study are available on request from the corresponding author. The data are not publicly available due to privacy or ethical restrictions.

Conflicts of Interest: The authors declare no conflict of interest.

References

1. Ghorbani Gorji, S.; Smyth, H.E.; Sharma, M.; Fitzgerald, M. Lipid oxidation in mayonnaise and the role of natural antioxidants: A review. *Trends Food Sci. Technol.* **2016**, *56*, 88–102. [CrossRef]
2. Di Mattia, C.; Balestra, F.; Sacchetti, G.; Neri, L.; Mastrocola, D.; Pittia, P. Physical and structural properties of extra-virgin olive oil based mayonnaise. *LWT-Food Sci. Technol.* **2015**, *62*, 764–770. [CrossRef]
3. Geng, M.; Hu, T.; Zhou, Q.; Taha, A.; Qin, L.; Lv, W.; Xu, X.; Pan, S.; Hu, H. Effects of different nut oils on the structures and properties of gel-like emulsions induced by ultrasound using soy protein as an emulsifier. *Int. J. Food Sci. Technol.* **2021**, *56*, 1649–1660. [CrossRef]
4. Raudsepp, P.; Brüggemann, D.A.; Lenferink, A.; Otto, C.; Andersen, M.L. Oxidative stabilization of mixed mayonnaises made with linseed oil and saturated medium-chain triglyceride oil. *Food Chem.* **2014**, *152*, 378–385. [CrossRef]
5. Eidhin, D.N.; O'Beirne, D. Oxidative stability of camelina oil in salad dressings, mayonnaises and during frying. *Int. J. Food Sci. Technol.* **2010**, *45*, 444–452. [CrossRef]
6. Symoniuk, E.; Ratusz, K.; Krygier, K. Oxidative stability and the chemical composition of market cold-pressed linseed oil. *Eur. J. Lipid Sci. Technol.* **2017**, *119*, 1700055. [CrossRef]
7. Ratusz, K.; Symoniuk, E.; Wroniak, M.; Rudzińska, M. Bioactive compounds, nutritional quality and oxidative stability of cold-pressed camelina (*Camelina sativa* L.) oils. *Appl. Sci.* **2018**, *8*, 2606. [CrossRef]
8. Siger, A.; Nogala-Kalucka, M.; Lampart-Szczapa, E. The content and antioxidant activity of phenolic compounds in cold-pressed plant oils. *J. Food Lipids* **2008**, *15*, 137–149. [CrossRef]
9. Grajzer, M.; Szmalcel, K.; Kuźmiński, Ł.; Witkowski, M.; Kulma, A.; Prescha, A. Characteristics and antioxidant potential of cold-pressed oils—Possible strategies to improve oil stability. *Foods* **2020**, *9*, 1630. [CrossRef]
10. Ozdemir, N.; Kantekin-Erdogan, M.N.; Tat, T.; Tekin, A. Effect of black cumin oil on the oxidative stability and sensory characteristics of mayonnaise. *J. Food Sci. Technol.* **2018**, *55*, 1562–1568. [CrossRef]
11. Thanonkaew, A.; Wongyai, S.; Decker, E.A.; McClements, D.J. Formation, antioxidant property and oxidative stability of cold pressed rice bran oil emulsion. *J. Food Sci. Technol.* **2015**, *52*, 6520–6528. [CrossRef] [PubMed]
12. Rong, Y.; Chen, L.; Zhu, T.; Song, Y.; Yu, M.; Shan, Z.; Sands, A.; Hu, F.B.; Liu, L. Egg consumption and risk of coronary heart disease and stroke: Dose-response meta-analysis of prospective cohort studies. *BMJ* **2013**, *346*, e8539. [CrossRef] [PubMed]
13. Li, Y.; Zhou, C.; Zhou, X.; Li, L. Egg consumption and risk of cardiovascular diseases and diabetes: A meta-analysis. *Atherosclerosis* **2013**, *229*, 524–530. [CrossRef] [PubMed]
14. Shin, J.Y.; Xun, P.; Nakamura, Y.; He, K. Egg consumption in relation to risk of cardiovascular disease and diabetes: A systematic review and meta-analysis. *Am. J. Clin. Nutr.* **2013**, *98*, 146–159. [CrossRef]
15. Gibin, D.; Simonetto, A.; Zanini, B.; Gilioli, G. A framework assessing the footprints of food consumption. An application on water footprint in Europe. *Environ. Impact Assess. Rev.* **2022**, *93*, 106735. [CrossRef]
16. Gonzales-González, V.A.; Díaz, A.M.; Fernández, K.; Rivera, M.F. Prevalence of food allergens sensitization and food allergies in a group of allergic Honduran children. *Allergy Asthma Clin. Immunol.* **2018**, *14*, 23. [CrossRef]
17. Alu'datt, M.H.; Rababah, T.; Alhamad, M.N.; Ereifej, K.; Gammoh, S.; Kubow, S.; Tawalbeh, D. Preparation of mayonnaise from extracted plant protein isolates of chickpea, broad bean and lupin flour: Chemical, physiochemical, nutritional and therapeutic properties. *J. Food Sci. Technol.* **2017**, *54*, 1395–1405. [CrossRef]
18. Aluko, R.E.; McIntosh, T. Limited enzymatic proteolysis increases the level of incorporation of canola proteins into mayonnaise. *Innov. Food Sci. Emerg. Technol.* **2005**, *6*, 195–202. [CrossRef]
19. Sørensen, A.M.; Nielsen, N.S.; Hyldig, G.; Jacobsen, C. Influence of emulsifier type on lipid oxidation in fish oil-enriched light mayonnaise. *Eur. J. Lipid Sci. Technol.* **2010**, *112*, 1012–1023. [CrossRef]
20. Rahmati, K.; Mazaheri Tehrani, M.; Daneshvar, K. Soy milk as an emulsifier in mayonnaise: Physico-chemical, stability and sensory evaluation. *J. Food Sci. Technol.* **2014**, *51*, 3341–3347. [CrossRef]
21. Karshenas, M.; Goli, M.; Zamindar, N. Substitution of sesame and peanut defatted-meal milk with egg yolk and evaluation of the rheological and microstructural properties of low-cholesterol mayonnaise. *Food Sci. Technol. Int.* **2019**, *25*, 633–641. [CrossRef] [PubMed]
22. Ali, M.R.; EL Said, R.M. Assessment of the potential of Arabic gum as an antimicrobial and antioxidant agent in developing vegan "egg-free" mayonnaise. *J. Food Saf.* **2020**, *40*, e12771. [CrossRef]
23. Raikos, V.; Hayes, H.; Ni, H. Aquafaba from commercially canned chickpeas as potential egg replacer for the development of vegan mayonnaise: Recipe optimisation and storage stability. *Int. J. Food Sci. Technol.* **2020**, *55*, 1935–1942. [CrossRef]
24. He, Y.; Shim, Y.Y.; Mustafa, R.; Meda, V.; Reaney, M.J.T. Chickpea cultivar selection to produce aquafaba with superior emulsion properties. *Foods* **2019**, *8*, 685. [CrossRef]
25. Damian, J.J.; Huo, S.; Serventi, L. Phytochemical content and emulsifying ability of pulses cooking water. *Eur. Food Res. Technol.* **2018**, *244*, 1647–1655. [CrossRef]
26. He, Y.; Meda, V.; Reaney, M.J.T.; Mustafa, R. Aquafaba, a new plant-based rheological additive for food applications. *Trends Food Sci. Technol.* **2021**, *111*, 27–42. [CrossRef]
27. Muehlbauer, F.J.; Sarker, A. Economic importance of chickpea: Production, value, and world Trade. In *The Chickpea Genome*; Varshney, R.K., Thudi, M., Muehlbauer, F., Eds.; Compendium of Plant Genomes; Springer International Publishing: Cham, Switzerland, 2017; pp. 5–12. [CrossRef]

28. Buhl, T.F.; Christensen, C.H.; Hammershøj, M. Aquafaba as an egg white substitute in food foams and emulsions: Protein composition and functional behavior. *Food Hydrocoll.* **2019**, *96*, 354–364. [CrossRef]
29. Meurer, M.C.; de Souza, D.; Ferreira Marczak, L.D. Effects of ultrasound on technological properties of chickpea cooking water (aquafaba). *J. Food Eng.* **2020**, *265*, 109688. [CrossRef]
30. He, Y.; Purdy, S.K.; Tse, T.J.; Tar'an, B.; Meda, V.; Reaney, M.J.T.; Mustafa, R. Standardization of aquafaba production and application in vegan mayonnaise analogs. *Foods* **2021**, *10*, 1978. [CrossRef]
31. Cheung, L.; Wanasundara, J.; Nickerson, M.T. The effect of pH and nacl levels on the physicochemical and emulsifying properties of a cruciferin protein isolate. *Food Biophys.* **2014**, *9*, 105–113. [CrossRef]
32. Pearce, K.N.; Kinsella, J.E. Emulsifying properties of proteins: Evaluation of a turbidimetric technique. *J. Agric. Food Chem.* **1978**, *26*, 716–723. [CrossRef]
33. *PN-75-A-04018:1975/Az3:2002*; Agri-Food Products—Determination of Nitrogen Content by the Kjeldahl Method and the Conversion of Protein. Polish Committee for Standardization: Warszawa, Poland, 2002.
34. *ISO 5508*; Animal and Vegetable Fats and Oils—Analysis by Gas Chromatography of Methyl Esters of Fatty Acid. International Organization for Standardization: Geneva, Switzerland, 1996.
35. *ISO 5509*; Animal and Vegetable Fats and Oils—Preparation of Methyl Esters of Fatty Acids. International Organization for Standardization: Geneva, Switzerland, 2000.
36. Szydłowska-Czerniak, A.; Tułodziecka, A.; Momot, M.; Stawicka, B. Physicochemical, antioxidative, and sensory properties of refined rapeseed oils. *J. Am. Oil Chem. Soc.* **2019**, *96*, 405–419. [CrossRef]
37. Alizadeh, L.; Abdolmaleki, K.; Nayebzadeh, K.; Shahin, R. Effects of tocopherol, rosemary essential oil and ferulago angulata extract on oxidative stability of mayonnaise during its shelf life: A comparative study. *Food Chem.* **2019**, *285*, 46–52. [CrossRef] [PubMed]
38. *ISO 3960*; Animal and Vegetable Fats and Oils—Determination of Peroxide Value–Iodometric (Visual) Endpoint Determination. International Organization for Standardization: Geneva, Switzerland, 2017.
39. *ISO 6885*; Animal and Vegetable Fats and Oils—Determination of Anisidine Value. International Organization for Standardization: Geneva, Switzerland, 2016.
40. *ISO 660*; Animal and Vegetable Fats and Oils—Determination of Acid Value and Acidity. International Organization for Standardization: Geneva, Switzerland, 2020.
41. Yan, W.; Qiao, L.; Gu, X.; Li, J.; Xu, R.; Wang, M.; Reuhs, B.; Yang, Y. Effect of high pressure treatment on the physicochemical and functional properties of egg yolk. *Eur. Food Res. Technol.* **2010**, *231*, 371–377. [CrossRef]
42. Suhag, R.; Dhiman, A.; Thakur, D.; Kumar, A.; Upadhyay, A. Physico-chemical and functional properties of microfluidized egg yolk. *J. Food Eng.* **2020**, *294*, 110416. [CrossRef]
43. Karaca, A.C.; Low, N.; Nickerson, M. Emulsifying properties of chickpea, faba bean, lentil and pea proteins produced by isoelectric precipitation and salt extraction. *Food Res. Int.* **2011**, *44*, 2742–2750. [CrossRef]
44. Mustafa, R.; He, Y.; Shim, Y.Y.; Reaney, M.J.T. Aquafaba, wastewater from chickpea canning, functions as an egg replacer in sponge cake. *Int. J. Food Sci. Technol.* **2018**, *53*, 2247–2255. [CrossRef]
45. Xu, Y.; Obielodan, M.; Sismour, E.; Arnett, A.; Alzahrani, S.; Zhang, B. Physicochemical, functional, thermal and structural properties of isolated kabuli chickpea proteins as affected by processing approaches. *Int. J. Food Sci. Technol.* **2017**, *52*, 1147–1154. [CrossRef]
46. Bouyer, E.; Mekhloufi, G.; Rosilio, V.; Grossiord, J.-L.; Agnely, F. Proteins, polysaccharides, and their complexes used as stabilizers for emulsions: Alternatives to synthetic surfactants in the pharmaceutical field? *Int. J. Pharm.* **2012**, *436*, 359–378. [CrossRef]
47. Günal-Köroğlu, D.; Turan, S.; Capanoglu, E. Interaction of lentil protein and onion skin phenolics: Effects on functional properties of proteins and in vitro gastrointestinal digestibility. *Food Chem.* **2022**, *372*, 130892. [CrossRef]
48. Chung, C.; Sher, A.; Rousset, P.; Decker, E.A.; McClements, D.J. Formulation of food emulsions using natural emulsifiers: Utilization of quillaja saponin and soy lecithin to fabricate liquid coffee whiteners. *J. Food Eng.* **2017**, *209*, 1–11. [CrossRef]
49. Bors, M.D.; Semeniuc, C.A.; Socaci, S.; Vlaic, R.; Moldovan, O.P.; Pop, A.V.; Tofa, M. The influence of variety and processing on the total phenolic content and antioxidant activity of mustard. *Rom. Biotechnol. Lett.* **2017**, *22*, 12827–12833.
50. Xu, B.; Chang, S.K.C. Effect of soaking, boiling, and steaming on total phenolic contentand antioxidant activities of cool season food legumes. *Food Chem.* **2008**, *110*, 1–13. [CrossRef] [PubMed]
51. Rabiej-Kozioł, D.; Krzemiński, M.P.; Szydłowska-Czerniak, A. Synthesis of steryl hydroxycinnamates to enhance antioxidant activity of rapeseed oil and emulsions. *Materials* **2020**, *13*, 4536. [CrossRef]
52. De Bruno, A.; Romeo, R.; Gattuso, A.; Piscopo, A.; Poiana, M. Functionalization of a vegan mayonnaise with high value ingredient derived from the agro-industrial sector. *Foods* **2021**, *10*, 2684. [CrossRef]
53. Bolarin, F.M.; Oke, M.O. Effect of addition of moringa (*Moringa oleifera* L.) seed flour on the quality characteristics of mayonnaise. *IJMRAP* **2018**, *1*, 11–17. [CrossRef]
54. McClements, D.J. *Food Emulsions: Principles, Practices, and Techniques*, 3rd ed.; CRC Press: Boca Raton, FL, USA, 2015. [CrossRef]

55. Yildirim, M.; Sumnu, G.; Sahin, S. Rheology, particle-size distribution, and stability of low-fat mayonnaise produced via double emulsions. *Food Sci. Biotechnol.* **2016**, *25*, 1613–1618. [CrossRef]
56. Armaforte, E.; Hopper, L.; Stevenson, G. Preliminary investigation on the effect of proteins of different leguminous species (cicer arietinum, vicia faba and lens culinarius) on the texture and sensory properties of egg-free mayonnaise. *LWT-Food Sci. Technol.* **2021**, *136*, 110341. [CrossRef]

foods

MDPI

Article

Improvement of Oil Valorization Extracted from Fish By-Products Using a Handheld near Infrared Spectrometer Coupled with Chemometrics

Sonia Nieto-Ortega [1,*], Idoia Olabarrieta [1], Eduardo Saitua [1], Gorka Arana [2], Giuseppe Foti [1] and Ángela Melado-Herreros [1]

[1] AZTI, Food Research, Basque Research and Technology Alliance (BRTA), Parque Tecnológico de Bizkaia, Astondo Bidea, Edificio 609, 48160 Derio, Spain; iolabarrieta@azti.es (I.O.); saitua@azti.es (E.S.); gfoti@azti.es (G.F.); amelado@azti.es (Á.M.-H.)

[2] Department of Analytical Chemistry, University of the Basque Country UPV/EHU, 48080 Bilbao, Spain; gorka.arana@ehu.eus

* Correspondence: snieto@azti.es; Tel.: +34-667-174-323

Abstract: A handheld near infrared (NIR) spectrometer was used for on-site determination of the fatty acids (FAs) composition of industrial fish oils from fish by-products. Partial least square regression (PLSR) models were developed to correlate NIR spectra with the percentage of saturated fatty acids (SFAs), monounsaturated fatty acids (MUFAs), polyunsaturated fatty acids (PUFAs) and, among them, omega-3 (ω-3) and omega-6 (ω-6) FAs. In a first step, the data were divided into calibration validation datasets, obtaining good results regarding R^2 values, root mean square error of prediction (RMSEP) and bias. In a second step, all these data were used to create a new calibration, which was uploaded to the handheld device and tested with an external validation set in real time. Evaluation of the external test set for SFAs, MUFAs, PUFAs and ω-3 models showed promising results, with R^2 values of 0.98, 0.97, 0.97 and 0.99; RMSEP (%) of 0.94, 1.71, 1.11 and 0.98; and bias (%) values of −0.78, −0.12, −0.80 and −0.67, respectively. However, although ω-6 models achieved a good R^2 value (0.95), the obtained RMSEP was considered high (2.08%), and the bias was not acceptable (−1.76%). This was corrected by applying bias and slope correction (BSC), obtaining acceptable values of R^2 (0.95), RMSEP (1.09%) and bias (−0.05%). This work goes a step further in the technology readiness level (TRL) of handheld NIR sensor solutions for the fish by-product recovery industry.

Keywords: no-waste; omega-3; circular economy; smart sensors; reuse; fish oil industry; recovery; chemometrics; lipid profile

Citation: Nieto-Ortega, S.; Olabarrieta, I.; Saitua, E.; Arana, G.; Foti, G.; Melado-Herreros, Á. Improvement of Oil Valorization Extracted from Fish By-Products Using a Handheld near Infrared Spectrometer Coupled with Chemometrics. *Foods* **2022**, *11*, 1092. https://doi.org/10.3390/foods11081092

Academic Editors: Marco Poian, Francesco Caponio and Antonio Piga

Received: 25 February 2022
Accepted: 8 April 2022
Published: 10 April 2022

Publisher's Note: MDPI stays neutral with regard to jurisdictional claims in published maps and institutional affiliations.

Copyright: © 2022 by the authors. Licensee MDPI, Basel, Switzerland. This article is an open access article distributed under the terms and conditions of the Creative Commons Attribution (CC BY) license (https://creativecommons.org/licenses/by/4.0/).

1. Introduction

Worldwide fisheries production and global per capita fish consumption have highly incremented in recent years [1]. The industrialization of the fish sector has brought not only a huge development but also an increase in the number of by-products generated during fish processing [2]. It is estimated that more than 70% of total fish captures are processed, of the processed products, about 50% result in solid waste and by-products [3]. These by-products are usually composed of viscera, heads, cut-offs, skin and fish that is damaged or unsuitable for human consumption [4]. Moreover, an additional source of by-products is represented by unwanted, non-targeted fish species (by-catches) that cannot be commercialized for direct human consumption [5]. These large quantities of unused fish products create serious pollution and environmental problems. Therefore, their correct reuse must become a priority for fish-processing countries and companies [6].

Most of these by-products should not be considered waste or less valuable materials [7], as they have great potential to be reused for higher-value applications [8]. Due to their high nutritive value, it is possible to give them a second life [7]. These secondary products can be processed into products such as fish sausages, pâté, cakes, gelatin, soups, sauces or snacks (i.e., the consumption of small fish bones with a minimum amount of meat as snacks, which is actually done in some countries [9]).

The production of omega-3 (ω-3)-rich fish oils represents an opportunity for valorizing fish by-products [3] and to achieve the zero-waste goal. The estimated amount of oil present in fish by-products varies from 2% to 30% of the total composition, depending on many factors, such as the fat content of the fish species and the distribution of fat in the fish body [10]. Fish oil is usually a good source of long-chain polyunsaturated fatty acids (PUFAs) [11], in particular docosahexaenoic acid (DHA) and eicosapentaenoic acid (EPA) [12]. It can be reused to generate products of high added value for the pharmaceutical industry and as raw material for food supplements [13]. Therefore, the characterization of the fatty acids (FAs) profile of fish oils is essential because EPA and DHA levels determine the destination of the product and therefore its market value [12]. Such characterization is crucial when fish oil is obtained from canning industry by-products, where it is mixed with vegetable oils, which may change the oil FAs profiles, reducing the ω-3 proportion.

Nowadays, the most common technique used to analyze the FAs profile of fish oil is gas chromatography with a flame ionization detector (GC-FID) [14], a complex technique that is relatively slow and generates toxic waste [15]. Thus, a simpler and faster technique capable of providing a response in real time would allow companies to quickly assess the FAs profile of oil and determine its most convenient destination. In this sense, near-infrared spectroscopy (NIRS) represents a valid alternative to GC-FID or other more traditional methods, as it is a rapid, non-destructive technique [16,17] that has been used in recent years in industry for quality control and process monitoring [18–20]. Furthermore, recent advances have allowed for a significant reduction in the size and cost of such devices, making them suitable for on-site determination [21].

This is not the first time that NIR has been used for the evaluation of the lipid profile of fish derivates. Some authors, such as Bekhit et al. and van der Merwe et al. [22,23], have studied NIR to analyze PUFAs in ω-3 supplements. Others, including dos Santos et al., analyzed the ω-3 and omega-6 (ω-6) content directly in fish fillets [24]. Other techniques, such as FT-NIR, were used by Karunathilaka et al. and Cascant et al. [14,25] to analyze omega-3 supplements and salmon. Only a few authors have used NIR to directly analyze fish oils [11] or used portable spectrometers [26]. Most research has been developed with big laboratory equipment and/or using processed fish pharmaceutical supplements or fish fillets, which prevents their use for this application in an actual industrial environment in the short term. More efforts are still needed to elevate the low technology readiness levels (TRLs) of such studies to be useful for the by-products industry. To the best of our knowledge, this is the first study demonstrating the scalability to industrial TRLs of NIR technology for measurement of the lipid profile of fish oil directly extracted from fish by-products.

Therefore, the principal objective of this study is to assess the potential of a portable device based on NIRS in combination with a partial least square regression (PLSR) analysis to characterize the FAs profile of fish oils in a rapid and non-destructive way. Thus, the device was not only calibrated to determine the ω-3 and ω-6 content but also to measure the complete fish oil profile, determining the saturated fatty acids (SFAs), monounsaturated fatty acids (MUFAs) and polyunsaturated fatty acids (PUFAs) levels. The main objective of this work was to demonstrate the high level of maturity of a handheld NIR spectroscopy sensor in combination with chemometrics for the rapid characterization of fish oil in the fish by-product industry. This technique could enable a fast and accurate classification of processed products in the appropriate market category with economic benefits for the company and increased efficiency.

2. Materials and Methods

2.1. Samples and Reagents

2.1.1. Oil Mixture Preparation

Samples were supplied by a local company, which collects and reuses fishing surplus and fish industry by-products from different industrial activities. Eight fish oil samples (named with consecutive letters from A to H) obtained from fish by-products were used to make 269 different mixtures. The origin of the fish species of the oils, as well as the industries and processes from which they came, were unknown. These samples were divided in two sets: calibration (172 mixtures) and validation (97 mixtures). For external validation purposes and to ensure the robustness of the calibration, 29 new mixtures were made. The set of mixtures used for this aim was composed of three out of eight of the previous fish oils, together with a new oil (I) and two additional commercial fish oil supplements (named Supplement A and Supplement B) (Table 1).

Table 1. Samples of oil mixtures used in each dataset.

	Calibration Set	Validation Set	External Validation Set
Number of mixtures	172	97	29
Oils and supplements used	A, B, C, D, E, F and G	B, E, F, G and H	B, E, F, I, Supplement A and Supplement B

The volume of the prepared oil mixtures was at least 3 mL. Therefore, different volumes of the initial oils were taken and mixed using automatic pipettes. The minimum amount of oil used for the mixtures was 0.1 mL, and the maximum was 2.9 mL. For some mixtures, only 2 oils were used, and the maximum number of oils used in a mixture was 6. The percentage of oil in each mixture was formulated so that the range of the mixtures covered all possible variability. All samples were filtered with Whatman grade 1 filter paper before analysis.

2.1.2. Reagents

The reagents used for the methylation process of the FAs were methanol, sodium chloride, hydrochloric acid, phenolphthalein (Thermo Fisher Scientific™, Roskilde, Denmark) and sodium methylate (ACROS organics™, part of Thermo Fisher Scientific™, Geel, Belgium). For the chromatographic analysis, n-Hexane (Thermo Fisher Scientific™, Roskilde, Denmark) was used as a solvent.

2.2. Reference Analysis

GC-FID was employed as the reference method to analyze the fat profile of the fish oils. To extract the FAs from the oils and transform them into fatty acid methyl esters (FAMEs), the methylation process described in Commission Regulation (EC) No. 796/2002 (2002), method B, was used with some modifications [27]. In this procedure, 80 mg of sample were transferred to a flat-bottom flask, where 8 mL of sodium methylate in methanol (0.6 mol/L) and some pumice stones were added. The mixture was boiled with a reflux condenser for 10 min. Once the mixture was chilled, two drops of phenolphthalein were incorporated, and a solution of hydrochloric acid in methanol (3.5%) was added until the solution became colorless, a sign of complete acidification. The sample mixture was boiled again under the same conditions, and when cooled, 8 mL of n-hexane was added with 5 mL of a concentrate solution of sodium chloride, shaking the mixture vigorously for 1 min. Finally, the same concentrate solution of sodium chloride was added to elevate the organic phase, which contained the FAMEs, and it was transferred to a gas chromatograph vial before injection.

The solution with the FAMEs was analyzed in a gas chromatograph (Agilent 5890 from Agilent Technologies Inc., Santa Clara, CA, USA) with a DB 23 column (60 m × 0.25 mm id × 0.25 μm from J&W scientific, Santa Clara, CA, USA), a flame ionization detector (FID) and helium as carrier gas (at 30 psi and a flow rate of 1.2 mL/min). To conduct the chromatographic analysis, 2 μL of sample was injected in split mode (split of 80 mL/min) at 220 °C. The initial temperature of the chromatograph oven was 40 °C, which was maintained for 3 min. The temperature was increased at a rate of 25 °C per minute, up to 125 °C, where it was maintained for 2 min. Next, the temperature was increased again, this time at a rate of 4 °C per minute, and maintained at 180 °C for 1 min. The last temperature increase was at a rate of 1 °C per minute, up to 215 °C, where it was maintained for 10 min. Finally, the temperature of the detector was increased to 250 °C. Each GC-FID analysis was conducted 3 times.

Data from the chromatogram were collected with ChemStation Software (version A.10.02) from Agilent, (Santa Clara, CA, USA). The FAMEs of the oils, which were equal to their respective FAs, were identified with FAMEs chromatographic external standards from Sigma-Aldrich (PUFA No. 3 From Menhaden Oil and Supelco 37 Component FAME Mix). Then, the area under each FA peak was integrated in the chromatogram, and the percentage of the total oil represented by each area was calculated. Afterwards, the percentage of FAs that belonged to the same group was summed to obtain the final percentage of all the categories (SFAs, MUFAs, PUFAs, ω-3 and ω-6) for each oil. The FAs of each group are presented in Table 2.

Table 2. Identified FAs in each category.

FAs Group	Fatty Acids
SFAs	Myristic (14:0), Palmitic (16:0), Stearic (18:0), Arachidic (20:0)
MUFAs	Palmitoleic (16:1), Oleic (18:1), Gadoleic (20:1), Erucic (22:1)
PUFAs	Linoleic (18:2), Gamma-linolenic (18:3), Stearidonic (18:4), Arachidonic (20:4), EPA (20:5), Clupanodonic (22:5), DHA (22:6).
ω-3	Alpha-linoleic (18:3), Stearidonic (18:4), EPA (20:5), Clupanodonic (22:5), DHA (22:6)
ω-6	Linoleic (18:2), Arachidonic (20:4)

SFAs: saturated fatty acids, MUFAs: monounsaturated fatty acids, PUFAs: polyunsaturated fatty acids, ω-3: omega-3 fatty acids, ω-6: omega-6 fatty acids.

Due to the complexity of the chromatographic method (time-consuming) and the need to analyze many samples to create a robust chemometric model, only the initial pure oils were analyzed (from A to I), and the composition of each oil mixture was calculated afterwards. To ensure that the composition of the mixtures was correct, a few were chosen randomly and analyzed.

2.3. NIRS Data Acquisition

A compact, handheld NIR spectrometer device was used (MicroNIR OnSite, developed by VIAVI Solutions Inc., Monza, Italy), working from 900 to 1650 nm, with a resolution of 6 nm. Samples were scanned in transflectance mode, with a special accessory for liquids (MicroNIR side-view vial holder by VIAVI) in small glass vials. A dark measurement (acquired with the lamp turned off) and a white diffuse reflectance standard (a white reference with 99% reflectance) were used for calibration. Each spectrum was the average of 100 scans, with an integration time of 8.2 ms. All spectra were taken in duplicate. For the external validation, the average of two spectra was considered for each sample.

2.4. Model Building

Models were developed for each FAs category: SFAs, MUFAs. PUFAs, ω-3 and ω-6. Therefore, the spectral data were considered as X, whereas the data obtained from the chromatographic analysis (the percentage of each FAs category for each model) were considered as Y.

X data were preprocessed before the multivariate analysis. Several methods were tested, such as standard normal variate without and with detrend (SNV and SNVd), multiplicative scatter correction (MSC), Saviztky–Golay first and second derivatives (with different polynomial orders and windows) and combinations of all of them. This step was necessary because it eliminates the irrelevant information that cannot be correctly processed [28], and it improves the regression [29]. X and Y data were mean centered in all cases before creation of the models.

To correlate the NIR spectra and the reference data (SFAs, MUFAs, PUFAs, ω-3 and ω-6 percentage), several partial least square regression (PLSR) models were developed [30]. For each developed model, two steps were followed:

- 1st step: The five models were built using Matlab R2013a equipped with the PLS_toolbox (version 8.2.1) (The Mathworks, Natick, USA). For the calibration, $n_{c.1}$ = 172 samples were used, and one model was developed for each FAs category. In all cases, a venetian blinds cross-validation (CV) with 10 data splits and 2 samples per blind was carried out. Then, the model was validated using the validation set ($n_{v.1}$ = 97).
- 2nd step: All the data used in the previously developed models ($n_{c.1}$ and $n_{v.1}$) were used to create a new dataset, which was used as calibration dataset ($n_{c.2}$ = 269). Then, a random CV with 20 segments and 27 samples per segment was carried out. These models were uploaded into the MicroNIR OnSite to directly predict an external dataset (n_{test} = 29) in real time in the place of analysis and without the necessity of extracting the data from the spectrometer and analyzing it afterwards in a computer. To build the mentioned calibration model, The Unscrambler® X 10.5.1 software was used (CAMO Software AS, Oslo, Norway).

As figures of merit of the models, the coefficient of determination (R^2), the root mean square error (RMSE) and the bias value were calculated for the CV and the prediction. To study the distribution of the oil mixtures used in each dataset, their mean, standard deviation, minimum and maximum values were calculated and expressed in percentage.

3. Results and Discussion

3.1. Determination of Fatty Acid Profiles of the Samples by Reference Analysis

3.1.1. Fatty Acid Composition of the Initial Oils

Table 3 displays the composition (% of the FAs categories) of the nine initial oils from which the mixtures were made and the two commercial supplements of ω-3.

Table 3. Composition (%) of the initial oils with their standard deviation.

Oils	SFAs	MUFAs	PUFAs	ω-3	ω-6
A	27.84 ± 0.32	42.05 ± 0.08	30.11 ± 0.25	27.50 ± 0.23	2.61 ± 0.03
B	14.26 ± 0.08	53.76 ± 0.12	31.98 ± 0.05	14.18 ± 0.05	17.80 ± 0.02
C	29.27 ± 0.30	25.60 ± 0.14	45.13 ± 0.30	42.24 ± 0.31	2.90 ± 0.00
D	17.47 ± 0.19	49.26 ± 0.03	33.27 ± 0.21	17.70 ± 0.13	15.57 ± 0.09
E	21.09 ± 0.26	44.94 ± 0.52	33.98 ± 0.30	22.27 ± 0.10	11.70 ± 0.32
F	29.66 ± 0.19	25.58 ± 0.09	44.76 ± 0.15	41.88 ± 0.14	2.89 ± 0.02
G	17.51 ± 0.21	49.10 ± 0.16	33.39 ± 0.07	17.87 ± 0.05	15.52 ± 0.03
H	18.51 ± 0.40	48.75 ± 0.19	32.74 ± 0.22	17.43 ± 0.14	15.31 ± 0.07
I	18.10 ± 0.08	49.18 ± 0.31	32.73 ± 0.24	17.50 ± 0.08	15.23 ± 0.17
Supplement A	14.34 ± 0.02	56.56 ± 0.30	29.10 ± 0.32	14.25 ± 0.39	14.85 ± 0.08
Supplement B	29.50 ± 0.02	25.32 ± 0.03	45.19 ± 0.05	42.08 ± 0.07	3.11 ± 0.01

For each sample, three replications were performed. SFAs: saturated fatty acids, MUFAs: monounsaturated fatty acids, PUFAs: polyunsaturated fatty acids, ω-3: omega-3 fatty acids, ω-6: omega-6 fatty acids.

Oil samples show high variability in their FAs profiles. This suggests that the oils collected for this study might have different origins and could come from different kinds of fish, production methods or various types of processing industries. This sample variability highlights the importance of determining the lipid profile of fish oils, as the percentage of the different groups, especially ω-3, varies significantly between samples.

On the one hand, oils A, C and F and Supplement B showed the typical seawater fish oil composition regarding PUFAs, where most of the PUFAs come from ω-3 FAs [31–33]. In these samples, PUFAs represented between 30% and 46% of the total FAs of the oils. They had an elevated ω-3 content, which almost corresponded with all the PUFAs in the samples, and a lower content of ω-6. Considering their composition, these samples may come from a process where only seawater fish is involved, i.e., fish fillet processing [31,32].

On the other hand, samples B, D, E, G, H and I and supplement A had PUFAs content between 25% and 34%, which is also typical in fish oil [31]. However, these samples presented a higher content of ω-6 FAs than the previous set, in which ω-3 FAs were predominant. This is due to a high level of linoleic acid (18:2) (data not shown), which may have two explanations: on the one hand, it might be due to the fish species from which the oil was obtained, i.e., this PUFAs profile is characteristic of freshwater fish [34], which has a higher ω-6 content in comparison with seawater fish [32,33]. On the other hand, it might be due to the type of fish processing industry from which the samples originated. In the canning industry, by-products of fish oil are mixed with vegetable oils, such as sunflower oil, which is rich in ω-6 PUFAs (linoleic acid) [35,36].

3.1.2. Fatty Acid Profile of the Oil Mixtures

The results of the characterization and the statistics of the oil samples in the different sets of data used in the models are shown in Table 4.

Table 4. Results of the characterization of all the samples.

		Dataset	*n*	Mean ± SD (%)	Minimum (%)	Maximum (%)
SFAs	1st Step	Calibration	172	22.0 ± 3.8	14.3	29.7
		Validation	97	24.1 ± 3.8	14.4	29.7
	2nd Step	Calibration	269	22.7 ± 3.8	14.3	29.7
		External Validation	29	20.5 ± 4.8	14.3	29.7
MUFAs	1st Step	Calibration	172	41.4 ± 6.9	25.6	53.8
		Validation	97	37.2 ± 7.4	25.6	53.6
	2nd Step	Calibration	269	40.1 ± 7.1	25.6	53.8
		External Validation	29	43.7 ± 9.2	25.3	56.6
PUFAs	1st Step	Calibration	172	36.5 ± 3.6	30.1	45.1
		Validation	97	38.6 ± 3.6	32.0	44.8
	2nd Step	Calibration	269	37.2 ± 3.6	30.1	45.1
		External Validation	29	35.8 ± 4.4	29.1	45.2
ω-3	1st Step	Calibration	172	26.2 ± 6.9	14.2	42.2
		Validation	97	30.0 ± 7.4	14.3	41.9
	2nd Step	Calibration	269	27.4 ± 7.0	14.2	42.2
		External Validation	29	23.7 ± 9.0	14.2	42.1
ω-6	1st Step	Calibration	172	10.3 ± 3.8	2.6	17.8
		Validation	97	8.6 ± 3.9	2.9	17.7
	2nd Step	Calibration	269	9.8 ± 3.8	2.6	17.8
		External Validation	29	12.1 ± 4.7	2.9	17.8

n: number of samples, SD: standard deviation, SFAs: saturated fatty acids, MUFAs: monounsaturated fatty acids, PUFAs: polyunsaturated fatty acids, ω-3: omega-3 fatty acids, ω-6: omega-6 fatty acids.

MUFAs constitute the majority group in most cases representing: in the 1st step, 41.4% of total FAs composition on the calibration set and, in the 2nd step, the 40.1% of the calibration set and the 43.7% of the external validation set. However, PUFAs are the majority group in the validation set of the 1st step, with a percentage of 38.6%. On the contrary, ω-6 is the least common group in the four sets of samples, with percentages of 10.3% and 8.6% in the calibration and validation set of the first step, respectively, and 9.8% and 12.1% in the calibration and external validation set of the second step, respectively.

3.2. Performance of the PLSR Models of the Target Oils

3.2.1. Model Results

The CV and validation results of the five models developed for each category of FAs in the first step are shown in Table 5.

Table 5. Principal statistics of the five models developed in the first step.

		X Preprocessing	Y Preprocessing	LV	R2	RMSE (%)	Bias (%)
SFAs	CV Validation	2nd derivative (order 2, window 5) + Mean Center	Mean Center	5	0.98 0.98	0.57 0.68	-2×10^{-3} −0.40
MUFAs	CV Validation	SNV + Mean Center	Mean Center	3	0.99 0.97	0.74 1.27	-3×10^{-4} 0.25
PUFAs	CV Validation	SNV + 2nd derivative (order 2, window 15) + Mean Center	Mean Center	5	0.97 0.96	0.65 0.85	2×10^{-4} −0.49
ω-3	CV Validation	SNV + 2nd derivative (order 2, window 15) + Mean Center	Mean Center	6	0.99 0.99	0.48 0.60	-2×10^{-3} −0.26
ω-6	CV Validation	MSC (using the mean of the spectra as reference) + 1st derivative (order 2 window 5) + Mean Center	Mean Center	6	0.96 0.95	0.78 0.90	0.02 −0.34

LV: latent variables, CV: cross-validation, SNV: standard normal variate, MSC: multiplicative scatter correction, RMSE: root mean square error, SFAs: saturated fatty acids, MUFAs: monounsaturated fatty acids, PUFAs: polyunsaturated fatty acids, ω-3: omega-3 fatty acids, ω-6: omega-6 fatty acids.

In Table 5, all the models developed for the validation of SFAs, MUFAs, PUFAs, ω-3 and ω-6 achieved good results, with R^2_{val} values of 0.98, 0.97, 0.96, 0.99 and 0.95; small errors of 0.68, 1.27, 0.85, 0.60 and 0.90; and a low bias value of −0.40, 0.25, −0.49, −0.26, −0.34, respectively.

On the other hand, the results of the five models developed in the second step (CV and external validation) are shown in Table 6.

Table 6. Principal statistics calculated for the five models developed in the second step.

		X Pretreatment	Y Pretreatment	LV	R2	RMSE (%)	Bias (%)
SFAs	CV External validation	2nd derivative (order 2, window 5) + Mean Center	Mean Center	5	0.98 0.98	0.60 0.94	-4×10^{-3} −0.78
MUFAs	CV External validation	SNV + Mean Center	Mean Center	3	0.99 0.97	0.77 1.71	5×10^{-4} −0.12
PUFAs	CV External validation	SNV + 2nd derivative (order 2, window 15) + Mean Center	Mean Center	5	0.97 0.97	0.65 1.11	2×10^{-3} −0.80
ω-3	CV External validation	SNV + 2nd derivative (order 2, window 15) + Mean Center	Mean Center	6	0.99 0.99	0.71 0.98	-5×10^{-6} −0.67
ω-6	CV External validation	MSC (using the mean of the spectra as reference) + 1st derivative (order 2 window 5) + Mean Center	Mean Center	6	0.96 0.95	0.74 2.09	-1×10^{-4} −1.76

LV: latent variables, CV: cross-validation, SNV: standard normal variate, MSC: multiplicative scatter correction, RMSE: root mean square error, SFAs: saturated fatty acids, MUFAs: monounsaturated fatty acids, PUFAs: polyunsaturated fatty acids, ω-3: omega-3 fatty acids, ω-6: omega-6 fatty acids.

In this case (Table 6), models for SFAs, MUFAs, PUFAs and ω-3 achieved good results in the external validation set regarding R^2 (0.98, 0.97, 0.97 and 0.99), RMSEP (0.94%, 1.71%, 1.11% and 0.98%) and bias (−0.78%, −0.12%, −0.30% and −0.67%), respectively.

Although the ω-6 model achieved good results in terms of R^2, the RMSEP and the bias in the validation showed high values: 2.09% and −1.76%, respectively. This is very common in quantitative NIRS and may be due to block effects occurring between measuring conditions [37]. In this case, there are two possible reasons for these effects. (i) The measurement conditions: all the measurements were performed in a laboratory under controlled temperature; therefore, the authors believe they might have a small effect. (ii) The possibly different origins of the oils, including different fish species and different processing industries. Seawater fish, the most consumed type of fish, is naturally low in ω-6 FAs, with most PUFAs resulting from the presence of ω-3 FAs [38,39]. However, as stated in Section 3.1.1, some of the fish oil samples had a higher content of ω-6 FAs. This finding could result from: (i) the presence of vegetable oils mixed with the fish oil, which is plausible if some of the samples came from the canning industry or (ii) the presence of samples from industries where the raw material is freshwater fish. However, the model can be corrected using techniques such as bias and slope correction (BSC) [40]. Applying this technique to the external test set (Figure 1), the following results are obtained: $R^2 = 0.95$; RMSEP = 1.09%; bias = −0.05%.

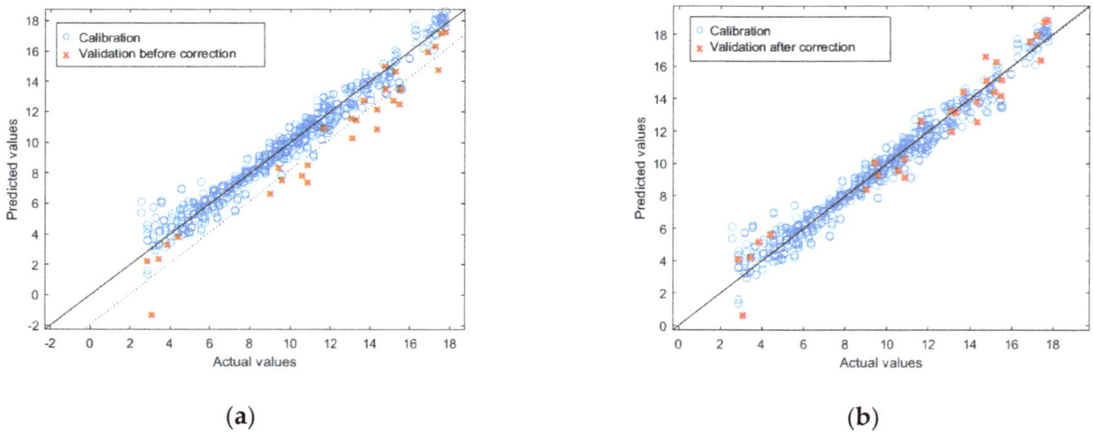

Figure 1. Results of the external validation of the PLSR for prediction of ω-6 before (**a**) and after (**b**) bias and slope correction.

These results are in accordance with those of other studies found in the literature that studied the fish oil profile of different matrices. In dietary supplements, Hespanhol et al. [26] and Bekhit et al. [22] obtained similar R^2 values (0.97 and 0.98, respectively) for ω-3 prediction, although their models were less complex, with one and two latent variables (LVs), respectively. The differences in complexity may be due to the fact that in the present study, the fish oil was analyzed directly from by-products with no previous processing (cleaning, refining, etc.), as it was made with dietary supplements. The results from the MUFAs, ω-3 and ω-6 models are similar to those obtained by Karunathilaka et al. [14] in dietary supplements, with RMSEP values of 1.03, 1.42 and 0.93, respectively. In other matrices, such as the model system created by Afseth et al. (using 70 different mixtures of protein, water and oil blends) [41], the error obtained for SFAs, MUFAs and PUFAs was similar to our results, with RMSEP values of 1.20, 0.80 and 0.60, respectively.

The good results achieved by the SFAs, MUFAs, PUFAs and ω-3 models in external validation and in the ω-6 models after the BSC suggest that the models can predict new samples from different fish oil industries. Furthermore, the ω-6 model could be improved with the addition of new samples of different origins, which would correct the bias and slope deviation.

3.2.2. Spectral Information of the Models

Raw spectra of the oil mixtures used during the experiment are shown in Figure 2.

Although information is usually hidden in the NIR spectrum, characteristic absorption bands from oil samples are observed in the raw spectra (Figure 2) at 900, 1020, 1200 and 1400 nm. The first two weak peaks observed are around 900 and 1020 nm. The former corresponds to the C-H stretching third overtone of CH_3, whereas the latter is a combination of the C-H stretching first overtone and the C-H deformation second overtone, again from CH_3 [11]. The first strong peak at 1200 nm is due to the second overtone of the stretching mode of C-H bonds in various chemical groups [42,43]. The second strong peak, localized between 1300 and 1500 nm, is caused by the combination of the stretching and deformation first overtone of C-H in CH, CH_2 and CH_3 [11].

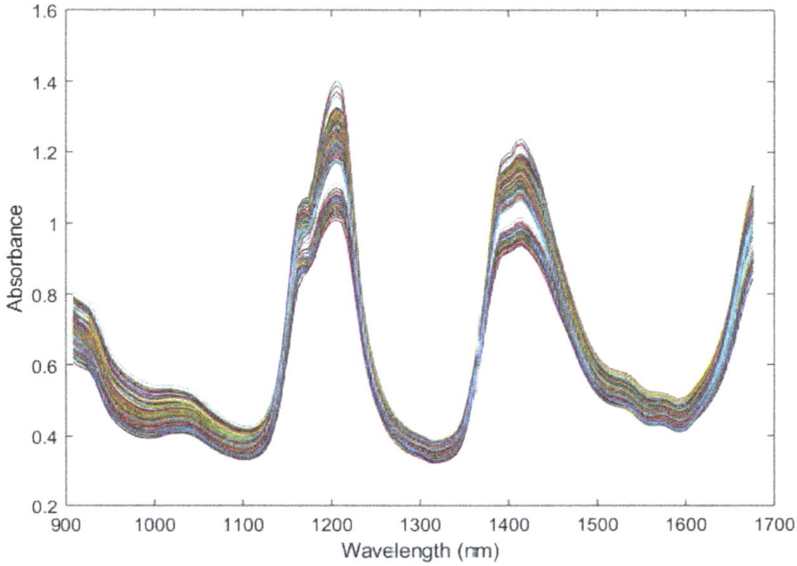

Figure 2. Raw spectra of the oil mixtures used during the experiment.

The loadings corresponding to the first and second latent variables (LV1 and LV2) of the five models developed in the second step, which contain information about all the data used in the experiments, are shown in Figure 3. LV1 retains the greatest amount of variance in most of the models, except for the SFAs model, wherein LV2 retains the most information. The large peaks in the loadings of the models resemble the main peaks of the raw spectra.

NIR absorption peaks related to the FAs information are associated with the vibrations of C-H and CH_2 [44]. Although they are usually above 1700 nm in the spectra, where two important regions are located at 1720 and 2143 nm [45], the presence of other bands related to C-H overtones at shorter wavelengths makes possible the measurement of oils with devices whose spectral range covers only wavelengths lower than 1700 nm, as demonstrated by Basri et al. [46].

As can be seen in Figure 3a–e, LV1 and LV2 of all the models show important peaks in the region between 1050 and 1300 nm. This region corresponds to the second overtone of C-H stretching, and it is one of the most important regions to determine FAs with this technology [42–44].

LV1 of PUFAs, ω-3 and ω-6 (Figure 3c–e) and LV2 in all the models (Figure 3a–e) show peaks in the region between 1300 and 1500 nm (Figure 3a,c–e). This absorption region is caused by the combination of the stretching and deformation of the first overtone of C-H in CH, CH_2 and CH_3 [11].

The increase found in the region between 1600 and 1670 nm can be seen in LV1 of PUFAs, ω-3 and ω-6 (Figure 3c–e) and in LV2 of MUFAs, PUFAs, ω-3 and ω-6 (Figure 3b–e). According to Hourant et al. [47], the wavelengths between 1600 and 1780 nm are related to the first overtone of the C-H group in -CH_3, and the peak that is starting to grow may correspond with the first part of that region. On the contrary, LV1 of SFAs and MUFAs (Figure 3a,b) and LV2 of SFAs (Figure 3a) present a peak with a maximum around 1600 nm. This region of the spectra is related to the C-H first overtone of = CH_2, which acquires its maximum at 1620 nm [48].

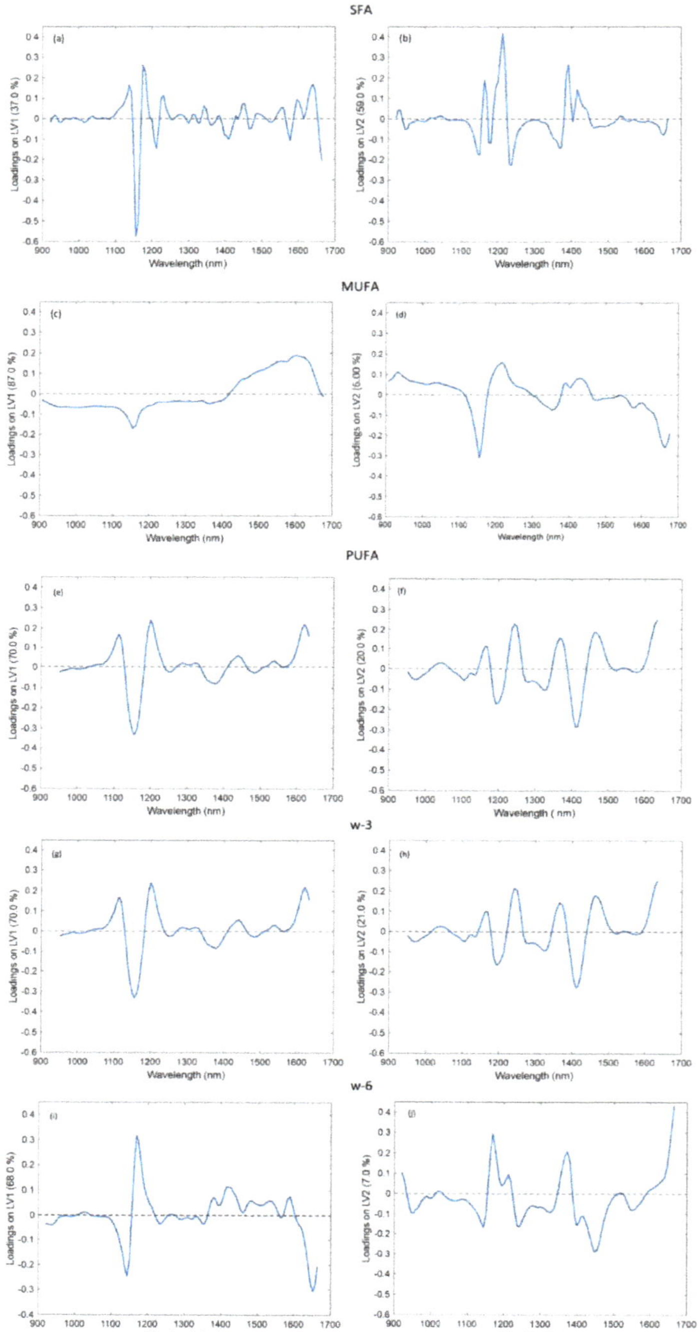

Figure 3. LV1 and LV2 of the second-step models. (**a**,**b**) SFAs, (**c**,**d**) MUFAs, (**e**,**f**) PUFAs, (**g**,**h**) ω-3, (**i**,**j**) ω-6.

The similarity in shape between PUFAs and ω-3 loadings suggests that they are closely related (Figure 3c,d). ω-6 loadings also present peaks at similar wavelengths (Figure 3e) as PUFAs and ω-3 loadings. This result was expected because fish PUFAs are mostly composed of ω-3 and ω-6 FAs [49], as can be seen in Table 3.

4. Conclusions

This study demonstrates the possibility of using a handheld NIR spectrometer as an alternative to GC-FID to determine fish oil fat composition on-site in a fast and non-destructive way. NIR spectroscopy, coupled with chemometrics, can predict concentrations of SFAs, MUFAs, PUFAs and ω-3 FAs with good results, with the SFAs and ω-3 models performing best in external validation (R^2 of 0.98 and 0.99, RMSEP = 0.94% and 0.98%, and BIAS = 0.78% and −0.67%, respectively, in the test set).

Although the technique produced a high error of prediction and bias in the ω-6 FAs model (RMSEP = 2.09% and Bias = −1.76%), this was corrected with the application of BSC, obtaining an R^2 of 0.95, an RMSEP of 1.09% and a bias of −0.05%, which could be improved in the future with the addition of new oil samples to the model.

The results presented in this study demonstrate that NIR spectroscopy is a mature technology capable of rapidly and efficiently determining the quality of oils extracted from fish by-products, which makes it suitable for industrial applications. This will allow food industries to rapidly and efficiently determine the quality and commercial value of oil coming from fish by-products.

Author Contributions: Conceptualization, I.O., E.S. and Á.M.-H.; methodology, S.N.-O., I.O., E.S. and Á.M.-H.; validation, S.N.-O., I.O., E.S. and Á.M.-H.; formal analysis, S.N.-O. and Á.M.-H.; investigation, S.N.-O., I.O., E.S. and Á.M.-H.; data curation, S.N.-O., E.S. and Á.M.-H.; writing—original draft preparation, S.N.-O., I.O., G.A., G.F. and Á.M.-H.; writing—review and editing, S.N.-O., I.O., G.A., G.F. and Á.M.-H.; visualization, S.N.-O., I.O. and Á.M.-H. project administration, I.O.; funding acquisition, I.O. and Á.M.-H. All authors have read and agreed to the published version of the manuscript.

Funding: This research part of the "Gaitik -Monitoring of quality parameters for process automation: PAT technology for the improvement of production systems" project, funded by the Basque Government—Department of Economic Development, Sustainability and Environment—Vice. Dept. of Agriculture, Fishing and Food Policy, Directorate of Quality and Food Industries.

Data Availability Statement: The datasets generated for this study are available on request from the corresponding author.

Acknowledgments: The authors greatly acknowledge the Basque Government—Department of Economic Development, Sustainability and Environment.—Vice. Dept. of Agriculture, Fishing and Food Policy, Directorate of Quality and Food Industries for the funding of the project and for the scholarship of Sonia Nieto-Ortega. They also acknowledge the company Barna (Mundaka, Spain) for providing fish oil samples for this research project. This paper is contribution n° 1096 from AZTI, Food Research, Basque Research and Technology Alliance (BRTA).

Conflicts of Interest: The authors declare no conflict of interest. The funders had no role in the design of the study; in the collection, analyses, or interpretation of data; in the writing of the manuscript, or in the decision to publish the results.

References

1. Nawaz, A.; Li, E.; Irshad, S.; Hhm, H.; Liu, J.; Shahbaz, H.M.; Ahmed, W.; Regenstein, J.M. Improved effect of autoclave processing on size reduction, chemical structure, nutritional, mechanical and in vitro digestibility properties of fish bone powder. *Adv. Powder Technol.* **2020**, *31*, 2513–2520. [CrossRef]
2. Gehring, C.K.; Gigliotti, J.C.; Moritz, J.S.; Tou, J.C.; Jaczynski, J. Functional and nutritional characteristics of proteins and lipids recovered by isoelectric processing of fish by-products and low-value fish: A review. *Food Chem.* **2011**, *124*, 422–431. [CrossRef]
3. Rubio-Rodríguez, N.; de Diego, S.M.; Beltrán, S.; Jaime, I.; Sanz, M.T.; Rovira, J. Supercritical fluid extraction of fish oil from fish by-products: A comparison with other extraction methods. *J. Food Eng.* **2012**, *109*, 238–248. [CrossRef]
4. Rustad, T.; Storrø, I.; Slizyte, R. Possibilities for the utilisation of marine by-products. *Int. J. Food Sci. Technol.* **2011**, *46*, 2001–2014. [CrossRef]

5. Iñarra, B.; Bald, C.; Cebrián, M.; Peral, I.; Llorente, R.; Zufía, J. Evaluation of unavoidable unwanted catches valorisation options: The Bay of Biscay case study. *Mar. Policy* **2020**, *116*, 103680. [CrossRef]
6. Rodriguez, Y.E.; Pereira, N.A.; Haran, N.S.; Mallo, J.C.; Fernandez-Gimenez, A.V. A new approach to fishery waste revalorization to enhance Nile tilapia (*Oreochromis niloticus*) digestion process. *Aquac. Nutr.* **2017**, *23*, 1351–1361. [CrossRef]
7. Simat, V.; Vlahovic, J.; Soldo, B.; Skroza, D.; Ljubenkov, I.; Mekinic, I.G. Production and Refinement of Omega-3 Rich Oils from Processing By-Products of Farmed Fish Species. *Foods* **2019**, *8*, 125. [CrossRef]
8. Aspevik, T.; Oterhals, Å.; Rønning, S.B.; Altintzoglou, T.; Wubshet, S.G.; Gildberg, A.; Afseth, N.K.; Whitaker, R.D.; Lindberg, D. Valorization of Proteins from Co- and By-Products from the Fish and Meat Industry. In *Chemistry and Chemical Technologies in Waste Valorization*; Lin, C.S.K., Ed.; Springer International Publishing: Cham, Switzerland, 2018; pp. 123–150. [CrossRef]
9. FAO. The State of World Fisheries and Aquaculture. 2020. Available online: https://www.fao.org/3/ca9229en/ca9229en.pdf (accessed on 9 February 2022).
10. Ramakrishnan, V.V.; Ghaly, A.E.; Brooks, M.S.; Budge, S.M. Extraction of oil from mackerel fish processing waste using Alcalase Enzyme. *Enzym. Eng.* **2013**, *2*, 1000115. [CrossRef]
11. Wu, D.; Chen, X.J.; Cao, F.; Sun, D.W.; He, Y.; Jiang, Y.H. Comparison of Infrared Spectroscopy and Nuclear Magnetic Resonance Techniques in Tandem with Multivariable Selection for Rapid Determination of omega-3 Polyunsaturated Fatty Acids in Fish Oil. *Food Bioprocess Technol.* **2014**, *7*, 1555–1569. [CrossRef]
12. Cheng, J.H.; Sun, D.W.; Liu, G.X.; Chen, Y.N. Developing a multispectral model for detection of docosahexaenoic acid (DHA) and eicosapentaenoic acid (EPA) changes in fish fillet using physarum network and genetic algorithm (PN-GA) method. *Food Chem.* **2019**, *270*, 181–188. [CrossRef]
13. Al Khawli, F.; Pateiro, M.; Dominguez, R.; Lorenzo, J.M.; Gullon, P.; Kousoulaki, K.; Ferrer, E.; Berrada, H.; Barba, F.J. Innovative Green Technologies of Intensification for Valorization of Seafood and Their By-Products. *Mar. Drugs* **2019**, *17*, 689. [CrossRef]
14. Karunathilaka, S.R.; Choi, S.H.; Mossoba, M.M.; Yakes, B.J.; Bruckner, L.; Ellsworth, Z.; Srigley, C.T. Rapid classification and quantification of marine oil omega-3 supplements using ATR-FTIR, FT-NIR and chemometrics. *J. Food Compos. Anal.* **2019**, *77*, 9–19. [CrossRef]
15. Hernandez-Martinez, M.; Gallardo-Velazquez, T.; Osorio-Revilla, G.; Almaraz-Abarca, N.; Ponce-Mendoza, A.; Vasquez-Murrieta, M.S. Prediction of total fat, fatty acid composition and nutritional parameters in fish fillets using MID-FTIR spectroscopy and chemometrics. *LWT Food Sci. Technol.* **2013**, *52*, 12–20. [CrossRef]
16. Alexandrakis, D.; Downey, G.; Scannell, A.G.M. Rapid Non-destructive Detection of Spoilage of Intact Chicken Breast Muscle Using Near-infrared and Fourier Transform Mid-infrared Spectroscopy and Multivariate Statistics. *Food Bioprocess Technol.* **2012**, *5*, 338–347. [CrossRef]
17. Leme, L.M.; Nakamura, F.; Tanamati, A.A.C.; Valderrama, P.; Marco, P.H. Fast non-invasive screening to detect fraud in oil capsules. *LWT Food Sci. Technol.* **2019**, *109*, 179–185. [CrossRef]
18. Siesler, H.W. Basic Principles of Near-Infrared Spectroscopy. In *Handbook of Near-Infrared Analysis*, 3rd ed.; Burns, D.A., Ciurczak, E.W., Eds.; CRC Press: Boca Raton, FL, USA, 2008; pp. 7–20.
19. Salguero-Chaparro, L.; Baeten, V.; Fernández-Pierna, J.A.; Peña-Rodríguez, F. Near infrared spectroscopy (NIRS) for on-line determination of quality parameters in intact olives. *Food Chem.* **2013**, *139*, 1121–1126. [CrossRef]
20. Ríos-Reina, R.; García-González, D.L.; Callejón, R.M.; Amigo, J.M. NIR spectroscopy and chemometrics for the typification of Spanish wine vinegars with a protected designation of origin. *Food Control* **2018**, *89*, 108–116. [CrossRef]
21. Marques, E.J.N.; de Freitas, S.T.; Pimentel, M.F.; Pasquini, C. Rapid and non-destructive determination of quality parameters in the 'Tommy Atkins' mango using a novel handheld near infrared spectrometer. *Food Chem.* **2016**, *197*, 1207–1214. [CrossRef]
22. Bekhit, M.Y.; Grung, B.; Mjøs, S.A. Determination of Omega-3 Fatty Acids in Fish Oil Supplements Using Vibrational Spectroscopy and Chemometric Methods. *Appl. Spectrosc.* **2014**, *68*, 1190–1200. [CrossRef]
23. Van der Merwe, S.; Manley, M.; Wicht, M. Enhancing near infrared spectroscopy models to identify omega-3 fish oils used in the nutraceutical industry by means of calibration range extension. *J. Near Infrared Spectrosc.* **2018**, *26*, 245–261. [CrossRef]
24. Dos Santos, D.A.; Coqueiro, A.; Gonçalves, T.R.; Carvalho, J.C.; Bezerra, J.S., Jr.; Matsushita, M.; de Oliveira, C.A.L.; Março, P.H.; Valderrama, P.; Ribeiro, R.P. Omega-3 and Omega-6 Determination in Nile Tilapia's Fillet Based on MicroNIR Spectroscopy and Multivariate Calibration. *J. Braz. Chem. Soc.* **2020**, *31*, 1883–1890. [CrossRef]
25. Cascant, M.M.; Breil, C.; Fabiano-Tixier, A.S.; Chemat, F.; Garrigues, S.; de la Guardia, M. Determination of fatty acids and lipid classes in salmon oil by near infrared spectroscopy. *Food Chem.* **2018**, *239*, 865–871. [CrossRef]
26. Hespanhol, M.C.; Souza, J.C.; Pasquini, C. Feasibility of a portable, low-cost near-infrared spectrophotometer for the quality screening of omega-3 dietary supplements. *J. Pharm. Biomed. Anal.* **2020**, *189*, 113436. [CrossRef]
27. EUR-lex. Commission Regulation (EC) No 796/2002. Available online: https://eur-lex.europa.eu/eli/reg/2002/796/oj (accessed on 9 July 2021).
28. Nicolai, B.M.; Beullens, K.; Bobelyn, E.; Peirs, A.; Saeys, W.; Theron, K.I.; Lammertyn, J. Nondestructive measurement of fruit and vegetable quality by means of NIR spectroscopy: A review. *Postharvest Biol. Technol.* **2007**, *46*, 99–118. [CrossRef]
29. Rinnan, Å.; van den Berg, F.; Engelsen, S.B. Review of the most common pre-processing techniques for near-infrared spectra. *TrAC Trends Anal. Chem.* **2009**, *28*, 1201–1222. [CrossRef]
30. Wold, S.; Sjöström, M.; Eriksson, L. PLS-regression: A basic tool of chemometrics. *Chemom. Intell. Lab. Syst.* **2001**, *58*, 109–130. [CrossRef]

31. Özogul, Y.; Özogul, F. Fatty acid profiles of commercially important fish species from the Mediterranean, Aegean and Black Seas. *Food Chem.* **2007**, *100*, 1634–1638. [CrossRef]
32. Özogul, Y.; Özogul, F.; Alagoz, S. Fatty acid profiles and fat contents of commercially important seawater and freshwater fish species of Turkey: A comparative study. *Food Chem.* **2007**, *103*, 217–223. [CrossRef]
33. Li, G.; Sinclair, A.J.; Li, D. Comparison of Lipid Content and Fatty Acid Composition in the Edible Meat of Wild and Cultured Freshwater and Marine Fish and Shrimps from China. *J. Agric. Food Chem.* **2011**, *59*, 1871–1881. [CrossRef]
34. Rahnan, S.A.; Huah, T.S.; Nassan, O.; Daud, N.M. Fatty acid composition of some Malaysian freshwater fish. *Food Chem.* **1995**, *54*, 45–49. [CrossRef]
35. Moret, S.; Purcaro, G.; Conte, L.S. Polycyclic aromatic hydrocarbons in vegetable oils from canned foods. *Eur. J. Lipid Sci. Technol.* **2005**, *107*, 488–496. [CrossRef]
36. Ganesan, K.; Sukalingam, K.; Xu, B. Impact of consumption and cooking manners of vegetable oils on cardiovascular diseases—A critical review. *Trends Food Sci. Technol.* **2018**, *71*, 132–154. [CrossRef]
37. Melado-Herreros, A.; Nieto-Ortega, S.; Olabarrieta, I.; Gutiérrez, M.; Villar, A.; Zufía, J.; Gorretta, N.; Roger, J.-M. Postharvest ripeness assessment of 'Hass' avocado based on development of a new ripening index and Vis-NIR spectroscopy. *Postharvest Biol. Technol.* **2021**, *181*, 111683. [CrossRef]
38. Özogul, Y.; Özogul, F.; Çiçek, E.; Polat, A.; Kuley, E. Fat content and fatty acid compositions of 34 marine water fish species from the Mediterranean Sea. *Int. J. Food Sci. Nutr.* **2009**, *60*, 464–475. [CrossRef]
39. Kocatepe, D.; Turan, H. Proximate and Fatty Acid Composition of Some Commercially Important Fish Species from the Sinop Region of the Black Sea. *Lipids* **2012**, *47*, 635–641. [CrossRef]
40. Osborne, B.G.; Fearn, T. Collaborative evaluation of universal calibrations for the measurement of protein and moisture in flour by near infrared reflectance. *Int. J. Food Sci. Technol.* **1983**, *18*, 453–460. [CrossRef]
41. Afseth, N.K.; Segtnan, V.H.; Marquardt, B.J.; Wold, J.P. Raman and near-infrared spectroscopy for quantification of fat composition in a complex food model system. *Appl. Spectrosc.* **2005**, *59*, 1324–1332. [CrossRef]
42. Downey, G.; McIntyre, P.; Davies, A.N. Detecting and Quantifying Sunflower Oil Adulteration in Extra Virgin Olive Oils from the Eastern Mediterranean by Visible and Near-Infrared Spectroscopy. *J. Agric. Food Chem.* **2002**, *50*, 5520–5525. [CrossRef]
43. Garrido-Varo, A.; Sánchez, M.-T.; De la Haba, M.-J.; Torres, I.; Pérez-Marín, D. Fast, low-cost and non-destructive physico-chemical analysis of virgin olive oils using near-infrared reflectance spectroscopy. *Sensors* **2017** *17*, 2642. [CrossRef]
44. Cheng, J.H.; Sun, D.W. Partial Least Squares Regression (PLSR) Applied to NIR and HSI Spectral Data Modeling to Predict Chemical Properties of Fish Muscle. *Food Eng. Rev.* **2017**, *9*, 36–49. [CrossRef]
45. Martín, J.F.G. Optical path length and wavelength selection using Vis/NIR spectroscopy for olive oil's free acidity determination. *Int. J. Food Sci. Technol.* **2015**, *50*, 1461–1467. [CrossRef]
46. Basri, K.N.; Hussain, M.N.; Bakar, J.; Sharif, Z.; Khir, M.F.A.; Zoolfakar, A.S. Classification and quantification of palm oil adulteration via portable NIR spectroscopy. *Spectrochim. Acta Part A* **2017**, *173*, 335–342. [CrossRef] [PubMed]
47. Hourant, P.; Baeten, V.; Morales, M.T.; Meurens, M.; Aparicio, R. Oil and Fat Classification by Selected Bands of Near-Infrared Spectroscopy. *Appl. Spectrosc.* **2000**, *54*, 1168–1174. [CrossRef]
48. Shenk, J.S.; Workman, J.J., Jr.; Westerhaus, M.O. Application of NIR spectroscopy to agricultural products. In *Handbook of Near-Infrared Analysis*, 3rd ed.; Burns, D.A., Ciurczak, E.W., Eds.; CRC Press: Boca Raton, FL, USA, 2008; pp. 365–404.
49. Aubourg, S.P. Lipid compounds. In *Hanbook of Seafood and Seafood Products Analysis*, Nollet, L.M.L., Toldrá, F., Eds.; CRC Press: Boca Raton, FL, USA, 2010; pp. 69–86.

foods

MDPI

Article

Vine Shoots as a Source of *Trans*-Resveratrol and ε-Viniferin: A Study of 23 Italian Varieties

Mirella Noviello [1], Antonio Francesco Caputi [1], Giacomo Squeo [1], Vito Michele Paradiso [2], Giuseppe Gambacorta [1] and Francesco Caponio [1,*]

[1] Department of Soil, Plant and Food Science (DISSPA), University of Bari Aldo Moro, Via Amendola, 165/a, I-70126 Bari, Italy; mirella.noviello@uniba.it (M.N.); antonio.caputi1@uniba.it (A.F.C.); giacomo.squeo@uniba.it (G.S.); giuseppe.gambacorta@uniba.it (G.G.)

[2] Department of Biological and Environmental Sciences and Technologies, University of Salento, S.P. 6, Lecce-Monteroni, I-73100 Lecce, Italy; vito.paradiso@unisalento.it

* Correspondence: francesco.caponio@uniba.it

Abstract: Stilbenes are a family of phenolic secondary metabolites that are known for their important roles in plant protection and human health. Numerous studies show that vine shoots, one of the most abundant winery wastes, could be used as a source of bioactive compounds such as stilbenes. The predominant stilbenoids in vine shoots are *trans*-resveratrol (Rsv) and ε-viniferin (Vf), whose content varies depending on numerous intrinsic and extrinsic factors. The present work investigates the influence of pre-treatment and variety on stilbene concentration in vine shoots. Vine shoots of the Primitivo and Negroamaro varieties were submitted to four different trials before stilbene extraction (untreated, dried at 50 °C for 24 h, dried at 70 °C for 15 min, and dried at 80 °C for 10 min). The results showed that the heat pre-treatments had a slight impact on the total phenol and stilbene content. In contrast, the variety variable had a stronger impact on stilbene concentration, ranging from 2700 to 6400 mg kg^{-1} DW for untreated vine shoots of 23 Italian varieties. In all vine shoots, the most abundant stilbene compound was Rsv and the highest content was found in vine shoots of the Nero di Troia (5298.1 mg kg^{-1} DW) and Negroamaro (5249.4 mg kg^{-1} DW) varieties.

Keywords: stilbene; vine shoots; viticulture waste; *trans*-resveratrol; ε-viniferin; Italian varieties

Citation: Noviello, M.; Caputi, A.F.; Squeo, G.; Paradiso, V.M.; Gambacorta, G.; Caponio, F. Vine Shoots as a Source of *Trans*-Resveratrol and ε-Viniferin: A Study of 23 Italian Varieties. *Foods* **2022**, *11*, 553. https://doi.org/10.3390/foods11040553

Academic Editor: Maria Cecilia do Nascimento Nunes

Received: 16 December 2021
Accepted: 14 February 2022
Published: 15 February 2022

Publisher's Note: MDPI stays neutral with regard to jurisdictional claims in published maps and institutional affiliations.

Copyright: © 2022 by the authors. Licensee MDPI, Basel, Switzerland. This article is an open access article distributed under the terms and conditions of the Creative Commons Attribution (CC BY) license (https://creativecommons.org/licenses/by/4.0/).

1. Introduction

Stilbenoids are a natural family of polyphenolic compounds that exist both as monomers and as oligomers, with a diphenyl ethylene group oriented in cis or trans configurations [1]. These compounds have gained interest not only for their several biological activities, but also for their complex structural conformation [2]. Numerous studies show that the beneficial properties of stilbenes for human health include protective effects against cancer (as they inhibit cell proliferation) [3], diabetes [4], neurodegenerative diseases such as Alzheimer's disease [5], and coronary heart disease [6]. They are also used as multifunctional ingredients in cosmetics [7]. Recently, the possibility of developing drugs against COVID-19 using natural stilbene compounds has been evaluated [8]. In addition, stilbenes are used in agriculture as alternative anti-phytopathogenic substances [9,10].

Stilbenes are mainly synthesized by plants as phytoalexins in response to biotic and abiotic stress (e.g., pathogens, ultraviolet irradiation, heavy metal ions, mechanical damage, frost, thermal treatment, or ozone) [11]. Their distribution is very heterogeneous in the plant kingdom [12]. In fact, stilbenes have been isolated and identified in at least 72 plant species belonging to 31 genera and 12 families, including *Vitaceae*, in which these compounds are present in lignified stem tissue, in grape berries and in wines [11,13,14]. Several reviews have indicated that winery wastes and by-products are rich in stilbenes, which have been extracted and applied in multiple fields based on their beneficial properties [15,16].

The cultivation of vines is widespread: in 2020, the world area under vine cultivation for all purposes (wine and juices, table grapes, and raisins) was estimated at 7.3 million hectares (Mha), of which 3.3 Mha are in the European Union. Italy has an area under vine cultivation of 719 thousand hectares, an increase of over 0.8% from 2019 [17]. Consequently, the wine-growing sector produces many and various wastes, generated from agricultural practices (e.g., vine shoots, leaves, stems) as well as from the winemaking process (e.g., grape stalks, pomace, wine lees). In particular, vine shoots (also called grapevine canes) are the most significant vine waste material from a quantitative point of view, with a weight of 2–5 tonnes per hectare per year, depending on density of plantation, climate, vigour of the vine, and other agronomical factors [18].

Vine shoots have a very low economic value; in fact, they are burned [19] or incorporated into the soil to promote the degradation of organic matter and reduce the need for organic fertilizers [20]. Some other possible applications of this material include the production of pulp paper [21], solid biofuels [22], or the possibility of obtaining activated carbon [23]. Recently, attention has shifted to the possibility of using vine shoots in the agri-food industry, in a circular economy perspective. One of the possible applications studied is their use as an alternative to oak chips as an enological coadjuvant to improve the sensorial profile of wines, [24,25]. Moreover, recent studies have shown that vine shoots are rich in bioactive compounds, such as stilbenes, that make this by-product an untapped source of these compounds with important antioxidant, anti-microbial, and anti-aging properties and multiple possible applications [15]. Up to 41 stilbenes have been found in vine shoots and among these, *trans*-resveratrol (Rsv) and ε-viniferin (Vf) are the most abundant [15,26]. Several studies tested stilbene-enriched vine shoot extracts as a preservative in wine in order to reduce the use of SO_2 in winemaking [15,27].

The concentration and composition of stilbenes in vine shoots are subject to extreme variability due to many intrinsic and extrinsic factors. These factors include the variety and geographical area of origin [28–30], vineyard age [31], or climate conditions [32]. Considering the variety analysed in literature, vine shoots of Pinot Noir and Gewurztraminer have been reported as those with the highest content of stilbenes [15,18,29,30]. On the other hand, the extrinsic factors include the extraction method [26], storage time and temperature of the vine shoots, or various pre-treatments, such as the cutting length or thermal treatments, before stilbenes extraction [32–38]. Despite the available reports, the literature does not clarify univocally the effects of these heat treatments on stilbene quantities [26,39,40]. Moreover, it is well known that the low-temperature/long-time heat treatments, mostly adopted for vine shoots, generally led to a higher reduction of the nutritional values of foods than the high-temperature/short-time heat treatments [41]. A previous work proved that treatments applied to Coratina olive cultivar leaves at high temperatures and short times did not cause a reduction of the phenolic compounds [42]. Consequently, investigations on the effect of the temperature-time conditions are necessary to preserve these compounds and increase the extraction yields.

The aim of this study was twofold: (i) select the most appropriate vine shoots treatment before stilbene extraction (untreated, dried at 50 °C for 24 h, dried at 70 °C for 15 min, dried at 80 °C for 10 min) using two testing varieties (Primitivo and Negroamaro); (ii) study the variability of the total phenolic content and the Rsv and Vf amounts of vine shoots from 23 Italian varieties. To the best of our knowledge, the stilbene contents of vine shoots from these Italian varieties has not been studied yet.

2. Materials and Methods

2.1. Plant Materials

Vine shoots of 23 varieties of *Vitis vinifera* L. were selected: Aglianico (AG), Bianco d'Alessano (BA), Bombino Bianco (BB), Bombino Nero (BN), Ciliegiolo (CI), Fiano Bianco d'Avellino (FB), Italia (IT), Malvasia Bianca (MB), Malvasia Nera di Brindisi (MN), Maresco Bianco (MA), Minutolo Bianco (MI), Montepulciano (MO), Negroamaro (NE), Nero di Troia (NT), Notardomenico (ND), Ottavianello (OT), Palieri (PA), Primitivo (PR), Sangiovese (SA),

Susumaniello (SU), Trebbiano (TR), Verdeca (VE), and Vittoria (VI). All vine shoots were sampled during winter (February 2021) from a varietal collection located in Locorotondo (Puglia, Italy; coordinates: longitude 17°13′3.741″ E, latitude 40°45′42.763″ N) grown under the same conditions. The vineyard was planted in 1985 on a sub-alkaline medium-textured soil. About 10 kg of vine shoots from each variety, sampled from large batches and representative of these, were collected and stored intact under controlled conditions (darkness, at 15 ± 3 °C) for 6 weeks [34]. Then, two different representative subsamples of about 1 kg for each variety were considered for the subsequent analyses. Table 1 shows additional information about the varieties chosen in this work.

Table 1. The grapevine variety name, grape colour, usual use, and acronym used in the text.

Variety	Colour	Use	Acronym
Aglianico	red	wine	AG
Bianco d'Alessano	white	wine	BA
Bombino Bianco	white	wine	BB
Bombino Nero	red	wine	BN
Ciliegiolo	red	wine	CI
Fiano Bianco d'Avellino	white	wine	FB
Italia	white	table	IT
Malvasia Bianca	white	wine	MB
Malvasia Nera di Brindisi	red	wine	MN
Maresco Bianco	white	wine	MA
Minutolo Bianco	white	wine	MI
Montepulciano	red	wine	MO
Negroamaro	red	wine	NE
Nero di Troia	red	wine	NT
Notardomenico	red	wine	ND
Ottavianello	red	wine	OT
Palieri	red	table	PA
Primitivo	red	wine	PR
Sangiovese	red	wine	SA
Susumaniello	red	wine	SU
Trebbiano	white	wine	TR
Verdeca	white	wine	VE
Vittoria	white	table	VI

2.1.1. Evaluation of Treatments Impact

Vine shoots from the Primitivo and Negroamaro varieties were manually cut (particle size about 5 cm long), cut crosswise, heat-treated (as described below), ground (particle size ranging from 0.2–4 mm) using a hammer mill (Dietz-Motoren KG, Elektromotorenfabrik, 7319 Dettingen-teck, Germany), and immediately submitted to extraction and analyses. Four different treatments of the vine shoots were tested before stilbene extraction (untreated, dried at 50 °C for 24 h, dried at 70 °C for 15 min, dried at 80 °C for 10 min). A thermostatic oven (TFC 120 forced air oven, ArgoLab) was used for the drying process. The moisture content of the vine shoots was measured using a thermobalance (Ladwag MAC 110/NP, Radwag, Poland).

2.1.2. Evaluation of Variety Impact

The stilbene contents of the untreated vine shoots from 23 Italian varieties were assessed. The vine shoots were manually cut (particle size around 5 cm long), cut crosswise, ground (particle size ranging from 0.2–4 mm) using a hammer mill (Dietz-Motoren KG, Elektromotorenfabrik, 7319 Dettingen-teck, Germany), and immediately submitted to extraction and analyses. The moisture content of the vine shoots was measured using a thermobalance (Ladwag Mac 110/NP, Radwag, Poland).

2.2. Extraction Procedure

The extraction of the stilbenes from the vine shoots was carried out according to Vergara et al. [29], with some modifications. Briefly, an aliquot of vine shoots (2 g) was added with 16 mL of an ethanol/water solution (80:20 *v*/*v*) and sonicated in an ultrasonic bath (CP104 Standard Ultrasonic Cleaning Machine, CEIA, Padova, Italy) at room temperature and 50 Hz for 5 min. The extract was centrifuged (SL 16R Centrifuge, Thermo Scientific, MA, USA) at 10,000× *g* for 5 min, the supernatant was separated, filtered through Whatman filter paper (GE Healthcare, Milan, Italy) (67 g m^{-2}), and then filtered using nylon filters of 0.45 μm (Sartorius Stedim Biotech Gmbh, Göttingen, Germany) and used for chemical characterization. Extractions were carried out in duplicate for each condition tested.

2.3. Extract Characterization

2.3.1. Total Phenolic Content Determination

The total phenol content was determined according to the Folin–Ciocalteu method [43]. To 980 μL of H_2O Milli-Q, 20 μL of appropriately diluted extract, 100 μL of Folin–Ciocalteu reagent were added. After 3 min, 800 μL of 7.5% Na_2CO_3 were added and then the sample was stored in the dark for 60 min. The absorbance was read at 720 nm (Cary 60 UV-Vis, Agilent Technologies, Mulgrave, Australia). The results were expressed as mg of gallic acid equivalents (GAE) per g of dry weight sample (mg GAE g^{-1} DW). Each sample was analysed in duplicate.

2.3.2. Antioxidant Activity Evaluation

The DPPH (2,2-diphenyl-1-picrylhydrazyl) assay was performed on the extracts according to the procedure of Tarantino et al. [44]. Each extract (50 μL) was combined with 950 μL DPPH solution (0.08 mM in ethanol). The decrease in absorbance was read at 517 nm using a Cary 60 UV-Vis spectrophotometer (Agilent, Cernusco, Milan, Italy). The results were expressed in μmol Trolox equivalents g^{-1} dry weight for all vine shoot samples (μmol TE g^{-1} DW). All determinations were carried out in duplicate. Antioxidant activity was also determined by ABTS-TEAC assay [44]. For spectrophotometry, the reaction took place directly in cuvettes by adding 50 μL of each sample to 950 μL of final $ABTS^{\bullet+}$ solution. After 8 min, the decrease in absorbance was measured at 734 nm, using a Cary 60 UV-Vis spectrophotometer (Agilent, Cernusco, Milan, Italy). The results were expressed in μmol TE g^{-1} dry weight for all vine shoot samples (μmol TE g^{-1} DW). Each sample was analysed in duplicate.

2.3.3. Quantification of Rsv and Vf by HPLC-DAD

The analysis of the stilbenes was performed according to the method of Ewald et al. [38] using high-performance liquid chromatography (UltiMate 3000 HPLC, Thermo scientific, Munich, Germany) that included an HPG-3200RS binary pump, WPS-3000RS/TRS autosampler, TCC-3000RS column oven, and a DAD-3000RS photodiode array detector. HPLC separation was achieved on Acclaim™ 120 C18 columns (120 Å 3 × 150 mm, 3 μm) maintained at 25 °C using a mobile phase consisting of 1% aqueous acetic acid (*v*/*v*) (A) and methanol (B). The separation was carried out at 25 °C with a flow rate of 0.6 mL min^{-1} under the following conditions: 0 min (20% B), 10 min (20% B) 6.5 min (37% B), 12.6 min (50% B), and 21.0 min (100% B). Under these conditions, Rsv and Vf were eluted with a retention time of 14.7 min and 17.8 min and monitored at 306 and 324 nm, respectively. Calibration curves were prepared using the endotoxin standards (Sigma-Aldrich, Steinheim, Germany) of Rsv (R^2 = 0.9993) and Vf (R^2 = 0.9994) in the concentration range 1–500 mg L^{-1}. The amount of Rsv and Vf found in each extract was expressed as mg of compound kg^{-1} of DW. Each sample was analysed in duplicate.

2.4. Statistical Analysis

Minitab17 (Minitab Inc., State College, PA, USA) was used for the statistical analysis of all results, reported as mean ± standard deviation (SD) of two replications. To evaluate the differences between samples, one-way ANOVA was applied. The Fisher LSD test was employed for the post-hoc comparisons of the means. Correlation between variables was determined by Pearson's correlation coefficient test. Statistical significance was set at $p < 0.05$ level.

3. Results and Discussion

3.1. Evaluation of the Pre-Treatments

3.1.1. Total Phenolic and Antioxidant Activity

Table 2 reports the mean values, standard deviation, and results of the statistical analysis of the total phenolic contents and the antioxidant activity measured in the vine shoot extracts subjected to the different investigated treatments. Several studies have shown that vine shoots are rich in phenolic compounds [45–49]. Concerning the vine shoots of the Primitivo variety, the results showed that the extracts, irrespective of the pre-treatments, contained similar amounts of TPC, except for vine shoots treated at 50 °C for 24 h, in which a significant reduction was observed (18.4 ± 0.1 mg g^{-1} DW). Instead, in regards the vine shoot extracts of the Negroamaro variety, except for the treatment at 80 °C for 10 min, the other two applied heat treatments reduced the TPC. In particular, the treatment at 70 °C for 15 min reduced TPC by 11.3% with respect to the untreated vine shoots, which had the highest content (21.2 ± 0.1 vs. 23.9 ± 0.1 mg g^{-1} DW, respectively).

Table 2. The TPC (total phenolic content), DPPH (antioxidant activity determined by the DPPH assay), and ABTS (antioxidant activity determined by the ABTS assay) of the vine shoot extracts from the Primitivo and Negroamaro varieties. Results are expressed as mean ± standard deviation (n = 2); different letters for each variety in the same column indicate a significant difference according to the Fisher test ($p < 0.05$).

Sample	TPC (mg GAE g^{-1} DW)	DPPH (µmol TE g^{-1} DW)	ABTS (µmol TE g^{-1} DW)
Primitivo			
Untreated	20.1 ± 0.1 ^{a}	85.0 ± 4.1 ^{a}	115.6 ± 1.6 ^{a}
50 °C–24 h	18.4 ± 0.1 ^{b}	80.6 ± 1.3 ^{ab}	97.0 ± 0.7 ^{d}
70 °C–15 min	21.1 ± 0.6 ^{a}	78.2 ± 0.1 ^{b}	101.2 ± 0.8 ^{c}
80 °C–10 min	20.3 ± 0.6 ^{a}	81.8 ± 0.9 ^{ab}	110.8 ± 0.8 ^{b}
Negroamaro			
Untreated	23.9 ± 1.0 ^{a}	79.5 ± 0.2 ^{a}	136.5 ± 0.8 ^{a}
50 °C–24 h	21.8 ± 0.1 ^{bc}	57.9 ± 0.5 ^{d}	86.4 ± 0.9 ^{c}
70 °C–15 min	21.2 ± 0.1 ^{c}	63.2 ± 0.4 ^{c}	79.2 ± 0.7 ^{d}
80 °C–10 min	22.8 ± 0.1 ^{ab}	65.0 ± 0.7 ^{b}	88.7 ± 0.7 ^{b}

3.1.2. Stilbene Composition

The stilbene concentration (Rsv and Vf) as affected by each treatment is shown in Table 3. First of all, we determined that the variety influenced the stilbene content. In fact, untreated Negramaro vine shoot extracts contained a higher concentration of Rsv compared to Primitivo (5249.4 vs. 1861.3 mg kg^{-1} DW), while the latter had a higher concentration of Vf (1531.6 vs. 600.1 mg kg^{-1} DW).

It was evident that the drying treatment accounted for some variations in the Rsv and Vf concentrations, according to what we observed for TPC (Table 2). With respect to the Primitivo vine shoots, the drying at 50 °C for 24 h determined the reduction of Rsv (1663.8 ± 16.3 mg kg^{-1} DW) and Vf (1356.8 ± 10.0 mg kg^{-1} DW) when compared to the untreated vine shoots (1861.3 ± 9.8 mg kg^{-1} DW for Rf and 1531.6 ± 89.1 mg kg^{-1} DW for Vf). Minor differences were observed when comparing the other two treatments with the untreated sample: the Rsv concentration increased by only 6.6% after the treatment at 70 °C for 15 min and decreased slightly after the treatment at 80 °C for 10 min (1763.4 ± 98.3 mg kg^{-1} DW); after the treatment at 70 °C for 15 min and 80 °C for 10 min, the concentration of Vf increased by 13% and 5.6%, respectively. Thus, no significant differences were found between the concentrations of Vf after these two treatments. In regards to Negroamaro, significant differences were found among the treatments, with the untreated sample showing the highest Rsv (5249.4 ± 129.8 mg kg^{-1} DW) and Vf concentrations (600.1 ± 79.0 mg kg^{-1} DW) (Table 3).

Table 3. The stilbene concentrations in Primitivo and Negramaro vine shoot extracts. The means and standard deviation (n = 2) are represented in the same column and different letters for each variety indicate significant differences ($p < 0.05$).

	Stilbene Concentrations (mg kg^{-1} DW)	
Sample	***Trans*-Resveratrol**	**ε-Viniferin**
Primitivo		
Control	1861.3 ± 9.8 ^{ab}	1531.6 ± 89.1 ^{ab}
50 °C–24 h	1663.8 ± 16.3 ^{c}	1356.8 ± 10.0 ^{c}
70 °C–15 min	1983.8 ± 12.9 ^{a}	1731.2 ± 56.0 ^{a}
80 °C–10 min	1763.4 ± 98.3 ^{bc}	1617.5 ± 13.1 ^{a}
Negroamaro		
Control	5249.4 ± 129.8 ^{a}	600.1 ± 79.0 ^{a}
50 °C–24 h	4471.1 ± 73.9 ^{b}	451.5 ± 1.0 ^{c}
70 °C–15 min	4626.0 ± 37.7 ^{b}	455.7 ± 9.8 ^{c}
80 °C–10 min	4925.4 ± 14.4 ^{ab}	525.2 ± 38.2 ^{b}

Overall, these results suggest that the heat pre-treatments either left unchanged or caused a decrease in the stilbene concentration. In particular, the treatments with lower temperatures and longer times led to a significant reduction in Rsv and Vf. Most likely, the use of high temperatures may promote the degradation of some compounds, as reported by Piñeiro et al. [39]. In that case, in most of the selected vine shoots (from 15 grape cane varieties), the total stilbene concentration was significantly higher for freeze-dried extracts than for oven-dried extracts (40 °C for 15 days). However, our results are in contrast with those from Sánchez-Gómez et al. [40], who showed that the thermal treatment led to Rsv concentrations from 6 to 14 times higher than those in the control/no heat treated samples, depending on the vine variety (Airén and Moscatel grape canes).

To the best of our knowledge, this is the first report on the effect of these time-temperature drying parameters on the vine shoot stilbenes contents of Italian vine varieties.

3.2. *Evaluation of Different Italian Varieties*

Considering the results previously obtained, no heat treatment was applied to characterize the stilbene contents in the vine shoots of the investigated Italian varieties. Indeed, in absence of clear advantages, the heat-treatment results are a waste of energy, incompatible with the requests for sustainable processes.

3.2.1. Total Phenolic Content and Antioxidant Activity

The total phenolic contents of the vine shoots are given in Table 4. Vine shoots from the Sangiovese variety showed the lowest TPC, which was approximately 60% lower than Palieri, the variety with the highest content. These results agree with previous studies [50,51]. In fact, Çetin et al. [51], in evaluating the chemical composition of ten different Turkish grape canes varieties, showed that the total phenolic content changed significantly according to the varieties (in a range from 25.36 ± 1.62 to 36.56 ± 2.67 mg GAE g^{-1} DW). Similarly, according to Dorosh et al. [50], the amount of total phenolic content in Tinta Roriz vine shoot extracts (32.6 ± 2.1 mg GAE g^{-1} DW) was 1.6 fold higher than the value obtained from the Touriga Nacional variety (20.1 ± 0.6 mg GAE g^{-1} DW), for the same extraction time and ultrasound extraction technique. These results were in agreement with those from a previous study that also presented a summary table showing the results from selected published papers examining the phenolic compounds of vine shoots extractions [48].

Table 4. The total phenolic content and antioxidant activity of vine shoot extracts from 23 different Italian varieties. Means and standard deviation (n = 2) are represented in the same column and data followed by different letters indicate statistically significant differences according to the Fisher test ($p < 0.05$). For sample codes, see Table 1.

Sample	TPC (mg GAE g^{-1} DW)	DPPH (μmol TE g^{-1} DW)	ABTS (μmol TE g^{-1} DW)
AG	18.0 ± 0.3 jk	72.2 ± 2.1 hijk	102.2 ± 2.8 hi
BA	24.4 ± 0.2 de	90.6 ± 7.2 c	137.2 ± 11.7 bc
BB	17.8 ± 0.2 jk	69.4 ± 2.4 ijk	116.2 ± 1.2 efg
BN	22.5 ± 1.8 efg	92.9 ± 1.8 bc	135.0 ± 3.2 bc
CI	19.6 ± 0.5 hij	37.4 ± 0.8 l	34.2 ± 0.7 k
FB	20.4 ± 0.1 ghi	86.3 ± 0.0 cde	79.7±1.4 j
IT	27.6 ± 1.1 bc	84.9 ± 2.9 cdef	144.1 ± 16.4 b
MB	19.6 ± 1.3 hij	75.6 ± 2.1 ghij	75.7 ± 1.4 j
MN	21.3 ± 1.2 fgh	76.0 ± 8.0 ghij	121.8 ± 7.5 def
MA	23.0 ± 1.5 def	87.9 ± 1.9 cd	125.9 ± 0.2 cde
MI	22.4 ± 0.7 efg	77.4 ± 3.9 fghi	132.6 ± 4.2 cd
MO	28.6 ± 0.8 b	111.6 ± 0.8 a	156.4 ± 0.8 a
NE	23.9 ± 1.0 de	79.5 ± 0.2 efgh	136.5 ± 0.8 bc
NT	24.2 ± 2.0 de	81.2 ± 3.9 defg	103.0 ± 4.3 hi
ND	29.3 ±1.7 b	115.2 ± 1.2 a	112.6 ± 2.7 fgh
OT	18.8 ± 1.0 ijk	70.4 ± 2.3 ijk	108.4 ± 2.0 ghi
PA	36.9 ± 2.2 a	112.1 ± 0.6 a	131.5 ± 3.9 cd
PR	20.1 ± 0.1 ghij	85.0 ± 4.1 cdef	115.6 ± 1.6 efg
SA	14.7 ± 0.2 l	67.2 ± 1.4 k	108.4 ± 1.3 ghi
SU	22.4 ± 1.3 efg	91.7 ± 7.7 c	99.7 ± 3.8 i
TR	18.9 ± 0.1 hijk	72.0 ± 7.6 hijk	108.0 ± 3.5 ghi
VE	17.4 ± 1.4 k	69.2 ± 5.9 jk	108.4 ± 7.2 ghi
VI	25.1 ± 1.3 cd	100.5 ± 1.9 b	111.8 ± 8.6 fgh

Table 4 shows the antioxidant properties of the extracts from the vine shoots of the Italian varieties evaluated. The antioxidant activity showed statistically significant differences among the varieties with the same tendency as that previously described for TPC. As reported in Table 5 and as previously demonstrated, antioxidant activity correlates with the total phenolic content of grape cane extracts [52]. These results were quite consistent with those provided by the DPPH assay, since the Palieri, Montepulciano, and Notardomenico vine shoot extracts showed the highest antioxidant capacity (112.1 ± 0.6, 111.6 ± 0.8, and 115.2 ± 1.2 μmol TE g^{-1} DW, respectively) and, at the same time, the highest total phenolic content. Additionally, according to the ABTS assay, the Montepulciano vine shoot extracts had the highest antioxidant activity (156.4 ± 0.8 μmol TE g^{-1} DW). It is very difficult to compare the results obtained from this characterisation with those from other studies because most of those used different assays to evaluate the antioxidant activity. Nevertheless, some researchers have compared the antioxidant activity of vine shoot extracts of different varieties [45,49,50]. For example, Guerrero et al. [18] found significant differences in the antioxidant activity of the vine shoots from 22 grape varieties (including *Vitis vinifera sativa*, *Vitis vinifera sylvestris*, and hybrid direct producers), measured using the oxygen radical absorbance capacity (ORAC) assay (range from 1700 to 5300 μmol, Trolox equivalent g^{-1} DW).

3.2.2. Stilbene Composition

Figure 1 shows the stilbene concentration of the vine shoot extracts of the investigated varieties while the Rsv, Vf, and total stilbenes concentrations (mean ± standard deviation) are reported in Table S1 (Supplementary Material). The mean total concentration of stilbenes, approximately 4500 mg kg^{-1} DW, varied greatly depending on the variety, with values ranging between 2700 and 6400 mg kg^{-1} DW for Verdeca and Palieri, respectively, with 2.4-fold higher results for the latter. Nevertheless, the Palieri, Montepulciano, and Italia varieties presented the highest total stilbene concentration. In contrast, the Verdeca, Bianco d'Alessano, and Trebbiano varieties presented the lowest total stilbene concentration. In previous studies, a wide variability (from 2.5 to 4-fold) of total polyphenol amounts was already observed among different vine shoot varieties [18,30,37].

Table 5. The Pearson's correlation coefficients between the TPC, ABTS, and DPPH in 23 vine shoot extracts.

	TPC	ABTS	DPPH
TPC	1	-	-
ABTS	0.450 (p = 0.002)	1	-
DPPH	0.760 (p < 0.001)	0.606 (p < 0.001)	1

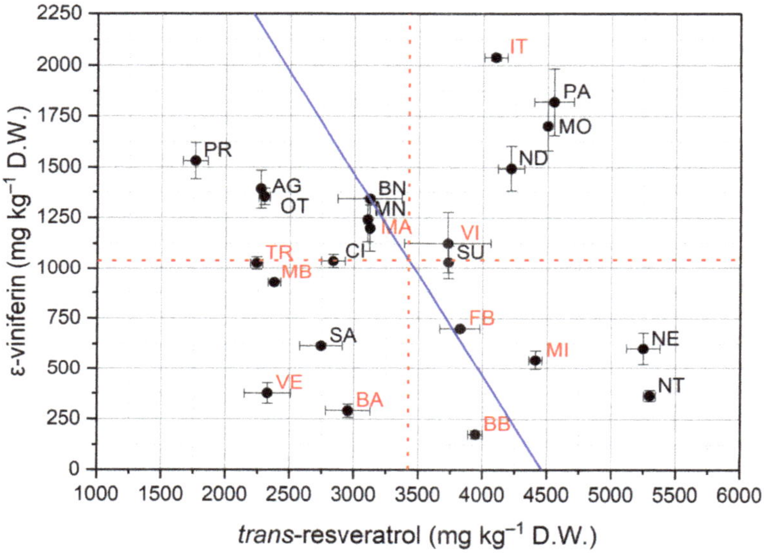

Figure 1. The stilbene contents in vine shoots from 23 different Italian varieties. Red dotted lines—the mean contents of *trans*-resveratrol (Rsv) and ε-viniferin (Vf); blue solid line—the mean content of *trans*-resveratrol + ε-viniferin. Black labels indicate black grape varieties; red labels indicate white grape varieties. For sample codes, see Table 1.

The major stilbene compounds found in all the collected samples were Rsv (mean of 3422.2 mg kg^{-1} DW), followed by Vf (mean of 1040.0 mg kg^{-1} DW). An example of HPLC-DAD chromatograms of the Palieri vine shoot extract is provided in the Supplementary Material Figure S1. These results agree with those observed in the studies by Vergara et al. [29] and Gorena et al. [33], in which the major stilbene compound found in most grape cane extracts considered were Rsv. In contrast, according to Guerrero et al. [18,32] and Lambert et al. [30], Vf was the most abundant compound in vine shoots of different *Vitis* varieties.

The highest mean concentration of Rsv was determined for the Nero di Troia (5298.1 ± 45.2 mg kg^{-1} DW) and Negroamaro vine shoots (5249.4 ± 129.8 mg kg^{-1} DW), followed by the Montepulciano and Palieri varieties. On the other hand, the Primitivo vine shoots (1861.3 ± 9.8 mg kg^{-1} DW) showed the lowest concentration of Rsv, about 64.9% less than Nero di Troia. There are many studies showing the differences between the stilbene contents in vine shoots from different varieties and species of vines [26,32], but there are no studies concerning the variation in vine shoots of these Italian varieties. Nevertheless, comparable concentrations of Rsv were found in vine shoots of different Chilean varieties, in which Gewurztraminer (mean 4628 ± 568 mg kg^{-1} DW) and Pinot noir varieties (mean 3676 ± 353 mg kg^{-1} DW) were determined to contain high levels of this compound [29]. Recently, Zwingelstein et al. [31] showed that vine shoots of the Mondeuse variety contained higher levels of Rsv (3759–4636 mg kg^{-1} DW) than those of the Jacquère variety (2259–2994 mg kg^{-1} DW). Lower concentrations were found by Zhang et al. [28], in which the *Vitis Vinifera* vine shoots grown in China exhibited an Rsv content ranging from 664 to 1751 mg kg^{-1}.

In regard to Vf, Figure 1 clearly reveals that most vine shoot extracts of red berry varieties had a concentration of Vf above the average. Nevertheless, the highest concentration was found in the vine shoot extracts of the Italia variety (2038.4 ± 15.8 mg kg^{-1} DW), when compared to other varieties. The Bombino Bianco variety (175.9 ± 19.6 mg kg^{-1} DW) showed a concentration 91.37% lower than that of the Italia variety. These results agreed with those observed in the studies by Guerrero et al. [18] in which the highest concentration of Vf, found in Gewurztraminer (2810.4 mg kg^{-1} DW), was similar to that found in this study. Similarly, according to Lambert al. [30], the most abundant stilbenoid in grape canes of sixteen *Vitis Vinifera* varieties was Vf (mean of 2171 mg kg^{-1} DW).

To evaluate the correlation between the TPC and the concentration of Rsv and Vf, the Pearson's correlation coefficient test was applied (Table 6). A significant correlation between TPC and Rsv and between TPC and Vf was observed, whilst no correlation was found between the two considered stilbenes (p = 0.697). A clear explanation for this absence of correlation is difficult to determine, considering that several sources of variability could affect the stilbene synthesis and outcome. From the genetic point of view, stilbene synthase (STS) is the key enzyme in the stilbene biosynthetic pathway, and grapevines contain a large number of STS genes [53,54]. Moreover, as reported in a recent review [11], the expression of these genes also varies according to environmental stress. At the same time, the specific varieties affect the accumulation of stilbenes, even under the same environmental conditions [32,34]. Vf is an oligomer of Rsv that accumulates in plants by oxidative coupling, affected by different biotic and abiotic stresses [11,55]. Thus, it could be supposed that Vf accumulation is independent of the original Rsv content, yet much more correlated to environmental the stresses on the plant material.

Table 6. The Pearson's correlation coefficients between the TPC, *trans*-resveratrol, and ε-viniferin in 23 vine shoot extracts.

	TPC	*Trans*-Resveratrol	ε-Viniferin
TPC	1	-	-
trans-resveratrol	0.626 (p < 0.001)	1	-
ε-viniferin	0.515 (p < 0.001)	−0.059 (p = 0.697)	1

4. Conclusions

Vine shoots are a rich source of bioactive compounds, with Rsv and Vf stilbenes characterised as the most important. The amounts of these stilbenes in the vine shoots could be strongly affected by both extrinsic and intrinsic factors. Our results showed that the heat pre-treatment of the plant material had a negligible effect on the concentration of TPC, Rsv, and Vf. On the other hand, the genotype had a strong influence on Rsv and Vf accumulation. The results of this work confirmed the possibility of obtaining extracts particularly rich in Rsv from Italian vine shoots, assigning an important economic value to a waste product with zero cost.

Thanks to its many applications, resveratrol has great potential in the future market. A recent report shows that the global resveratrol market will reach USD 99.4 million by the end of 2026 [56]. However, its price also depends on the costs of the raw materials and the entire extraction process. Considering this last point, the outcomes of this work impart useful insights proving that there is no need to consume energy for vine shoot pre-treatment, decreasing the general costs. However, more studies are needed to confirm these observations and to investigate the concentration of Rsv and other stilbene compounds in the same vine shoot varieties from different geographical areas or in other Italian varieties.

Supplementary Materials: The following supporting information can be downloaded at: https://www.mdpi.com/article/10.3390/foods11040553/s1, Figure S1: Stilbenes HPLC-DAD chromatogram of cultivar Palieri vine-shoots extract detected at 306 nm (a) and 324 nm (b); Table S1: Stilbene concentrations (mg kg^{-1} DW) in vine-shoots from 23 different Italian varieties. Means and standard deviation (n = 2) are represented in the same column and data followed by different letters indicate statistically significant differences according to Fisher's LSD test ($p < 0.05$). For sample codes see Table 1 of the main text.

Author Contributions: Conceptualization, M.N., A.F.C. and F.C.; methodology, M.N. and A.F.C.; software, V.M.P.; formal analysis, M.N. and A.F.C.; writing—original draft preparation, M.N., A.F.C. and F.C.; writing—review and editing, M.N., A.F.C., G.S., V.M.P., G.G. and F.C.; supervision, F.C.; funding acquisition, F.C. All authors have read and agreed to the published version of the manuscript.

Funding: This research was funded by the Ministero dell'Istruzione, dell'Università e della Ricerca-programmi di ricerca 2017 (2017JTNK78) and "GOOD-BY-WASTE. Obtain GOOD products—exploit BY-products—reduce WASTE" (CUP H98D19000940006).

Data Availability Statement: Not applicable.

Conflicts of Interest: The authors declare no conflict of interest.

References

1. Gindro, K.; Alonso-Villaverde, V.; Viret, O.; Spring, J.-L.; Marti, G.; Wolfender, J.-L.; Pezet, R. Stilbenes: Biomarkers of Grapevine Resistance to Disease of High Relevance for Agronomy, Oenology and Human Health. In *Plant Defence: Biological Control*; Mérillon, J.M., Ramawat, K.G., Eds.; Springer: Dordrecht, The Netherlands, 2012; pp. 25–54. ISBN 9789400719330.
2. Pecyna, P.; Wargula, J.; Murias, M.; Kucinska, M. More than Resveratrol: New Insights into Stilbene-Based Compounds. *Biomolecules* **2020**, *10*, 1111. [CrossRef] [PubMed]
3. Kiskova, T.; Kubatka, P.; Büsselberg, D.; Kassayova, M. The Plant-Derived Compound Resveratrol in Brain Cancer: A Review. *Biomolecules* **2020**, *10*, 161. [CrossRef]
4. Kubyshkin, A.; Shevandova, A.; Petrenko, V.; Fomochkina, I.; Sorokina, L.; Kucherenko, A.; Gordienko, A.; Khimich, N.; Zyablitskaya, E.; Makalish, T.; et al. Anti-inflammatory and antidiabetic effects of grape-derived stilbene concentrate in the experimental metabolic syndrome. *J. Diabetes Metab. Disord.* **2020**, *19*, 1205–1214. [CrossRef] [PubMed]
5. Vion, E.; Page, G.; Bourdeaud, E.; Paccalin, M.; Guillard, J.; Bilan, A.R. Trans ε-viniferin is an amyloid-β disaggregating and anti-inflammatory drug in a mouse primary cellular model of Alzheimer's disease. *Mol. Cell. Neurosci.* **2018**, *88*, 1–6. [CrossRef] [PubMed]
6. Dyck, G.J.B.; Raj, P.; Zieroth, S.; Dyck, J.R.B.; Ezekowitz, J.A. The Effects of Resveratrol in Patients with Cardiovascular Disease and Heart Failure: A Narrative Review. *Int. J. Mol. Sci.* **2019**, *20*, 904. [CrossRef]
7. Anna Malinowska, M.; Billet, K.; Drouet, S.; Munsch, T.; Unlubayir, M.; Tungmunnithum, D.; Giglioli-Guivarc'H, N.; Hano, C.; LaNoue, A. Grape Cane Extracts as Multifunctional Rejuvenating Cosmetic Ingredient: Evaluation of Sirtuin Activity, Tyrosinase Inhibition and Bioavailability Potential. *Molecules* **2020**, *25*, 2203. [CrossRef]
8. Wahedi, H.M.; Ahmad, S.; Abbasi, S.W. Stilbene-based natural compounds as promising drug candidates against COVID-19. *J. Biomol. Struct. Dyn.* **2020**, *39*, 1–10. [CrossRef]
9. Rayne, S.; Karacabey, E.; Mazza, G. Grape cane waste as a source of trans-resveratrol and trans-viniferin: High-value phytochemicals with medicinal and anti-phytopathogenic applications. *Ind. Crop. Prod.* **2008**, *27*, 335–340. [CrossRef]
10. Santra, H.K.; Banerjee, D. Natural Products as Fungicide and their Role in Crop Protection. In *Natural Bioactive Products in Sustainable Agriculture*; Springer: Singapore, 2020; pp. 131–219. ISBN 978-981-15-3023-4.
11. Valletta, A.; Iozia, L.M.; Leonelli, F. Impact of Environmental Factors on Stilbene Biosynthesis. *Plants* **2021**, *10*, 90. [CrossRef]
12. Rivière, C.; Pawlus, A.D.; Mérillon, J.-M. Natural stilbenoids: Distribution in the plant kingdom and chemotaxonomic interest in Vitaceae. *Nat. Prod. Rep.* **2012**, *29*, 1317–1333. [CrossRef]
13. Benbouguerra, N.; Hornedo-Ortega, R.; Garcia, F.; El Khawand, T.; Saucier, C.; Richard, T. Stilbenes in grape berries and wine and their potential role as anti-obesity agents: A review. *Trends Food Sci. Technol.* **2021**, *112*, 362–381. [CrossRef]
14. Goufo, P.; Singh, R.K.; Cortez, I. A Reference List of Phenolic Compounds (Including Stilbenes) in Grapevine (*Vitis vinifera* L.) Roots, Woods, Canes, Stems, and Leaves. *Antioxidants* **2020**, *9*, 398. [CrossRef] [PubMed]
15. Aliaño-González, M.J.; Richard, T.; Cantos-Villar, E. Grapevine Cane Extracts: Raw Plant Material, Extraction Methods, Quantification, and Applications. *Biomolecules* **2020**, *10*, 1195. [CrossRef] [PubMed]

16. Troilo, M.; Difonzo, G.; Paradiso, V.; Summo, C.; Caponio, F. Bioactive Compounds from Vine Shoots, Grape Stalks, and Wine Lees: Their Potential Use in Agro-Food Chains. *Foods* **2021**, *10*, 342. [CrossRef]
17. OIV State of the World Vitivinicultural Sector in 2020. Available online: http://www.oiv.int/public/medias/7298/oiv-state-of-the-vitivinicultural-sector-in-2019.pdf (accessed on 12 November 2021).
18. Guerrero, R.F.; Biais, B.; Richard, T.; Puertas, B.; Waffo-Teguo, P.; Merillon, J.-M.; Cantos-Villar, E. Grapevine cane's waste is a source of bioactive stilbenes. *Ind. Crop. Prod.* **2016**, *94*, 884–892. [CrossRef]
19. Peralbo-Molina, Á.; de Castro, M.D.L. Potential of residues from the Mediterranean agriculture and agrifood industry. *Trends Food Sci. Technol.* **2013**, *32*, 16–24. [CrossRef]
20. Gómez, S.J.; Causapé, M.C.; Martínez, A.A. Distribution of nutrients in anaerobic digestion of vine shoots. *Bioresour. Technol.* **1993**, *45*, 93–97. [CrossRef]
21. Jiménez, L.; Angulo, V.; Ramos, E.; De la Torre, M.; Ferrer, J. Comparison of various pulping processes for producing pulp from vine shoots. *Ind. Crop. Prod.* **2006**, *23*, 122–130. [CrossRef]
22. Mendívil, M.A.; Muñoz, P.; Morales, M.P.; Juárez, M.C.; García-Escudero, E. Chemical characterization of pruned vine shoots from La Rioja (Spain) for obtaining solid bio-fuels. *J. Renew. Sustain. Energy* **2013**, *5*, 033113. [CrossRef]
23. Corcho-Corral, B.; Olivares-Marín, M.; Fernández-González, C.; Gómez-Serrano, V.; Macías-García, A. Preparation and textural characterisation of activated carbon from vine shoots (*Vitis vinifera*) by H3PO4—Chemical activation. *Appl. Surf. Sci.* **2005**, *252*, 5961–5966. [CrossRef]
24. Cebrián-Tarancón, C.; Sánchez-Gómez, R.; Cabrita, M.J.; García, R.; Zalacain, A.; Alonso, G.L.; Salinas, M.R. Winemaking with vine-shoots. Modulating the composition of wines by using their own resources. *Food Res. Int.* **2019**, *121*, 117–126. [CrossRef] [PubMed]
25. Fanzone, M.; Catania, A.; Assof, M.; Jofré, V.; Prieto, J.; Gil Quiroga, D.; Sottano, J.L.; Sari, S. Application of Vine-Shoot Chips during Winemaking and Aging of Malbec and Bonarda Wines. *Beverages* **2021**, *7*, 51. [CrossRef]
26. Zwingelstein, M.; Draye, M.; Besombes, J.-L.; Piot, C.; Chatel, G. Viticultural wood waste as a source of polyphenols of interest: Opportunities and perspectives through conventional and emerging extraction methods. *Waste Manag.* **2019**, *102*, 782–794. [CrossRef] [PubMed]
27. Gutiérrez-Escobar, R.; Fernández-Marín, M.I.; Richard, T.; Fernández-Morales, A.; Carbú, M.; Cebrian-Tarancón, C.; Torija, M.J.; Puertas, B.; Cantos-Villar, E. Development and characterization of a pure stilbene extract from grapevine shoots for use as a preservative in wine. *Food Control* **2020**, *121*, 107684. [CrossRef]
28. Zhang, A.; Fang, Y.; Li, X.; Meng, J.; Wang, H.; Li, H.; Zhang, Z.; Guo, Z. Occurrence and Estimation of trans-Resveratrol in One-Year-Old Canes from Seven Major Chinese Grape Producing Regions. *Molecules* **2011**, *16*, 2846–2861. [CrossRef]
29. Vergara, C.; Von Baer, D.; Mardones, C.; Wilkens, A.; Wernekinck, K.; Damm, A.; Macke, S.; Gorena, T.; Winterhalter, P. Stilbene Levels in Grape Cane of Different Cultivars in Southern Chile: Determination by HPLC-DAD-MS/MS Method. *J. Agric. Food Chem.* **2012**, *60*, 929–933. [CrossRef]
30. Lambert, C.; Richard, T.; Renouf, E.; Bisson, J.; Waffo-Téguo, P.; Bordenave, L.; Ollat, N.; Mérillon, J.-M.; Cluzet, S. Comparative Analyses of Stilbenoids in Canes of Major *Vitis vinifera* L. Cultivars. *J. Agric. Food Chem.* **2013**, *61*, 11392–11399. [CrossRef]
31. Zwingelstein, M.; Draye, M.; Besombes, J.-L.; Piot, C.; Chatel, G. trans-Resveratrol and trans-ε-Viniferin in Grape Canes and Stocks Originating from Savoie Mont Blanc Vineyard Region: Pre-extraction Parameters for Improved Recovery. *ACS Sustain. Chem. Eng.* **2019**, *7*, 8310–8316. [CrossRef]
32. Guerrero, R.F.; Aliaño-González, M.J.; Puertas, B.; Richard, T.; Cantos-Villar, E. Comparative analysis of stilbene concentration in grapevine shoots of thirteen Vitis during a three-year study. *Ind. Crop. Prod.* **2020**, *156*, 112852. [CrossRef]
33. Gorena, T.; Sáez, V.; Mardones, C.; Vergara, C.; Winterhalter, P.; von Baer, D. Influence of post-pruning storage on stilbenoid levels in *Vitis vinifera* L. canes. *Food Chem.* **2014**, *155*, 256–263. [CrossRef]
34. Houillé, B.; Besseau, S.; Courdavault, V.; Oudin, A.; Glévarec, G.; Delanoue, G.; Guérin, L.; Simkin, A.J.; Papon, N.; Clastre, M.; et al. Biosynthetic Origin of E-Resveratrol Accumulation in Grape Canes during Postharvest Storage. *J. Agric. Food Chem.* **2015**, *63*, 1631–1638. [CrossRef] [PubMed]
35. Cebrián, C.; Sánchez-Gómez, R.; Salinas, M.R.; Alonso, G.L.; Zalacain, A. Effect of post-pruning vine-shoots storage on the evolution of high-value compounds. *Ind. Crop. Prod.* **2017**, *109*, 730–736. [CrossRef]
36. Billet, K.; Houillé, B.; Besseau, S.; Mélin, C.; Oudin, A.; Papon, N.; Courdavault, V.; Clastre, M.; Giglioli-Guivarc'H, N.; Lanoue, A. Mechanical stress rapidly induces E-resveratrol and E-piceatannol biosynthesis in grape canes stored as a freshly-pruned byproduct. *Food Chem.* **2018**, *240*, 1022–1027. [CrossRef] [PubMed]
37. Billet, K.; Unlubayir, M.; Munsch, T.; Malinowska, M.A.; de Bernonville, T.D.; Oudin, A.; Courdavault, V.; Besseau, S.; Giglioli-Guivarc'H, N.; Lanoue, A. Postharvest Treatment of Wood Biomass from a Large Collection of European Grape Varieties: Impact on the Selection of Polyphenol-Rich Byproducts. *ACS Sustain. Chem. Eng.* **2021**, *9*, 3509–3517. [CrossRef]
38. Ewald, P.; Delker, U.; Winterhalter, P. Quantification of stilbenoids in grapevine canes and grape cluster stems with a focus on long-term storage effects on stilbenoid concentration in grapevine canes. *Food Res. Int.* **2017**, *100*, 326–331. [CrossRef]
39. Piñeiro, Z.; Marrufo-Curtido, A.; Serrano, M.J.; Palma, M. Ultrasound-Assisted Extraction of Stilbenes from Grape Canes. *Molecules* **2016**, *21*, 784. [CrossRef]
40. Sánchez-Gómez, R.; Zalacain, A.; Alonso, G.; Salinas, M. Effect of toasting on non-volatile and volatile vine-shoots low molecular weight phenolic compounds. *Food Chem.* **2016**, *204*, 499–505. [CrossRef]

41. Singh, R.P.; Heldman, D.R. *Introduction to Food Engineering*, 3rd ed.; Gulf Professional Publishing: Oxford, UK, 2001; ISBN 9780080574493.
42. Difonzo, G.; Russo, A.; Trani, A.; Paradiso, V.M.; Ranieri, M.; Pasqualone, A.; Summo, C.; Tamma, G.; Silletti, R.; Caponio, F. Green extracts from Coratina olive cultivar leaves: Antioxidant characterization and biological activity. *J. Funct. Foods* **2017**, *31*, 63–70. [CrossRef]
43. Difonzo, G.; Aresta, A.; Cotugno, P.; Ragni, R.; Squeo, G.; Summo, C.; Massari, F.; Pasqualone, A.; Faccia, M.; Zambonin, C.; et al. Supercritical CO_2 Extraction of Phytocompounds from Olive Pomace Subjected to Different Drying Methods. *Molecules* **2021**, *26*, 598. [CrossRef]
44. Tarantino, A.; Difonzo, G.; Lopriore, G.; Disciglio, G.; Paradiso, V.M.; Gambacorta, G.; Caponio, F. Bioactive compounds and quality evaluation of 'Wonderful' pomegranate fruit and juice as affected by deficit irrigation. *J. Sci. Food Agric.* **2020**, *100*, 5539–5545. [CrossRef]
45. Delgado-Torre, M.P.; Ferreiro-Vera, C.; Priego-Capote, F.; Pérez-Juan, P.M.; de Castro, M.D.L. Comparison of Accelerated Methods for the Extraction of Phenolic Compounds from Different Vine-Shoot Cultivars. *J. Agric. Food Chem.* **2012**, *60*, 3051–3060. [CrossRef] [PubMed]
46. Sánchez-Gómez, R.; Zalacain, A.; Alonso, G.L.; Salinas, M.R. Vine-Shoot Waste Aqueous Extracts for Re-use in Agriculture Obtained by Different Extraction Techniques: Phenolic, Volatile, and Mineral Compounds. *J. Agric. Food Chem.* **2014**, *62*, 10861–10872. [CrossRef] [PubMed]
47. Sánchez-Gómez, R.; Sánchez-Vioque, R.; Santana-Méridas, O.; Martín-Bejerano, M.; Alonso, G.; Salinas, M.; Zalacain, A. A potential use of vine-shoot wastes: The antioxidant, antifeedant and phytotoxic activities of their aqueous extracts. *Ind. Crop. Prod.* **2017**, *97*, 120–127. [CrossRef]
48. Moreira, M.M.; Barroso, M.F.; Porto, J.V.; Ramalhosa, M.J.; Švarc-Gajić, J.; Estevinho, M.L.M.F.; Morais, S.; Delerue-Matos, C. Potential of Portuguese vine shoot wastes as natural resources of bioactive compounds. *Sci. Total Environ.* **2018**, *634*, 831–842. [CrossRef] [PubMed]
49. Farhadi, K.; Esmaeilzadeh, F.; Hatami, M.; Forough, M.; Molaei, R. Determination of phenolic compounds content and antioxidant activity in skin, pulp, seed, cane and leaf of five native grape cultivars in West Azerbaijan province, Iran. *Food Chem.* **2016**, *199*, 847–855. [CrossRef]
50. Dorosh, O.; Moreira, M.M.; Rodrigues, F.; Peixoto, A.F.; Freire, C.; Morais, S.; Delerue-Matos, C. Vine-Canes Valorisation: Ultrasound-Assisted Extraction from Lab to Pilot Scale. *Molecules* **2020**, *25*, 1739. [CrossRef]
51. Çetin, E.S.; Altinöz, D.; Tarçan, E.; Baydar, N.G. Chemical composition of grape canes. *Ind. Crop. Prod.* **2011**, *34*, 994–998. [CrossRef]
52. Karacabey, E.; Mazza, G. Optimisation of antioxidant activity of grape cane extracts using response surface methodology. *Food Chem.* **2010**, *119*, 343–348. [CrossRef]
53. Vannozzi, A.; Dry, I.B.; Fasoli, M.; Zenoni, S.; Lucchin, M. Genome-wide analysis of the grapevine stilbene synthase multigenic family: Genomic organization and expression profiles upon biotic and abiotic stresses. *BMC Plant Biol.* **2012**, *12*, 130. [CrossRef]
54. Parage, C.; Tavares, R.; Rety, S.; Baltenweck-Guyot, R.; Poutaraud, A.; Renault, L.; Heintz, D.; Lugan, R.; Marais, G.A.; Aubourg, S.; et al. Structural, Functional, and Evolutionary Analysis of the Unusually Large Stilbene Synthase Gene Family in Grapevine. *Plant Physiol.* **2012**, *160*, 1407–1419. [CrossRef]
55. Dubrovina, A.S.; Kiselev, K.V. Regulation of stilbene biosynthesis in plants. *Planta* **2017**, *246*, 597–623. [CrossRef] [PubMed]
56. Global Resveratrol Market Research Report 2020. Available online: https://www.industryresearch.co/global-resveratrol-market-15064120 (accessed on 8 December 2021).

foods

MDPI

Article

Sustainable Drying and Green Deep Eutectic Extraction of Carotenoids from Tomato Pomace

Celeste Lazzarini [1], Enrico Casadei [1], Enrico Valli [1,2,*], Matilde Tura [3], Luigi Ragni [1,2], Alessandra Bendini [1,2] and Tullia Gallina Toschi [2,3]

[1] Department of Agricultural and Food Sciences (DISTAL), Alma Mater Studiorum—Università di Bologna, Piazza Goidanich 60, 47521 Cesena, Italy; celeste.lazzarini3@unibo.it (C.L.); enrico.casadei15@unibo.it (E.C.); luigi.ragni@unibo.it (L.R.); alessandra.bendini@unibo.it (A.B.)
[2] Interdepartmental Centre for Industrial Agrofood Research (CIRI Agroalimentare), Alma Mater Studiorum—Università di Bologna, Via Quinto Bucci 336, 47521 Cesena, Italy; tullia.gallinatoschi@unibo.it
[3] Department of Agricultural and Food Sciences (DISTAL), Alma Mater Studiorum—Università di Bologna, Viale Fanin, 40, 40127 Bologna, Italy; matilde.tura2@unibo.it
* Correspondence: enrico.valli4@unibo.it; Tel.: +39-0547-338116

Abstract: The extraction of molecules with high added value plays an important role in the recovery of food waste. This work aimed to valorize tomato pomace, a by-product composed of skin and seeds, through extraction of carotenoids, especially lycopene and β-carotene. The tomato pomace was dried using three different methods (freeze-drying, heat drying, and non-thermal air-drying) to reduce its weight, volume, and water activity and to concentrate the carotenoid fraction. These drying approaches were compared considering the extractive potential. Three solvent mixtures were compared, a traditional one (n-hexane:acetone) and two green deep eutectic solvent mixtures (ethyl acetate:ethyl lactate and menthol:lactic acid) in combination with different drying procedures. The extract obtained using ethyl acetate:ethyl lactate with non-thermal air-drying showed the highest contents of lycopene and β-carotene (75.86 and 3950.08 μg/g of dried sample, respectively) compared with the other procedures.

Keywords: tomato pomace; lycopene; β-carotene; extraction; sustainability; food by-products; deep eutectic solvents; non-thermal drying

Citation: Lazzarini, C.; Casadei, E.; Valli, E.; Tura, M.; Ragni, L.; Bendini, A.; Gallina Toschi, T. Sustainable Drying and Green Deep Eutectic Extraction of Carotenoids from Tomato Pomace. *Foods* **2022**, *11*, 405. https://doi.org/10.3390/foods11030405

Academic Editors: Francesco Caponio, Marco Poiana and Antonio Piga

Received: 24 December 2021
Accepted: 24 January 2022
Published: 30 January 2022

Publisher's Note: MDPI stays neutral with regard to jurisdictional claims in published maps and institutional affiliations.

Copyright: © 2022 by the authors. Licensee MDPI, Basel, Switzerland. This article is an open access article distributed under the terms and conditions of the Creative Commons Attribution (CC BY) license (https://creativecommons.org/licenses/by/4.0/).

1. Introduction

The cultivated tomato, *Solanum lycopersicum* L., is the world's most highly consumed vegetable, thanks to its versatility and its status as an ingredient in a large variety of different foods [1]. In 2018, it was estimated that nearly 5 billion hectares are cultivated to produce tomatoes, with an average production of more than 182 billion tons worldwide [2].

Tomato skin and seeds are usually undesirable parts for the preparation of most tomato derivates, and thus there is the need for separation from the pulp. The produced by-product accounts for 1.5 to 5% of the initial weight of the fruit, which is a concerning amount of material considering its widespread cultivation [3–6]. This tomato by-product, called tomato pomace, is rich in fiber and other important compounds such as sugars, proteins, pectins, fats, and vitamins [3]. The peel has a higher content of fibers, carotenoids, and phenols than seeds, which mainly consist of oil and proteins [4,7–10].

Tomatoes and tomato products meet consumer demands in terms of cost, convenience, availability, and taste, while they also deliver beneficial health effects, being easily included in a large variety of culturally diverse dishes [11]. Carotenoids are naturally present in tomatoes; among these is lycopene, which is mainly responsible for the red color of the fruit with a claimed nutraceutical effect [12], and it also appears to act as a powerful antioxidant, preventing the action of free radicals and carcinogenic cells [13,14]. Tomato can play an important role in valorization of by-products, not only for feedstock and production of biofuels [15–17], but also through the extraction of important compounds, including lycopene [18]. In fact, lycopene can be used in cosmetic preparations due to its properties of reducing lipid oxidation and preventing damage related to the action of UV rays [18]. The antioxidant activity of carotenoids can also be applied to the food supplement field [19] and in the meat industry against lipid oxidation and as colorants, thus improving the oxidative stability of meat products [20]. Indeed, natural antioxidants are safe to be consumed and can be generally obtained with fewer rough chemical processes rather than synthetic ones; they can be extracted and applied in different food sectors in addition to the meat industry [21]. The food industry nowadays prefers "green" solvents for the extraction of these compounds, due to their non-toxic features, food safety, and recycling possibilities [22].

Moreover, colors play an important role in the marketing success of any food product and can often influence consumer preferences [23,24]; following the trend of more interest in natural products, even natural colorants are preferred as a healthier alternative to synthetic ones [25].

Lycopene is responsible for the bright reddish color of tomato and tomato-based products: its color is due to its chemical characteristics and its eleven linear conjugated double bonds in a polyene chain, which is able to absorb almost all visible light radiation while reflecting low-frequency wavelengths [26]. Due to its strong shade and absence of toxicity, this pigment is used for a wide range of applications in food industries [27].

β-carotene is also an important natural colorant, which is responsible for the orange color of tomato and other vegetables such as carrots [28]; natural pigments are responsible for the appearance of fruits and vegetables and for their visual attractiveness, and their consumption is associated with health benefits, such as the decreased risk of developing of some diseases [28,29].

Lycopene has gained increasing importance thanks to its unique properties; in this context, the extraction of carotenoids is also fundamental to take advantage of their full potential [30]. According to FAOSTAT [31], the total production of food as the major source of lycopene (namely tomatoes, pumpkins, and melons) has increased annually as has the associated waste production, which are usually lycopene-rich parts [32]. The global nutraceutical market reached a value of USD 289.8 billion in 2021, including 1.5 billion for carotenoids in 2017 and USD 2.0 billion in 2022 [33,34], while lycopene accounts for about 7% of the total market [35].

β-carotene has also gained increasing importance thanks to its antioxidant properties [36]; it also has other interesting properties such as protection against cardiovascular disease and cancer, improvement of the immune system, filtration of UV-light, and anti-inflammatory activities [37].

As for lycopene, β-carotene has also seen a major demand, reaching high prices (more than 500 EUR per kg), making tomato pomace a profitable source of this important biological compound [38,39]. Moreover, traditional solvent extraction may have low efficiency with consumption of large amounts of solvent and time [40–42]; moreover, the solvents may be harmful for both the environment and human health [43,44]. Organic solvents can easily enter body parts and organs, where they are converted via osmotic processes into water-soluble forms, which sometimes can be more toxic than the parent compound [45].

Organic solvents are highly volatile and are more likely to be inhaled via respiration, affecting the lungs and other organs of operators; therefore, replacement with greener and less toxic solvents is becoming increasingly important due to increasing health and environmental concerns [45,46]. Moreover, new emerging technologies, such as ultrasound-assisted extraction, can promote lycopene extraction while preserving this important compound for human and environmental health [32].

In particular, to avoid the potentially deleterious effects of traditional solvents, new classes of reagents have become more popular, such as deep eutectic solvents (DES), a particular class of solvents considered to be environmentally friendly [47]. A DES mixture is made of two or more compounds that are typically solid at room temperature, but when mixed at particular ratios, change into a liquid [48]. DES can be easily synthesized by combining a hydrogen bond acceptor (HBA) and a hydrogen bond donor (HBD), and their combination results in changes in the physical characteristics of the mixture (e.g., melting point, solubility) [47].

Therefore, their biodegradability is extraordinarily high, and their toxicity is nonexistent or very low. Because of their minimal ecological footprint, low cost of their constituents, tunability of their physicochemical properties, and ease of preparation, DESs are successfully and progressively replacing often hazardous and volatile organic compounds (VOCs) in many fields of science [49].

DESs have several properties that make them suitable for different types of extraction: they are not derived from petroleum, while they are cheap, easy to prepare, and biodegradable and have high purity [50]. In addition, one of the major advantages related to the use of DESs is their tunability: the physical characteristics of a large number of eutectic mixtures can be varied (such as viscosity, density, etc.) by simply changing one or more of the components of the mixture [47].

Concerning green extraction, a variety of techniques can be considered, one of which is supercritical fluid extraction (SFE). This technique is increasingly used, thanks to its versatility that allows one to fine-tune the solvent; this can be done by modifying its polarity, e.g., with ethanol, according to the polarity of the target compounds. Moreover, when using CO_2, since it is volatile at ambient conditions, it can be easily separated from the extract, with benefits for health and the environment [51]. The use of supercritical CO_2 can successfully extract thermolabile compounds (generally at temperatures around 60 °C), such as carotenoids, considering the inert characteristics of this solvent, namely non-explosive and non-toxic characteristics [51]. Technologies such as supercritical fluid extraction require particular equipment that is comprehensive of pumps, stainless steel extraction vessels, pipes, etc., and therefore upscaling of this technology would be very expensive. However, once the facilities are present, the required solvent is easily available and inexpensive [52]. Generally, due to high costs, this technique is frequently used to extract valuable compounds such as the those used in the food, pharmaceutical, and cosmetic industries [52].

Drying methods are fundamental to enhance extraction yields and improve durability, especially considering the high water content of by-products typical of the food industry; one of the most common is freeze-drying, which is one of the best techniques to prevent the spoilage of samples and to preserve composition [53]. Even if freeze-drying is able to drastically reduce the water content of dried material, it requires large amounts of energy and is also expensive, especially depending on the material to be dried; it has been estimated that the cost of freeze-drying is nearly eight times higher than the cost of air drying, which can be applied for high-value products and compounds [54,55].

Other drying techniques involve the use of heat, such as oven drying, which involves exposure to high temperatures with adverse effects on the nutritional and chemical properties [56]. The use of an innovative prototype of non-thermal air-drying has the potential to overcome the drawbacks related to the use of thermal energy.

This research tested three different solvent mixtures on a tomato pomace composed of peels and seeds, comparing their extractive potential in terms of lycopene and β-carotene content. Furthermore, three drying methods, namely freeze-drying, heat drying in a conventional oven, and nonthermal air-drying with the use of a prototype, were applied to favor the action of the solvents used. This study thus investigated the possible substitutions of traditional extraction techniques and drying methods with more sustainable ones with comparable extractive potentials, which is an essential aspect for any industrial application.

2. Materials and Methods

The raw material used was tomato pomace, made of skins and seeds, from the industrial production of tomato purée. After collection—performed in an Italian company, named La Cesenate Conserve Alimentari S.p.a. located in Cesena, immediately after production of the purée—pomace was stored, packed in plastic bags containing 5 kg each, in a freezer at −40 °C until extraction tests in order to preserve its characteristics.

2.1. Water Content and Water Activity (a_w)

Preliminary analyses of the physical characteristics of the tomato by-product were performed. To quantify the water content, a gravimetric method was used by weighing 5 g of tomato pomace and drying it with the use of a traditional lab oven (M20-VN, PID system, Monza Brianza, Italy) at 105 °C. The samples were kept in the oven until their weight was stable (around 8 h), and the analysis was carried out in three replicates.

The water content was calculated with the following equation:

$$Water\ content\ \% = \frac{m_h - m_d}{m_h} 100 \quad (1)$$

where m_h refers to the weight of the humid tomato pomace and m_d of the dried material.

For a_w, the analysis was done after defrosting at 20 °C the tomato pomace in specific plastic containers, placing the product on the bottom of the disk in such a way to cover its entire surface, and quantified using Aqualab (Series 3, Decagon Devices Inc., Pullman, WA, USA).

Both water content and water activity analyses were repeated in three replicates for the raw tomato pomace and on each dried by-product.

2.2. Drying Treatments

To enhance the effectiveness of the solvent mixtures to extract carotenoids from the by-product (see Section 2.4), the following procedures were used. For (lab-scale) freeze-drying (Heto PowerDry LL3000 Freeze Dryer, Thermo Fisher Scientifics, Warminster, United Kingdom), the by-product was stored for 12 h at −40 °C and was then freeze-dried within a 27 h treatment cycle (the samples were coded as "L"). For air-drying treatment (sample code "E"), a prototype previously developed at Interdepartmental Centre for Industrial Agrofood Research (CIRI Agro, Cesena, Italy) that does not use thermal energy was used, thus preserving the by-product from degradation. The prototype is composed of a pseudo stadium-shaped chamber in which the material is set. The movement of a blade rotor, together with a flux of compressed air, allows ventilation, closed-loop circulation, and crunching of the tomato pomace. The overall external dimensions of the chamber, made of stainless steel, are 80 cm × 50 cm, with a height of 25 cm, while the four-blade helical rotor has a diameter of 24 cm. It is rotated by a 0.37 kW three-phase asynchronous motor. The rotation speed can be regulated using an inverter up to its maximum nominal value of 2790 min^{-1}. The compressed air flow is injected tangentially to promote circulation of the air-product flow. The drying procedure took 2.5 h, in which the by-product was mixed to guarantee homogeneity every 30 min.

The procedure was also carried out through a traditional heating laboratory oven (PID system M20-VN) (sample code "S") set at 85 °C, chosen as a temperature to preserve as much as possible the characteristics of tomato pomace based on the literature [57]. To reach a comparable a_w with respect to the air-dried material, the process took 2 h.

2.3. Carotenoids Extraction Methods

The extraction methods described herein use different categories of solvents, both traditional and green. For traditional solvents, an *n*-hexane-acetone (1:1 % *v*/*v*) (*n*-hexane purchased from Sigma Aldrich, with purity ≥97%, Darmstadt, Germany and acetone purchased from Carlo Erba Reagents, with purity ≥99.8%, Cornaredo, Milan, Italy) mixture was tested (the extracts were coded as "T"), while DES DL-menthol and lactic acid solution (1:8 molar ratio) (code "ML") (DL menthol from Sigma Aldrich, purity ≥98%, Darmstadt, Germany, and lactic acid Sigma Aldrich, purity ≥85 %, Darmstadt, Germany), ethyl acetate (Sigma Aldrich, purity ≥99.8%, Darmstadt, Germany), and ethyl lactate (Sigma Aldrich, purity ≥98%, Darmstadt, Germany) solution (30% of ethyl acetate and 70% of ethyl lactate) (code "G") were used as the DES solvents. The procedure follows previously developed analytical protocols and applied in the literature on the same matrix [50,58].

To compare the three extraction procedures, it is fundamental to adopt the same conditions in terms of temperature, time, and solvent/sample volume/mass ratio: according to the literature, the parameters that favor carotenoid extraction are 63 °C for 20 min in ultrasonic bath (2209 S3 Ultrasonic Cleaner, Milan, Italy), with 100 mL/mg solvent/sample, both for *n*-hexane:acetone solution and ethyl acetate:ethyl lactate, but 120 mg/L for the DL-menthol:lactic acid solution [50,58]. As reported by Silva and colleagues [50,58], the solvents were prepared following the following ratios: 1:1 % *v/v* *n*-hexane:acetone, 70:30 *v/v* ethyl acetate:ethyl lactate, 8:1 mol/mol DL-menthol:lactic acid.

After preparation of the three above-mentioned solvent mixtures, 0.2 g of raw by-product was weighed for each sample, doubling the quantities suggested by the operation procedures tested by Silva and colleagues [50,58] due to the high moisture content of tomato pomace. When the extraction was performed on treated by-product (freeze-dried, air-dried and heat-dried), 0.1 g of by-product was considered. Next, 10 mL of *n*-hexane:acetone was added to a test tube for a trial, 10 mL of ethyl acetate:ethyl lactate for another, and lastly 12 mL of DL-menthol:lactic acid for another, in three replicates each; each solvent mixture was heated at 60 °C before the interaction with the by-product. After this pre-heating, the solvent mixture was mixed with the tomato pomace using a vortex (Vortex agitator Zx3, VELP Scientifica Srl, Monza Brianza, Italy) for 10 s and placed in an ultrasound water bath at 60 °C (±1 °C) for 20 min.

Finally, the glass tubes were removed from the water bath and centrifuged (MPW M-SCIENCE, MPW Med. Instruments, Warsaw, Poland) at 2916 rpm for 10 min to obtain the separation of the solid and liquid phases. The liquid extract was withdrawn and transferred into another glass tube for spectrophotometric analyses.

2.4. Spectrophotometric Analyses

The quantification of carotenoid content was performed by spectrophotometric analysis (spectrophotometer UV-1800, Shimadzu, Kyoto, Japan) using specific calibration curves for lycopene (Supelco, purity ≥90% Darmstadt, Germany) and β-carotene (Sigma Aldrich, purity ≥97%, Darmstadt, Germany) constructed by dissolving the related standards at different concentrations in dichloromethane solutions. Dichloromethane (λ cut-off ≤ 233 nm) (Sigma Aldrich, purity ≥99.9%, Darmstadt, Germany) was chosen because it can dissolve both carotenoids; moreover, preliminary tests were performed to check the spectra between 250 and 600 nm of lycopene and β-carotene solutions dissolved in dichloromethane and in the other solvents (*n*-hexane/acetone, ethyl acetate/ethyl lactate and menthol/lactic acid). The resulting peaks were very similar for each solution. Absorbance peaks were identified for lycopene at 477 nm and 481 nm, and for β-carotene at 461 nm. Both wavelengths were considered for lycopene (477 nm for the extractions using the green solvent solutions according to the literature [58] and 481 nm for the extracts obtained with the conventional one according to the absorbance peak obtained from the traditional extract). The calibration curve of lycopene at 481 nm has the equation y = 0.2481x (R^2 = 0.9957), the one at 477 nm y = 0.2393x (R^2 = 0.9829), and the curve of β-carotene y = 0.0042x + 0.0265 (R^2 = 0.9928).

The absorbance of each extract was determined using a UV-spectrophotometer at the abovementioned wavelengths against the respective blank (the different solvent mixtures) and linear regression equations ($R^2 > 0.99$) were obtained for the three calibration curves and used to determine the lycopene and β-carotene content (μg/g) in each extract; the absorbance of the solvent (blank) concentrations was expressed as μg carotenoids per g of by-product weighted. A regular check was carried out for the accuracy and reproducibility of the absorbance and wavelength scales as well as for stray light.

Color of the extracts was also determined. A quartz cuvette was filled with the extract, and the transmittance was read on a Jasco dual beam spectrophotometer model V-550 UV-Vis (Jasco, Tokyo, Japan). Results were expressed using the CIEL*a*b* scale.

2.5. Data Processing and Statistical Analysis

Data processing and calculation were carried out with Microsoft® spreadsheet program 2016 (Microsoft Corp., Redmond, WA, USA). Analysis of variance (Analysis of variance (one-way ANOVA, Tukey's HSD, $p < 0.05$) and PCA were performed with XLSTAT (Addinsoft Corp., Paris, France).

3. Results and Discussion

3.1. Water Content and Water Activity (a_w) in the Raw and Treated By-Products

The analyses of moisture level and water activity are relevant to assess how each treatment described in Section 2.3 can preserve the by-product by drying it out. As seen in Table 1, the initial moisture content was 63% while the water activity was 0.99. The freeze-dried by-product had the lowest level of moisture content and a_w, followed by the non–heat-dried by-product obtained with the prototype. This highlights the satisfactory action of the air-drying prototype, since it allows the level of a_w to decrease below 0.7, while the lowest level of a_w was reached with the freeze-drier. According to the literature, at an a_w lower than 0.75, bacterial growth is inhibited [59]. Finally, the heat-dried by-product had a_w values comparable to those obtained with the prototype, thus supporting possible inhibition of bacterial activity despite the higher moisture content observed for this treatment compared to the two other drying methods; these results are similar to those found in the literature for comparable products [60].

Table 1. Mean (±SD) moisture content (%) and mean a_w for each tomato by-product, both raw and treated, as described in Section 2.3. By-product: raw (P), freeze-dried (L), non-thermal air-dried (E), and heat-dried (S).

Sample	Mean Moisture Content% ± SD	Mean a_w
Raw (P)	63.4 ± 1.7	0.99
Freeze-dried (L)	3.5 ± 0.1	0.30
Air-dried (E)	9.9 ± 0.2	0.67
Heat-dried (S)	20.3 ± 0.2	0.69

3.2. Lycopene and β-Carotene Content in the Extracts

The carotenoid content in extracts strongly differ in relation to both the solvent used and the drying method (Table 2).

Concerning the solvent mixture, the one with the highest extraction potential was the traditional one, acetone:*n*-hexane, which had some practical issues related to evaporation of the solvent during the heating/ultrasound-assisted extraction. This could be useful in a closed system, with possible solvent recovery and extract concentration, but in an open system, it is not sustainable. The use of the DES mixture, composed of ethyl acetate:ethyl lactate, showed promising results in terms of amount of lycopene extracted. In fact, the DES eutectic mixture (samples with code "G") was the second-best option for extraction of carotenoids from the raw by-product, and in particular lycopene, compared to the ultrasound-assisted extraction conducted with the use of traditional solvents (code "T" in Table 2).

In fact, from the raw by-product treated with the greener solvent solution, it is possible to obtain extracts with 27.44 µg/g of lycopene (sample G), which is significantly lower than the 34.11 µg/g obtained with acetone:*n*-hexane (sample T). The use of DES allows practical issues to be overcome related to the high volatility of the traditional organic solution. A similar result was seen for β-carotene with 1510.19 µg/g (sample G) obtained with ethyl acetate:ethyl lactate compared with 2117.64 µg/g from the traditional extraction (sample T).

Table 2. Carotenoid content in the extract obtained with the three different solvent solutions in combination with the three drying methods. The samples were named using the combination of codes, identifying the different solvent solution, acetone:n-hexane (1:1 % v/v) (T), ethyl acetate:ethyl lactate (30:70 %v/v) (G), and DL-menthol:lactic acid (8:1 mol/mol) (ML), and the three drying techniques, freeze-drying (L), non-thermal air-drying (E), and heat drying (S) (e.g., MLL: extract obtained from the use of DL-menthol:lactic acid from the freeze-dried by-product).

Extract	Lycopene Content (µg/g) ± SD	β-Carotene Content (µg/g) ± SD
T	34.11 $^{a;B}$ ± 1.11	2117.64 $^{a;B}$ ± 100.46
G	27.44 $^{b;C}$ ± 2.50	1510.19 $^{b;B}$ ± 79.51
ML	12.32 $^{c;C}$ ± 1.78	492.46 $^{c;B}$ ± 87.88
TL	96.55 $^{a;A}$ ± 10.39	5717.46 $^{a;A}$ ± 710.70
GL	56.28 $^{b;B}$ ± 6.07	3000.45 $^{b;B,C}$ ± 26.58
MLL	48.89 $^{b;B}$ ± 1.89	2273.87 $^{b;A}$ ± 92.58
TE	81.20 $^{a;A}$ ± 4.20	4777.59 $^{a;A}$ ± 256.04
GE	75.86 $^{a;A}$ ± 10.94	3950.08 $^{a,b;A}$ ± 597.49
MLE	82.86 $^{a;A}$ ± 11.18	2923.02 $^{b;A}$ ± 583.24
TS	15.87 $^{a;B}$ ± 0.21	507.29 $^{a;C}$ ± 48.56
GS	1.80 $^{c;D}$ ± 0.83	ND
MLS	11.22 $^{b;C}$ ± 0.45	78.27 $^{b;B}$ ± 21.61
Solvent mixture	<0.0001	<0.0001
Drying method (treatment)	<0.0001	<0.0001
Solvent mixture * Drying method (treatment)	<0.0001	<0.0001

Different lowercase letters indicate significant differences (one-way ANOVA, Tukey's HSD, $p < 0.05$) among the extracts obtained by the raw and differently treated by-products using the same solvent mixtures. Different capital letters indicate significant differences (one-way ANOVA, Tukey's HSD, $p < 0.05$) among the extracts, obtained by the raw and the treated by-product with the same drying technology, or non-treatment, using different solvent mixtures. * ND: not detected.

On the other hand, the DL-menthol:lactic acid solution showed difficulties in the interaction between the solvents in the mixture and the by-product due to its viscosity, thus lowering the extraction potential, with 12.32 µg/g and 492.46 µg/g, respectively, for lycopene and β-carotene (sample ML).

Moreover, on the dried by-product, both the drying method and the choice of the extraction mixture had a strong influence on the lycopene and β-carotene content in the final extracts (Table 2). In particular, concerning the drying methods, the prototype (E) showed good potential while preserving the carotenoid content, as shown in Table 3. The use of non-thermal air-drying was superior to freeze-drying in terms of extractions with the DES mixture for both lycopene and β-carotene. The nature of the prototype allows the by-product to be pulverized, and this appears extremely useful, rendering the subsequent interaction between solvent and smashed by-product more effective. Indeed, by reducing the particle size of the raw material, it is possible to enhance the extraction of lycopene and β-carotene; in fact, according to the literature, the extraction potential is inversely proportional to the particles dimension [61]. Indeed, the concentration of lycopene in the extracts obtained from the three different solvent mixtures of the by-product dried with the use of the prototype are comparable: (i) 81.20 µg/g with acetone:n-hexane (sample "TE"), (ii) 75.86 µg/g with ethyl acetate:ethyl lactate (sample "GE"), and (iii) 82.86 µg/g with menthol:lactic acid (sample MLE) (Table 2). These results confirm the high potential of the unconventional non-thermal drying procedure. In fact, it allows the by-product to be uniformly pulverized, thus enhancing the extraction potential, thanks to the synergic action of the flux of air and the rotation of the blade rotor, while preserving the most important biological compounds. For this reason, with non-thermal air-drying, the greener solvents performed at their best in terms of extractive capacity.

Table 3. CIE coordinates of the color space regarding the different samples of tomato by-product extracts. L* indicated the lightness, a* the red/green coordinate, and b* the yellow/blue. Higher values of a* are redder, while higher values of b are more yellow rather than blue.

Sample	a*	b*	L*
T	−6.01 b;C	22.28 a;A	95.31 b;A
TL	−4.85 b;B,C	25.85 a;A	91.82 b;B
TE	−4.32 b;B	22.48 a;A	93.43 a;B
TS	−2.22 a;A	7.91 a;B	96.20 a;A
G	−5.17 a,b;C	18.74 a;A	95.23 b;A
GL	−2.32 a;A,B	7.72 b;B,C	95.27 a;A
GE	−3.17 a;B	9.73 b;B	95.14 a;A
GS	−1.34 a;A	3.10 b;C	95.51 a;A
ML	−4.16 a;C	10.42 b;A	96.23 a;A
MLL	−3.00 a;B	9.17 b;A	95.66 a;B
MLE	−3.15 a;B	9.71 b;A	95.35 a;B
MLS	−1.38 a;A	3.86 b;B	96.44 a;A
Solvent mixture	<0.0001	<0.0001	<0.0001
Drying method (treatment)	<0.0001	<0.0001	<0.0001
Solvent mixture * Drying method (treatment)	<0.0001	0.005	0.010

Different lowercase letters indicate significant differences (one-way ANOVA, Tukey's HSD, $p < 0.05$) among the extracts obtained by the raw and the differently treated by-products using the same solvent mixtures. Different capital letters indicate significant differences among the extracts, obtained by the raw and the treated by-product with the same drying technology using different solvent mixtures. * ND: not detected.

The lycopene concentration of the by-product obviously has a decisive influence on the yield. In fact, Silva and colleagues [58], adopting the same extractive conditions, reported a higher lycopene content (1334.8 μg/g of dried material). Different storage conditions of the by-product in the factory before sampling and diverse tomato variety appear to be responsible for this variability. To verify if it is convenient to extract carotenoids and lycopene from a by-product of the tomato canning industry, the starting point is an evaluation of the quality and quantity of these components in the by-product itself. In addition, the lycopene content in extracts is lower than that of β-carotene; this is due to the previously mentioned storage and processing conditions that can lead to the degradation of lycopene, since its degradation rate is higher than that of β-carotene [62].

Concerning heat drying with the use of the oven (samples coded with "S"), the tests performed show that the lycopene and β-carotene content in the extracts were significantly lower than those found when applying the other drying treatments (Table 2). The main cause may be temperature, which is responsible for the degradation of carotenoids, and, in particular, the total decay of these compounds occurs at 100–145 °C according to the literature, even if a proportion is also affected by the increase in temperature, up to 80 °C for 2 h, used in the oven [57]. It is clear that heat treatments should be avoided when tomato pomace is used, independently of the solvent used for the subsequent extraction, since the carotenoid yield is dramatically reduced (Table 2).

3.3. Color and Principal Component Analysis

The treatment to dry out the by-product and the choice of solvent mixture also showed a significant impact on the color of the extracts, in terms of L*a*b* (Table 3).

Based on some considerations, in terms of quality, explained in Section 3.2, the different extracts are clearly discriminated in a Principal Component Analysis (PCA) biplot (Figure 1), which takes into account, besides the determination of lycopene and β-carotene contents, the results of color determination.

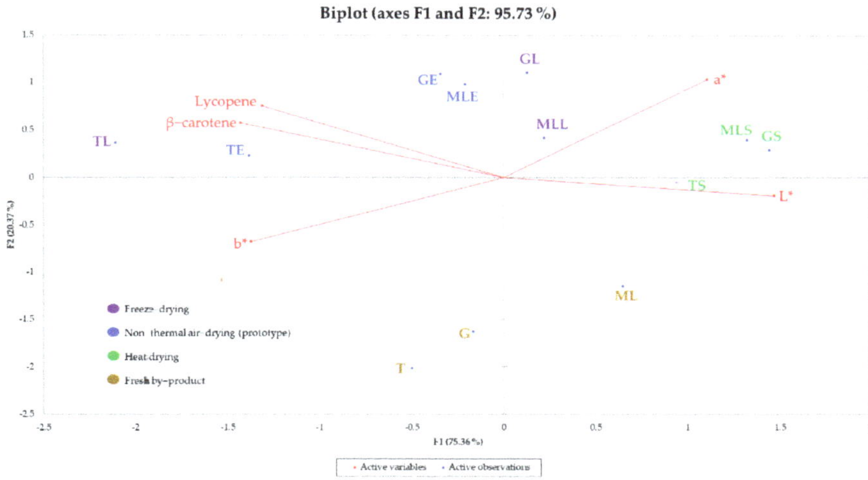

Figure 1. PCA biplot built with the results of the color analysis and the lycopene and β-carotene content for all the samples under consideration.

As shown in Figure 1, the lycopene and β-carotene contents were negatively correlated with the L* (r = −0.807 and r = −0.905, respectively, $p \leq 0.05$). On the other hand, b* was positively correlated with the β-carotene content (r = 0.651, $p \leq 0.05$). The variable b*, if positive, is considered as a yellow index [63]. Extracts obtained from freeze-dried and non-thermal air-dried tomato pomace showed a higher level of carotenoids vs. fresh and heat dried (see Table 2) and higher percentages of the yellow and red components (see Table 3). The differences in color among the samples, also related to the content of lycopene and β-carotene, should be taken into consideration for the formulation of different cosmetic, food, and pharmaceutical products, such as the type of the solvent used for the extraction (e.g., food grade), its residual amount in a completely dried extract, or other possible restrictions through laws.

4. Conclusions

This study demonstrates advancements regarding the action of green solvents on food by-products, in combination with the use of an innovative non-thermal drying method. The results, in particular the lycopene extraction from the non–thermal-dried tomato by-product, confirm ethyl acetate:ethyl lactate mixture as an effective alternative to the traditional solvents.

Regarding the drying methods tested, it is difficult to compare the differences in energy consumption, since the non-thermal drying prototype was produced only at a lab scale. However, its upgrade to an industrial or semi-industrial plant would surely bring significant advantages in terms of sustainability, compared to heating, which is typical of traditional dryers, or long-time freezing at low temperatures for freeze-drying, both of which require high energy consumption. This preliminary work has highlighted the possibility of drying and the potential of non-conventional extraction techniques to extract carotenoids; the use of HPLC to quantify them would be needed in further studies to confirm these findings.

Furthermore, the synergy between a sustainable drying process and the green solvent used for the extraction appears to be a valid strategy to reduce energy consumption and, at the same time, sustain the environment.

In particular, for the tomato by-product and lycopene and β-carotene extraction, the valorization can be part of a project of industrial symbiosis, where the two technological phases—(i) concentration/stabilization and possible packaging of the by-product to guarantee a defined shelf-life and preserve it from spoilage and degradation and (ii) extraction of the fractions/molecules of interest—could be held in two different factories or even industrial chains. The first (stabilization/packaging) could become a new final phase of the tomato supply chain, and the second (extraction) could be conducted elsewhere in a biorefinery or in a specific industrial environment, putting in place tailored flux diagrams for cosmetic, food supplement, and/or pharmaceutical applications.

Author Contributions: Conceptualization, T.G.T., A.B. and E.V.; methodology, C.L., E.C. and E.V.; validation, T.G.T.; formal analysis, C.L., M.T. and L.R.; investigation, C.L., M.T. and L.R.; resources A.B. and T.G.T.; data curation, C.L., M.T. and L.R.; writing—original draft preparation, C.L., M.T. and E.C.; writing—review and editing, C.L., M.T., E.C., E.V., L.R., A.B. and T.G.T.; visualization C.L., M.T., L.R. and E.C.; supervision, T.G.T., A.B. and E.V.; project administration T.G.T. All authors have read and agreed to the published version of the manuscript.

Funding: This research received no external funding.

Data Availability Statement: Data is contained within the article.

Acknowledgments: The authors gratefully acknowledge Claudio Bellini, plant manager of La Cesenate Conserve Alimentari S.p.a., for providing the tomato pomace.

Conflicts of Interest: The authors declare no conflict of interest.

References

1. Organization for Economic Co-operation and Development (OECD). *Safety Assessment of Transgenic Organism in the Environment, OECD Consensus Documents*; Harmonisation of Regulatory Oversight in Biotechnology Series; OECD Publishing: Paris, France, 2017; Volume 7. [CrossRef]
2. FAOSTAT. Available online: http://www.fao.org/faostat/en/#home (accessed on 25 May 2021).
3. Palomo, I.; Concha-Meyer, A.; Lutz, M.; Said, M.; Saez, B.; Vasquez, A.; Fuentes, E. Chemical characterization and antiplatelet potential of bioactive extract from tomato pomace (by-product of tomato paste). *Nutrients* **2019**, *11*, 456. [CrossRef] [PubMed]
4. Kartal, C.; Kaplan Turkoz, B.; Otles, S. Prediction, identification and evaluation of bioactive peptides from tomato seed proteins using in silico approach. *J. Food Meas. Charact.* **2020**, *14*, 1865–1883. [CrossRef]
5. Silva, Y.P.A.; Borba, B.C.; Reis, M.G.; Caliari, M.; Ferreira, T.A.P. Tomato industrial waste as potential source of nutrients. In Proceedings of the X CIGR Section VI International Technical Symposium Food: In the Tree That Sustains Life, Gramado, Brasil, 24–27 October 2016.
6. Vidyarthi, S.K.; Simmons, C.W. Characterization and management strategies for process discharge streams in California industrial tomato. *Sci. Total Environ.* **2020**, *723*, 137976. [CrossRef]
7. Valdez-Morales, M.; Espinosa-Alonso, L.G.; Espinoza-Torres, L.C.; Delgado-Vargas, F.; Medina-Godoy, S. Phenolic content and antioxidant and antimutagenic activities in tomato peel, seeds, and by products. *J. Agric. Food Chem.* **2014**, *62*, 5281–5289. [CrossRef] [PubMed]
8. Sokolov, S.; Deyneka, I.; Katanaeva, Y.; Ugrekhelidze, N.; Yatskova, E.; Kulikova, N. The effectiveness evaluation of the carotenoids extraction from the tomatoes by means of absorption spectrophotometery method. In Proceedings of the E3S Web of Conferences, Online, 29 June 2020; Volume 175, p. 01011.
9. Pandya, D.; Akbari, S.; Bhatt, H.; Joshi, D.C. Standardization of solvents extraction process for lycopene extraction from tomato pomace. *J. Appl. Biotechnol. Bioeng.* **2017**, *2*, 00019. [CrossRef]
10. Lu, Z.; Wang, J.; Gao, R.; Ye, F.; Zhao, G. Sustainable valorisation of tomato pomace: A comprehensive review. *Trends Food Sci. Technol.* **2019**, *86*, 172–187. [CrossRef]
11. Sharma, S.; Katoch, V.; Kumar, S.; Chatterjee, S. Functional relationship of vegetable colors and bioactive compounds: Implications in human health. *J. Nutr. Biochem.* **2021**, *92*, 108615. [CrossRef]
12. Ali, M.Y.; Sina, A.A.I.; Khandker, S.S.; Neesa, L.; Tanvir, E.M.; Kabir, A.; Khalil, M.I.; Gan, S.H. Nutritional composition and bioactive compounds in tomatoes and their impact on human health and disease: A review. *Foods* **2021**, *10*, 45. [CrossRef]
13. Mai, Z.; Ngan, R.; Ng, W.; Yuen, K.; Ip, D.; Chan, Y.; Lee, A.; Ho, S.; Lung, M.; Lam, T. Dietary fiber intake from fresh and preserved food and risk of nasopharyngeal carcinoma: Observational evidence from a Chinese population. *Nutr. J.* **2021**, *20*, 110. [CrossRef]
14. Li, N.; Wu, X.; Zhuang, W.; Xia, L.; Chen, Y.; Wu, C.; Rao, Z.; Du, L.; Zhao, R.; Yi, M.; et al. Tomato and lycopene and multiple health outcomes: Umbrealla review. *Food Chem.* **2021**, *343*, 128396. [CrossRef]
15. Correddu, F.; Lunesu, M.; Buffa, G.; Atzori, A.; Nudda, A.; Battacone, G.; Pulina, G. Can agro-industrial by-products rich in polyphenols be advantageously used in the feeding and nutrition of dairy small ruminants? *Animals* **2020**, *10*, 131. [CrossRef] [PubMed]

16. Hassan, S.S.; Williams, G.A.; Jaiswal, A.K. Lignocellulosic biorefineries in Europe: Current state and prospects. *Trends Biotechnol.* **2019**, *37*, 231–234. [CrossRef]
17. Moreno, A.D.; Duque, A.; Gonzalez, A.; Ballesteros, I.; Negro, M.J. Valorization of greenhouse horticulture waste from a biorefinery perspective. *Foods* **2021**, *10*, 114. [CrossRef] [PubMed]
18. Wathoni, N.; Haerani, A.; Yuniarsih, N.; Haryanti, R. A review on herbal cosmetic in Indonesia. *Int. J. App. Pharm.* **2018**, *10*, 13–16. [CrossRef]
19. Naviglio, D.; Sapio, L.; Langella, C.; Ragone, A.; Illiano, M.; Naviglio, S.; Gallo, M. Beneficial effects and perspective strategies for lycopene food enrichment: A systematic review. *Syst. Rev. Pharm.* **2019**, *10*, 383–392.
20. Dominguez, R.; Gullòn, P.; Pateiro, M.; Minekata, P.; Zhang, W.; Lorenzo, J. Tomato as potential source of natural additives for meat industry. A review. *Antioxidants* **2020**, *9*, 73. [CrossRef] [PubMed]
21. Kumar, P.; Chatli, M.; Mehta, N.; Malav, O.; Verma, A.; Kumar, D.; Rathour, M. Antioxidant and antimicrobial efficacy of sapota powder in pork patties stored under different packaging conditions. *Korean J. Food Sci. Anim. Resour.* **2018**, *38*, 593–605.
22. Awad, A.M.; Kumar, P.; Ismail-Fitry, M.R.; Josuoh, S.; Ab Aziz, M.F.; Sazili, A.Q. Green extractions of bioactive compounds from plant biomass and their application in meat as natural antioxidant. *Antioxidants* **2021**, *10*, 1465. [CrossRef]
23. Carocho, M.; Morales, P.; Ferreira, I.C.F.R. Natural food additives: Quo vadis? *Trends Food Sci. Technol.* **2015**, *45*, 284–295. [CrossRef]
24. Lopez, C.J.; Caleja, C.; Prieto, M.A.; Barreiro, M.F.; Barros, L.; Ferreira, I.C F.R. Optimization and comparison of heat and ultrasound assisted extraction techniques to obtain anthocyanin compounds from *Arbutus unedo* L. Fruits. *Food Chem.* **2018**, *264*, 81–91. [CrossRef]
25. Neri-Numa, I.A.; Pessoa, M.G.; Paulino, B.N.; Pastore, G.M. Genipin: A natural blue pigment for food and health purposes. *Trends Foods Sci. Technol.* **2017**, *67*, 271–279. [CrossRef]
26. Caseiro, M.; Ascenso, A.; Costa, A.; Creagh-Flynn, J.; Johnson, M.; Simones, S. Lycopene in human health. *LWT Food Sci. Technol.* **2020**, *127*, 109323. [CrossRef]
27. Manzoor, M.; Singh, J.; Gani, A.; Noor, N. Valorization of natural colors as health-promoting bioactive compounds: Phytochemical profile, extraction techniques, and pharmacological perspectives. *Food Chem.* **2021**, *362*, 130141. [CrossRef] [PubMed]
28. Gosh, S.; Sarkar, T.; Das, A.; Chakraborty, R. Natural colorants from plant pigments and their encapsulation: An emerging window for the food industry. *LWT Food Sci. Technol.* **2022**, *153*, 112527. [CrossRef]
29. Ghosh, S.; Sarkar, T.; Das, A.; Chakraborty, R. Micro and nanoencapsulation of natural colors: A holistic view. *Appl. Biochem. Biotechnol.* **2021**, *193*, 3787–3811. [CrossRef]
30. Vieira, D.; Caliari, M.; de Souza, E.; Soares Junior, M. Methods for a pigments extraction and determination of color in tomato for processing cultivars. *Food Sci. Technol.* **2020**, *40*, 11–17. [CrossRef]
31. FAOSTAT. Food and Agriculture Organization of the United Nations. Available online: https://www.fao.org/faostat/en/#data/QC/visualize (accessed on 25 May 2021).
32. Deng, Y.; Zhao, S.; Yang, X.; Hou, F.; Fan, L.; Wang, W.; Xu, E.; Cheng, H.; Guo, M.; Liu, D. Evaluation of extraction technologies of lycopene: Hindrance of extraction, effects on isomerization and comparative analysis—A review. *Trends Food Sci. Technol.* **2021**, *115*, 285–296. [CrossRef]
33. BBC Research. Neutraceuticals: Global Markets to 2026. Available online: https://www.bccresearch.com/market-research/food-and-beverage/nutraceuticals-global-markets.html (accessed on 3 December 2021).
34. BBC Research. The Global Market for Carotenoids. Available online: https://www.bccresearch.com/market-research/food-and-beverage/the-global-market-for-carotenoids.html (accessed on 3 December 2021).
35. Saini, R.K.; Moon, S.H.; Keum, Y. An updated review on use of tomato pomace and crustacean processing waste to recover commercially vital carotenoids. *Food Res. Int.* **2018**, *108*, 516–529. [CrossRef]
36. Junior, S.S.; Casagrande, J.G.; Toledo, C.A.L.; Ponce, F.S.; Ferreira, F.S.; Zanuzo, M.R.; Diamante, M.S.; Lima, G.P.P. Selection of thermotolerant Italian tomato cultivars with high fruit yield and nutritional quality for the consumer taste grown under protected cultivation. *Sci. Hortic.* **2022**, *291*, 19. [CrossRef]
37. Jalali-Jivan, M.; Fathi-Achachlouei, B.; Ahmadi-Gavlighi, H.; Jafari, S. Improving the extraction efficiency and stability of β-carotene from carrot by enzyme-assisted green nanoemulsification. *Innov. Food Sci. Emerg. Technol.* **2021**, *74*, 18. [CrossRef]
38. Junker-Frohn, L.; Luck, M.; Schmittgen, S.; Wensing, J.; Carraresi, L.; Thiele, B.; Groher, T.; Reimer, J.; Broring, S.; Noga, G.; et al. Tomato's green gold: Bioeconomy potential of residual leaf biomass as a novel source for the secondary metabolite rutin. *ACS Omega* **2019**, *4*, 19071–19080. [CrossRef] [PubMed]
39. Almeida, P.V.; Rodrigues, R.P.; Garspar, M.C.; Braga, M.E.M.; Quina, M.J. Integrated management of residues from tomato production: Recovery of value-added compounds and biogas production in the biorefinery context. *J. Environ. Manag.* **2021**, *299*, 111. [CrossRef] [PubMed]
40. Bhattacharjee, C.; Saxena, V.K.; Dutta, S. Insights into effectiveness of tight ultrafiltration and frozen storage in bioactive compound retention in watermelon juice concentrate. *J. Food Process Eng.* **2020**, *43*, 117. [CrossRef]
41. Deng, Q.; Zinoviadou, K.G.; Galanakis, C.M.; Orlien, V.; Vorobiev, E.; Lebovka, N.; Barba, F.J. The effects of conventional and non-conventional processing on glucosinolates and its derived forms, isothiocynates: Extraction, degradation and applications. *Food Eng. Rev.* **2015**, *7*, 357–381. [CrossRef]

42. Wen, L.; Zhang, Z.; Sun, D.; Sivagnanam, S.P.; Tiwari, B.K. Combination of emerging technologies for the extraction of bioactive compounds. *Crit. Rev. Food Sci. Nutr.* **2020**, *60*, 1826–1841. [CrossRef]
43. Koteswararao, P.R.; Tulasi, S.L.; Pavani, Y. Impact of solvents on environmental pollution. National Seminar on Impact of Toxic Metals, Minerals and Solvents leading to Environmental lution. *J. Chem. Pharm. Sci.* **2014**, *3*, 132–135.
44. Naviglio, D.; Pizzolongo, F.; Ferrara, L.; Aragòn, A.; Santini, A. Extraction of pure lycopene from industrial tomato by-products in water using a new high-pressure process. *J. Sci. Food Agric.* **2008**, *88*, 2414–2420. [CrossRef]
45. Joshi, D.; Adhikari, N. An overview on common organic solvents and their toxicity. *J. Pharm. Res. Int.* **2019**, *28*, 118. [CrossRef]
46. Płotka-Wasylka, J.; Rutkowska, M.; Owczarek, K.; Tobiszewski, M.; Namieśnik, J. Extraction with environmentally friendly solvents. *TrAC Trends Anal. Chem.* **2017**, *91*, 12–25. [CrossRef]
47. Torres-Valenzuela, L.S.; Ballesteros-Gomez, A.; Rubio, S. Green solvents for the extraction of high added-value compounds from agri-food waste. *Food Eng. Rev.* **2020**, *12*, 83–100. [CrossRef]
48. Sekharan, T.R.; Chandira, R.M.; Tamilvanan, S.; Rajesh, S.C.; Venkateswarlu, B.S. Deep eutectic solvents as an alternative to other harmful solvents. *Biointerface Res. Appl. Chem.* **2022**, *12*, 847–860.
49. Khandelwal, S.; Tailor, Y.; Kumar, M. Deep eutectic solvents (DESs) as eco-friendly and sustainable solvent/catalyst system in organic trasformations. *J. Mol. Liq.* **2016**, *215*, 345–386. [CrossRef]
50. Silva, Y.P.A.; Ferreira, T.A.P.C.; Jiao, G.; Brooks, M.S. Sustainable approach for lycopene extraction from tomato processing by-product using hydrophobic eutectic solvents. *J. Food Sci. Technol.* **2019**, *56*, 1649–1654. [CrossRef] [PubMed]
51. Aniceto, J.; Rodrigues, V.; Portugal, I.; Silva, C. Valorization of tomato residues by supercritical fluid extraction. *Processes* **2021**, *10*, 125. [CrossRef]
52. Getachew, A.; Jacobsen, C.; Holdt, S. Emerging technologies for the extraction of marine phenolics: Opportunities and challenges. *Mar. Drugs* **2020**, *18*, 122. [CrossRef] [PubMed]
53. Clausen, I.C.; Ustad, T.S.; Strommen, I.; Walde, P.M. Atmospheric freeze drying: A review. *Dry. Technol.* **2010**, *25*, 947–957. [CrossRef]
54. Ratti, C. Hot and freeze-drying of high-value foods: A review. *J. Food Eng.* **2001**, *49*, 311–319. [CrossRef]
55. Rybak, K.; Parniakov, O.; Samborska, K.; Wiktor, A.; Witrowa-Rajchert, D.; Nowacka, M. Energy and quality aspects of freeze-drying preceded by traditional and novel pre-treatment methods as exemplified by red bell pepper. *Sustainability* **2021**, *13*, 116. [CrossRef]
56. Singh, A.P.; Mandal, R.; Shojaei, M.; Singh, A.; Kowalczewski, P.L.; Ligaj, M.; Pawlicz, J.; Jarzebski, M. Novel drying methods for sustainable upcycling of brewers' spent grains as a plant protein source. *Sustainability* **2020**, *12*, 3660. [CrossRef]
57. Gheonea, I.; Aprodu, I.; Enachi, E.; Horincar, G.; Bolea, C.A.; Bahrim, G.E.; Rapeanu, G.; Stanciuc, N. Investigations on thermostability of carotenoids from tomato peels in oils using a kinetic approach. *J. Food Process. Preserv.* **2020**, *44*, 19.
58. Silva, Y.P.A.; Ferreira, T.A.P.C.; Celli, G.B.; Brooks, M.S. Optimization of lycopene extraction from tomato processing waste using an eco-friendly ethyl lactate-ethyl acetate solvent: A green valorization approach. *Waste Biomass Valorization* **2019**, *10*, 2851–2861. [CrossRef]
59. Sperber, W.H. Influence of water activity on foodborne bacteria: A review. *J. Food Prot.* **1983**, *46*, 142–150. [CrossRef] [PubMed]
60. Kiranoudis, C.T.; Maroulis, Z.B.; Tsami, E.; Marinos-Kouris, D. Equilibrium moisture content and heat of desorption of some vegetables. *J. Food Eng.* **1993**, *20*, 55–74. [CrossRef]
61. MacHmudah, S.; Winardi, S.; Sasaki, M.; Goto, M.; Kusumoto, N.; Hayakawa, K. Lycopene extraction from tomato peel by-product containing tomato seed using supercritical carbon dioxide. *J. Food Eng.* **2012**, *108*, 290–296. [CrossRef]
62. Henry, L.; Catignani, G.; Schwartz, S. Oxidative degradation kinetics of lycopene, lutein, and 9-cis and all trans β-carotene. *JAOCS* **1998**, *75*, 823–829. [CrossRef]
63. Morales-Soriano, E.; Panozzo, A.; Ugás, R.; Grauwet, T.; Van Loey, T.; Hendrickx, M. Carotenoid profile basic structural indicators of native Peruvian chili peppers. *Eur. Food Res. Technol.* **2019**, *245*, 717–732. [CrossRef]

foods

MDPI

Article

Extraction of Natural Gum from Cold-Pressed Chia Seed, Flaxseed, and Rocket Seed Oil By-Product and Application in Low Fat Vegan Mayonnaise

Taha Hijazi [1], Salih Karasu [1,*], Zeynep Hazal Tekin-Çakmak [1] and Fatih Bozkurt [1,2]

1 Department of Food Engineering, Faculty of Chemical and Metallurgical Engineering, Davutpasa Campus, Yildiz Technical University, Istanbul 34349, Turkey; hijazi8319@gmail.com (T.H.); zhazaltekin@gmail.com (Z.H.T.-Ç.); fbozkurt@yildiz.edu.tr (F.B.)

2 Department of Food Engineering, Engineering and Architecture Faculty, Muş Alparslan University, Muş 49250, Turkey

* Correspondence: skarasu@yildiz.edu.tr; Tel.: +90-212-383-46-23

Abstract: This study involves the modeling of rheological behavior of the gum solution obtained from cold-pressed chia seed (CSG), flaxseed (FSG), and rocket seed (RSG) oil by-products and the application of these gums in a low-fat vegan mayonnaise formulation as fat replacers and emulsifier. CSG, FSG, and RSG solutions showed shear-thinning flow behavior at all concentrations. The K values ranged between 0.209 and 49.028 Pa·s^n for CSG, FSG, and RSG solutions and significantly increased with increased gum concentration. The percentage recovery for the G′ was significantly affected by gum type and concentrations. CSG, FSG, and RSG showed a solid-like structure, and the storage modulus (G′) was higher than the loss modulus (G″) in all frequency ranges. The rheological characterization indicated that CSG, FSG, and RSG could be evaluated as thickeners and gelling agents in the food industry. In addition, the rheological properties, zeta potential, and particle size and oxidative stability (at 90 °C) of low-fat vegan mayonnaise samples prepared with CSG, FSG, and RSG were compared to samples prepared with guar gum (GG), Arabic gum (AG), and xanthan gum (XG). As a result, CSG, FSG, and RSG could be utilized for low-fat vegan mayonnaise as fat and egg replacers, stabilizers, and oxidative agents. The results of this study indicated that this study could offer a new perspective in adding value to flaxseed, chia seed, and rocket seed cold-press oil by-product.

Keywords: cold-pressed oil by-product; gum; thixotropic behavior; low-fat vegan mayonnaise; thickeners; gelling agents

Citation: Hijazi, T.; Karasu, S.; Tekin-Çakmak, Z.H.; Bozkurt, F. Extraction of Natural Gum from Cold-Pressed Chia Seed, Flaxseed, and Rocket Seed Oil By-Product and Application in Low Fat Vegan Mayonnaise. *Foods* **2022**, *11*, 363. https://doi.org/10.3390/foods11030363

Academic Editors: Marco Poiana, Francesco Caponio and Antonio Piga

Received: 15 December 2021
Accepted: 19 January 2022
Published: 27 January 2022

Publisher's Note: MDPI stays neutral with regard to jurisdictional claims in published maps and institutional affiliations.

Copyright: © 2022 by the authors. Licensee MDPI, Basel, Switzerland. This article is an open access article distributed under the terms and conditions of the Creative Commons Attribution (CC BY) license (https://creativecommons.org/licenses/by/4.0/).

1. Introduction

Edible oils contain ubiquitous dietary ingredients that may contain high quantities of bioactive compounds, such as sterols, tocopherols, and unsaturated fatty acids [1]. The cold-pressed extracted oils are not subjected to any refining and chemical processes. Also, cold-pressed oil by-products can be used in the food industry as they are not exposed to a solvent such as hexane. The moisture is removed before the processing of cold-pressed oil by-products and the oil is removed during production. The obtained byproducts are rich in proteins and polysaccharides. Today, although cold-press oil byproducts are an important source in terms of nutrition and technology, they have not taken their place in the food industry sufficiently. However, the development of new uses for cold-pressed oil byproducts of the edible oil industry by converting them to value-added products would prevent their disposal as waste and would encourage sustainable and competitive industrial supply [2]. Cold-pressed oil byproducts of seeds, such as flaxseed, chia seed, and rocket seed have an important potential in terms of natural gum. These byproducts have high gum content and can be easily extracted from food.

Flaxseed (*Linum usitatissimum* L.) contains rich dietary fiber (~30% by weight), unsaturated fat (~40% by weight), and bioactive protein/peptides (~20% by weight) [3,4]. Cold-pressed flaxseed oil byproducts (FOB) contained 46.37% of carbohydrate, 27.67% of protein, 9.15% of oil, and 2.87% of ash on a wet basis [5]. Flaxseed gum (FG) is a natural polysaccharide and protein blend derived from flaxseed. FG is rich in soluble dietary fiber (3–9 wt% of the flaxseed) and can be used as a thickener, a stabilizer, a gelling agent, and an emulsifier [6–8].

Chia seed (*Salvia hispanica* L.) is rich in carbohydrates (~42%, mainly dietary fiber), proteins (17–24%), oil (25–40%), and omega-3/omega-6 fatty acids (60–80% of total oil) [9]. Chia seed polysaccharide (CSP), also known as chia seed gum (CSG), is a water-soluble anionic heteropolysaccharide isolated from the chia seed coat [10]. Chia seed contains about 5% mucilage, which can also act as soluble fiber [11]. Cold-pressed chia seed oil by-products obtained from the oil industry can be considered as a natural source of oil substitutes due to its high polysaccharide and protein content [12]. Chia seed gum can be used in a variety of industrial applications, including dietary fiber supplements, fat replacer, gelling agents, thickeners, stabilizers, emulsifiers, bulking agents, and film/coating agents [13].

Rocket seed (*Eruca sativa*) contains a variety of health-promoting phytochemical compounds, such as polyphenols, fibers, and glucosinolates. It has been used in the industry for oil production due to its high oil contents (20.0%) and Erucic acid used for the manufacture of a wide range of industrial products, e.g., plasticizers, surfactants, detergents, coatings, and polyesters. [14]. The rocket seed has a significant amount of total carbohydrate (23.1%), crude fibers (20.4%), and protein (31.0%) [15]. Thanks to their carbohydrate and protein content, rocket seeds have reasonable gum content with outstanding functional properties. Rocket seed gum (RSG) has high carbohydrate (80.38%) and low protein (5.81%) contents so the purification method used when obtaining gum could be appropriate. The protein content of gum is a crucial parameter determining its emulsion, foaming, and film-forming capacity. When comparing commercial gums, the protein content of RSG was higher than that of xanthan gum (2.125%), lower than that of guar gum (8.19%), and close to that of locust bean gum (5.2–7.4%) [16].

Mayonnaise is a semi-solid oil in water (O/W) emulsion that mainly consists of vegetable oil (70–80%), egg yolk, vinegar, and salt [17]. Egg yolk has an important place among the components to ensure the stability of mayonnaise and contributes to the overall organoleptic properties of the final product [18]. However, there has been a growing tendency toward substituting eggs with plant-based components, particularly in the creation of mayonnaise analogs, due to health and environmental concerns, recently. Egg yolk is an important ingredient that affects oil droplet distribution and emulsion stability in emulsion products, such as mayonnaise. Therefore, it is very difficult to find an alternative stabilizing agent to egg yolk in vegan mayonnaise production. In recent years, the number of studies on alternatives to egg yolk has been increasing. In this study, low-fat vegan mayonnaise samples were prepared using CSG, FSG, RSG, AG, GG, and XG, and rheological properties were compared.

There are many studies on the rheological properties of different gum solutions. However, the researchers focused more on the steady and dynamic rheological properties of the gums. The number of studies on the thixotropic properties of natural gums is limited. In this sense, there are important gaps in the literature. In this study, natural gums were produced from the cold-pressed chia seed, flaxseed, and rocket seed oil by-products. This study will also demonstrate the thixotropic behavior of a wide range of gums using specific rheological tests such as 3-ITT. Also, gums produced from cold-pressed oil by-products can be used as an alternative emulsifier for oil in water (O/W) emulsions such as mayonnaise and salad dressings to obtain a more stable O/W emulsion having expected quality. In this study, by-product gums were utilized as egg yolk replacers to produce vegan mayonnaise.

2. Materials and Methods

In this study, cold-pressed oil byproducts obtained from chia seed (*Salvia hispanica* L.), flaxseed (*Linum usitatissimum* L.), and rocket seed (*Eruca sativa*) were supplied from ONEVA Food Co. (Istanbul, Turkey) for gum extraction. Ethanol and all other chemicals and standards to be used during gum extraction are of analytical quality and purchased from Merck (Darmstadt, Germany) or Sigma-Aldrich (St. Louis, MO, USA). Sunflower oil and sunflower lecithin, distilled water, CSG, FSG, RSG, guar gum (GG), Arabic gum (AG), and xanthan gum (XG) were used for low-fat vegan mayonnaise preparation.

2.1. Gum Extraction

Gums were extracted from cold-pressed oil by-products (chia seed, flaxseed, and rocket seed) according to the procedure described by Naji-Tabasi et al. [19]. The gums were removed from the by-products by soaking with water followed by ethanol precipitation. Five hundred by-products are weighed and added to 10 L of distilled water and mixed with a magnetic heated stirrer at 80 °C for 2 h to obtain gel structures that come out of the byproducts. The extracted gum solution from the byproducts was passed through to the sieve and then purified by mixing with three volumes of 96% ethanol to precipitate gum and left in the oven at 30 °C for 1 day to dry. The seed gums are dried, packaged at 4 °C, and stored in dry conditions.

2.2. Gum Characterization

The moisture contents of CSG, FSG, and RSG were determined by drying the sample at 105 °C for 5 h in a conventional oven (Memmert UF-110, Schwabach, Germany) [20]. The ash contents were determined by burning of the organic material of the sample at 550 °C in a muffle furnace (WiseTherm-Daihan FH-03, Seoul, Korea) for 6 h [21]. The protein contents were determined by the Kjeldahl method using the Behr Kjeldahl unit (Unit-S5, Ahlen, Germany) with the conversion factor of 6.25 [22]. The fat contents were analyzed by Soxhlet extraction using hexane as a solvent [23].

The monosaccharide (glucose, galactose, mannose, and xylose) compositions of CSG, FSG, and RSG were determined using an HPLC tool (Agilent 1260 Infinity) equipped with Rezex ROA-Organic Acid H+ (New Column, 300 × 7.8 mm). A 250-mg gum sample was heated with 2 M of H_2SO_4 for 2 h at 120 °C. Then, the hydrolyzed mixture was cooled, the pH of the solution was adjusted to 7 using 5 M of NaOH and the final volume was adjusted to 10 mL with distilled water. Then, it was filtered through a 0.45-µm syringe filter and injected into the column [24].

ATR-FTIR (attenuated total reflection-Fourier transform infrared) spectroscopy (Bruker Tensor 27, Borken, Germany) equipped with a KBr beam diffuser and DLaTGS detector was used to characterize CSG, FSG, and RSG. ATR-FTIR spectra of CSG, FSG, and RSG with wavenumbers ranging from 400 to 4000 cm^{-1} were acquired with 16 scans per spectrum and 2 cm^{-1} resolutions. Deconvolution was applied to all spectra by interpretation of changes in the overlapped amide I band (1600–1700 cm^{-1}) using Origin 2020b software to determine changes in the secondary structure of hydrolysates [25].

2.3. Preparation of Gum Solutions

Gum solutions were prepared at concentrations of 1.0–2.0% using flaxseed and chia seed gums, and 1.0–5.0% for rocket seed gum. First of all, natural gums dissolve in water at a certain concentration and were expected to mix in a magnetic stirrer for 12 h to fully hydrate. Each measurement was repeated three times.

2.4. Rheological Properties

The flow behavior and thixotropic and dynamic rheological properties of gum solutions were determined using a temperature-controlled rheometer (MCR 302, Anton Paar, Graz, Austria). All rheological measurements were carried out at 25 °C.

2.5. Flow Behavior Rheological Properties

The flow behavior characteristics of the gum solutions were determined in the range of 0–100 shear rate (1/s) using a parallel plate configuration and a 0.5-mm gap between the rheometer probe and the sample plate. Apparent viscosity values corresponding to shear stress and shear rate were recorded. The flow behavior rheological properties were modeled using the Herchel–Bulkley model by nonlinear regression.

$$\tau = K(\gamma)^n \tag{1}$$

In Equation (1), τ is the shear stress (Pa), K is the consistency coefficient (Pa·s^n), γ is the shear rate (s^{-1}), and n is the flow behavior index (dimensionless).

2.6. Thixotropic Properties

2.6.1. Constant Shear Rate

Shear stress or $\eta 50$ values were used to determine the time-dependent behavior of gum solutions at 25 °C. The experimental data could be fitted to the Weltman model (Equation (2)) and second-order structural kinetic model (Equation (3)), with n = 2 [26,27].

$$\sigma = a + b \ln t \tag{2}$$

In Equation (2), σ is the shear stress at time t (Pa), a is the initial shear stress (Pa), b is the time coefficient of thixotropic breakdown (Pa), and t is the time of shearing (s).

$$\left[\frac{\eta - \eta_e}{\eta_0 - \eta_e}\right]^{(1-n)} = (n-1)kt + 1 \tag{3}$$

In Equation (3), η_0 is the initial apparent viscosity at t = 0 (structured state), η_e is the equilibrium apparent viscosity as $t \to \infty$ (non-structured state), $k = k(\gamma)$ is the rate constant, and n is the order of the structure breakdown reaction assumed as 2 in the second-order structural kinetic model.

2.6.2. 3-ITT Rheological Properties

The 3-ITT rheological properties were determined in the range of 0.5/s constant shear rate and 150/s variable shear rate, respectively. When the values were selected, the linear viscoelastic region was taken into account, and the linear viscoelastic region of the samples ends at 50 s^{-1}. Gum solutions were subjected to a very low shear rate (0.5/s) for 100 s during the first-time interval. In the second time interval, it was subjected to the specified shear force for 40 s. In the third time interval, the dynamic rheological behavior in the second time interval was determined by exposing the samples to the low shear rate in the first time interval. The second-order structural kinetic model was used, and G_0, G_e, and k values were calculated by the following Equation (4):

$$\left[\frac{G' - G_e}{G_0 - G_e}\right]^{1-n} = (n-1)kt + 1 \tag{4}$$

In Equation (4), G_0 (the initial values of the storage modulus in the first interval), G_e (the equilibrium storage modulus as $t \to \infty$), k (the rate constant of recovery of the sample) and in this model n = 2, are specified.

In Equations (5) and (6), G_i (at the initial state of the product), G_0 (after deformation applied G' value), and G_e (after recovery of sample G' value) values are characterized by deformation equation [28]:

$$\%\, Deformation = \left(\frac{(G_i - G_0)}{G_i}\right) \times 100 \tag{5}$$

The recovery degrees of CSG, FSG, and RSG is determined by Equation (6),

$$\% \; Recovery = \left(\frac{G_e}{G_i}\right) \times 100 \quad (6)$$

2.6.3. Dynamical Rheological Properties

The dynamic rheological properties of the gum solutions were carried out using a parallel plate configuration. First, the amplitude sweep test was performed with a strain value of 0.1% to determine the direct viscoelastic region. The frequency sweep test was applied in the range of 0.1–10 Hz and 0.1–64 (ω) angular velocity in the primary viscoelastic region. The values of storage modulus (G') and loss modulus (G'') were measured against angular velocity and frequency. The parameters related to dynamic rheological properties were determined using the Oswald-de Waele model and nonlinear regression [29].

$$G' = K'(\omega)^{n'} \quad (7)$$

$$G'' = K''(\omega)^{n'} \quad (8)$$

Equations (7) and (8), G' value corresponds to storage modulus (Pa), G'' value to lose modulus (Pa), ω value to angular velocity value (s^{-1}), n' and n'' values to flow behavior index values, and K' and K'' values to consistency coefficient (Pa·s^n).

2.7. Vegan Mayonnaise Preparation and Analysis

The different types of gum (CSG (2%), FSG (2%), RSG (5%), Arabic gum (AG:1%), guar gum (GG:1%), and xanthan gum (XG:0.4%) were used in vegan mayonnaise production. Firstly, the gums were dispersed at 25 °C in water at different ratios. Afterward, the gum was hydrated by stirring at 1000 rpm in a magnetic stirrer for 6 h. The obtained dispersion was combined with sunflower oil (30%) and lecithin (1%) and homogenized for 3 min utilizing Ultra Turrax (Daihan, HG15D) at 10,000 rpm. Finally, low-fat vegan mayonnaise samples were obtained.

2.7.1. Rheological Analysis

All rheological analyses, i.e., flow behavior properties, dynamic rheological properties, and 3-ITT rheological properties of vegan mayonnaise prepared with byproduct gums, were determined using a temperature-controlled rheometer (MCR 302, Anton Paar, Austria) at 25 °C.

2.7.2. Zeta Potential and Particle Size

The zeta(ζ-)-potential, the oil particle size (d_{32}), and polydispersity index (PdI) were measured by a particle size meter (Nanosizer, Malvern Instruments, Malvern, Worcestershire, UK) with electrophoresis and dynamic light scattering system in the continuous phase of the mayonnaise samples. Until the measurement, the samples were diluted 500 times with ultrapure water and then homogenized by mixing in an ultrasonic water bath. This procedure was repeated in triplicate for each sample by using the Zeta potential measurement, and the averages of the values and the standard deviations were determined.

2.7.3. Oxidative Stability (OXITEST)

The oxidative stability of the mayonnaise samples was tested using the OXITEST Device (Velp Scientifica, Usmate, Italy) according to AKSOY et al. [30]. All samples were weighed before the oxidative stability analysis started. Firstly, 20 g of each mayonnaise sample were weighed into the sample cells. The device temperature and the oxygen pressure were adjusted at 90 °C and 6 bars, respectively. The induction period (IP) value obtained by the OXITEST system was used to evaluate the oxidative stability values of the samples.

2.8. Statistical Analysis

The statistical analysis was carried out using the Statistica software program (StatSoft, Inc., Tulsa, OK, USA). All the rheological analyses were conducted in triplicate. The standard deviation and mean values were presented. ANOVA was conducted to determine the differences in rheological parameters of gum solutions. Duncan, multiple comparison tests at 95% significance level were used to determine the effect of analysis parameters.

3. Results and Discussion

3.1. Characterization of Gum

Table 1 showed physicochemical properties of CSG, FSG, and RSG. CSG consisted of 73.59% carbohydrate, 9.45% protein, 1.01% fat, 9.60% moisture, 6.26% ash (%*w*/*w*). FSG contained 78.56% carbohydrate, 11.38% protein, 2.16% fat, 9.08% moisture, 7.89% ash (%*w*/*w*) while RSG consist of 60.48% carbohydrate, 21.00% protein, 1.94% fat, 9.95% moisture, 6.63% ash (%*w*/*w*).

Table 1. Physicochemical properties of gum.

	CSG	FSG	RSG
Carbohydrate (%*w*/*w*)	73.59 ± 0.29 [b]	78.56 ± 0.06 [a]	70.48 ± 0.20 [c]
Protein (%*w*/*w*)	9.45 ± 0.07 [b]	11.38 ± 0.10 [a]	11.00 ± 0.01 [a]
Fat (%*w*/*w*)	1.01 ± 0.02 [b]	2.16 ± 0.10 [a]	1.94 ± 0.04 [a]
Moisture (%*w*/*w*)	9.60 ± 0.18 [s]	9.08 ± 0.06 [b]	9.95 ± 0.08 [a]
Ash (%*w*/*w*)	6.26 ± 0.14 [b]	7.89 ± 0.11 [a]	6.63 ± 0.05 [b]
Monosaccharides (%)			
Glucose	20.46 ± 0.16 [a]	20.27 ± 0.15 [a]	10.26 ± 0.06 [b]
Galactose	6.70 ± 0.07 [c]	19.91 ± 0.03 [b]	22.08 ± 0.08 [a]
Mannose	4.09 ± 0.01 [b]	0.41 ± 0.01 [c]	39.12 ± 0.12 [a]
Xylose	36.15 ± 0.15 [b]	38.75 ± 0.05 [a]	-

CSG: chia seed byproduct gum, FSG: flaxseed byproduct gum, RSG: rocket seed byproduct gum. The different lowercase letter in the same line indicates statistical significance ($p < 0.05$).

The protein content of gums is a very important parameter affecting the emulsification and foaming ability of gums. The protein content of the obtained gums was found to be higher than other natural and commercial gums [16,31]. RSG is rich in galactose and mannose, while FSG and CSG are rich in xylose and glucose. It can be said that RSG has a galactomannan structure as in many gums. The ratio of mannose to galactose was found to be 1.77 in RSG. The mannose to galactose ratio strongly affected the technological properties of gums such as their cold-water solubility, thickening, gelling, and crystalizing properties. The mannose galactose ratio was lower than guar gum (2:1) and LBG (4:1), and higher than fenugreek gum (1:1) [32,33]. In a similar study, xylose and glucose were reported to be higher than mannose and galactose from studies on the sugar profile of mucilage and gums obtained from flaxseed and chia seeds [34–37].

The FTIR spectrum of CSG, FSG, and RSG is illustrated in Figure 1. The results revealed that all gum samples showed a similar FTIR spectrum with some minor differences. This difference might be due to their protein content and different sugar and organic acid composition. Characteristic bands varying between 3500–3100 cm^{-1} are attributed to the hydroxyl (-OH) stretch that forms the gross structure of carbohydrates [38]. The bands between 3000–2800 cm^{-1} represent -C-H stretching of the aromatic rings and the methyl group (CH3) [38]. The bands at 1654 cm^{-1} and 1618 cm^{-1} for chia seed and its mucilage are assigned to a mannose ring [39].The bands at around 1597 cm^{-1} and 1422 cm^{-1} are due to the carboxyl groups of uronic acid residues in the gum polysaccharide or the presence of protein in the gum samples [40]. This result is compatible with the protein content of the gums. The wavenumber between 950 cm^{-1} and 1200 cm^{-1} is generally considered the fingerprint region of polysaccharides where the main chemical groups in polysaccharides are identified. The band at around 1030 cm^{-1} is assigned to C-O-C stretching of 1→4 glycosidic bonds [41]. Also, the strong absorption at 1014 cm^{-1} shows the stretching vibration of the C–N [42]. The band at 864 cm^{-1} is assigned to the β-anomeric C-H deformation and glycosidic linkages of glucopyranose and xylopyranose units [38]. The FTIR spectrum obtained is very similar to many mucilage and gums studied in the literature. Hadad and Goli [43] observed the FTIR spectrum of flaxseed mucilage gave the absorption peaks at 3321, 2922, 1612, 1410, and 1050 cm^{-1}. Darwish and El-Sohaimy [44] reported the absorption peaks of chia seed mucilage at 1739, 1539, 1444, 1419, 1157, 1058, and 618 cm^{-1}.

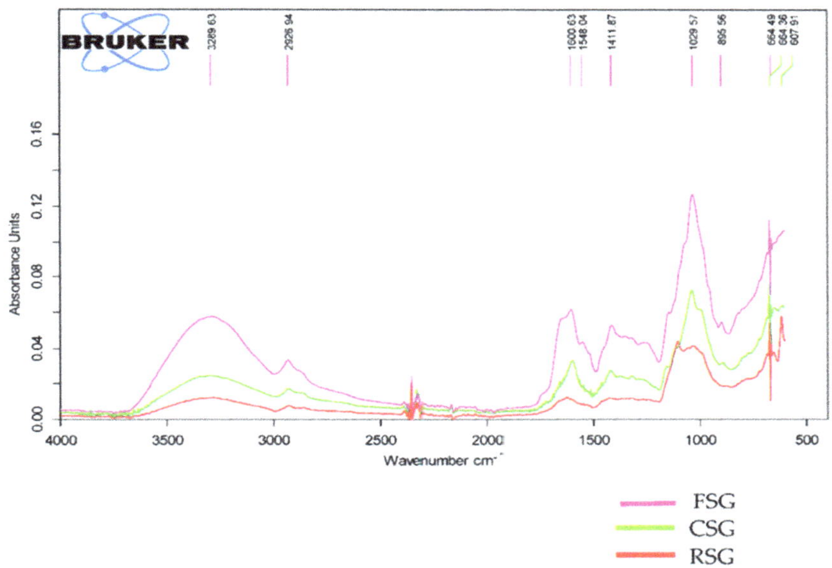

Figure 1. FTIR spectra of the FSG (Flaxseed byproduct gum), CSG (Chia seed byproduct gum), and RSG (Rocket seed byproduct gum) in the spectral region between 400 and 4000 cm^{-1}.

3.2. Flow Behavior Rheological Properties of the CSG, FSG, and RSG Solutions

Figure 2 exhibits the flow properties of CSG (1.0, 1.5, and 2.0 %, *w*/*w*), FSG (1.0, 1.5, and 2.0 %, *w*/*w*), and RSG (1.0, 1.5, 2.0, 3.0, and 5.0 %, *w*/*w*) solutions with different concentrations over the range of shear rate from 0.01 to 100 s^{-1} at 25 °C. The decrease in viscosity by an increase in shear rate indicated a non-Newtonian shear-thinning (pseudoplastic) flow behavior for CSG, FSG, and RSG solutions at all concentrations. The decrease in the viscosity values of the samples due to the increasing shear rate can be explained by the breaking of the weak bonds between the molecules in the product as a result of the applied force and the weakening of the interaction between the components [45,46]. As predicted, the constant shear viscosity increased with the increase of gum concentrations. The most used hydrocolloids in the food industry, such as locust bean gum [47] and xanthan gum [48], showed shear-thinning rheological behavior. Similar shear-thinning flow behavior of gum dispersions was previously reported by Sanchez et al. [49], Marcotte et al. [50], Yamazaki et al. [51], Razavi et al. [52], and Chaharlang and Samavati [53]. CSG and FSG dispersions showed high viscosity and pronounced shear-thinning behavior at low concentrations (1–2%); however, more RSG exhibited weak pseudo-plasticity at lower concentrations. This might be explained by the varied origins of these gums. The gums with high molecular weight and strong intermolecular interactions have high viscosity solutions [54]. The orientation effect is another explanation for the shear thinning behavior. As the shear rate increases, the randomly distributed polymer molecules become more and more aligned in the flow direction, resulting in reduced contact between neighboring polymer chains [55].

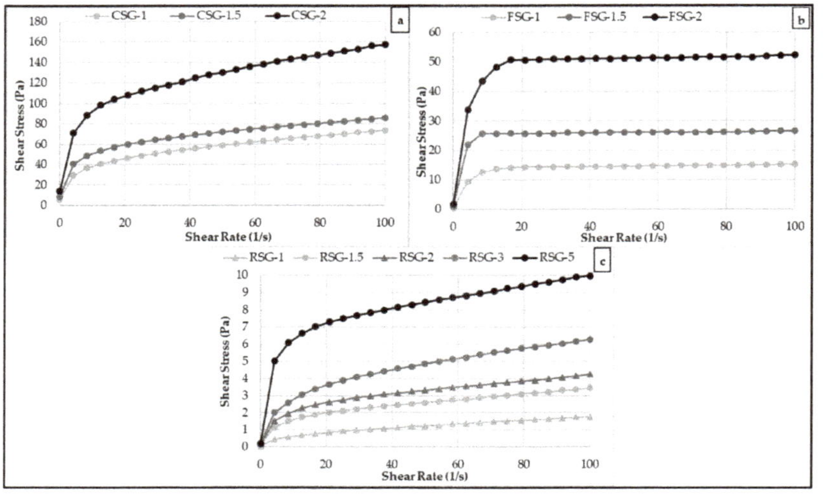

Figure 2. Steady shear rheological properties of the FSG (flaxseed byproduct gum), CSG (chia seed byproduct gum), and RSG (rocket seed byproduct gum) solutions with different concentrations ((**a**) CSG (1–2%), (**b**) FSG (1–2%), (**c**) RSG (1–5%)).

Shear-thinning gums are commonly used to enhance the texture and rheological properties of food products. The high-shear processing operations, such as pumping and filling cause a reduction in the apparent viscosity of the solution. However, the high apparent viscosity produces a pleasant tongue feel during consumption [56].

The power-law model parameters (the consistency index (K) and the flow behavior index (n)) are shown in Table 2. The power-law model was successfully applied for the modeling of flow behavior properties of CSG, FSG, and RSG solutions (The coefficients of determination: $R^2 > 0.96$). Many previous investigations have shown that the power-law model was the suitable model for describing the flow behavior of gum solutions [47,50,56,57].

Table 2. Steady shear power-law parameters of gum solutions.

	Gum (%)	K (Pa·s^n)	*n*	R^2
FSG-1	1.0	8.256 ± 0.043 ^{C}	0.143 ± 0.000	0.942
FSG-1.5	1.5	15.699 ± 0.120 ^{B}	0.143 ± 0.001	0.972
FSG-2	2.0	28.731 ± 0.341 ^{A}	0.145 ± 0.001	0.956
CSG-1	1.0	18.578 ± 0.087 ^{C}	0.296 ± 0.011	0.998
CSG-1.5	1.5	27.889 ± 0.092 ^{B}	0.243 ± 0.003	0.995
CSG-2	2.0	49.028 ± 0.0359 ^{A}	0.252 ± 0.001	0.994
RSG-1	1.0	0.209 ± 0.001 ^{E}	0.460 ± 0.004	0.999
RSG-1.5	1.5	0.694 ± 0.015 ^{D}	0.342 ± 0.003	0.997
RSG-2	2.0	0.985 ± 0.002 ^{C}	0.313 ± 0.001	0.996
RSG-3	3.0	1.217 ± 0.003 ^{B}	0.305 ± 0.000	0.998
RSG-5	5.0	3.553 ± 0.021 ^{A}	0.224 ± 0.001	0.987

CSG: chia seed byproduct gum, FSG: flaxseed byproduct gum, RSG: rocket seed byproduct gum. A different uppercase letter in the same column indicates statistical significance ($p < 0.05$).

The K values were found as 0.209–49.028 Pa·s^n. The increase of gum concentrations increased K values for all types of gum solutions. At the same concentration, CSG had the highest K values, followed by FSG and RSG. The n values of CSG, FSG, and RSG solutions at all concentrations were less than 1, indicating that all gums showed a non-Newtonian shear-thinning behavior (Table 2). All the gum solutions had strong shear-thinning behavior with *n* values as low as 0.143–0.460, indicating a significant deviation from Newtonian behavior, and they, like many other shear-thinning hydrocolloids, have a high viscosity and pleasant mouthfeel at low shear rates. CSG, FSG, and RSG samples exhibited shear-thinning behavior like locust bean gum [47], monoi gum [57], guar gum [58], and xanthan gum [48], which was explained to be due to weak bonds formed as a result of shearing [59]. When the n number is smaller than 0.6, it has been observed that non-Newtonian behavior becomes relevant [60]. The increase of gum concentrations caused decreasing n values. These results demonstrated that different concentrations of CSG, FSG, and RSG solutions affected the steady shear properties of CSG, FSG, and RSG solutions.

Figure 3 shows the viscosity of CSG, FSG, and RSG solutions with different concentrations over the range of shear rate from 0.01 to 100 s^{-1}. As seen in Figure 2, the viscosity of the CSG, FSG, and RSG samples decreased with an increase in the shear rate at all concentrations due to the pseudoplastic behavior of gum solutions. The increase in shear rate led to a breakdown of molecular bonds, and therefore molecules became regular and internal friction decreased. As a result, the viscosity of CSG, FSG, and RSG solutions decreased. As the gum became entangled in the solution, the viscosity of the gum solutions increased with gum concentration, as can be seen in Figure 3. The higher viscosity value in CSG-2 (a solution containing 2% chia gum) compared to other gum solutions was associated with the strong interactions in hydrogen bonds [61].

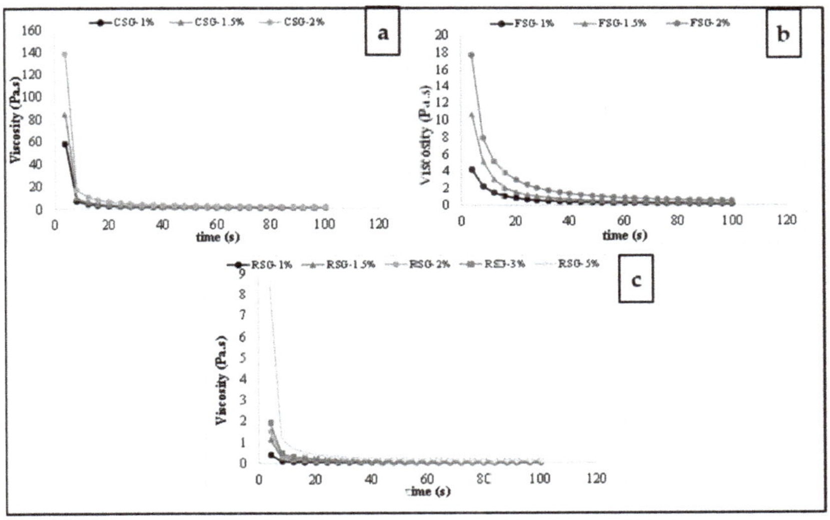

Figure 3. The viscosity of CSG: chia seed byproduct gum, FSG: flaxseed byproduct gum, RSG: rocket seed byproduct gum solutions with different concentrations ((**a**) CSG (1–2%), (**b**) FSG (1–2%), (**c**) RSG (1–5%)).

3.3. *Thixotropic Properties*

3.3.1. Constant Shear Rate

CSG, FSG, and RSG solutions were sheared at a constant shear rate of 0.5 s^{-1} at 25 °C in the shear decay test. At 25 °C, Figure 4 depicts the change in shear stress of CSG, FSG, and RSG solutions as a function of time. Following that, the data were fitted to two different models: the Weltmann [62] model and the second-order structural model [63].

The computed parameters and their related determination coefficients are shown in Table 3. Both models could accurately reflect the time-dependency of shear stress values at all gum concentration levels with good R^2 values.

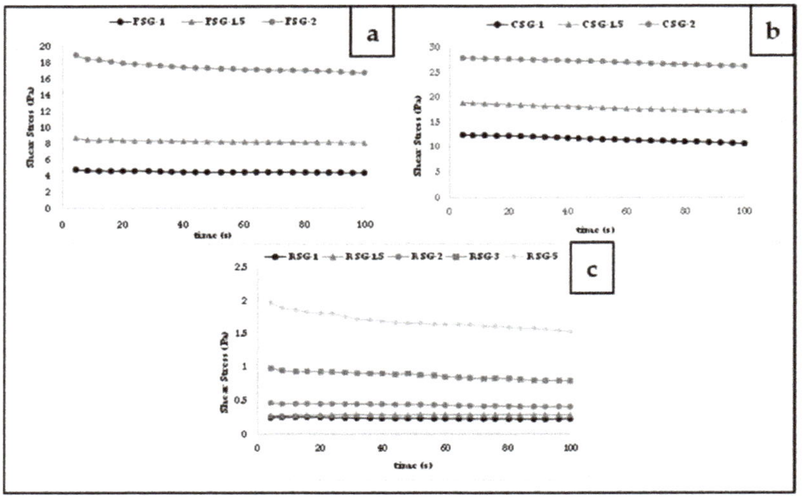

Figure 4. Shear stress vs. time at a constant shear rate (0.5 s^{-1}) for CSG: chia seed byproduct gum, FSG: flaxseed byproduct gum, RSG: rocket seed byproduct gum solutions with different concentrations ((**a**) CSG (1–2%), (**b**) FSG (1–2%), (**c**) RSG (1–5%)).

Table 3. Weltman and second-order structural kinetic model parameters defining the time-dependent flow behavior of gum solutions at 25 °C.

		Weltman Model			**Second-Order Structural Model**				
	Gum (%)	**A (Pa)**	**-B (Pa)**	**R^2**	**k ∗ 1000**	**η_0 (Pa·s)**	**η_e (Pa·s)**	**η_0/η_e**	**R^2**
FSG-1	1.0	4.910 C	0.133 C	0.984	3.666 C	23.346 C	5.276 C	4.425 C	0.996
FSG-1.5	1.5	8.911 B	0.193 B	0.986	7.641 B	59.080 B	6.415 B	9.209 B	0.993
FSG-2	2.0	19.923 A	0.683 A	0.996	13.522 A	138.530 A	9.166 A	15.113 A	0.997
CSG-1	1.0	13.988 C	0.550 C	0.921	5.566 C	74.176 C	11.842 C	6.264 C	0.997
CSG-1.5	1.5	20.407 B	0.661 B	0.952	7.480 B	256.477 B	15.567 B	16.475 B	0.996
CSG-2	2.0	29.349 A	0.701 A	0.999	14.329 A	423.919 A	20.661 A	20.517 A	0.994
RSG-1	1.0	0.281 E	0.012 E	0.966	1.309 E	1.219 D	0.947 E	1.287 E	0.998
RSG-1.5	1.5	0.309 D	0.017 D	0.992	3.157 D	6.626 C	3.976 B	1.668 D	0.999
RSG-2	2.0	0.731 C	0.041 C	0.993	6.787 C	8.837 B	4.466 A	1.978 C	0.990
RSG-3	3.0	2.195 B	0.137 B	0.987	8.602 B	8.953 B	3.092 C	2.896 B	0.992
RSG-5	5.0	3.281 A	0.212 A	0.996	10.656 A	17.499 A	2.899 D	6.036 A	0.993

CSG: chia seed byproduct gum, FSG: flaxseed byproduct gum, RSG: rocket seed byproduct gum; constant shear rate (0.5 s^{-1}). A different uppercase letter in the same column indicates statistical significance ($p < 0.05$).

The second-order structural model provides information on the change in time-dependent flow characteristics caused by shearing, as well as the rate of breakdown, based on the sample's structured and nonstructured state. The rate constant (k) represents the rate of structural breakdown (thixotropy degree), whereas the initial-to-equilibrium viscosity ratio (η_0/η_e) offers a relative assessment of structural breakdown quantity [26]. The parameters of the second-order structural model are shown in Table 3. Gum type also affected the magnitudes of k values and η_0/η_e values. As can be expected, η_0 and η_e values of gum solutions were increased with the increase of gum concentrations. Moreover, the k values and η_0/η_e ratios of gum solutions were increased with the increase of gum concentrations. The highest k and η_0/η_e values belong to CSG solutions, indicating that CSG solutions showed a faster rate of thixotropic breakdown and the extent of thixotropy.

The Weltman model predicted well the relationship between the shear stress and shearing time of gum solutions, and its parameters (A and B) were utilized to examine the effect of temperature on stress decay behavior (Table 3). Parameter A represents the shear stress threshold for structural breakdown, while the time coefficient-B represents the amount of structural breakdown caused by applied shear [64]. Negative B values indicate how soon the apparent viscosity achieves equilibrium [60]. Lower B values imply that the structure of a product is less affected by stirring. Therefore, RSG and FSG were affected less than CSG by stirring due to their lower B values. The Weltman model's coefficients were affected by the gum concentrations and applied constant shear rate [65].

3.3.2. Three Interval Thixotropic Time Test (3-ITT)

Figure 5 showed the structural recovery of CSG, FSG, and RSG solutions by 3-ITT, which simulates the sudden and nonlinear deformation of gum solutions. The structural recovery tendency of CSG, FSG, and RSG solutions increased with the increase of gum concentrations. The lowest concentrations of gum solutions caused the lowest structural recovery. These results showed that the structural breakdown observed in the second time interval due to high sudden shear force could be easily recovered as the gum concentration increased [66].

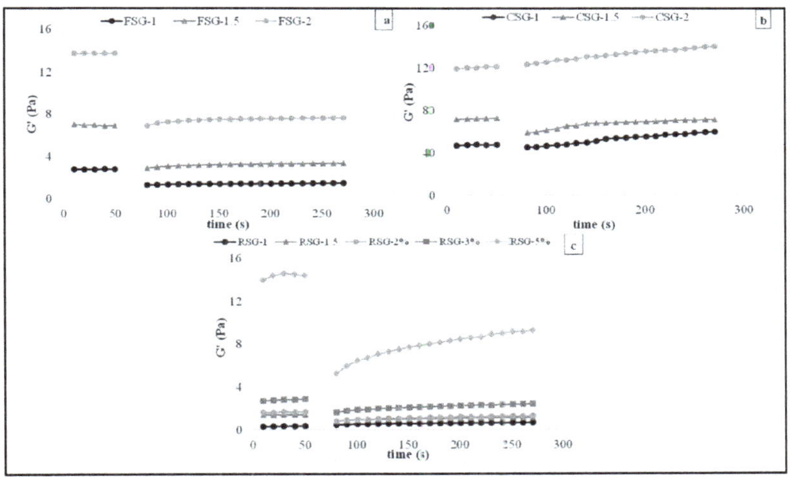

Figure 5. 3-ITT rheological properties of CSG: chia seed byproduct gum, FSG: flaxseed byproduct gum, RSG: rocket seed byproduct gum solutions with different concentrations ((**a**) CSG (1–2%), (**b**) FSG (1–2%), (**c**) RSG (1–5%)).

Table 4 indicates the structural deformation and recovery ratio determined by fitting 3-ITT rheological data using a second-order structural model. The thixotropic constant (k), initial storage modulus (G_0), equilibrium storage modulus (G_e), and the ratio of G_e and G_0 (G_e/G_0) were calculated by the second-order structural kinetic model. The G_e/G_0 value represents the recovery percentage; the larger it is numerically, the faster it can be evaluated to tend to recover. G_e/G_0 values were between 1.124 and 1.750. CSG had the highest G_e, G_0, and G_e/G_0 values, indicating that CSG solutions were the higher recoverable character. As seen in Table 4, the k value indicates the thixotropic rate of samples, and a higher k indicates a higher recovery rate. For each type of gum, the higher gum content showed a higher k value. Samples containing the highest gum concentration showed the highest k and G_e/G_0, indicating that the sample showed the highest thixotropic behavior and viscoelastic solid character. Also, FSG-2 had the highest k value.

Table 4. Second-order structural kinetic model parameters for 3-ITT.

	G_0	G_e	G_e/G_0	k × 1000	R^2	% D	% R
FSG-1	1.285 ± 0.011 C	1.542 ± 0.107 C	1.200 C	7.41 C	0.970	53.79	55.47
FSG-1.5	2.694 ± 0.049 B	3.390 ± 0.012 B	1.259 B	35.56 B	0.998	61.35	48.64
FSG-2	6.368 ± 0.124 A	8.200 ± 0.300 A	1.288 A	56.71 A	0.997	53.52	59.85
CSG-1	38.788 ± 1.561 C	59.525 ± 0.565 C	1.535 C	33.27 C	0.909	16.41	128.29
CSG-1.5	44.234 ± 0.095 B	75.015 ± 0.352 B	1.696 B	38.99 B	0.990	38.05	61.95
CSG-2	76.001 ± 0.219 A	133.000 ± 0.435 A	1.750 A	41.80 A	0.948	36.13	111.76
RSG-1	0.461 ± 0.003 E	0.518 ± 0.019 E	1.124 E	11.00 E	0.983	17.67	256.47
RSG-1.5	0.744 ± 0.014 D	0.842 ± 0.005 D	1.132 D	14.91 D	0.980	46.47	89.33
RSG-2	0.800 ± 0.020 C	0.914 ± 0.259 C	1.143 C	7.08 C	0.983	50.62	99.60
RSG-3	1.268 ± 0.209 B	1.557 ± 0.083 B	1.228 B	8.56 B	0.996	47.89	105.48
RSG-5	4.617 ± 0.308 A	5.869 ± 0.438 A	1.271 A	9.47 A	0.998	65.59	81.91

CSG: chia seed byproduct gum, FSG: flaxseed byproduct gum, RSG: rocket seed byproduct gum. % D: % Deformation and % R: %Recovery. A different uppercase letter in the same column indicates statistical significance.

3.4. Viscoelastic Behavior of the CSG, FSG, and RSG Solutions

A frequency sweep test was utilized for the determining viscoelastic behavior of gum solutions. Figure 6 indicated the dynamic viscoelastic characteristics of CSG, FSG, and RSG solutions with different concentrations. As seen, the changes in storage (G′) and loss (G″) values for gum solutions were demonstrated as a function of angular frequency (ω) at 25 °C. The structure of the gum solutions may be shown in dynamic mechanical spectra, which reveal the frequency dependence of storage modulus G′ and loss modulus G″. There were no cross points between G′ and G″ values, revealing that G′ values were greater than G″ across the entire frequency range studied. As a result, gum solutions had typical gel-like behavior within the experimental frequency range at 25 °C. This viscoelastic behavior was in good agreement with that reported by Chaisawang and Suphantharika [67] and KUTLU et al. [68].

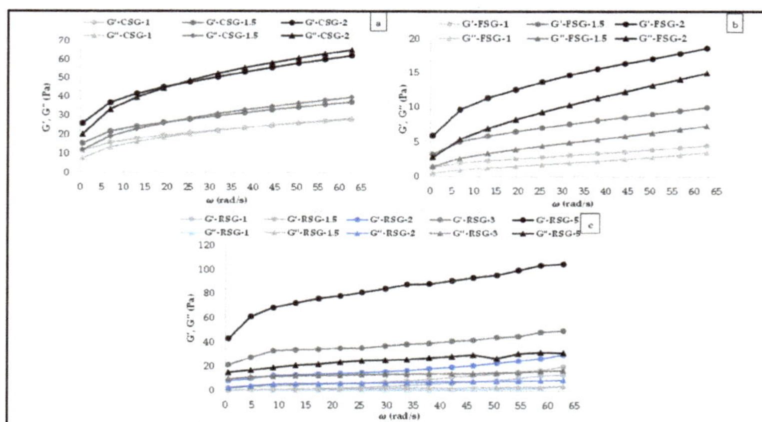

Figure 6. Viscoelastic behavior of the (**a**): CSG: chia seed byproduct gum (1–2%), (**b**): FSG: flaxseed byproduct gum (1–2%), (**c**): RSG: rocket seed byproduct gum (1–5%) solutions with different concentrations.

Non-linear regression was used to analyze experimental G′ and G″ values vs. ω, and the computed magnitudes of slopes (n′ and n″), intercept (K′ and K″), and coefficient of determination (R^2) values are shown in Table 5. K″ values were determined to be higher than K′ values, indicating that gum solutions exhibited liquid-like characteristics. Similar results were reported in previous studies on different gum solutions [66–68] and different food products [69,70]. All of the solutions exhibited gel-like behavior due to the positive slopes (n′ and n″ > 0). The n′ value may be used to determine the strength and type of the gel; n′ = 0 indicates a covalent gel, while n′ > 0 indicates a physical gel. Low n′ values (around zero) indicate that G′ does not vary with frequency, but n′ values close to 1 indicate that the system acts like a viscous gel.

Table 5. Power-law parameters of dynamic rheological properties of the CSG, FSG, and RSG solutions.

	K′ (Pa·s)	n′	R^2	K″ (Pa·s)	n″	R^2
FSG-1	0.949 ± 0.002 C	0.358 ± 0.002	0.9994	0.217 ± 0.004 C	0.656 ± 0.069	0.9981
FSG-1.5	2.933 ± 0.025 B	0.285 ± 0.015	0.9975	1.049 ± 0.006 B	0.459 ± 0.010	0.9880
FSG-2	5.666 ± 0.031 A	0.279 ± 0.005	0.9997	2.216 ± 0.033 A	0.453 ± 0.006	0.9985
CSG-1	10.679 ± 0.031 C	0.326 ± 0.002	0.9978	7.328 ± 0.012 C	0.326 ± 0.002	0.9917
CSG-1.5	14.655 ± 0.093 B	0.304 ± 0.001	0.9990	11.013 ± 0.307 B	0.304 ± 0.011	0.9964
CSG-2	24.849 ± 0.101 A	0.212 ± 0.000	0.9992	19.543 ± 0.076 A	0.287 ± 0.000	0.9807
RSG-1	0.011 ± 0.001 E	0.118 ± 0.103	0.9993	0.007 ± 0.000 E	0.117 ± 0.079	0.9989
RSG-1.5	0.074 ± 0.002 D	0.809 ± 0.098	0.989	0.060 ± 0.003 D	0.801 ± 0.106	0.988
RSG-2	0.151 ± 0.002 C	0.679 ± 0.045	0.9987	0.150 ± 0.034 C	0.721 ± 0.009	0.9983
RSG-3	0.232 ± 0.001 B	0.508 ± 0.002	0.9899	0.219 ± 0.019 B	0.720 ± 0.032	0.9983
RSG-5	2.178 ± 0.041 A	0.202 ± 0.000	0.9982	0.634 ± 0.092 A	0.213 ± 0.001	0.9910

CSG: chia seed byproduct gum (1–2%), FSG: flaxseed byproduct gum (1–2%), RSG: rocket seed byproduct gum (1–5%); K′ and K″: consistency coefficient (Pa·s^n); n′ and n″: flow behavior index values; R^2: determination of coefficient. A different uppercase letter in the same column indicates statistical significance.

3.5. Rheological Properties, Zeta Potential, and Particle Size and Oxidative Stability of Low-Fat Vegan Mayonnaise Samples with Different Types of Gum

3.5.1. Rheological Properties

According to the rheological characterization results, RSG, FSG, and CSG gave the highest K, K′, recovery values and thixotropic degree at the highest concentrations used. RSG, FSG, and CSG were compared with commercial gums at these concentrations. In determining the concentrations of commercial gums in mayonnaise production, the concentrations that gave successful results in the literature were taken into account. The steady shear rheogram of vegan mayonnaise samples was shown in Figure 7, which was absolute evidence of shear-thinning behavior (pseudoplastic behavior) of vegan mayonnaise. This typical behavior was reported for vegan mayonnaise by other researchers [71,72]. It can be concluded from Figure 7 that the mayonnaise sample that formulated the CG and FG had the highest shear-thinning behavior, followed by GG.

The viscoelastic characteristics of the vegan mayonnaise samples obtained by frequency sweep measurement were characterized and shown in Figure 7. As seen, G′ was greater than G″ across the measured frequency range and both G′ and G″ were hardly influenced by frequency change (Figure 6). It can be concluded that all vegan mayonnaise samples showed more solid-like properties. This conclusion was conducted by [73] for mayonnaise formulated with micronized konjac gel. CG-VM samples had the highest G′ and G″ values.

Figure 7 indicated the 3-ITT rheological properties of the vegan mayonnaise samples. Due to deformations during high-speed mixing and homogenization, as well as during consumption, such as when the packed food is shaken or pressed, thixotropic behavior is crucial for O/W emulsions, notably in vegan mayonnaise with low oil content. As can be observed in Figure 7, all samples showed thixotropic behavior in the third interval. Mayonnaise samples lost their viscoelastic characteristics after severe shear deformation but recovered them after a second period. These findings suggested that all mayonnaise samples may maintain their viscoelastic character throughout food processing involving a large amount of abrupt deformation, such as homogenization or pumping, as well as consumption under shaking and squeezing. This is the ideal flow behavior for mayonnaise.

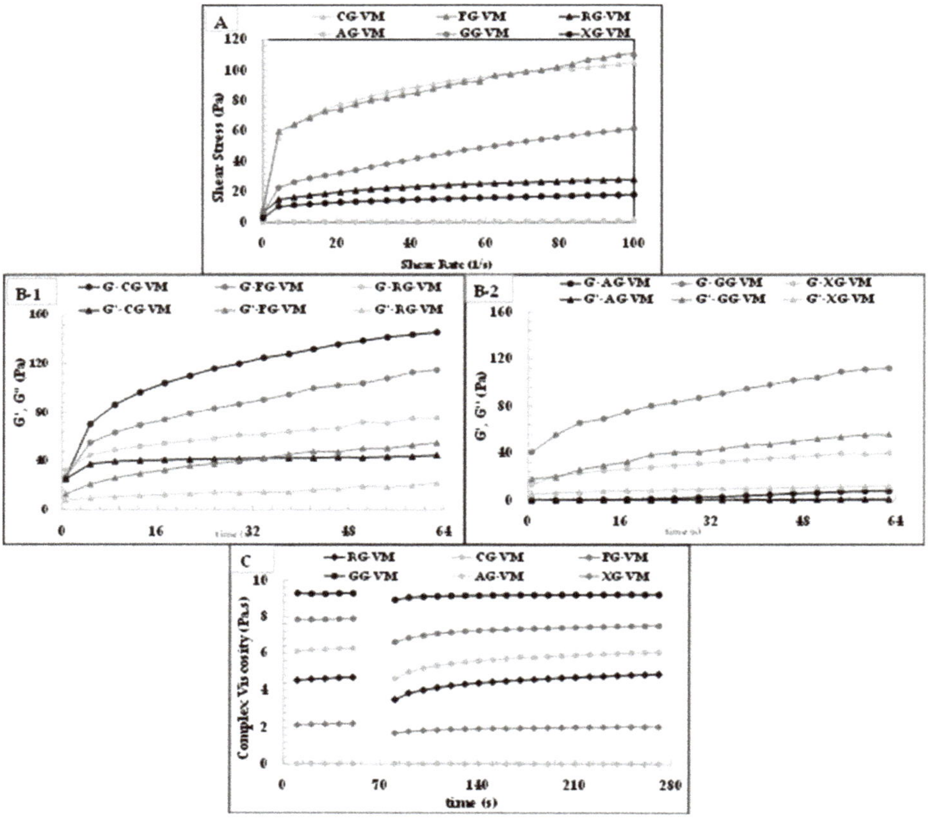

Figure 7. Rheological properties of vegan mayonnaise samples formulated with different gums. FG-VM: flaxseed oil byproduct gum vegan mayonnaise, CG-VM: chia seed oil byproduct gum vegan mayonnaise, RG-VM: rocket seed oil by-product gum vegan mayonnaise, GG-VM: guar gum vegan mayonnaise, XG-VM: xanthan gum vegan mayonnaise, AG-VM: gum Arabic vegan mayonnaise. (**A**) Steady shear rheological properties of the low-fat vegan mayonnaise samples, (**B-1**) Viscoelastic behavior of the low-fat vegan mayonnaise samples formulated with CG, FG, and RG, (**B-2**) Viscoelastic behavior of the low-fat vegan mayonnaise samples formulated with AG, GG, and XG, (**C**) 3-ITT rheological properties of the low-fat vegan mayonnaise samples.

Table 6 presented the rheological properties of low-fat vegan mayonnaise samples prepared with a different type of gum. The flow parameters of the flow index (n), consistency index (K), and coefficient of determination (R^2) were shown in Table 6. The n values were 0.208–0.769, indicating that all mayonnaise samples expected for AG-VM were pseudoplastic fluids ($n < 1$). The n values were reported as lower than 1 for some commercial mayonnaises and vegan mayonnaise samples [74–77]. K values of vegan mayonnaise samples were determined between 0.007 and 38.582 Pa·s^n. CG-VM and FG-VM samples had similar and highest K values, explaining the strongest non-Newtonian behavior.

Table 6 indicated the dynamic power-law parameters of vegan mayonnaise samples. K′ and K″ values of samples were found as 0.012–43.317 Pa·s^n and 0.006–30.423 Pa·s^n, respectively. For the generation of dense viscoelastic interfacial networks at the air/water interface, neutral protein-polysaccharide complexes are recommended. These networks can minimize a thin-gas film's permeability while also promoting foam stability, resulting in much lower interfacial area loss and air bubble coalescence rates. While unevenly charged protein-polysaccharide solutions can stabilize electrostatic repulsion forces between droplet surfaces and induce stability against flocculation and creaming of emulsions, they can also cause flocculation and creaming.

Table 6. Rheological properties of low-fat vegan mayonnaise samples * with a different type of gum.

	Gum (%)	K (Pa·s^n)	R^2		
CG-VM	2.0	38.582 ± 0.183 ^{A}	0.985		
FG-VM	2.0	35.913 ± 0.238 ^{B}	0.986		
RG-VM	5.0	10.647 ± 0.232 ^{C}	0.999		
AG-VM	1.0	0.007 ± 0.002 ^{E}	0.997		
GG-VM	1.0	10.888 ± 0.313 ^{C}	0.994		
XG-VM	0.4	7.050 ± 0.069 ^{D}	0.994		
	K′ (Pa·s^n)	n′	K″ (Pa·s^n)	n″	R^2
CG-VM	43.317 ± 0.098 ^{A}	0.299 ± 0.000	30.423 ± 0.032 ^{A}	0.094 ± 0.005	0.9973
FG-VM	31.321 ± 0.291 ^{C}	0.309 ± 0.001	11.688 ± 0.020 ^{C}	0.370 ± 0.016	0.9960
RG-VM	32.511 ± 0.310 ^{C}	0.191 ± 0.009	5.534 ± 0.197 ^{D}	0.301 ± 0.002	0.9835
AG-VM	0.012 ± 0.001 ^{E}	0.992 ± 0.031	0.006 ± 0.000 ^{F}	1.201 ± 0.021	0.9993
GG-VM	36.752 ± 0.180 ^{B}	0.264 ± 0.002	12.489 ± 0.129 ^{B}	0.361 ± 0.024	0.9790
XG-VM	12.721 ± 0.201 ^{D}	0.274 ± 0.008	4.364 ± 0.082 ^{E}	0.238 ± 0.007	0.9941
	G_0	G_e	G_e/G_0	k × 1000	R^2
CG-VM	64.613	104.270	1.614 ^{A}	60.789 ^{B}	0.9844
FG-VM	53.781	70.591	1.313 ^{D}	37.57 ^{C}	0.9990
RG-VM	38.071	47.951	1.260 ^{E}	24.987 ^{E}	0.9874
AG-VM	0.258	0.180	0.698 ^{F}	4.32 ^{F}	0.9938
GG-VM	65.192	97.138	1.490 ^{B}	68.876 ^{A}	0.9974
XG-VM	14.478	20.176	1.394 ^{C}	29.90 ^{D}	0.9981

CG-VM: vegan mayonnaise contained chia seed byproduct gum, FG-VM: vegan mayonnaise contained flaxseed byproduct gum, RG-VM: vegan mayonnaise contained rocket seed byproduct gum, AG-VM: vegan mayonnaise contained Arabic gum, GG-VM: vegan mayonnaise contained guar gum, XG-VM: vegan mayonnaise contained xanthan gum; K, K′, and K″: consistency coefficient (Pa·s^n); n, n′, and n″: flow behavior index values; G_0: the initial values of the storage modulus; G_e: the equilibrium storage modulus; k: the rate constant of recovery of the sample; R^2: determination of coefficient. * low-fat vegan mayonnaise samples contain 30% vegetable oil and 1% lecithin. A different uppercase letter in the same column indicates statistical significance.

3.5.2. Zeta Potential and Particle Size

Zeta (ζ) potential is an important parameter that shows whether O/W emulsions can remain stable for a long time. As the ζ-potential value moves away from 0, in other words, the system having a negative or positive charge is a positive indicator for the long stability of the product. Table 7 showed that the ζ-potential values of the samples were found between (−42.80) and (−31.90) mV. The first interpretation we can make by looking at these values is that the ζ-potential value of all samples is higher than 0 or that the samples are stable products to a certain degree. The absolute ζ-potential values of our vegan mayonnaise samples were similar except for AG-VM, indicating that the samples can remain stable in these formulations for a long time. The fact that gum forms a compact structure by reducing the mobility of the mobile phase in increasing the stability of mayonnaise samples, thus reducing the action potential of the oil molecules in this tight structure and restricting the interaction of the droplets play a primary role [77]. The fact that the oil droplets have a certain electrical potential and interact with each other thanks to the electrostatic repulsive force, preventing flocculation. The zeta potential of samples prepared with gums obtained from byproducts is similar to commercial gums, indicating that the gums obtained from these byproducts can be successfully applied in an emulsion product, such as mayonnaise and salad dressing.

Table 7. The zeta potential and particle size of vegan mayonnaise samples with a different type of gum.

	ζ-Potential (mV)	PdI	d_{32} (μm)
CG-VM	−42.0 ± 1.58 ^{A}	0.507 ± 0.121 ^{D}	5418.00 ± 268.88 ^{C}
FG-VM	−40.4 ± 1.79 ^{A}	0.701 ± 0.050 ^{A}	8581.33 ± 284.39 ^{A}
RG-VM	−39.5 ± 1.21 ^{A}	0.302 ± 0.000 ^{E}	3689.67 ± 282.13 ^{D}
AG-VM	−31.9 ± 0.62 ^{B}	0.249 ± 0.048 ^{F}	1240.00 ± 102.77 ^{E}
GG-VM	−42.8 ± 1.04 ^{A}	0.539 ± 0.092 ^{C}	5851.33 ± 114.01 ^{C}
XG-VM	−41.2 ± 0.89 ^{A}	0.613 ± 0.074 ^{B}	6720.32 ± 160.99 ^{B}

CG-VM: vegan mayonnaise contained chia seed byproduct gum, FG-VM: vegan mayonnaise contained flaxseed byproduct gum, RG-VM: vegan mayonnaise contained rocket seed byproduct gum, AG-VM: vegan mayonnaise contained Arabic gum, GG-VM: vegan mayonnaise contained guar gum, XG-VM: vegan mayonnaise contained xanthan gum. ζ-Potential: zeta-potential (mV); PdI: polydispersity index; d_{32}: the oil particle size (μm). A different uppercase letter in the same column indicates statistical significance.

The oil particle size (d_{32}) and PdI value of the samples were found as 1240.00–8581.33 μm, and 0.249–0.701, respectively. The researchers emphasized that the oil particle diameter decreased significantly as the polysaccharide concentration increased [78,79]. As can be seen, the mayonnaise samples containing AG and RG exhibited lower particle size and PdI value as compared to other mayonnaise samples. Also, all samples have a sufficient zeta potential value.

3.5.3. Oxidative Stability (at 90 °C)

The oxidative stability of the vegan mayonnaise samples was determined with the Oxitest device, and the IP values of the samples were recorded. Table 8 showed the IP values of the samples. The IP values varied between 4:38 and 13:44 h. The sample prepared with xanthan gum showed the lowest IP value. The IP values of the samples prepared with gums obtained from byproducts were higher than the vegan mayonnaise samples prepared with xanthan gum. The IP value of the sample prepared with RSG was much higher than other samples. The very high IP value of the sample prepared with RSG can be explained by the RSG concentration. Due to the low consistency of RSG, it was used at the level of 5%. During the extraction of RSG, antioxidant products, such as phenolic compounds, may have transferred to gum solution and caused higher IP value. The higher IP value of FSG and RSG than xanthan gum. This result indicated that gums obtained from byproducts could exhibit a favorable condition in terms of oxidative stability as well as providing desirable consistency.

Table 8. The induction period (IP (h)) of vegan mayonnaise samples with a different type of gum at 90 °C.

Samples	IP (h)
CG-VM	6:15 ± 0.02 ^{B}
FG-VM	5:35 ± 0.08 ^{D}
RG-VM	13:44 ± 0.12 ^{A}
AG-VM	6:00 ± 0.01 ^{C}
GG-VM	5:27 ± 0.04 ^{D}
XG-VM	4:38 ± 0.03 ^{E}

CG-VM: vegan mayonnaise contained chia seed byproduct gum, FG-VM: vegan mayonnaise contained flaxseed byproduct gum, RG-VM: vegan mayonnaise contained rocket seed byproduct gum, AG-VM: vegan mayonnaise contained Arabic gum, GG-VM: vegan mayonnaise contained guar gum, XG-VM: vegan mayonnaise contained xanthan gum. IP: induction period (h). A different uppercase letter in the same column indicates statistical significance.

4. Conclusions

In recent years, cold-pressed oil consumption has increased due to the bioactive components in its structure. After the consumption of cold-pressed oil, a byproduct rich in protein, carbohydrates, and bioactive components emerges. Adding value to these byproducts is an important issue for the cold press oil industry. In this study, the production of gum from cold-pressed flax, chia, and rocket seed oil wastes and the use of the obtained gum in vegan mayonnaise were investigated. FSG and CSG are rich in xylose and glucose, while RSG is rich in galactose and mannose. While FGS and CSG exhibited high pseudoplastic, viscoelastic solid, and recoverable structure at low concentrations, RSG exhibited these properties at high concentrations. The potential for use as stabilizers and emulsifiers in the production of FSG, CSG, and RSG vegan mayonnaise has been investigated. The results of the study showed that FSG, CSG, and RSG can provide desired rheological and microstructural properties and oxidative stability in the production of reduced-fat vegan mayonnaise. The results of this study indicated that this study could offer a new perspective in adding value to flaxseed, chia seed, and rocket seed cold-press oil by-product.

Author Contributions: Conceptualization, S.K., Z.H.T.-Ç. and F.B.; methodology, S.K.; software, S.K. and Z.H.T.-Ç.; validation, Z.H.T.-Ç.; formal analysis, Z.H.T.-Ç.; investigation, S.K., Z.H.T.-Ç. and F.B.; data curation, Z.H.T.-Ç., F.B. and T.H.; writing—original draft preparation, S.K., F.B., T.H. and Z.H.T.-Ç.; writing—review and editing, S.K. and Z.H.T.-Ç.; visualization, S.K. and Z.H.T.-Ç.; supervision, S.K.; funding acquisition, T.H. All authors have read and agreed to the published version of the manuscript.

Funding: This research received no external funding.

Data Availability Statement: The data presented in this study are available on request from the corresponding author.

Conflicts of Interest: The authors declare no conflict of interest.

References

1. Parker, T.D.; Adams, D.A.; Zhou, K.; Harris, M.; Yu, L. Fatty Acid Composition and Oxidative Stability of Cold-pressed Edible Seed Oils. *J. Food Sci.* **2003**, *68*, 1240–1243. [CrossRef]
2. Mildner-Szkudlarz, S.; Różańska, M.; Siger, A.; Kowalczewski, P.Ł.; Rudzińska, M. Changes in chemical composition and oxidative stability of cold-pressed oils obtained from by-product roasted berry seeds. *LWT* **2019**, *111*, 541–547. [CrossRef]
3. Bekhit, A.E.-D.A.; Shavandi, A.; Jodjaja, T.; Birch, J.; Teh, S.; Ahmed, I.A.M.; Al-Juhaimi, F.Y.; Saeedi, P.; Bekhit, A.A. Flaxseed: Composition, detoxification, utilization, and opportunities. *Biocatal. Agric. Biotechnol.* **2018**, *13*, 129–152. [CrossRef]
4. Wu, S.; Wang, X.; Qi, W.; Guo, Q. Bioactive protein/peptides of flaxseed: A review. *Trends Food Sci. Technol.* **2019**, *92*, 184–193. [CrossRef]
5. Tekin, Z.H.; Karasu, S. Cold-pressed flaxseed oil by-product as a new source of fat replacers in low-fat salad dressing formulation: Steady, dynamic and 3-ITT rheological properties. *J. Food Process. Preserv.* **2020**, *44*, e14650. [CrossRef]
6. Chen, C.; Huang, X.; Wang, L.-J.; Li, D.; Adhikari, B. Effect of flaxseed gum on the rheological properties of peanut protein isolate dispersions and gels. *LWT* **2016**, *74*, 528–533. [CrossRef]
7. Chen, H.-H.; Xu, S.-Y.; Wang, Z. Interaction between flaxseed gum and meat protein. *J. Food Eng.* **2007**, *80*, 1051–1059. [CrossRef]
8. Rashid, F.; Ahmed, Z.; Hussain, S.; Huang, J.-Y.; Ahmad, A. *Linum usitatissimum* L. seeds: Flax gum extraction, physicochemical and functional characterization. *Carbohydr. Polym.* **2019**, *215*, 29–38. [CrossRef]
9. Timilsena, Y.P.; Wang, B.; Adhikari, R.; Adhikari, B. Preparation and characterization of chia seed protein isolate–chia seed gum complex coacervates. *Food Hydrocoll.* **2016**, *52*, 554–563. [CrossRef]
10. Timilsena, Y.P.; Adhikari, R.; Kasapis, S.; Adhikari, B. Rheological and microstructural properties of the chia seed polysaccharide. *Int. J. Biol. Macromol.* **2015**, *81*, 991–999. [CrossRef]
11. Suri, S.; Passi, S.J.; Goyat, J. Chia seed (*Salvia hispanica* L.)—A new age functional food. In Proceedings of the 4th International Conference on Recent Innovations in Science Engineering and Management, New Delh, India, 20 March 2016; pp. 286–299.
12. Akcicek, A.; Karasu, S. Utilization of cold pressed chia seed oil waste in a low-fat salad dressing as natural fat replacer. *J. Food Process Eng.* **2018**, *41*, e12694. [CrossRef]
13. Punia, S.; Dhull, S.B. Chia seed (*Salvia hispanica* L.) mucilage (a heteropolysaccharide): Functional, thermal, rheological behaviour and its utilization. *Int. J. Biol. Macromol.* **2019**, *140*, 1084–1090. [CrossRef] [PubMed]
14. Koubaa, M.; Mhemdi, H.; Sanlaville, Q.; Vorobiev, E. Recovery of oil, erucic acid, and phenolic compounds from rapeseed and rocket seeds. *Chem. Eng. Technol.* **2016**, *39*, 1431–1437. [CrossRef]

15. Nail, T.; Ali, M.; Salim, E. Phytochemical studies on Sudanese rocket (*Eruca sativa*) seeds and oil constituents. *Am. J. Phytomed. Clin. Ther.* **2017**, *5*, 1–5.
16. Sherahi, M.H.; Fathi, M.; Zhandari, F.; Hashemi, S.M.B.; Rashidi, A. Structural characterization and physicochemical properties of *Descurainia sophia* seed gum. *Food Hydrocoll.* **2017**, *66*, 82–89. [CrossRef]
17. Depree, J.A.; Savage, G.P. Physical and flavour stability of mayonnaise. *Trends Food Sci. Technol.* **2001**, *12*, 157–163. [CrossRef]
18. Harrison, L.J.; Cunningham, F.E. Factors influencing the quality of mayonnaise: A review. *J. Food Qual.* **1985**, *8*, 1–20. [CrossRef]
19. Naji-Tabasi, S.; Razavi, S.M.A.; Mohebbi, M.; Malaekeh-Nikouei, B. New studies on basil (*Ocimum bacilicum* L.) seed gum: Part I—Fractionation, physicochemical and surface activity characterization. *Food Hydrocoll.* **2016**, *52*, 350–358. [CrossRef]
20. Yebeyen, D.; Lemenih, M.; Feleke, S. Characteristics and quality of gum arabic from naturally grown *Acacia senegal* (Linne) Willd. trees in the Central Rift Valley of Ethiopia. *Food Hydrocoll.* **2009**, *23*, 175–180. [CrossRef]
21. Shabani, H.; Askari, G.; Jahanbin, K.; Khodaeian, F. Evaluation of physicochemical characteristics and antioxidant property of *Prunus avium* gum exudates. *Int. J. Biol. Macromol.* **2016**, *93*, 436–441. [CrossRef]
22. Sibaja-Hernández, R.; Román-Guerrero, A.; Sepúlveda-Jiménez, G.; Rodríguez-Monroy, M. Physicochemical, shear flow behaviour and emulsifying properties of *Acacia cochliacantha* and *Acacia farnesiana* gums. *Ind. Crops Prod.* **2015**, *67*, 161–168. [CrossRef]
23. Rincón, F.; Muñoz, J.; Ramírez, P.; Galán, H.; Alfaro, M.C. Physicochemical and rheological characterization of *Prosopis juliflora* seed gum aqueous dispersions. *Food Hydrocoll.* **2014**, *35*, 348–357. [CrossRef]
24. Razavi, S.M.A.; Cui, S.W.; Guo, Q.; Ding, H. Some physicochemical properties of sage (*Salvia macrosiphon*) seed gum. *Food Hydrocoll.* **2014**, *35*, 453–462. [CrossRef]
25. Wang, J.; Su, Y.; Jia, F.; Jin, H. Characterization of casein hydrolysates derived from enzymatic hydrolysis. *Chem. Cent. J.* **2013**, *7*, 62. [CrossRef] [PubMed]
26. Abu-Jdayil, B. Modelling the time-dependent rheological behavior of semisolid foodstuffs. *J. Food Eng.* **2003**, *57*, 97–102. [CrossRef]
27. Van Hecke, E.; Nguyen, P.U.; Clausse, D.; Lanoisellé, J.L. Flow behaviour of carrot puree: Modelling the influence of time, temperature and potato flakes addition. *Int. J. Food Sci. Technol.* **2012**, *47*, 177–185. [CrossRef]
28. Toker, O.S.; Karasu, S.; Yilmaz, M.T.; Karaman, S. Three interval thixotropy test (3ITT) in food applications: A novel technique to determine structural regeneration of mayonnaise under different shear conditions. *Food Res. Int.* **2015**, *70*, 125–133. [CrossRef]
29. Yoo, B.; Rao, M.A.; Steffe, J.F. Yield stress of food dispersions with the vane method at controlled shear rate and shear stress. *J. Texture Stud.* **1995**, *26*, 1–10. [CrossRef]
30. Aksoy, F.S.; Tekin-Cakmak, Z.H.; Karasu, S.; Aksoy, A.S. Oxidative stability of the salad dressing enriched by microencapsulated phenolic extracts from cold-pressed grape and pomegranate seed oil by-products evaluated using OXITEST. *Food Sci. Technol.* **2021**, *in press*. [CrossRef]
31. Bhushette, P.R.; Annapure, U.S. Physicochemical, functional and rheological investigation of Soymida febrifuga exudate gum. *Int. J. Biol. Macromol.* **2018**, *111*, 1116–1123. [CrossRef]
32. Hellebois, T.; Tsevdou, M.; Soukoulis, C. Chapter Five—Functionalizing and bio-preserving processed food products via probiotic and synbiotic edible films and coatings. In *Advances in Food and Nutrition Research*; da Cruz, A.G., Prudencio, E.S., Esmerino, E.A., da Silva, M.C., Eds.; Academic Press: Cambridge, MA, USA, 2020; Volume 94, pp. 161–221.
33. Kontogiorgos, V. Galactomannans (Guar, Locust Bean, Fenugreek, Tara). In *Encyclopedia of Food Chemistry*; Elsevier: Amsterdam, The Netherlands, 2018.
34. Oomah, B.D.; Kenaschuk, E.; Cui, S.; Mazza, G. Variation in the composition of water-soluble polysaccharides in flaxseed. *J. Agric. Food Chem.* **1995**, *43*, 1484–1488. [CrossRef]
35. Safdar, B.; Pang, Z.; Liu, X.; Jatoi, M.A.; Mehmood, A.; Rashid, M.T.; Ali, N.; Naveed, M. Flaxseed gum: Extraction, bioactive composition, structural characterization, and its potential antioxidant activity. *J. Food Biochem.* **2019**, *43*, e13014. [CrossRef] [PubMed]
36. Kaushik, P.; Dowling, K.; Adhikari, R.; Barrow, C.J.; Adhikari, B. Effect of extraction temperature on composition, structure and functional properties of flaxseed gum. *Food Chem.* **2017**, *215*, 333–340. [CrossRef] [PubMed]
37. Timilsena, Y.P.; Adhikari, R.; Kasapis, S.; Adhikari, B. Molecular and functional characteristics of purified gum from Australian chia seeds. *Carbohydr. Polym.* **2016**, *136*, 128–136. [CrossRef] [PubMed]
38. Cerqueira, M.A.; Souza, B.W.S.; Simões, J.; Teixeira, J.A.; Domingues, M.R.M.; Coimbra, M.A.; Vicente, A.A. Structural and thermal characterization of galactomannans from non-conventional sources. *Carbohydr. Polym.* **2011**, *83*, 179–185. [CrossRef]
39. Goh, K.K.T.; Matia-Merino, L.; Chiang, J.H.; Quek, R.; Soh, S.J.B.; Lentle, R.G. The physico-chemical properties of chia seed polysaccharide and its microgel dispersion rheology. *Carbohydr. Polym.* **2016**, *149*, 297–307. [CrossRef]
40. Vinod, V.T.P.; Sashidhar, R.B.; Suresh, K.I.; Rama Rao, B.; Vijaya Saradhi, U.V.R.; Prabhakar Rao, T. Morphological, physico-chemical and structural characterization of gum kondagogu (*Cochlospermum gossypium*): A tree gum from India. *Food Hydrocoll.* **2008**, *22*, 899–915. [CrossRef]
41. Fonseca, P.R.M.S.; Dekker, R.F.H.; Barbosa, A.M.; Silveira, J.L.M.; Vasconcelos, A.F.D.; Monteiro, N.K.; Aranda-Selverio, G.; Da Silva, M.D.L.C. Thermal and Rheological Properties of a Family of Botryosphaerans Produced by *Botryosphaeria rhodina* MAMB-05. *Molecules* **2011**, *16*, 7488–7501. [CrossRef]
42. Xie, J.-H.; Liu, X.; Shen, M.-Y.; Nie, S.-P.; Zhang, H.; Li, C.; Gong, D.-M.; Xie, M.-Y. Purification, physicochemical characterisation and anticancer activity of a polysaccharide from Cyclocarya paliurus leaves. *Food Chem.* **2013**, *136*, 1453–1460. [CrossRef]

43. Hadad, S.; Goli, S.A.H. Fabrication and characterization of electrospun nanofibers using flaxseed (*Linum usitatissimum*) mucilage. *Int. J. Biol. Macromol.* **2018**, *114*, 408–414. [CrossRef]
44. Darwish, A.; El-Sohaimy, S. Functional Properties of Chia Seed Mucilage Supplemented In Low Fat Yoghurt. *Alex. Med. J.* **2018**, *39*, 450–459. [CrossRef]
45. Bortnowska, G.; Balejko, J.; Schube, V.; Tokarczyk, G.; Krzemińska, N.; Mojka, K. Stability and physicochemical properties of model salad dressings prepared with pregelatinized potato starch. *Carbohydr. Polym.* **2014**, *111*, 624–632. [CrossRef] [PubMed]
46. Utrilla-Coello, R.G.; Rodríguez-Huezo, M.E.; Carrillo-Navas, H.; Hernández-Jaimes, C.; Vernon-Carter, E.J.; Alvarez-Ramirez, J. In vitro digestibility, physicochemical, thermal and rheological properties of banana starches. *Carbohydr. Polym.* **2014**, *101*, 154–162. [CrossRef] [PubMed]
47. Dakia, P.A.; Blecker, C.; Robert, C.; Wathelet, B.; Paquot, M. Composition and physicochemical properties of locust bean gum extracted from whole seeds by acid or water dehulling pre-treatment. *Food Hydrocoll.* **2008**, *22*, 807–818. [CrossRef]
48. Zhong, L.; Oostrom, M.; Truex, M.J.; Vermeul, V.R.; Szecsody, J.E. Rheological behavior of xanthan gum solution related to shear thinning fluid delivery for subsurface remediation. *J. Hazard. Mater.* **2013**, *244*, 160–170. [CrossRef]
49. Sanchez, C.; Renard, D.; Robert, P.; Schmitt, C.; Lefebvre, J. Structure and rheological properties of acacia gum dispersions. *Food Hydrocoll.* **2002**, *16*, 257–267. [CrossRef]
50. Marcotte, M.; Taherian Hoshahili, A.R.; Ramaswamy, H.S. Rheological properties of selected hydrocolloids as a function of concentration and temperature. *Food Res. Int.* **2001**, *34*, 695–703. [CrossRef]
51. Yamazaki, E.; Kurita, O.; Matsumura, Y. High viscosity of hydrocolloid from leaves of *Corchorus olitorius* L. *Food Hydrocoll.* **2009**, *23*, 655–660. [CrossRef]
52. Razavi, S.M.A.; Taheri, H.; Quinchia, L.A. Steady shear flow properties of wild sage (*Salvia macrosiphon*) seed gum as a function of concentration and temperature. *Food Hydrocoll.* **2011**, *25*, 451–458. [CrossRef]
53. Chaharlang, M.; Samavati, V. Steady shear flow properties of *Cordia myxa* leaf gum as a function of concentration and temperature. *Int. J. Biol. Macromol.* **2015**, *79*, 56–62. [CrossRef]
54. Yaseen, E.I.; Herald, T.J.; Aramouni, F.M.; Alavi, S. Rheological properties of selected gum solutions. *Food Res. Int.* **2005**, *38*, 111–119. [CrossRef]
55. Adeli, M.; Samavati, V. Studies on the steady shear flow behavior and chemical properties of water-soluble polysaccharide from Ziziphus lotus fruit. *Int. J. Biol. Macromol.* **2015**, *72*, 580–587. [CrossRef] [PubMed]
56. Koocheki, A.; Taherian, A.R.; Bostan, A. Studies on the steady shear flow behavior and functional properties of *Lepidium perfoliatum* seed gum. *Food Res. Int.* **2013**, *50*, 446–456. [CrossRef]
57. Vardhanabhuti, B.; Ikeda, S. Isolation and characterization of hydrocolloids from monoi (*Cissampelos pareira*) leaves. *Food Hydrocoll.* **2006**, *20*, 885–891. [CrossRef]
58. Torres, M.D.; Hallmark, B.; Wilson, D.I. Effect of concentration on shear and extensional rheology of guar gum solutions. *Food Hydrocoll.* **2014**, *40*, 85–95. [CrossRef]
59. Wu, Y.; Ding, W.; Jia, L.; He, Q. The rheological properties of tara gum (*Caesalpinia spinosa*). *Food Chem.* **2015**, *168*, 366–371. [CrossRef]
60. Koocheki, A.; Mortazavi, S.A.; Shahidi, F.; Razavi, S.M.A.; Taherian, A.R. Rheological properties of mucilage extracted from *Alyssum homolocarpum* seed as a new source of thickening agent. *J. Food Eng.* **2009**, *91*, 490–496. [CrossRef]
61. Fijan, R.; Šostar-Turk, S.; Lapasin, R. Rheological study of interactions between non-ionic surfactants and polysaccharide thickeners used in textile printing. *Carbohydr. Polym.* **2007**, *68*, 708–717. [CrossRef]
62. Weltmann, R.N. Breakdown of Thixotropic Structure as Function of Time. *J. Appl. Phys.* **1943**, *14*, 343–350. [CrossRef]
63. Dzuy Nguyen, Q.; Jensen, C.T.B.; Kristensen, P.G. Experimental and modelling studies of the flow properties of maize and waxy maize starch pastes. *Chem. Eng. J.* **1998**, *70*, 165–171. [CrossRef]
64. Alvarez, M.D.; Canet, W. Time-independent and time-dependent rheological characterization of vegetable-based infant purees. *J. Food Eng.* **2013**, *114*, 449–464. [CrossRef]
65. Koocheki, A.; Razavi, S.M. Effect of Concentration and Temperature on Flow Properties of *Alyssum homolocarpum* Seed Gum Solutions: Assessment of Time Dependency and Thixotropy. *Food Biophys.* **2009**, *4*, 353–364. [CrossRef]
66. Razmkhah, S.; Razavi, S.M.A.; Mohammadifar, M.A. Purification of cress seed (*Lepidium sativum*) gum: A comprehensive rheological study. *Food Hydrocoll.* **2016**, *61*, 358–368. [CrossRef]
67. Chaisawang, M.; Suphantharika, M. Pasting and rheological properties of native and anionic tapioca starches as modified by guar gum and xanthan gum. *Food Hydrocoll.* **2006**, *20*, 641–649. [CrossRef]
68. Kutlu, G.; Akcicek, A.; Bozkurt, F.; Karasu, S.; Tekin-Cakmak, Z.H. Rocket seed (*Eruca sativa* Mill) gum: Physicochemical and comprehensive rheological characterization. *Food Sci. Technol.* **2021**, *in press*. [CrossRef]
69. Toker, O.S.; Karaman, S.; Yuksel, F.; Dogan, M.; Kayacier, A.; Yilmaz, M.T. Temperature dependency of steady, dynamic, and creep-recovery rheological properties of ice cream mix. *Food Bioprocess Technol.* **2013**, *6*, 2974–2985. [CrossRef]
70. Yılmaz, M.T.; Karaman, S.; Cankurt, H.; Kayacier, A.; Sagdic, O. Steady and dynamic oscillatory shear rheological properties of ketchup–processed cheese mixtures: Effect of temperature and concentration. *J. Food Eng.* **2011**, *103*, 197–210. [CrossRef]
71. Worrasinchai, S.; Suphantharika, M.; Pinjai, S.; Jamnong, P. β-Glucan prepared from spent brewer's yeast as a fat replacer in mayonnaise. *Food Hydrocoll.* **2006**, *20*, 68–78. [CrossRef]

72. Rudra, S.G.; Hanan, E.; Sagar, V.; Bhardwaj, R.; Basu, S.; Sharma, V. Manufacturing of mayonnaise with pea pod powder as a functional ingredient. *J. Food Meas. Charact.* **2020**, *14*, 2402–2413. [CrossRef]
73. Li, J.; Wang, Y.; Jin, W.; Zhou, B.; Li, B. Application of micronized konjac gel for fat analogue in mayonnaise. *Food Hydrocoll.* **2014**, *35*, 375–382. [CrossRef]
74. Dickie, A.M.; Kokini, J.L. An Improved Model for Food Thickness from non-Newtonian Fluid Mechanics in the Mouth. *J. Food Sci.* **1983**, *48*, 57–61. [CrossRef]
75. Juszczak, L.; Fortuna, T.; Kośla, A. Sensory and rheological properties of Polish commercial mayonnaise. *Food/Nahrung* **2003**, *47*, 232–235. [CrossRef] [PubMed]
76. Izidoro, D.; Sierakowski, M.-R.; Waszczynskyj, N.; Haminiuk, C.W.; de Paula Scheer, A. Sensory evaluation and rheological behavior of commercial mayonnaise. *Int. J. Food Eng.* **2007**, *3*, 1–15. [CrossRef]
77. Singla, N.; Verma, P.; Ghoshal, G.; Basu, S. Steady state and time dependent rheological behaviour of mayonnaise (egg and eggless). *Int. Food Res. J.* **2013**, *20*, 2009–2016.
78. Moschakis, T.; Murray, B.; Dickinson, E. Microstructural evolution of viscoelastic emulsions stabilised by sodium caseinate and xanthan gum. *J. Colloid Interface Sci.* **2005**, *284*, 714–728. [CrossRef] [PubMed]
79. Khalloufi, S.; Corredig, M.; Goff, H.; Alexander, M. Flaxseed gums and their adsorption on whey protein-stabilized oil-in-water emulsions. *Food Hydrocoll.* **2009**, *23*, 611–618. [CrossRef]

foods

MDPI

Article

Byproducts (Flour, Meals, and Groats) from the Vegetable Oil Industry as a Potential Source of Antioxidants

Mihaela Multescu [1,2,*], Ioana Cristina Marinas [3], Iulia Elena Susman [1,2] and Nastasia Belc [1]

1 National Institute of Research and Development for Food Bioresources—IBA Bucharest, 020323 Bucharest, Romania; iulia.susman@bioresurse.ro (I.E.S.); nastasia.belc@bioresurse.ro (N.B.)
2 Faculty of Biotechnology, University of Agronomic Sciences and Veterinary Medicine of Bucharest, 011464 Bucharest, Romania
3 Research Institute of the University of Bucharest—ICUB, 050095 Bucharest, Romania; ioana.cristina.marinas@gmail.com
* Correspondence: mihaela.multescu@gmail.com

Abstract: The present study presents the use of photochemiluminescence assay (PCL) and 2,2 diphenyl-1-picryl-hydrazyl (DPPH), 2,2′-azinobis-(3-ethylbenzothiazoline-6-sulfonic acid (ABTS), the ferric reducing antioxidant power (FRAP), and cupric ion reducing antioxidant capacity (CUPRAC) methods for the measurement of lipid-soluble antioxidant capacity (ACL) of 14 different byproducts obtained from the vegetable oil industry (flour, meals, and groats). The research showed that the analyzed samples contain significant amounts of phenolic compounds between 1.54 and 74.85 mg gallic acid per gram of byproduct. Grape seed flour extract had the highest content of total phenolic compounds, 74.85 mg GAE/g, while the lowest level was obtained for the sunflower groats, 1.54 mg GAE/g. DPPH values varied between 7.58 and 7182.53 mg Trolox/g of byproduct, and the highest antioxidant capacity corresponded to the grape seed flour (7182.53 mg Trolox/g), followed by walnut flour (1257.49 mg Trolox/g) and rapeseed meals (647.29 mg Trolox/g). Values of ABTS assay of analyzed samples were between 0 and 3500.52 mg Trolox/g of byproduct. Grape seed flour had the highest value of ABTS (3500.52 mg Trolox/g), followed by walnut flower (1423.98) and sea buckthorn flour (419.46). The highest values for FRAP method were represented by grape seed flour (4716.75 mg Trolox/g), followed by sunflower meals (1350.86 mg Trolox/g) and rapeseed flour (1034.92 mg Trolox/g). For CUPRAC assay, grape seed flour (5936.76 mg Trolox/g) and walnut flour (1202.75 mg Trolox/g) showed the highest antioxidant activity. To assess which method of determining antioxidant activity is most appropriate for the byproducts analyzed, relative antioxidant capacity index (RACI) was calculated. Depending on the RACI value of the analyzed byproducts, the rank of antioxidant capacity ranged from −209.46 (walnut flour) to 184.20 (grape seed flour). The most sensitive methods in developing RACI were FRAP (r = 0.5795) and DPPH (r = 0.5766), followed by CUPRAC (r = 0.5578) and ABTS (r = 0.4449), respectively. Strong positive correlations between the antioxidant capacity of lipid-soluble compounds measured by PCL and other methods used for determining antioxidant activity were found (r > 0.9). Analyses have shown that the different types of byproducts obtained from the vegetable oil industry have a high antioxidant activity rich in phenolic compounds, and thus their use in bakery products can improve their nutritional quality.

Keywords: byproducts; vegetable oil industry; phenolics; flavonoids; antioxidant activity; photochemiluminescence

Citation: Multescu, M.; Marinas, I.C.; Susman, I.E.; Belc, N. Byproducts (Flour, Meals, and Groats) from the Vegetable Oil Industry as a Potential Source of Antioxidants. *Foods* **2022**, *11*, 253. https://doi.org/10.3390/foods11030253

Academic Editors: Marco Poiana, Francesco Caponio and Antonio Piga

Received: 9 December 2021
Accepted: 13 January 2022
Published: 18 January 2022

Publisher's Note: MDPI stays neutral with regard to jurisdictional claims in published maps and institutional affiliations.

Copyright: © 2022 by the authors. Licensee MDPI, Basel, Switzerland. This article is an open access article distributed under the terms and conditions of the Creative Commons Attribution (CC BY) license (https://creativecommons.org/licenses/by/4.0/).

1. Introduction

During the process of obtaining vegetable oils, considerable amounts of waste and byproducts are generated. These byproducts from the vegetable oil industry are important due to their high value-added substances, and they represent an excellent source of bioactive components, such as antioxidants. Byproducts such as flour, meals, and groats resulting from the vegetable oil industry are considered economic resources due to the antioxidant compounds, which have attracted interest in making functional products with a higher nutritional value, satisfying consumer demand for such products [1].

Sea buckthorn berries are rich in substances with biological activity [2]. Sea buckthorn is rich in nutrients and antioxidants, so it is used as a dietary supplement [3]. Ghendov-Mosanu et al. [2] have determined the characteristics of sea buckthorn flour and investigated its effects on sensory, physicochemical, and antioxidant properties. The results obtained showed that sea buckthorn flour is a good source of ascorbic acid, polyphenols, and flavonoids. This byproduct added in different concentrations in wheat bread has extended the shelf life to 72 h and has improved antioxidant activity. Increasing the concentration of sea buckthorn flour in wheat bread was directly proportional to its benefits. Hemp is rich in protein, fat, carbohydrates, and fiber. It contains significant amounts of macroelements, such as P, K, Mg, Na, and Ca. Hemp flour is a good source of bioactive compounds, especially polyphenols [4]. Rusu et al. [4] have nutritionally characterized the bread with the addition of hemp flour in different concentrations. Studies have shown that the addition of hemp flour to bread improves the nutritional properties.

Walnuts are a significant source of vegetable protein and amino acids. They contain over 50% oil rich in polyunsaturated fatty acids. Walnuts are an excellent source of phytochemicals, such as phenolic compounds, carotenoids, and tocopherols [5]. Almoraie et al. [6] determined the possibility of using walnut flour in the production of wheat bread. Bread samples with different concentrations of walnut flour have been shown to have a higher nutritional value than bread samples made from 100% wheat flour. Grape seed flour contains a significant number of polyphenols and presents high antioxidant capacity [7]. This byproduct can be an alternative in the production of various foods due to a high content of dietary fiber. Previous studies have shown that grape seed flour has been used in many products, such as cookies, pancakes, butter biscuits, and bread [8].

Sunflower has many nutritional components. Examples are sunflower flour, meal, groats, etc. [9]. Sunflower meal can be used in a wide range of bakery products due to the high concentration of antioxidants. Byproducts such as sunflower meal represent an excellent source of protein, essential amino acids, and fiber. It contains essential amino acids, vitamin B, and minerals, and has a high antioxidant property [10]. Grasso et al. [11] observed that sunflower meal could be used for the nutritional improvement of muffins. Rapeseed's meal is rich in phenolic compounds, tocopherols, vitamins B, calcium, magnesium, and presents high antioxidant activity [12]. Along with rapeseed meal, black sesame meal is a potential source of polyphenolics with high antioxidant status [13]. Flax seeds contain a high level of oil and are rich in polyunsaturated fatty acids. They are also an important source of soluble and insoluble fiber [14]. Numerous phytochemical compounds with antioxidant activity are found in flax seeds, including phenolic acids, flavonoids, and lignins [15]. Pourabedin et al. [16] observed that adding flaxseed flour to toast bread increases phenolic compounds. Milk thistle is known for its high content of bioactive compounds, mainly flavonoids with a powerful antioxidant character [17]. Milk thistle seeds contain the largest number of active substances, about 70–80% of silymarin flavonolignans and about 20–30% of substances with polyphenolic structures [18].

The byproducts (meal) obtained in the vegetable oils industry contain phenolic compounds with different chemical structures, such as tocopherols, carotenoids, flavonoids, lignins, phenolic acids, and tannins with high values of antioxidant capacity [19]. These are natural and cheap sources of antioxidants that could replace synthetic additives (BHT or BHA) [19]. It is known that antioxidants play an important role in preventing many diseases, such as cancer and cardiovascular disease [20]. The main biologically active compounds of the byproducts obtained from the vegetable oil industry are polyphenols and flavonoids. Phenolic compounds are a broad group of secondary metabolites that are spread throughout the plant. Polyphenolics have special properties for human health, including anti-inflammatory activity, enzyme inhibition, antimicrobial, antiallergic, reducing the risk of cardiovascular disease, and cytotoxic antitumor activity [21]. Antioxidants such as polyphenols are considered possible protective agents, reducing the oxidative damage caused by reactive oxygen species in the human body and delaying the progression of many chronic diseases, as well as the oxidation of low-density lipoproteins (LDL), which play an important role in atherosclerosis [22]. One of the major causes of atherosclerosis is the high cholesterol. Flavonoids represent a class of phenolic compounds, which are found naturally in plants. These bioactive compounds are also found in a variety of nutraceutical, pharmaceutical, medicinal, and cosmetic applications. This is attributed to their antioxidant, anti-inflammatory, antimutagenic, and anticarcinogenic properties, along with their ability to modulate the key function of cellular enzymes. Research on flavonoids has received an additional boost with the discovery of a low rate of cardiovascular mortality and the prevention of cardiovascular disease [23].

The aim of this study was to determine the total polyphenolic and flavonoid content, as well as to evaluate the antioxidant activity by different methods (DPPH, FRAP, CUPRAC, ABTS) and the antioxidant capacity by PCL-ACL of various byproducts obtained from the vegetable oil industry. It is necessary to find alternative strategies for the use of these byproducts in order to avoid their impact on the environment and to increase the profitability of plant resources. The main attraction of these byproducts obtained in the vegetable oil industry is the possibility to use them in the production of food products with high nutritional value.

2. Materials and Methods

2.1. Reagents and Standards

2,2-Diphenyl-1-picrylhydrazyl (DPPH), 2,2′-azino-bis(3-ethylbenzothiazoline-6-sulphonic acid) diammonium salt (ABTS), ferric reducing antioxidant potential (FRAP), 2,4,6-tripyridyl-s-triazine (TPTZ), (+)-catechin, gallic acid, and Trolox (6-hydroxy-2,5,7,8-tetrame thylchroman-2-carboxylic acid) were purchased from Sigma Chemical Co. (Switzerland). Folin–Ciocalteu's phenol reagent was purchased from Merck (Germany). All chemicals used were of analytical grade. Standard solutions were prepared with distilled deionized water.

2.2. Byproduct Materials

The following types of byproducts were considered: sea buckthorn flour, hemp flour, walnut flour, grape seed flour, rapeseed meals, sunflower meals, black sesame meals, red grape seed meals, golden flax meals, thistle meals, sesame groats, thistle groats, coriander groats, sunflower groats. The byproducts were obtained from the vegetable oil industry and provided by the Association of Operators in Organic Farming Bio-Romania (Romania). The samples were ground using a laboratory mill and kept in closed jars in a refrigerator at 4 °C.

2.3. Qualitative Analysis by ATR-FTIR

The FTIR spectrum for byproducts and dry extracts was recorded at room temperature using the Cary 630 FTIR Spectrometer in ATR mode (Agilent Technologies Inc., Santa Clara, CA, USA). The chosen measurement range was 4000–650 cm^{-1}, number of scans 400, resolution 4 cm^{-1}. The FTIR spectra of dry extract were produced on the alcoholic extract of byproducts dried previously in an oven at 25 °C (for 24 h) [24].

2.4. Extraction Procedure

An amount of 0.2 g of byproduct was weighed and brought into 20 mL of ethanol 96 %. The extracts were obtained by the ultrasound-assisted extraction method involving frequencies ranging from 20 kHz to 2000 kHz for 1h at room temperature. Then, the extracts were centrifuged for 10 min at 10,000 rpm to remove the secondary materials [25].

2.5. Determination of Phenol Content

Total phenol content was determined by the Folin–Ciocalteu method [26]. A total of 50 μL of extract was mixed with 10 μL Folin–Ciocalteu reagent, 90 μL distilled water, and 10 μL of saturated sodium carbonate. The 96-well plates were allowed to stand in the dark for 60 min for color development. Absorbance was measured at 765 nm using a FlexStation 3 UV-Vis (Molecular Devices, GA, USA) Spectrophotometer. A standard curve was prepared by using different concentrations (10–50 μg/mL) of gallic acid in the same condition with samples (R^2 = 0.9966). Total phenolic content was expressed as mg gallic acid equivalent/g of byproduct (mg GAE/g).

2.6. Determination of Flavonoid Content

Total flavonoid content (TFC) was assessed through the $AlCl_3$ method described by Woisky and Salatino [27]. Briefly, 0.1 mL sample/standard solution was mixed with 0.1 mL 10% sodium acetate and 0.12 mL 2.5% $AlCl_3$, the final volume being adjusted to 1 mL with 70% ethanol. The samples were then vortexed and incubated in the dark for 45 min. The absorbances were measured at 430 nm. A standard curve was plotted by using different concentrations (5–200 μg/mL) of quercetin (R^2 = 0.9980). Total flavonoid content was expressed as mg quercetin equivalent/g of byproduct (mg QE/g).

2.7. Determination of Antioxidant Activity through DPPH, CUPRAC, FRAP, and ABTS Methods

DPPH radical scavenging activity was determined based on the reduction in DPPH radical, according to Culetu et al. [28], with slight modifications. The reaction mixture consisted of 1 mL of sample and 6 mL of DPPH radical solution, which was incubated for 20 min in the dark. Then, the absorbance was measured at 517 nm. Antioxidant activity was calculated using a calibration curve (0.0156–0.0625 μg/mL) obtained with Trolox (R^2 = 0.9998). The results were expressed in mg Trolox/g of byproduct.

The CUPRAC method is based on the reduction of a cupric complex, neocuproin, by antioxidants in copper form. Copper ion reduction was performed according to a method described by Celik et al. [29]: 60 μL of sample/standard solutions of different concentrations were mixed with 50 μL $CuCl_2$ (10 mM), 50 μL neocuproin (7.5 mM), and 50 μL ammonium acetate buffer 1 M, pH = 7.00. After 30 min, the absorbance was measured at 450 nm. The stock Trolox solutions required for the calibration curve were 2 mM, and the working concentrations were between 0.125 and 2.0 mM (R^2 = 0.9977). The results were expressed in mg Trolox/g of byproduct.

FRAP assay—the determination of the antioxidant capacity of iron reduction was performed by the method described by Thaipong et al. [30]. The stock solutions included 300 mM acetate buffer (3.1 g $C_2H_3NaO_2$ $3H_2O$ and 16 mL $C_2H_4O_2$), pH 3.6, 10 mM 2,4,6-tripyridyl-s-triazine (TPTZ) solution in 40 mM HCl, and 20 mM FeCl3 6H2O solution. The fresh working solution was prepared by mixing 25 mL acetate buffer, 2.5 mL TPTZ solution, and 2.5 mL $FeCl_3$ $6H_2O$ solution, and then warming at 37 °C before using. After incubation, the absorbance was read at 593 nm. A 1 mM Trolox stock solution was used to plot the calibration curve, the concentration ranging between 25 and 250 μM Trolox (R^2 = 0.9962). The results were expressed in mg Trolox/g of byproduct.

Trolox equivalent antioxidant capacity (TEAC) assay was performed according to Re et al. [31] with slight modifications. A stable stock solution of $ABTS^+$ was produced by mixing a solution of 7 mM ABTS in 2.45 mM potassium persulphate. Then, the mixture was left standing in the dark at room temperature for 12–16 h before use. An $ABTS^+$ working solution was obtained by dilution with ethanol to an absorbance of around of 0.70. The reaction mixture consisted in 20 μL of sample/standard and 180 μL of $ABTS^+$ working solution and was incubated 30 min in the dark. The standard curve was linear between 20 and 200 μM Trolox (R^2 = 0.9975). The results were expressed in mg Trolox/g of byproduct.

2.8. Photochemiluminescence Assay

An amount of 0.05 g of samples was mixed with 30 mL methanol for 3 h [32]. The extractions were performed in tightly closed plastic tubes, vortexed, and centrifuged at 20 °C. The extraction was made in triplicate. The scavenging activity of byproduct samples was evaluated by a photochemiluminescence (PCL-ACL) method in which superoxide radical anions (O_2-) are generated from luminol. The extracts were dissolved in methanol. The reactions were carried out using kits for the determination of antioxidant capacity of lipid-soluble substances (Analytik Jena, Jena, Germany), mixing 2300 µL of methanol (reagent 1), 200 µL of buffer solution (reagent 2), 25 µL of luminol (reagent 3), and 10 µL of sample. Measurement was performed on a Photochem device with PCLsoft software (Analytik Jena). Trolox was used to prepare the calibration curve. The results are expressed as mg of Trolox equivalents per g of extract.

2.9. Statistical Analysis

Antioxidant assays were performed in at least three repetitions. Results are presented as means ± standard deviation (SD). To determine the relation between results of total phenolics and antioxidant assays, the Pearson correlation was used.

3. Results and Discussion

3.1. Qualitative Analysis by ATR-FTIR

Qualitative analysis of FTIR spectra provides a characteristic signature present for specific bonds of interest in antioxidants (phenolic, phenolic acids, flavonoids, hydroxycinnamic acids, etc.) by the presence of their molecular vibrations (stretching, bending, and torsion of chemical bonds) [33,34]. Therefore, FTIR spectra represent a molecular footprint of samples for both byproducts and dry extracts obtained by evaporation at room temperature in the dark. Because the amount of some of the dry extracts were low, it was not possible to perform FTIR spectra for all extracts. Figure 1 shows the comparative spectra for both byproducts and dry extracts to highlight specific bands of phenolic compounds.

Figure 1. FTIR spectra for sea buckthorn flour and its dry extract in the region between 650 and 4000 cm^{-1}.

It was possible to distinguish several peaks, which correspond to the functional groups and vibration modes of the polyphenolic components. The wide band between 3220 and 3249 cm^{-1} corresponded to the OH stretching modes and it could be attributed to polysaccharides and/or lignins. The peak at 3010 cm^{-1} was related to the C–H stretching vibration of the cis-double bond (=CH) groups. The asymmetric and symmetrical stretching vibrations of the CH_2 groups were found at 2924 cm^{-1} and 2853 cm^{-1}, respectively, in both dry extract and byproduct for hemp flour, seed flour harrows, sunflower meal, and sesame meal. They were mainly associated with the hydrocarbon chains of lipids or lignins [35]. The spectral band at 1744 cm^{-1} (for hemp flour, sunflower meal, and sesame meal) and the shoulder band at 1719 cm^{-1} (for sea buckthorn flour and walnut flour) was attributed to the absorption of C=O bonds of the ester groups and was related to the presence of fatty acids and glycerides, as well as pectins and lignins [36]. The bands around 1600 cm^{-1} (hemp flour, walnut flour, grape seed flour, and rapeseed meals) were associated with the extent of the carboxyl group and aromatic ring, for example, in pectins and phenolic compounds [37], but also with the bending vibrations of the OH groups. The footprint region from 1500 to 800 cm^{-1} was very rich in peaks from different ways of stretching, bending, swinging, scissoring, and twisting. This region was difficult to analyze due to its complexity, providing important information about organic compounds, such as carbohydrates, alcohols, and organic action, present in the samples. Aromatic C–C extending to ~1520 (seed flour harrows and rapeseed meals) and ~1443 cm^{-1} (walnut flour, grape seed flour, and sesame meal) was related to phenolic compounds [38]. Methyl radical bent in the plane at 1377 cm^{-1} (hemp flour, sunflower meal, and sesame meal) and C–O extending at ~1035 cm^{-1} (walnut flour, grape seed flour, and rapeseed meals) were related to polysaccharide structures [34,36,37]. The peak at 1143 cm^{-1} (identified for hemp flour and sesame meals) corresponded to the aromatic extent of C–H. The band at 765 cm^{-1} has been assigned to benzene ring 1,3-disubstituted, specific for phenolic compounds [38]. Therefore, in the studied extracts, specific bands of phenolic compounds were identified (Figures 1–4).

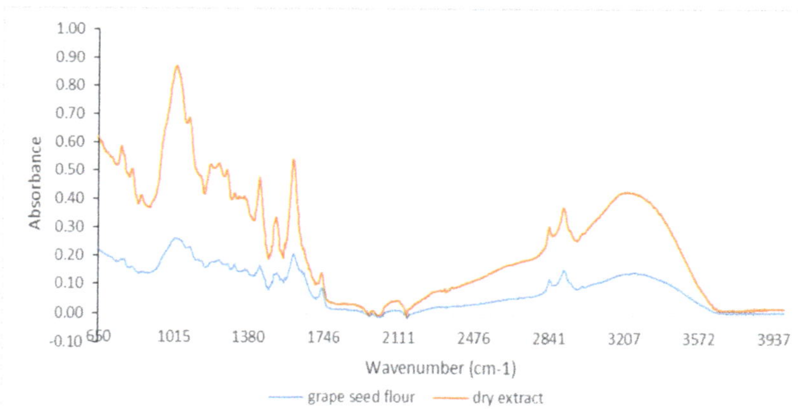

Figure 2. FTIR spectra for grape seed flour and its dry extract in the region between 650 and 4000 cm^{-1}.

The different byproducts from the vegetable oil industry contain significant amounts of phenolic compounds. Analyzed extracts contained between 1.54 and 74.85 mg GAE per gram of byproduct. Grape seed flour extract resulted in the highest amount of total phenolic compounds, 74.85 mg GAE/g. The lowest level was obtained for the sunflower groats, 1.54 mg GAE/g. These amounts were comparable with results described in the literature for other extracts of plant products. The byproducts of the wine industry, mainly fresh and fermented grape pomace, represent a potential source of natural phenolic substances. Grape seeds are richer in polyphenolic compounds and have a higher antioxidant activity than grape skins. The content of extractable phenolic compounds from seeds represents almost 70% of the total extractable phenols from grapes [39]. Rajakumari et al. [40] formulated a nanodispersion containing grape seed extract and analyzed its release profile and the antioxidant potential of the prepared formulations. They concluded that the formulations present a high antioxidant scavenging due to the content of phenolic compounds. The resulting value of total polyphenol content of the tested grape seed flour is higher than those obtained by Antonic et al. [8], 5.58 mg GAE/g. Lutterodt et al. [41] analyzed flours made from cold-pressed seeds of two grape cultivars. The obtained values were 5.93 and 6.66 mg GAE/g. In the same study, the results of TPC for another two cultivars were 67.9 and 89.6 mg GAE/g. These indicate that the TPC content is highly dependent on the cultivar.

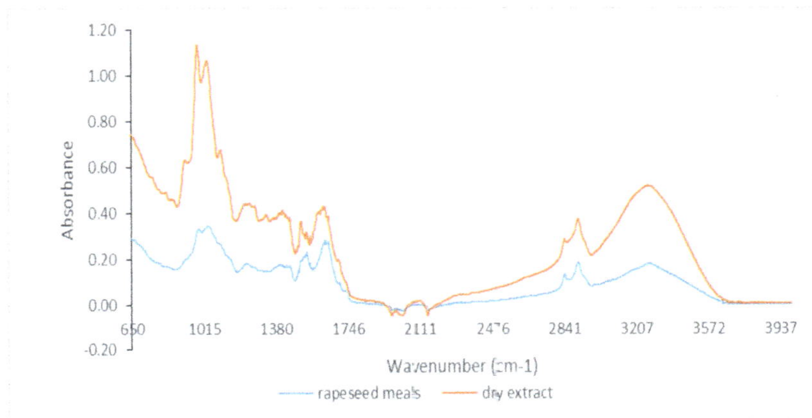

Figure 3. FTIR spectra for rapeseed meals and its dry extract in the region between 650 and 4000 cm^{-1}.

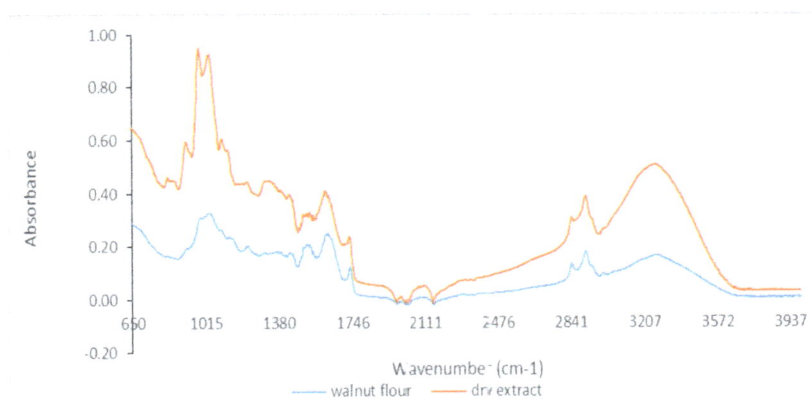

Figure 4. FTIR spectra for walnut flour and its dry extract in the region between 650 and 4000 cm^{-1}.

3.2. *Amount of Phenolic Compounds and Flavonoid Compounds*

The amount of total phenolic and flavonoid content in the byproduct extracts is shown in Table 1.

Gallic acid, 3,4-dihydroxybenzoic acid, (+)-catechin, 1,2-dihydroxybenzene, and syringic acid were the major phenolic compounds in hemp seed [42]. Phenolic content of hemp flour was 10 times higher than those obtained by Ertas et al. [43], who reported that raw hemp flour contains 0.405 mg GAE/g. In contradiction with our results, Mikulec et al. [44] reported that hemp flour is characterized by a polyphenol content of 0.98 mg GAE/g. Walnuts are recognized for their high antioxidant capacity, being associated with a high phenolic content. Phenolic compounds, hydrolyzable tannins, and flavanols are the major components in walnut flour [45]. Similar to our results, Burbano and Correa [46] reported that the total phenolic content of walnut flour is 10.9 ± 0.3 mg GAE/g. Santos et al. [47] studied the effect of roasting conditions on the composition and antioxidant properties of defatted walnut flour and they observed that walnut flour contains 20 mg GAE/g. Labuckas et al. [48] analyzed three walnut flour varieties with respect to total phenolic content (TPC). They observed that TPC varied between 16.3 and 23.7 mg GAE/g, the results being higher than those obtained in this study.

Table 1. Phenolic content and flavonoid content in the byproducts of the vegetable oil industry.

Sample Name	Total Phenolic Content (Mg GAE/g)	Total Flavonoid Content (Mg QE/g)
Sea buckthorn flour	11.52 ± 0.50	9.28 ± 0.06
Hemp flour	4.68 ± 0.13	4.12 ± 0.07
Walnut flour	13.75 ± 0.24	7.57 ± 0.09
Grape seed flour	74.85 ± 0.43	51.24 ± 1.46
Rapeseed meals	11.24 ± 0.65	9.46 ± 0.50
Sunflower meals	11.70 ± 0.35	7.92 ± 0.14
Black sesame meals	3.88 ± 0.20	nd
Red grape seed meals	6.84 ± 0.67	nd
Golden flax meals	3.80 ± 0.28	nd
Thistle meals	7.97 ± 0.48	nd
Sesame groats	4.31 ± 0.32	1.41 ± 0.25
Thistle groats	8.39 ± 0.19	4.99 ± 0.37
Coriander groats	1.73 ± 0.09	nd
Sunflower groats	1.54 ± 0.03	nd

Rapeseed meal is the byproduct of the rapeseed de-oiling process. Rapeseed contains the greatest number of phenolic compounds compared with other oilseed plants. The most significant phenolic compounds in rapeseed are sinapic acid derivatives, such as sinapine [49]. The rapeseed meal phenolic isolates contained between 0.42 and 6.94 mg GAE/g depending on the extraction solvent used [49]. According to Yang et al. [50], total phenolic compounds in rapeseed meals ranged from 38.50 to 63.95 mg GAE/g dry weight. These values are higher than those reported in the present study (11.24 mg GAE/g).

It is known that sunflower seeds are rich in phenolic compounds and the total phenolic content was in the range of 10–42 mg/g according to Kreps et al. [19]. Zilic et al. [51] investigated the total polyphenol content in three sunflower genotypes. They found that the polyphenol content ranged between 14.68 and 18.24 mg GAE/g for sunflower seed and 16.28 and 20.13 mg GAE/g for sunflower kernel, respectively. Chlorogenic and caffeic acids represented 70% of phenolic compounds in sunflower flour [19].

Elleuch et al. [52] have performed a detailed study on the chemical composition and the importance of each component in the seed coat. They studied the total phenolic content of raw sesame seed (0.88 mg GAE/g) and the raw seed coat (0.60 mg GAE/g). Sesame seed coat showed a relatively high polyphenol content (9.9 mg/g of seed coat dry matter) [53]. Compared with black sesame meals (3.88 mg GAE/g), it was observed that seed coat contains a higher amount of total polyphenols. Barthet et al. [54] determined antioxidant activity of flaxseed meal components. They reported that the amount of total phenolics extracted by the various solvents decreased from methanol to water extracts, the results varying between 98.2 mg GAE/g and 68.2 mg GAE/g, respectively. Our results are lower than those obtained by Barthet et al. [54].

The total phenolic content of milk thistle genotypes ranged from 2.06 to 3.60 mg GAE per g [17]. Stancheva et al. [55] reported that the total phenolic content of thistle was between 2.58 and 3.91 mg GAE/100 g. Phenolic content was higher in thistle meals (7.97 mg GAE/g) and thistle groats (8.39 mg GAE/g) than for milk thistle.

The flavonoid content represents approximately 80.56% of the total polyphenols for sea buckthorn flour extract, 88.03% for hemp flour extract, 55.05% for walnut flour extract, 68.46% for grape seed flour extract, 84.16% for rapeseed meals, 67.69% for sunflower meal extract, 32.41% for sesame groats, and 59.43% for thistle groats (Table 1). The results were correlated with the literature data; for example, the percentage of flavonoids was similar to that obtained by Nilova and Malyutenkova [56] for sea buckthorn flour extract (89.5%), and, for the hemp flour extract, the flavonoids content was significantly higher than those present in the literature, 0.29 mg catechin equivalent/g [42,57]. For walnut flour, a higher concentration of flavonoids was obtained than in Santos et al. [47]. Depending on the extraction solvent used, the content of flavonoids in rapeseed and sunflower varied between 0.74 and 4.19 mg catechin/g and 1.45 and 12.03 mg catechin/g, respectively [20]. In comparison with our results, rapeseed meal (11.24 mg QE/g) presented higher flavonoid content. Sunflower meals (11.70 mg QE/g) contain lower flavonoid levels than metanolic extract analyzed by Matthaus [20].

The variation in TPC and TFC could be due to the varietal differences, climate, harvest time, and other factors that affect the nutritional quality of the plants [58]. Moreover, the content of TPC and TFC depends on the extraction solvent. Khalil et al. [59] concluded that the extraction solvent of total phenolic contents and total flavonoid contents of pomegranate peel extracts is methanol, followed by ethanol and ethyl acetate. The vegetable oil industry generates substantial amounts of phenolic-rich byproducts, which could be valuable natural sources of antioxidants.

3.3. Comparison of the Antioxidant Activity of Selected byproducts

The response of antioxidants to different radical or oxidant sources may be different. Thus, no single method accurately reflects the mechanism of action of all radical sources or all antioxidant compounds in a complex system [60].

Evaluation of antioxidants in food is of great importance in today's context. Estimation of total antioxidant using different assays is important to get the overall antioxidant potential of any food matrix. The total antioxidant activity of the 14 byproducts obtained from the vegetable oil industry was estimated using four in vitro assays, namely, DPPH, ABTS, FRAP, and CUPRAC. The antioxidant capacity through PCL was also determined. The measurements of DPPH, ABTS, FRAP, and CUPRAC values are presented in Table 2

Table 2. Antioxidant activity of selected byproducts obtained in the vegetable oil industry.

Sample Name	DPPH	ABTS	FRAP	CUPRAC
Sea buckthorn flour	394.17 ± 1.50	419.46 ± 2.45	547.45 ± 13.14	503.43 ± 14.52
Hemp flour	139.59 ± 1.17	113.73 ± 13.47	285.81 ± 17.23	231.94 ± 18.74
Walnut flour	1257.49 ± 3.85	1423.98 ± 24.57	913.44 ± 15.19	1202.75 ± 23.99
Grape seed flour	7182.53 ± 6.12	3500.52 ± 66.45	4716.75 ± 131.88	5936.76 ± 96.42
Rapeseed meals	647.29 ± 1.36	406.55 ± 6.61	1034.92 ± 39.63	478.43 ± 30.88
Sunflower meals	628.58 ± 3.85	347.01 ± 20.97	1350.86 ± 72.20	510.49 ± 35.22
Black sesame meals	17.73 ± 1.04	48.81 ± 2.68	42.69 ± 2.75	63.31 ± 1.09
Red grape seed meals	200.77 ± 1.08	322.76 ± 27.81	119.92 ± 8.06	119.99 ± 5.54
Golden flax meals	9.25 ± 0.68	12.13 ± 0.61	61.54 ± 4.98	75.50 ± 8.02
Thistle meals	85.58 ± 3.29	292.81 ± 13.59	84.39 ± 6.18	125.75 ± 4.55
Sesame groats	7.58 ± 1.30	15.77 ± 5.05	66.21 ± 3.45	62.45 ± 2.38
Thistle groats	22.74 ± 1.74	293.14 ± 34.32	105.31 ± 12.41	112.54 ± 2.92
Coriander groats	17.64 ± 0.61	9.37 ± 2.22	26.47 ± 2.78	67.53 ± 0.95
Sunflower groats	55.06 ± 2.64	nd	34.46 ± 1.51	70.47 ± 2.96

All values are expressed as mg Trolox/g fresh weight.

These analyses can allow the identification and selection of byproducts in the vegetable oil industry that favor the enrichment of bakery products in antioxidants, which could lead to an increase in the quality of these products.

The total antioxidant activity using the DPPH method of tested samples was between 7.58 and 7182.53 mg Trolox/g of extract. Grape seed flour presented the highest antioxidant activity (7182.53 mg Trolox/g), while the value of DPPH in sesame groats was the lowest (7.58 mg Trolox/g). Scavenging activity of extracts of grape seed flour and walnut flour

(1257.49 mg Trolox/g) were significantly higher than that of sesame groats; thus, phenolic compounds of grape seed flour and walnut flour exhibited the strongest DPPH scavenging potency. Grape seeds are rich in gallic acid, vanilla acid, caffeic acid, ferulic acid, p-coumaric acid, chlorogenic acid, rutin, and quercetin [61]. According to Ross et al. [62], the DPPH radical scavenging ability of defatted grape seed flour was 0.67 mg Trolox/g dry matter, lower than the result from this study. Ross et al. [62] determined the effect of heating time and temperature on the DPPH antioxidant activity of 70% ethanol extracts of grape seed flour and they observed that, at a temperature of up to 150 °C, the values of DPPH activity of the grape seed flour did not significantly change. At a temperature above 180 °C, significant decreases in antioxidant activity were observed.

Walnut kernels are rich in gallic acid, caffeic acid, chlorogenic acid, ferulic acid, synaptic acid, salicylic acid, and ellagic acid [63]. Walnuts are recognized for their high antioxidant capacity when compared with other nuts, being normally associated with this property with a high phenolic content [64]. Studies have shown that caffeic acid and gallic acid exhibited high DPPH scavenging activities, 89.4% and 88.5%, respectively [65]. Sroka and Cisowski [66] confirmed the strong activity of gallic acid and pyrogallol on DPPH scavenging tests, 75% and 79.5%, respectively. Trandafir et al. [62] analyzed the effect of different solvents and extraction methods on antioxidant activity of full-fat and defatted walnut kernel. The results showed that values of antioxidant activity ranged between 17.25 and 49.25 mg Trolox/100g. They concluded that antioxidant capacity of full-fat walnut kernel and defatted walnut kernel is influenced by extraction method.

Chlorogenic and caffeic acids represent almost 70% of phenolic compounds in sunflower flour [67]. Numerous studies have demonstrated that sunflower meal presents high antioxidant capacity, which could be beneficial for further technological utilization [68]. Grasso et al. [69] investigated defatted sunflower seed flour and they reported an antioxidant radical scavenging value of 4.5 mg Trolox/g, lower than our concentration reported in the preset study. The antioxidant activity of phenolic acids and their esters depend partly on the number of hydroxyl groups in the molecule [70], which explains the high DPPH scavenging activity of caffeic acid and gallic acid with three hydroxyl groups. Kikuzaki et al. [71] also have reported similar compounds of scavenging ability towards DPPH radicals: caffeic acid, sinapic acid, ferulic acid, and p-coumaric acid. Chlorogenic and isochlorogenic acids, which are derivatives of caffeic and quinic acid, exhibited strong antioxidative properties, 93% and 86%, respectively [66]. Ferulic acid showed a DPPH scavenging capacity of 39.5%, having in its structure a single methoxy group in the meta position [65].

The values of DPPH reported in this study were higher for meal or flour than for groats. This is because, in all meal or flour samples, the polyphenol content was higher than in the groats, except black sesame meal and thistle meal, where the level of phenolic compounds was higher in the groats. It is generally predicted that DPPH radical scavenging activity is strongly affected by the content of phenolic compounds [72].

Because the contribution of the phenolic compounds to the overall antioxidant capacity is different, a correlation analysis was performed (Table 3). The results showed a strong positive correlation of TPC with antioxidant activity on DPPH radical, with a correlation coefficient of 0.9927. Moreover, a high correlation was recorded between DPPH and TFC (0.9811).

The antioxidant capacity is measured as the ability of compounds to decrease the color reacting with $ABTS^{+}$ radical and expressed relative to Trolox [73]. It is known that antioxidant activity depends on the content of the phenolics and the number and position of the hydroxyl groups of the aromatic ring binding site and the type of substituent. Similar to DPPH assay, the strongest antioxidant activity in the ABTS method was obtained by phenol with three hydroxyl groups, such as gallic acid and pyrogallol [74].

Table 3. The correlation coefficients between total phenolic content and flavonoids with DPPH, ABTS, CUPRAC, and FRAP.

	DPPH	ABTS	FRAP	CUPRAC
TPC	0.9927	0.9660	0.9752	0.9920
TFC	0.9811	0.9477	0.9825	0.9815

Values of $ABTS^+$ of analyzed byproducts ranged from 0 to 3500.52 mg Trolox/g. Grape seed flour had the highest value of ABTS assay (3500.52 mg Trolox/g), followed by walnut flower (1423.98) and sea buckthorn flour (419.46). Ross et al. [62] analyzed a sample of defatted grape seed flour and they observed a TEAC value of 4.68 mg Trolox/g dry matter, lower than our result. Grape seed contains a high amount of polyphenols with three hydroxyl groups, mainly flavonoids, including gallic acid, the monomeric flavan-3-ols catechin, epicatechin, gallocatechin, epigallocatechin, and epicatechin 3-O-gallate, and procyanidin dimers, trimers, and more highly polymerized procyanidins [75]. Moreover, the walnut flour contains a great content of phenolic compounds, such as hydrolyzable tannins. They are responsible for a high antioxidant activity in walnut flour. According to Pellegrini et al. [76], antioxidant capacity of walnut extracts was 119.91 µmol Trolox/g dry matter by ABTS method. Antioxidant capacity of extracts of the whole walnut determined by Arranz et al. [77] was 165.18 µmol Trolox/g dry matter. After the sample was defatted with petroleum ether, the antioxidant capacity was higher than whole walnut at 211.85 µmol Trolox/g dry matter. The authors concluded that the high fat content of walnut interferes in the determination of antioxidant capacity [77]. Seventeen phenolic acids were identified in sea buckthorn berries. Salicylic acid was the predominant phenolic acid in berries. Small quantities of p-hydroxybenzoic acid were detected [78]. Cyclic spermidine-alkaloid, feruloyl choline, kaempferol, and sinapine were identified as the main phenolic compounds in rapeseed meals [50].

The ABTS values for golden flax meals (12.13 mg Trolox/g), sesame groats (15.77 mg Trolox/g), and coriander groats (9.37 mg Trolox/g) were the lowest, suggesting small quantities of phenolic compounds. Barthet et al. [54] analyzed two samples of golden flax meals and brown flax meals. They reported ORAC values of 2.44 mg Trolox/g and 2.33 mg Trolox/g, respectively.

The correlation between TPC and TFC with ABTS method is shown in Table 3. The correlation coefficients in these cases are 0.9660 and 0.9477, respectively.

A wide range in FRAP values was observed among the analyzed byproducts, showing values of 34.46–4716.75 mg Trolox/g of byproduct. The highest value of FRAP was presented by grape seed flour (4716.75 mg Trolox/g), followed by sunflower meals (1350.86 mg Trolox/g) and rapeseed meals (1034.92 mg Trolox/g). Gougoulias and Mashev [79] analyzed 14 grape seeds varieties. The results of FRAP activity were lower than those presented in this article and ranged between 0.15 and 0.20 mg Trolox/g dry weight. In contradiction with our results, Antonic et al. [8] analyzed grape seed flour with respect to antioxidant activity and observed that FRAP activity was 0.056 mg Trolox/g. Ross et al. [62] found that the FRAP value of defatted grape seed flour was 0.93 mg Trolox/g dry matter. Szydłowska-Czerniak et al. [12] evaluated antioxidant capacity of rapeseed meal. Results ranged between 0.003 and 0.025 mg Trolox/g. Gvozdenac et al. [80] tested seed, hull, and kernel of different sunflower hybrids. The results of antioxidant activity (FRAP) in seed ranges between 76.5 and 83.0 mg Trolox/g, in hull between 4.5 and 8.4 mg Trolox/g, and in kernel between 116.1 and 1117.3 mg Trolox/g. According to Arranz et al. [77], antioxidant activity of aqueous organic extracts of defatted walnut through FRAP method was 114.92 µmol Trolox/g dry matter.

CUPRAC antioxidant assay delivered values in the range of 62.45–5936.75 mg Trolox/g. Grape seed flour presented the highest CUPRAC value (5936.76 mg Trolox/g), while the lowest value was obtained by sesame groats (62.45 mg Trolox/g). According to Bogoeva et al. [81], the antioxidant activity of 70% ethanolic extract of grape seed flour was evaluated and they obtained a CUPRAC value of 344.78 mg Trolox/g, lower than our result. Grasso et al. [69] investigated defatted sunflower seed flour with respect to antioxidant activity. They observed that defatted sunflower seed flour had higher antioxidant capacity measured by DPPH and CUPRAC assays compared to wheat flour. The CUPRAC value was 20 mg Trolox/g, lower than our result.

A high correlation coefficient was observed between TPC and TFC with FRAP and CUPRAC methods. The correlation coefficients in these cases were 0.9752, 0.9825, and 0.9920, 0.9815 respectively. Moreover, the analysis revealed a significant positive correlation of CUPRAC with FRAP (0.9716).

Zhou and Yu [82] determined the correlation coefficient for CUPRAC, FRAP, and total phenolics in cauliflower genotypes. The results reported presented a significant correlation coefficient of CUPRAC with FRAP (0.711), CUPRAC with total phenolics (0.470), and FRAP with total phenolics (0.504). Deng et al. [83] also observed that total phenolic content and the measured antioxidant properties were correlated with each other. Sun and Tanumihardjo [84] also revealed a significant linear correlation among different antioxidant capacities (FRAP and TEAC) in a study with 56 vegetables. They have also observed a positive correlation among total antioxidant capacities and total phenolic content.

In this study, most of the analyzed byproducts showed high values of antioxidant activity by the FRAP and CUPRAC method; this is probably due to other structures that can function as ligands.

3.4. Development of the Relative Antioxidant Capacity Index RACI

The relative antioxidant capacity index (RACI) was developed as a statistical perspective by integrating food antioxidant capacity data determined by the several methods used. For obtaining the RACI values, it is necessary to determine the standard scores for each method. Standard scores obtained using different methods of analysis will have a similar contribution to the central trend of the average and can be compared without interference from different units, scales, and distributions [84]. The trend of the RACI value matched with the standard score from the four methods (Table 4).

In order to create a ranking of the antioxidant capacities of the analyzed byproducts, RACI was calculated (Figure 5).

Table 4. Standard scores of antioxidant capacity and RACI for the analyzed byproducts.

Sample Name	DPPH	ABTS	FRAP	CUPRAC	RACI
Sea buckthorn flour	43.15	36.74	16.59	11.98	27.12
Hemp flour	−475.64	−43.23	−23.81	−24.77	−141.86
Walnut flour	−521.89	−75.00	−154.93	−86.04	−209.46
Grape seed flour	685.39	7.71	13.11	30.58	184.20
Rapeseed meal	−20.75	−40.69	9.07	−6.38	−14.69
Sunflower meal	65.56	−1.39	13.50	3.81	20.37
Black sesame meal	−95.45	−25.44	−27.02	−49.25	−49.29
Red grape seed meal	79.20	7.46	0.58	0.86	22.03
Golden flax meal	−123.80	−133.28	−6.40	−2.24	−66.43
Thistle meal	−2.14	14.73	−1.25	7.28	4.65
Sesame groats	−60.11	−13.85	−5.65	−9.78	−22.35
Thistle groats	−33.97	6.16	1.89	10.51	−3.85
Coriander groats	−36.89	−13.86	−4.92	28.83	−6.71
Sunflower groats	0.66	0.00	−12.50	5.79	−1.51

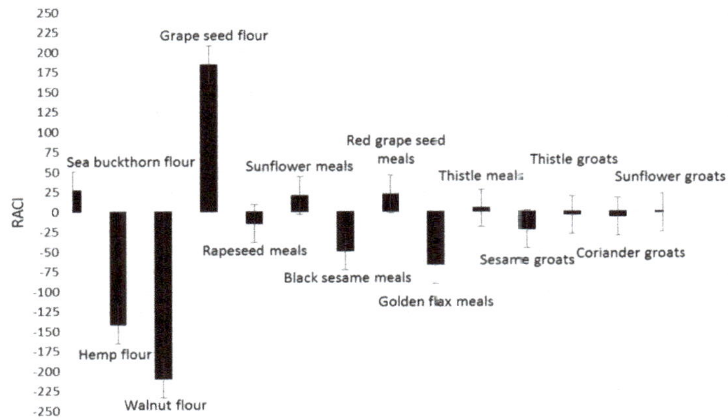

Figure 5. Relative antioxidant capacity index of 14 byproducts obtained in the vegetable oil industry.

RACI of analyzed byproducts is a scientific combination of data from different chemical methods with no unit limitation and no variance among methods. With this parameter, it is possible to make a much more precise comparison of the antioxidant capacities of the different food matrices. Based on the RACI value of the 14 byproducts obtained in the vegetable oil industry, the rank of antioxidant capacity ranged from −209.46 to 184.20. The highest relative antioxidant capacity index was observed in grape seed flour; the lowest was in walnut flour, with a value of −209.46. Each antioxidant activity used in this study was correlated with RACI. In this study, the most sensitive methods in developing RACI were FRAP and DPPH (r = 0.5795; r = 0.5766), followed by CUPRAC (r = 0.5578) and ABTS (r = 0.4449) (Figures 6–9).

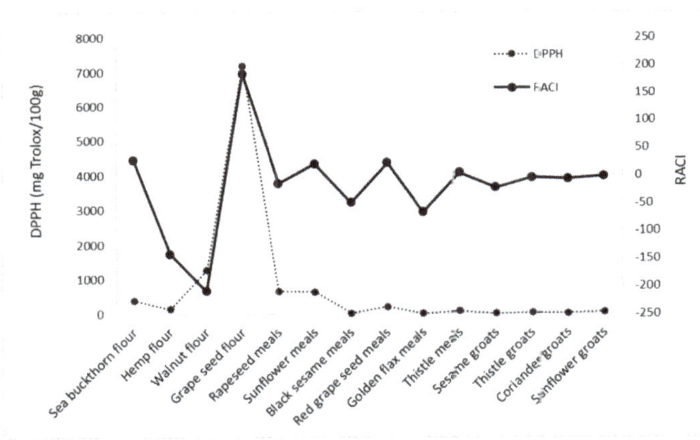

Figure 6. Correlation between RACI and DPPH activity.

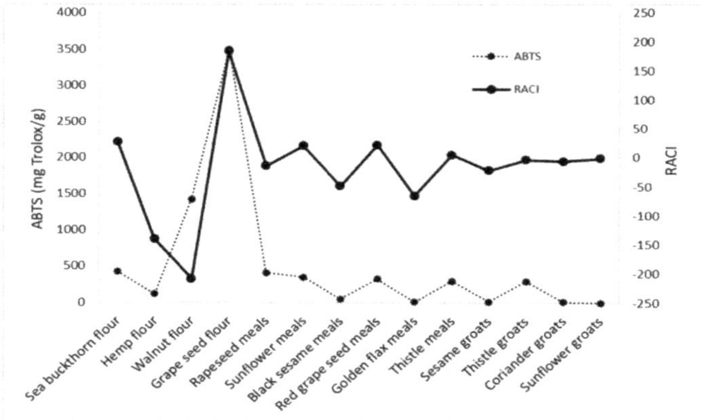

Figure 7. Correlation between RACI and ABTS activity.

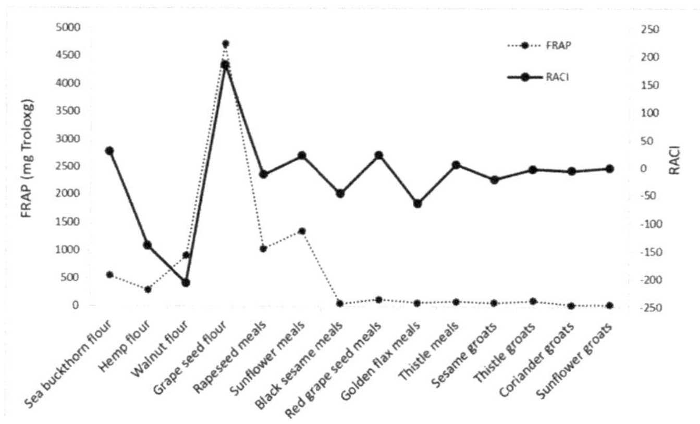

Figure 8. Correlation between RACI and FRAP activity.

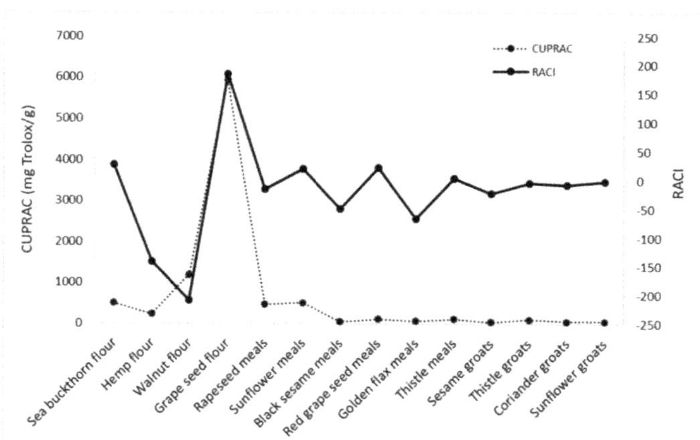

Figure 9. Correlation between RACI and CUPRAC activity.

3.5. Determination of Antioxidant Capacity Using the PHOTOCHEM Device in an ACL System

In the present study, the ACL concentration was expressed as equivalent antioxidant activity in mg Trolox per gram of extract

The inhibition of photochemiluminescence (PCL) of luminol (5-amino-2,3-dihydro-1,4-phthalazinedione) in methanol by constituents of the phenolic-containing byproduct extracts was monitored using the PHOTOCHEM (Figure 10).

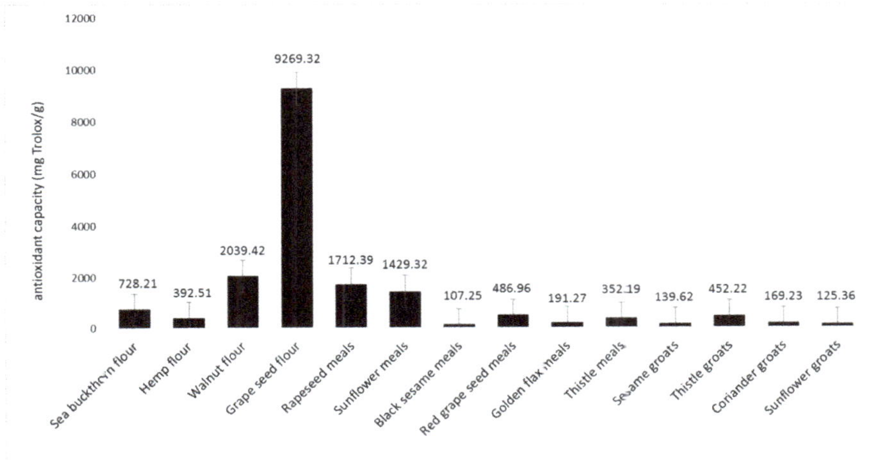

Figure 10. Antioxidant capacity using the PHOTOCHEM device in an ACL system.

Lipid soluble antioxidant capacity data of the 14 byproducts obtained in the vegetable oil industry were between 107.25 and 9269.32 mg Trolox/g.

As can be seen in Figure 10, the highest value was determined in the grape seed flour sample. This value exceeded other measured data several-fold. The walnut flour (2039.42 mg Trolox/g), rapeseed meals (1712.39 mg Trolox/g), and sunflower meals (1429.32 mg Trolox/g) were the next samples with ACL values above 1000 mg Trolox/g. Obtained data for the black sesame meals, golden flax meals, sesame groats, coriander groats, and sunflower groats were the lowest at just over 100 mg Trolox/g. Senila et al. [85] analyzed total antioxidant capacity of four types of seeds (sunflower, flax, hemp, and sesame) using the photochemiluminescence assay. The antioxidant capacity varied between 7.5 and 112.5 mg Trolox/g. The results were lower than those presented in this study.

Sielicka et al. [86] analyzed six samples of flaxseed, walnut, rapeseed, pumpkin seed, evening primrose, and black cumin cold-pressed in terms of lipid-soluble antioxidant capacity analysis. They used hexane as the extraction solvent. The values ranged between 1.0 and 7.7 mM Trolox/L oil. The antioxidant capacity of flaxseed, walnut, and rapeseed pressed cold oils presented the lowest values.

Strong positive correlations between the antioxidant capacity of lipid-soluble compounds measured by photochemiluminescence assay (PCL) and the other methods used for determining antioxidant activity were found (Table 5).

Table 5. Correlation coefficients between ACL and DPPH, ABTS, FRAP, and CUPRAC.

Method	DPPH	ABTS	FRAP	CUPRAC
ACL	0.9952	0.9735	0.9874	0.9930

It was observed that the highest correlation is between ACL and DPPH (r = 0.9952), followed by ACL and CUPRAC (0.9930). In order to determine the antioxidant capacity of lipid-soluble compounds of different cold-pressed oils, Sielicka et al. [86] used the photochemiluminescence and DPPH method. Similar to our results, they found a high correlation between these methods, with a correlation coefficient of 0.91.

All methods exhibited similar trends of antioxidant capacity of the analyzed byproducts obtained from the vegetable oil industry, but the obtained values differed, which might be attributed to the dissimilar measurement principles, as described by Prevc et al. [87]. In the PCL assay, optical excitation of a photosensitizer results in the generation of the superoxide radical, and then the ability to scavenge the radicals is evaluated by chemiluminescence. In contrast, the DPPH assay analyzes the ability to reduce the stable radical through determination of a decrease in absorbance. The CUPRAC assay measures the reducing power of antioxidants to convert cupric (Cu2+) to cuprous (Cu+) ion. The other cause of lower values of antioxidant capacity obtained by DPPH, CUPRAC, FRAP, and DPPH methods may be due to the use of ethanol instead of methanol.

It is known that the solvent used as an extraction medium can influence the degree of oxidation reaction [88]. Sielicka et al. [86] reported that pumpkin, sesame, rapeseed, and flaxseed oil samples dissolved in ethyl acetate presented much lower antioxidative capacity in the DPPH method than after solubilization of oils in other solvents.

4. Conclusions

In this study, the phenolic content, flavonoid content, and the lipid-soluble antioxidant capacity of 14 byproducts obtained in the vegetable oil industry were measured. Results confirm that the byproducts analyzed are a good source of many biological functional substances having considerable amounts of total phenolic content.

For determining the antioxidant activity, DPPH, ABTS, FRAP, and CUPRAC methods were used. The ACL method was used for determining the antioxidant capacity. The samples showed varied antioxidant capacities depending on the seed origin. In all methods performed, the highest antioxidant capacity was for the grape seed flour. Walnut flour, sunflower meals, and rapeseed meals are an excellent source of antioxidant substances with high antioxidant capacity. The polyphenol content and the antioxidant capacity of the byproducts from the vegetable oil industry are influenced by the variety and the method of obtaining the waste. The flour of the analyzed byproducts has a higher antioxidant activity than meals and groats. Photochemiluminescence analysis and DPPH, ABTS, FRAP, and CUPRAC assays were fully applicable to the evaluation of the antioxidant capacity of lipophilic fraction of byproducts obtained in vegetable oil industry samples, with correlation coefficients of 0.9952, 0.9735, 0.9874, and 0.9930, respectively. The results indicate that byproducts obtained from the vegetable oil industry (flour, meal, and groats) could be an inexhaustible source of phenolic compounds, especially flavonoids, with antioxidant properties as valuable functional ingredients with beneficial effects on human health. The byproducts obtained from the vegetable oil industry can be used as ingredients for new bakery products to improve their nutritional properties and antioxidant quality. Further studies are needed to determine the optimal concentration of the byproducts' addition into wheat flour in order to achieve an improvement in the nutritional and sensory properties and to increase the antioxidant capacity of the bakery products.

Author Contributions: Conceptualization, N.B.; methodology, M.M. and I.C.M.; software, M.M. and I.C.M.; validation, M.M. and I.C.M.; formal analysis, M.M., I.C.M. and I.E.S.; investigation, M.M., I.C.M. and I.E.S.; resources, N.B.; data curation, N.B. and M.M.; writing—original draft preparation, M.M. and I.C.M.; writing—review and editing, N.B.; visualization, M.M. and N.B.; supervision, N.B.; project administration, N.B.; funding acquisition, N.B. All authors have read and agreed to the published version of the manuscript.

Funding: This work was supported by a grant of the Romanian National Authority for Scientific Research and Innovation, CCDI-UEFISCDI, project number ERANET-COREORGANIC&SUSFOOD-PROVIDE-1, within PNCDI III.

Informed Consent Statement: Not applicable.

Data Availability Statement: The data presented in this study are available on request from the corresponding author. The data are not publicly available due to privacy or ethical restrictions.

Conflicts of Interest: The authors declare no conflict of interest.

References

1. Martins, Z.E.; Pinho, O.; Ferreira, I.M.P.L.V.O. Food industry by-products used as functional ingredients of bakery products. *Trends Food Sci. Technol.* **2017**, *67*, 106–128. [CrossRef]
2. Ghendov-Mosanu, A.; Cristea, E.; Patras, A.; Sturza, R.; Padureanu, S.; Deseatnicova, O.; Turculet, N.; Boestean, O.; Niculaua, M. Potential Application of Hippophae Rhamnoides in Wheat Bread Production. *Molecules* **2020**, *25*, 1272. [CrossRef] [PubMed]
3. Krejkarova, J.; Strakova, E.; Suchy, P.; Herzig, I.; Karaskova, K. Sea buckthorn (*Hippophae rhamnoides* L.) as a potential source of nutraceutics and its therapeutic possibilities—A review. *Acta Vet. Brno* **2015**, *84*, 257–268. [CrossRef]
4. Rusu, I.E.; Marc (Vlaic), R.A.; Muresan, C.C.; Mursan, A.E.; Muresan, V.; Pop, C.R.; Chis, M.S.; Man, S.M.; Filip, M.R.; Onica, B.-M. Hemp (*Cannabis sativa* L.) Flour-Based Wheat Bread as Fortified Bakery Product. *Plants* **2021**, *10*, 1558. [CrossRef]
5. Abdallah, I.B.; Tlili, N.; Martinez-Force, E. Content of carotenoids, tocopherols, sterols, triterpenic and aliphatic alcohols, and volatile compounds in six walnuts (*Juglans regia* L.) varieties. *Food Chem.* **2015**, *173*, 972–978. [CrossRef]
6. Almoraie, N.M. The Effect of Walnut Flour on the Physical and Sensory Characteristics of Wheat Bread. *Int. J. Food Sci.* **2019**, *2019*, 5676205. [CrossRef]
7. Yu, J.; Ahmedna, M. Functional components of grape pomace: Their composition, biological properties and potential applications. *Int. J. Food Sci.* **2013**, *48*, 221–237. [CrossRef]
8. Antonic, B.; Dordevic, D.; Jancikova, S.; Holeckova, D.; Tremlova, B.; Kulawik, P. Effect of Grape Seed Flour on the Antioxidant Profile, Textural and Sensory Properties of Waffles Textural and Sensory Properties of Waffles. *Processes* **2021**, *9*, 131. [CrossRef]
9. Adeleke, B.S.; Babalola, O.O. Oilseed crop sunflower (*Helianthus annuus*) as a source of food: Nutritional and health benefits. *Food Sci. Nutr.* **2020**, *8*, 4666–4684. [CrossRef]
10. Saleh, A.A.; El-Awady, A.; Amber, K.; Eid, Y.Z.; Alzawqari, M.H.; Selim, S.; Soliman, M.M.; Shukry, M. Effects of Sunflower Meal Supplementation as a Complementary Protein Source in the Laying Hen's Diet on Productive Performance, Egg Quality, and Nutrient Digestibility. *Sustainability* **2021**, *13*, 3557. [CrossRef]
11. Grasso, S.; Pintado, T.; Pérez-Jiménez, J.; Ruiz-Capillas, C.; Herrero, A.M. Characterisation of Muffins with Upcycled Sunflower Flour. *Foods* **2020**, *10*, 426. [CrossRef] [PubMed]
12. Szydłowska-Czerniak, A.; Amarowicz, R.; Szłyk, E. Antioxidant capacity of rapeseed meal and rapeseed oilsenriched with meal extract. *Eur. J. Lipid Sci. Technol.* **2010**, *112*, 750–760. [CrossRef]
13. Shahidi, F.; Liyana-Pathirana, C.M.; Wall, D.S. Antioxidant activity of white and black sesame seeds and their hull fractions. *Food Chem.* **2006**, *99*, 478–483. [CrossRef]
14. Bloedon, L.; Szapary, P. Flaxseed and cardiovascular risk. *Nutr. Rev.* **2004**, *62*, 18–27. [CrossRef]
15. Pouzo, L.B.; Descalzo, A.M.; Zaritzky, N.E.; Rossetti, L.; Pavan, E. Antioxidant status, lipid and color stability of aged beef from grazing steers supplemented with corn grain and increasing levels of flaxseed. *Meat Sci.* **2016**, *111*, 1–8. [CrossRef] [PubMed]
16. Pourabedin, M.; Aarabi, A.; Rahbaran, S. Effect of flaxseed flour on rheological properties, staling and total phenol of Iranian toast. *J. Cereal Sci.* **2017**, *76*, 173–178. [CrossRef]
17. Lucini, L.; Pellizzoni, M.; Ruzickova, G.; Molinari, G.P. Phytochemical constituents and in vitro radical scavenging activity of different aloe species. *Food Chem.* **2015**, *170*, 501–507. [CrossRef]
18. Křen, V.; Walterová, D. Silybin and silymarin—New effects and applications. *Biomed. Pap.* **2005**, *149*, 29–41. [CrossRef]
19. Kreps, F.; Vrbiková, L.; Schmidt, Š. Industrial Rapeseed and Sunflower Meal as Source of Antioxidants. *Int. J. Eng. Res. Appl.* **2004**, *4*, 45–54.
20. Matthäus, B. Antioxidant activity of extracts obtained from residues of different oilseeds. *J. Agric. Food Chem.* **2002**, *50*, 3444–3452. [CrossRef]
21. Shen, S.S.; Callaghan, D.; Juzwik, C.; Xiong, H.Q.; Huang, P.L.; Zhang, W.D. ABCG2 reduces ROS-mediated toxicity and inflammation: A potential role in Alzheimer's disease. *J. Neurochem.* **2010**, *114*, 1590–1604. [CrossRef] [PubMed]
22. Mullen, W.; Marks, S.C.; Crozier, A. Evaluation of phenolic compounds in commercial fruit juices and fruit drinks. *J. Agric. Food Chem.* **2007**, *55*, 3148–3157. [CrossRef]
23. Panche, A.N.; Diwan, A.D.; Chandra, S.R. Flavonoids: An overview. *J. Nutr. Sci.* **2016**, *5*, e47. [CrossRef] [PubMed]
24. Marinas, I.C.; Oprea, E.; Geana, E.I.; Tutunaru, O.; Gradisteanu Pircalabioru, G.; Zgura, I.; Chifiriuc, M.C. Valorization of Gleditsia triacanthos Invasive Plant Cellulose Microfibers and Phenolic Compounds for Obtaining Multi-Functional Wound Dressings with Antimicrobial and Antioxidant Properties. *Int. J. Mol. Sci.* **2021**, *22*, 33. [CrossRef]

25. Corbu, V.M.; Gheorghe, I.; Marinas, I.C.; Geană, E.I.; Moza, M.I.; Csutak, O.; Chifiriuc, M.C. Demonstration of Allium sativum Extract Inhibitory Effect on Biodeteriogenic Microbial Strain Growth, Biofilm Development, and Enzymatic and Organic Acid Production. *Molecules* **2021**, *26*, 7195. [CrossRef] [PubMed]
26. Singleton, L.V.; Orthofer, R.; Lamuela-Raventós, R.M. Analysis of total phenols and other oxidation substrates and antioxidants by means of folin-ciocalteu reagent. *Methods Enzymol.* **1999**, *299*, 152–178.
27. Woisky, R.G.; Salatino, A. Analysis of propolis: Some parameters and procedures for chemical quality control. *J. Apic. Res.* **1998**, *37*, 99–105. [CrossRef]
28. Culetu, A.; Fernandez-Gomez, B.; Ullate, M.; Del Castillo, M.D.; Andlauer, W. Effect of theanine and polyphenols enriched fractions from decaffeinated tea dust on the formation of Maillard reaction products and sensory attributes of breads. *Food Chem.* **2016**, *197*, 14–23. [CrossRef] [PubMed]
29. Celik, S.E.; Ozyürek, M.; Güçlü, K.; Apak, R. Determination of antioxidants by a novel on-line HPLC-cupric reducing antioxidant capacity (CUPRAC) assay with post-column detection. *Anal. Chim. Acta* **2010**, *674*, 79–88. [CrossRef] [PubMed]
30. Thaipong, K.; Boonprakob, U.; Crosby, K.; Cisneros-Zevallos, L.; Byrne, D.H. Comparison of ABTS, DPPH, FRAP, and ORAC assays for estimating antioxidant activity from guava fruit extracts. *J. Food Compos. Anal.* **2006**, *19*, 669–675. [CrossRef]
31. Re, R.; Pellegrini, N.; Proteggente, A.; Pannala, A.; Yang, M.; Rice-Evans, C. Antioxidant activity applying an improved ABTS radical cation decolorization assay. *Free Radic. Biol. Med.* **1999**, *26*, 1231–1237. [CrossRef]
32. Popov, I.N.; Lewin, G. Photochemiluminescent detection of antiradical activity; IV: Testing of lipid-soluble antioxidant. *J. Biochem. Biophys. Methods* **1996**, *31*, 1–8. [CrossRef]
33. Lupoi, J.S.; Singh, S.; Parthasarathi, R.; Simmons, B.A.; Henry, R.J. Recent innovations in analytical methods for the qualitative and quantitative assessment of lignin. *Renew. Sustain. Energy Rev.* **2015**, *49*, 871–906. [CrossRef]
34. Lucarini, M.; Durazzo, A.; Kiefer, J.; Santini, A.; Lombardi-Boccia, G.; Souto, E.B.; Romani, A.; Lampe, A.; Nicoli, S.F.; Gabrielli, P.; et al. Grape Seeds: Chromatographic Profile of Fatty Acids and Phenolic Compounds and Qualitative Analysis by FTIR-ATR Spectroscopy. *Foods* **2019**, *9*, 10. [CrossRef]
35. Kennedy, J.A.; Troup, G.J.; Pilbrow, J.R.; Hutton, D.R.; Hewitt, D.; Hunter, C.R.; Jones, G.P. Development of seed polyphenols in berries from Vitis vinifera L. cv. Shiraz. *Aust. J. Grape Wine Res.* **2000**, *6*, 244–254. [CrossRef]
36. Gao, Y.; Fangel, J.U.; Willats, W.G.T.; Vivier, M.A.; Moore, J.P. Dissecting the polysaccharide-rich grape cell wall changes during winemaking using combined high-throughput and fractionation methods. *Carbohydr. Polym.* **2015**, *133*, 567–577. [CrossRef] [PubMed]
37. Heredia-Guerrero, J.A.; Benítez, J.J.; Domínguez, E.; Bayer, I.; Cingolani, R.; Athanassiou, A.; Heredia, A. Infrared and Raman spectroscopic features of plant cuticles: A review. *Front. Plant Sci.* **2014**, *5*, 305. [CrossRef] [PubMed]
38. Wahyono, T.; Astuti, D.A.; Wiryawan, K.G.; Sugoro, I.; Jayanegara, A. Fourier Transform Mid-Infrared (FTIR) Spectroscopy to Identify Tannin Compounds in The Panicle of Sorghum Mutant Lines. *IOP Conf. Ser. Mater. Sci. Eng.* **2019**, *546*, 042045. [CrossRef]
39. Bosso, A.; Cassino, C.; Motta, S.; Panero, L.; Tsolakis, C.; Guaita, M. Polyphenolic Composition and In Vitro Antioxidant Activity of Red Grape Seeds as Byproducts of Short and Medium-Long Fermentative Macerations. *Foods* **2020**, *9*, 1451. [CrossRef]
40. Rajakumari, R.; Volova, T.; Oluwafemi, O.S.; Rajesh Kumar, S.; Thomas, S.; Kalarikkal, N. Grape seed extract-soluplus dispersion and its antioxidant activity. *Drug Dev. Ind. Pharm.* **2020**, *46*, 1219–1229. [CrossRef]
41. Lutterodt, H.; Slavin, M.; Whent, M.; Turner, E.; Yu, L.L. Fatty acid composition, oxidative stability, antioxidant and antiproliferative properties of selected cold-pressed grape seed oils and flours. *Food Chem.* **2011**, *128*, 391–399. [CrossRef]
42. Babiker, E.E.; Uslu, N.; Juhaimi, F.A.; Ahmed, I.A.M.; Ghafoor, K.; Özcan, M.M.; Almusallam, I.A. Effect of roasting on antioxidative properties, polyphenol profile and fatty acids composition of hemp (*Cannabis sativa* L.) seeds. *LWT* **2021**, *139*, 110537. [CrossRef]
43. Ertaş, N.; Aslan, M. Antioxidant and physicochemical properties of cookies containing raw and roasted hemo flour. *Acta Sci. Pol. Technol. Aliment.* **2020**, *19*, 177–184. [PubMed]
44. Mikulec, A.; Kowalski, S.; Sabat, R.; Skoczylas, Ł.; Tabaszewska, M.; Wywrocka-Gurgul, A. Hemp flour as a valuable component for enriching physicochemical and antioxidant properties of wheat bread. *LWT Food Sci. Technol.* **2019**, *102*, 164–172. [CrossRef]
45. Slatnar, A.; Mikulic-Petkovsek, M.; Stampar, F. Identifcation and quantifcation of phenolic compounds in kernels, oil and bagasse pellets of common walnut (*Juglans regia* L.). *Food Res Int.* **2015**, *67*, 255–263. [CrossRef]
46. Burbano, J.J.; Correa, M.J. Composition and Physicochemical Characterization of Walnut Flour, a By-product of Oil Extraction. *Plant Foods Hum. Nutr.* **2021**, *76*, 233–239. [CrossRef]
47. Santos, J.; Alvarez-Ortí, M.; Sena-Moreno, E. Efect of roasting conditions on the composition and antioxidant properties of defatted walnut four. *J. Sci. Food Agric.* **2018**, *98*, 1813–1820. [CrossRef] [PubMed]
48. Labuckas, D.O.; Maestri, D.M.; Perelló, M. Phenolics from walnut (*Juglans regia* L.) kernels: Antioxidant activity and interactions with proteins. *Food Chem.* **2008**, *107*, 607–612. [CrossRef]
49. Vuorela, S.; Meyer, A.S.; Heinonen, M. Impact of Isolation Method on the Antioxidant Activity of Rapeseed Meal Phenolics. *J. Agric. Food Chem.* **2004**, *52*, 8202–8207. [CrossRef]
50. Yang, S.C.; Arasu, M.V.; Chun, J.H.; Jang, Y.S.; Lee, Y.H.; Kim, I.H.; Lee, K.T.; Hong, S.T.; Kim, S.J. Identification and Determination of Phenolic Compounds in Rapeseed Mealsm(*Brassica napus* L.). *J. Agric. Chem. Environ.* **2015**, *4*, 14–23.
51. Žilić, S.; Maksimović Dragišić, J.; Maksimović, V.; Maksimović, M.; Basić, Z.; Crevar, M.; Stanković, G. The content of antioxidants in sunflower seed and kernel. *Helia* **2010**, *33*, 75–84. [CrossRef]

52. Elleuch, M.; Besbes, S.; Roiseux, O.; Blecker, C.; Attia, H. Quality characteristics of sesame seeds and by-products. *Food Chem.* **2007**, *103*, 641–650. [CrossRef]
53. Elleuch, M.; Bedigian, D.; Besbes, S.; Blecker, C.; Attia, H. Dietary fibre characteristics and antioxidant activity of sesame seed coats (Testae). *Int. J. Food Prop.* **2012**, *15*, 25–37. [CrossRef]
54. Barthet, V.J.; Klensporf-Pawlik, D.; Przybylski, R. Antioxidant activity of flaxseed meal components. *Can. J. Plant Sci.* **2014**, *94*, 593602. [CrossRef]
55. Stancheva, I.; Georgiev, G.; Geneva, M.; Ivanova, A.; Dolezal, M.; Tumova, L. Influence of foliar fertilization and growth regulator on milk thistle seed yield and quality. *J. Plant Nutr.* **2010**, *33*, 818–830. [CrossRef]
56. Nilova, L.; Malyutenkova, S. The possibility of using powdered sea-buckthorn in the development of bakery products with antioxidant properties. *Agron. Res.* **2018**, *16*, 1444–1456.
57. Teh, S.S.; Birch, E.J. Effect of ultrasonic treatment on the polyphenol content and antioxidant capacity of extract from defatted hemp, flax and canola seed cakes. *Ultrason. Sonochem.* **2014**, *21*, 346–353. [CrossRef] [PubMed]
58. Faugno, S.; Piccolella, S.; Sannino, M.; Principio, L.; Crescente, G.; Baldi, G.M.; Fiorentino, N.; Pacifico, S. Can agronomic practices and cold-pressing extraction parameters affect phenols and polyphenols content in hempseed oils? *Ind. Crops Prod.* **2019**, *130*, 511–519. [CrossRef]
59. Khalil, A.A.; Khan, M.R.; Shabbir, M.A. In vitro antioxidant activity and punicalagin content quantification of pomegranate peel obtained as agro-waste after juice extraction. *Pak. J. Agric. Sci.* **2018**, *55*, 197–201.
60. Prior, R.L.; Wu, X.; Schaich, K. Standardized Methods for the Determination of Antioxidant Capacity and Phenolics in Foods and Dietary Supplements. *J. Agric. Food Chem.* **2005**, *53*, 4290–4302. [CrossRef] [PubMed]
61. Tița, O.; Lengyel, E.; Stegarus, D.I.; Savescu, P.; Ciubara, A.B.; Constantinescu, M.A.; Tița, M.A.; Rața, D.; Ciubara, A. Identification and Quantification of Valuable Compounds in Red Grape Seeds. *Appl. Sci.* **2021**, *11*, 5124. [CrossRef]
62. Ross, C.F.; Hoye, C., Jr.; Fernandez-Plotka, V.C. Influence of Heating on the Polyphenolic Content and Antioxidant Activity of Grape Seed Flour. *J. Food Sci.* **2011**, *76*, C884–C890. [CrossRef] [PubMed]
63. Trandafir, I.; Cosmulescu, S.; Nour, V. Phenolic Profile and Antioxidant Capacity of Walnut Extract as Inflluenced by the Extraction Method and Solvent. *Int. J. Food Eng.* **2017**, *13*, 20150284. [CrossRef]
64. Bakkalbasi, E.; Meral, R.; Dogan, I.S. Bioactive compounds, physical and sensory properties of cake made with walnut press-cake. *J Food Qual.* **2015**, *38*, 422–430. [CrossRef]
65. Mathew, S.; Abraham, T.E.; Zakaria, Z.A. Reactivity of phenolic compounds towards free radicals under in vitro conditions. *J. Food Sci. Technol.* **2015**, *52*, 5790–5798. [CrossRef] [PubMed]
66. Sroka, Z.; Cisowski, W. Hydrogen peroxide scavenging, antioxidant and anti-radical activity of some phenolic acids. *Food Chem. Toxicol.* **2003**, *41*, 753–758. [CrossRef]
67. Sabir, M.A.; Sosulski, F.W.; Kernan, J.A. Phenolic constituents in sunflower flour. *J. Agric. Food Chem.* **1974**, *22*, 572–574. [CrossRef]
68. Weisz, G.M.; Kammerer, D.R.; Carle, R. Identification and quantification of phenolic compounds from sunflower (*Helianthus annuus* L.) kernels and shells by HPLC DAD/ESI-MS^n. *Food Chem.* **2009**, *115*, 758–765. [CrossRef]
69. Grasso, S.; Omoarukhe, E.; Wen, X.; Papoutsis, K.; Methven, L. The Use of Upcycled Defatted Sunflower Seed Flour as a Functional Ingredient in Biscuits. *Foods* **2019**, *8*, 305. [CrossRef] [PubMed]
70. Dziedzic, S.Z.; Hudson, B.J.F. Polyhydroxy chalcones and flavanones as antioxidants for edible oils. *Food Chem.* **1983**, *12*, 205–212. [CrossRef]
71. Kikuzaki, H.; Hisamoto, M.; Hirose, K.; Akiyama, K.; Taniguchi, H. Antioxidant Properties of Ferulic Acid and Its Related Compounds. *J. Agric. Food Chem.* **2002**, *50*, 2161–2168. [CrossRef]
72. Birasuren, B.; Kim, N.Y.; Jeon, H.L.; Kim, M.R. Evaluation of the antioxidant capacity and phenolic content of Agriophyllum pungens seed extracts from Mongolia. *Prev. Nutr. Food Sci.* **2013**, *18*, 188–195. [CrossRef] [PubMed]
73. Roginsky, V.; Lissi, E.A. Review of methods to determine chainbreaking antioxidant activity in food. *Food Chem.* **2005**, *92*, 235. [CrossRef]
74. Biskup, I.; Golonka, I.; Gamini, A.; Sroka, Z. Antioxidant activity of selected phenols estimated by ABTS and FRAP methods. *Postepy Hig. Med. Dosw.* **2013**, *67*, 958–963. [CrossRef]
75. Shi, J.; Yu, J.; Pohorly, J.; Kakuda, Y. Polyphenolics in Grape Seeds—Biochemistry and Functionality. *J. Med. Food* **2003**, *6*, 291–299. [CrossRef] [PubMed]
76. Pellegrini, N.; Serafini, M.; Salvatore, S.; Del Rio, D.; Bianchi, M.; Brighenti, F. Total antioxidant capacity of spices, dried fruits, nuts, pulses, cereals and sweets consumed in Italy assessed by three different in vitro assays. *Mol. Nutr. Food Res.* **2006**, *50*, 1030–1038. [CrossRef]
77. Arranz, S.; Perez-Jimenez, J.; Saura-Calixto, F. Antioxidant capacity of walnut (*Juglans regia* L.): Contribution of oil and defatted matter. *Eur. Food Res. Technol.* **2008**, *227*, 425–431. [CrossRef]
78. Zadernowski, R.; Naczk, M.; Czaplicki, S.; Rubinskiene, M.; Szalkiewicz, M. Composition of Phenolic Acids in Sea Buckthorn (*Hippophae rhamnoides* L.) Berries. *JAOCS* **2005**, *82*, 175–179. [CrossRef]
79. Gougoulias, N.; Mashev, N. Evaluation of polyphenols antioxidant activity of grape seeds (*V. vinifera*). *Oxid. Commun.* **2008**, *31*, 88–97.

80. Gvozdenac, S.M.; Prvulović, D.M.; Radovanović, M.N.; Ovuka, J.S.; Miklič, V.J.; Ačanski, J.M.; Tanasković, S.T.; Vukajlović, F.N. Life history of Plodia interpunctella Hübner on sunflower seeds: Effects of seed qualitative traits and the initial seed damage. *J. Stored Prod. Res.* **2018**, *79*, 89–97. [CrossRef]
81. Bogoeva, A.L.; Durakova, A.G. Sorption characteristics of full-fatted grape seeds flour of Bulgarian origin. *J. Agric. Food Res.* **2020**, *2*, 100026. [CrossRef]
82. Zhou, K.; Yu, L. Total phenolic contents and antioxidant properties of commonly consumed vegetables grown in Colorado. *LWT Food Sci. Tech.* **2006**, *39*, 1155–1162. [CrossRef]
83. Deng, G.F.; Lin, X.; Xu, X.R.; Gao, L.L.; Xie, J.F.; Li, H.B. Antioxidant capacities and total phenolic contents of 56 vegetables. *J. Funct. Foods* **2013**, *5*, 260–266. [CrossRef]
84. Sun, T.; Tanumihardjo, S.A. An integral approach to evaluate Food Antioxidant Capacity. *J. Food Sci.* **2007**, *72*, 159–165. [CrossRef]
85. Senila, L.; Neag, E.; Cadar, O.; Haydee Kovacs, M.; Becze, A.; Senila, M. Chemical, Nutritional and Antioxidant Characteristics of Different Food Seeds. *Appl. Sci.* **2020**, *10*, 1589. [CrossRef]
86. Sielicka, M.; Małecka, M.; Purłan, M. Comparison of the antioxidant capacity of lipid-soluble compounds in selected cold-pressed oils using photochemiluminescence assay (PCL) and DPPH method. *Eur. J. Lipid Sci. Technol.* **2014**, *116*, 388–394. [CrossRef]
87. Prevc, T.; Šegatin, N.; Ulrih, N.P.; Cigic, B. DPPH assay of vegetable oils and model antioxidants in protic and aprotic solvents. *Talanta* **2013**, *109*, 13–19. [CrossRef] [PubMed]
88. Schlesier, K.; Harwat, M.; Böhm, V.; Bitsch, R. Assessment of antioxidant activity by using different in vitro methods. *Free Radic. Res.* **2002**, *36*, 177–187. [CrossRef] [PubMed]

foods

MDPI

Article

Reuse of Wasted Bread as Soil Amendment: Bioprocessing, Effects on Alkaline Soil and Escarole (*Cichorium endivia*) Production

Claudio Cacace [1], Carlo Giuseppe Rizzello [2], Gennaro Brunetti [1], Michela Verni [1,*] and Claudio Cocozza [1]

1 Department of Soil, Plant, and Food Science, University of Bari, 70126 Bari, Italy; claudio.cacace@uniba.it (C.C.); gennaro.brunetti@uniba.it (G.B.); claudio.cocozza@uniba.it (C.C.)
2 Department of Environmental Biology, "Sapienza" University of Rome, 00185 Rome, Italy; carlogiuseppe.rizzello@uniroma1.it
* Correspondence: michela.verni@uniba.it

Abstract: In an era characterized by land degradation, climate change, and a growing population, ensuring high-yield productions with limited resources is of utmost importance. In this context, the use of novel soil amendments and the exploitation of plant growth-promoting microorganisms potential are considered promising tools for developing a more sustainable primary production. This study aimed at investigating the potential of bread, which represents a large portion of the global food waste, to be used as an organic soil amendment. A bioprocessed wasted bread, obtained by an enzymatic treatment coupled with fermentation, together with unprocessed wasted bread were used as amendments in a pot trial. An integrated analytical plan aimed at assessing (i) the modification of the physicochemical properties of a typical Mediterranean alkaline agricultural soil, and (ii) the plant growth-promoting effect on escarole (*Cichorium endivia* var. *Cuartana*), used as indicator crop, was carried out. Compared to the unamended soils, the use of biomasses raised the soil organic carbon content (up to 37%) and total nitrogen content (up to 40%). Moreover, the lower pH and the higher organic acid content, especially in bioprocessed wasted bread, determined a major availability of Mn, Fe, and Cu in amended soils. The escaroles from pots amended with raw and bioprocessed bread had a number of leaves, 1.7- and 1.4-fold higher than plants cultivated on unamended pots, respectively, showing no apparent phytotoxicity and thus confirming the possible re-utilization of such residual biomasses as agriculture amendments.

Keywords: wasted bread; bioprocessing; lactic acid bacteria; soil amendment

Citation: Cacace, C.; Rizzello, C.G.; Brunetti, G.; Verni, M.; Cocozza, C. Reuse of Wasted Bread as Soil Amendment: Bioprocessing, Effects on Alkaline Soil and Escarole (*Cichorium endivia*) Production. *Foods* **2022**, *11*, 189. https://doi.org/10.3390/foods11020189

Academic Editors: Marco Poiana, Francesco Caponio and Antonio Piga

Received: 14 December 2021
Accepted: 6 January 2022
Published: 11 January 2022

Publisher's Note: MDPI stays neutral with regard to jurisdictional claims in published maps and institutional affiliations.

Copyright: © 2022 by the authors. Licensee MDPI, Basel, Switzerland. This article is an open access article distributed under the terms and conditions of the Creative Commons Attribution (CC BY) license (https://creativecommons.org/licenses/by/4.0/).

1. Introduction

Many agricultural soils are characterized by a low content of organic matter (OM) that represents a limiting factor for crop growth and production [1]; moreover, the OM decomposition rate also increases with a warm climate and intensity of cultivation [2]. In the perspective of sustainable agriculture, the reuse of organic waste as soil amendments is a promising tool to recover soil fertility [3]. Several soil properties, such as pH, nutrient availability, structure and water infiltration, long-term carbon sequestration, and soil biological activity, are positively influenced after the addition of organic amendments [4].

A wide range of food waste and by-products with high content of OM represent an underutilized resource for agronomic applications [5], i.e., among some Mediterranean countries, bread and bakery products represent up to 20% of the total daily food waste produced by some surveyed consumers [6]. Melikoglu and Webb [7] estimated that the bread wasted daily worldwide is around hundreds of tons, and only a little quantity is reused mainly to feed livestock. The loss of bread occurs through the entire supply chain, not only at the household level: for example, during sandwich production, crusts and external layers removed from loaves represent up to 40% of the products [8]. Recently, the possibility of reusing wasted bread (WB) as a substrate for the cultivation of lactic acid bacteria (LAB) to be used as starters for the food industry was investigated [8]. Since LAB causes fast acidification through the production of organic acids, an acidified biomass, employed as a soil amendment, could be of interest for alkaline soils, such as Mediterranean ones. In such pH conditions, many essential plant nutrients are not available for crops, e.g., phosphorous precipitates as Ca phosphates [9], but the competition for the sorption sites between P and organic acids helps to increase P availability [10]. The same occurs for another important nutrient, iron, that in alkaline and oxygenated soils precipitates as iron oxides [11].

Recent scientific evidence confirmed LAB as plant growth-promoting microorganisms (PGPM); besides indirectly improving nutrient acquisition, they can act as biocontrol agents, improving the ability of the host plant to withstand biotic and abiotic stress, or by producing compounds that directly stimulate plant growth [12]. As for most PGPM, plant growth promotion is the simultaneous result of multiple biochemical mechanisms [12]. In addition, former *Lactobacillus* spp. is among the bacterial species able to bioaccumulate metals. Maintaining crop production within a context of land degradation, changing climate, and a growing population is of utmost significance. A rapid and more efficient transformation of the agricultural system that guarantees high-yield production with continually limited resources is therefore required.

In this framework, this study aimed at investigating the potential of wasted bread to be used as an organic soil amendment. A bioprocessed wasted bread (bWB), obtained by an enzymatic treatment coupled with fermentation and containing viable LAB cells at high cell density, together with a biomass of unprocessed wasted bread, were included in this study and used in a pot trial. As an effect of either the organic soil amendment or the viable microorganism supplementation, the physicochemical properties of a typical Mediterranean alkaline agricultural soil could be modified by bioprocessed wasted bread supplementation; additionally, a growth-promoting effect on escarole, used as an indicator crop, could be achieved. To confirm the above-mentioned hypothesis, an integrated analytical approach aimed at assessing the main characteristics of both soil and plants was carried out.

2. Materials and Methods

2.1. Bread-Based Amendments Preparation and Characterization

2.1.1. Raw Material, Enzymes, and Microorganisms

Wasted white wheat bread (surplus from the production) was kindly provided by the industrial bakery Vallefiorita Srl (Ostuni, Italy). Bread, having the following composition: moisture, 12.08%; proteins, 12.89%; fats, 1.55%; carbohydrates, 70.82%; and dietary fibers, 5.21%, was ground into small crumbs (<1 mm), mixed with distilled water (65%), and homogenized to obtain the WB.

Veron® Mac, a maltogenic amylase used in the bakery industry and purchased from AB Enzymes (Darmstadt, Germany), was added to the bread homogenate for bioprocessing. *Lactiplantibacillus plantarum* H64, belonging to the Culture Collection of the Department of Soil, Plant and Food Sciences (University of Bari, Italy), was used as a starter for biomass fermentation. The strain was routinely propagated on De Man, Rogosa, and Sharpe (MRS) (Oxoid, Basingstoke, Hampshire, UK) at 30 °C. When used for fermentation, cells grown until the late exponential phase of growth (circa 12h) were harvested by centrifugation at 9000× *g* for 10 min at 4 °C, washed twice in a sterile physiological solution (NaCl 0.9%, *w*/*v*), and resuspended in distilled water.

2.1.2. Bioprocessing

Bioprocessed wasted bread (bWB) was prepared by mixing ground bread (35%), distilled water (65%), amylase Veron® Mac at the concentration recommended by the manufacturer (3 mg/100 g), and the selected LAB strain at the final cell density of ca. 7 log cfu/g. The biomass was then incubated at 30 °C for 24 h, characterized, and used in pot trials. WB was also characterized and used as a control in all the experiments.

2.1.3. Characterization of Wasted Bread Biomasses

The pH of the biomasses was determined by a pH meter (Model 507, Crison, Milan, Italy) with a food penetration probe, and total titratable acidity (TTA) was determined according to the AACC method 02–31.01 [13] and expressed as the amount (mL) of 0.1 M NaOH necessary to reach pH of 8.4.

Presumptive LAB were enumerated, before and after fermentation, using MRS agar medium (Oxoid) supplemented with cycloheximide (0.1 g/L). Plates were incubated at 30 °C for 48 h, under anaerobiosis (AnaeroGen and AnaeroJar, Oxoid). WB and bWB were also characterized for the presence of yeasts, molds, and *Enterobacteriaceae*. Yeasts and molds were cultivated on Yeast Peptone Dextrose Agar medium (Sigma-Merck, Darmstadt, Germany), supplemented with 0.01% chloramphenicol, through pour and spread plate enumeration, respectively, and incubated at 25 °C whereas *Enterobacteriaceae* were determined on Violet Red Bile Glucose Agar (Oxoid) at 37 °C for 24 h.

Water/salt-soluble extracts (WSE) from wasted bread biomasses were prepared according to the method originally described by Osborne and modified by Weiss et al. [14] using 50 mM Tris–HCl (pH 8.8). After centrifugation, the supernatants were used to determine sugars, organic acids, peptides, and total free amino acid (TFAA) concentration.

Glucose and maltose were measured using the D-Fructose D-Glucose Assay Kit K-FRUGL and the Maltose-Sucrose-D-Glucose Assay Kit K-MASUG (Megazyme International Ireland Limited, Bray, Ireland), respectively, following the manufacturer's instructions.

Organic acids were quantified by High Performance Liquid Chromatography (HPLC), using an ÄKTA Purifier system (GE Healthcare, Buckinghamshire, UK) equipped with an Aminex HPX-87H column (ion exclusion, Biorad, Richmond, CA), as described by Rizzello et al. [15].

For the analysis of peptides, WSE were treated with trifluoroacetic acid (0.05% *wt/vol*), centrifuged (10.000× *g* for 10 min), and subjected to dialysis (cut-off 500 Da) to remove proteins and free amino acids, respectively. Then, peptide concentration was determined by the *o*-phtaldialdehyde method as described by Church et al. [16], and dialysates were analyzed through Reversed-Phase Fast Performance Liquid Chromatography (RP-FPLC) using a Resource RPC column and ÄKTA FPLC equipment with the UV detector operating at 214 nm (GE Healthcare Bio-Sciences AB, Uppsala, Sweden) as described by Rizzello et al. [15]. TFAA was analyzed by a Biochrom 30^+ series Automatic Amino Acid Analyzer (Biochrom Ltd., Cambridge Science Park, United Kingdom), equipped with a Li-cation-exchange column (4.6 × 200 mm internal diameter) [17].

WB and bWB were analyzed for moisture, ash content, pH, and electrical conductivity (EC) according to the methods previously proposed by Trinchera et al. [18]. In detail, the moisture, expressed as a percentage of the initial weight, was determined by drying samples at 105 °C overnight; the ash content, expressed as a percentage of the dry matter, was determined by combustion in a Controls 10-D1418/A muffle furnace at 550 °C for 12 h. The EC was measured using a Hanna Edge® EC instrument on sample/water extracts (1:10 *w/v*) after shaking for 30 min. Total N (TN) content was determined by the Kjeldahl method, while the organic carbon (OC) content was determined by dichromate oxidation and subsequent titration with ferrous sulfate according to Ciavatta et al. [19]. This method is suitable for samples characterized by high OC levels. The total P content was measured spectrophotometrically at 650 nm after incinerating biomass samples at 550 °C, suspending ashes in 10% hydrochloric acid solution, and developing the blue color in the filtered solution according to the Olsen method [20].

2.2. Pot Trial

2.2.1. Experimental Design

An alkaline soil, classified Calcic Luvisol according to IUSS Working Group WRB [21], was collected from a stone fruit orchard, air-dried, and used for the pot experiment. The particle size composition of the soil used for the pot trial was 173 ± 2, 356 ± 3, and

471 ± 3 g kg^{-1} of sand, silt, and clay, respectively, corresponding to clayey texture according to the Soil Survey Staff methodology [22].

For the pot trial, two experiments were carried out, one with plants and one without, aiming at assessing whether the changes observed in the soil features were a consequence of the amendment addition or the soil ecosystem interaction with the plant (rhizosphere effect). Treatments included in the pot experiment were: (i) not amended soil, without a plant (CTA); (ii) soil amended with WB, without a plant (WBA); (iii) soil amended with bWB, without a plant (bWBA); (iv) not amended soil, with a plant (CTP); (v) soil amended with WB, with a plant (WBP); (vi) soil amended with bWB and with a plant (bWBP). Pots (0.4 L each) were distributed in a completely randomized design with three replications for each treatment, for a total of 18 experimental pots. The trial was performed in a cold greenhouse at the University of Bari (South Italy). The amended pots received WB or bWB at a dose of about 25,000 kg ha^{-1}, according to the good local agricultural practices [3].

Thirty-day-old seedlings of *Cichorium endivia* var. *Cuartana*, a variety of escarole, were transplanted at the end of the first period of February 2020 and the trial was stopped at the beginning of April. The first irrigation was performed immediately after the transplanting for the rooting and establishment of the plants. During the trial, the temperature ranged from 5 °C at night to 23 °C at mid-day.

2.2.2. Soil Characterization

The soil was characterized at the beginning of the trial (T0) by means of pH, EC, TN, available phosphorous (P_{ava}), Mn, Fe and Cu, and OC content. The pH was measured in deionized water (pH_{H2O}) and in 1 M KCl (pH_{KCl}) suspensions at 1:2.5 soil to liquid ratio, whereas the electrical conductivity (EC) was measured in filtrates from a 1:2 soil to water ratio. The TN content was determined by the Kjeldahl method. The OC was measured by dichromate oxidation and ferrous sulfate titration according to the Walkley-Black method [23]. The P_{ava} was extracted with a 0.5 M $NaHCO_3$ solution and determined spectrophotometrically at 650 nm [20]. Diethylenetriaminepentaacetic acid (DTPA)-extractable fractions of Mn, Fe, and Cu were obtained from a 1:2 soil to DTPA solution. DTPA extracts were filtered by gravity through Whatman No. 42 filter paper, and the solutions were then analysed using an inductively coupled plasma iCAP 6000 Series ICP-OES Spectrometer (Thermo Electron Corporation, Walthman (MA), USA). The soil texture was identified using the Soil Survey Staff methodology [22].

At the end of the experiment, all soils were characterized again to investigate the effects of WB, bWB, and/or plants on the soil parameters with respect to T0.

2.2.3. Plant Characterization

To verify the effects of WB and bWB on the soil/plant system during the test, indirect measurements of the chlorophyll content were carried out using SPAD-502 (Konica Minolta, Japan). At the end of the test (50 days from transplanting), the total number of plant leaves and the fresh weight of the plant were determined to calculate the yield of each treatment. Moreover, leaf samples were analyzed for their P, Mn, Fe, and Cu content, aiming at verifying the effects of each treatment on leaf composition, the total P was obtained as described above for the wasted biomasses. The total Mn, Fe, and Cu content were determined using the microwave-assisted acid digestion method, adding a Suprapur® HNO_3:H_2O_2:HCl mixture (6:1:1, *v:v:v*) to each sample. At the end of the digestion, the samples were cooled, filtered through Whatman No. 42 filter paper, and diluted with distilled Milli-Q Reagent grade water and, finally analyzed by means of an inductively coupled plasma iCAP 6000 Series ICP-OES Spectrometer (Thermo Electron Corporation).

2.3. Statistical Analysis

Experimental data were tested against the normal distribution of variables (Shapiro—Wilk test) and the homogeneity of variance (Bartlett test) using R studio. The variables normally distributed with homogeneity of variances verified were subjected to an ANOVA and HSD test. Data not normally distributed were subjected to the Levene test and a no-parametric ANOVA analysis (Kruskal-Wallis test) and Dunn test.

3. Results

3.1. Amendment Characterization

In bWB, the initial cell density of presumptive LAB corresponded to the targeted inoculum and increased after 24 h of fermentation approximately 2 log cycles, reaching 9.35 ± 0.36 log cfu g^{-1}.

The amendments were also characterized for the presence of yeasts, molds, and *Enterobacteriaceae*. Yeasts and molds in WB were 3.8 ± 0.1 and 2.9 ± 0.0 log cfu g^{-1}, respectively, whereas *Enterobacteriaceae* were 4.1 ± 0.2 log cfu g^{-1}. After bioprocessing, all microbial species investigated were in a notably lower range, compared to WB. Yeasts and molds remained below 2 log cfu g^{-1}, whereas *Enterobacteriaceae* were not detected.

Relevant acidification was obtained after fermentation. The pH decreased from 5.92 ± 0.09 of WB to 3.89 ± 0.16 of bWB, with a production of 56 and 2 mmol kg^{-1} of lactic and acetic acid, respectively, which were detected in traces in unprocessed WB. As expected, the TTA value was significantly higher in bWB (8.46 ± 0.48 mL) compared to WB (1.54 ± 0.08 mL). Glucose and maltose, present in small amounts in WB (0.95 and 2.79 mg g^{-1}, respectively), were also found at higher concentrations in bWB (12.87 and 7.83 mg g^{-1}, respectively).

Peptide content in WB was 101.36 ± 4.29 mg g^{-1} and increased by ca. 38% during fermentation. This trend was confirmed by the FPLC chromatograms, where although no differences among peak area/total area ratios were observed at different percentages of acetonitrile, the total area and the number of detected peaks increased by roughly 20% in bWB compared to WB. On the contrary, TFAA slightly but significantly ($p < 0.05$) decreased after fermentation (ca. 700 mg g^{-1}). Nevertheless, γ-aminobutyric acid (GABA) content was higher in bWB (149.68 ± 1.36 mg g^{-1}) compared to WB (116.39 ± 3.12 mg g^{-1}).

WB bioprocessing slightly increased the OC content (7%) compared to unprocessed WB (Table 1). In contrast, WB biomass showed significantly higher EC and TN content than bWB, up to 6% and 9%, respectively. Accordingly, the C/N ratio was significantly higher in bWB than WB. Finally, even though bWB resulted in a numerically higher total P content (2150 ± 15 mg kg^{-1}) compared to WB (1716 ± 246 mg kg^{-1}), they were not significantly different.

Table 1. Chemical and physicochemical characteristics of the amendment. WB wasted bread; bWB, bioprocessed wasted bread (treated with amylase and fermented with *Lactiplantibacillus plantarum* H64).

Samples	Moisture (%)	Ash (%)	EC µS cm^{-1}	OC (%)	TN (%)	C/N	Total P mg kg^{-1}
WB	61.81 ± 6	0.73 ± 0.05	1950 ± 50 ^{a}	40.7 ± 1 ^{b}	2.47 ± 0.04 ^{a}	16.4 ± 0.07 ^{b}	1716 ± 246
bWB	65.04 ± 5	0.86 ± 0.06	1820 ± 60 ^{b}	43.7 ± 2 ^{a}	2.25 ± 0.02 ^{b}	19.4 ± 0.03 ^{a}	2150 ± 15
	ns	ns	*	*	*	**	ns ¥

Data are the means of three independent experiments ± standard deviations (n = 3). $^{a-b}$ Values in the same column followed by a different letter are significantly different according to the HSD test or Dunn test (¥). * Significant at $p \leq 0.05$; ** Significant at $p \leq 0.01$, ns: not significant.

3.2. Soil Characterization

The physicochemical properties of cultivated and uncultivated soils treated with the biomasses in comparison to CTP and the soil at the beginning of the trial (T0) are reported in Table 2.

Table 2. Chemical and physicochemical properties of cultivated (P) and uncultivated (A) pots. CT, control soil; WB, soil amended with wasted bread; bWB; soil amended with bioprocessed wasted bread (treated with amylase and fermented with *Lactiplantibacillus plantarum* H64).

Samples	pH_{H2O}	pH_{KCl}	EC $\mu S\ cm^{-1}$	OC $g\ kg^{-1}$	TN $g\ kg^{-1}$	P_{ava} $mg\ kg^{-1}$
			Uncultivated pots			
T0	8.20 ± 0.15^{a}	7.20 ± 0.08	200 ± 7^{b}	16.0 ± 0.45^{b}	1.60 ± 0.10^{bc}	45.5 ± 1.0^{ab}
CTA	8.20 ± 0.08^{a}	7.30 ± 0.08	319 ± 57^{b}	15.2 ± 0.51^{b}	1.5 ± 0.07^{c}	46.9 ± 1.9^{a}
WBA	7.70 ± 0.07^{b}	7.30 ± 0.04	805 ± 109^{a}	20.3 ± 1.47^{a}	2.1 ± 0.18^{a}	37.3 ± 2.5^{bc}
bWBA	7.70 ± 0.07^{b}	7.20 ± 0.02	764 ± 22^{a}	20.8 ± 0.23^{a}	1.9 ± 0.12^{ab}	36.1 ± 2.0^{c}
	***	ns	***	***	**	**
			Cultivated pots			
T0	8.20 ± 0.15^{a}	$7,20 \pm 0.08$	200 ± 7^{c}	16 ± 0.45^{b}	1.60 ± 0.10^{bc}	45.5 ± 1.0
CTP	8.07 ± 0.05^{a}	7.25 ± 0.01	417 ± 103^{b}	17.5 ± 0.49^{b}	1.31 ± 0.26^{c}	46.1 ± 2.2
WBP	7.67 ± 0.09^{b}	7.25 ± 0.11	685 ± 109^{a}	22.4 ± 1.23^{a}	1.96 ± 0.11^{ab}	48 ± 7.9
bWBP	7.57 ± 0.01^{b}	7.23 ± 0.07	786 ± 56^{a}	22.4 ± 0.83^{a}	2.17 ± 0.16^{a}	41.9 ± 1.8
	***	ns	***	***	**	ns ¥

Data are the means of three independent experiments ± standard deviations (n = 3). [a–c] Values in the same column, among cultivated or uncultivated pots data group, followed by a different letter are significantly different according to HSD test or Dunn test (¥). ** Significant at $p \leq 0.01$; *** Significant at $p \leq 0.001$; ns: not significant.

The pH_{H2O} of T0 and CTP was alkaline and ranged from 8.07 ± 0.05 to 8.20 ± 0.15, while bWB and WB supplementation significantly reduced the pH_{H2O} by roughly 8%, even if they did not show significant differences between each other. No significant differences were observed for the pH_{KCl} among all treatments as well (Table 2).

The EC value of cultivated pots of escarole (CTP) significantly increased (417 ± 103 $\mu S\ cm^{-1}$) compared to T0, and was further enhanced by the addition of the two biomasses. However, the two amended soils did not show significant differences between each other.

The soil was positively and significantly influenced by the amendments since the treated soils had higher OC and TN content, reaching up to 23% higher values at the end of the trial compared to the soil at T0, whereas CTP showed the lowest TN and OC content (Table 2). The P_{ava}, on the other hand, was not significantly influenced by the amendments.

As observed for the cultivated soils, the absence of the plants resulted in very similar trends of pH, EC, OC, and TN, meaning that those parameters were influenced mainly by the biomasses. The availability of P was significantly and negatively influenced by the treatments since, compared to CTA and the soil at the beginning of the trial, a decrease of up to 23% and 20%, respectively, was observed in uncultivated pots (Table 2). Nevertheless, P_{ava} content did not show any statistical difference among samples in cultivated soils.

The bioavailability of Mn, Fe, and Cu in soils with and without plants was also studied (Table 3). In the amended but not cultivated soils, biomass supplementation led to an increase in the availability of Mn and Fe, which were almost 3- and 2-fold higher compared to CTA, while no significant differences were observed for the available Cu among treatments.

The cultivated pots showed the higher availability of the selected elements, up to 36%, 34%, and 13% higher concentrations for Mn, Fe, and Cu, respectively, with bWBP showing the highest values between the amendments.

Table 3. Soil availability of selected micronutrients (mg kg^{-1}) in cultivated (P) and uncultivated (A) pots. CT, control soil; WB, soil amended with wasted bread; bWB; soil amended with bioprocessed wasted bread (treated with amylase and fermented with *Lactiplantibacillus plantarum* H64).

Samples	Mn	Fe	Cu
	Uncultivated pots		
CTA	8.06 ± 0.34 ^{b}	2.02 ± 0.05 ^{b}	1.23 ± 0.01
WBA	22.04 ± 6.20 ^{a}	2.88 ± 0.48 ^{a}	1.37 ± 0.07
bWBA	20.88 ± 1.90 ^{a}	2.85 ± 0.32 ^{ab}	1.49 ± 0.31
	** ¥	*	ns
	Cultivated pots		
CTP	10.72 ± 0.85 ^{b}	2.08 ± 0.08 ^{b}	1.24 ± 0.02 ^{b}
WBP	16.77 ± 1.65 ^{a}	2.87 ± 0.32 ^{a}	1.36 ± 0.05 ^{ab}
bWBP	16.68 ± 2.83 ^{a}	3.15 ± 0.21 ^{a}	1.43 ± 0.07 ^{a}
	*	**	*

Data are the means of three independent experiments ± standard deviations (n = 3). $^{a–b}$ Values in the same column, among cultivated or uncultivated pots data groups, followed by a different letter, are significantly different according to HSD. test or Dunn test (¥). * Significant at $p \leq 0.05$; ** Significant at $p \leq 0.01$, ns: not significant.

3.3. Plant Characterization

Table 4 reports the mean biometric features of plants at the end of the experiment. The escaroles from WB and bWB amended pots had a number of leaves 1.7- and 1.4- times higher, respectively, than plants cultivated on CTP pots. In particular, WBP and bWBP pots had a yield of 1.95 and 1.70 times higher than CTP, respectively.

Table 4. Biometric features of escarole plants at the end of the trial.

Samples	Number of Leaves per Plant	Treated/CTP Leaves Ratio	Average Head Escarole Fresh Weight (g)	Treated/CTP Yield Ratio
CTP	13 ± 1.15 ^{b}	-	6.6 ± 0.47 ^{b}	-
WBP	22 ± 3.78 ^{a}	1.7 ± 0.40	12.9 ± 0.95 ^{a}	1.95 ± 0.22
bWBP	19 ± 3.05 ^{ab}	1.4 ± 0.15	11.2 ± 1.36 ^{a}	1.70 ± 0.11
	*	ns	***	ns

Data are the means of three independent experiments ± standard deviations (n = 3). $^{a–b}$ Values in the same column followed by a different letter are significantly different according to the HSD test. * Significant at $p \leq 0.05$; *** Significant at $p \leq 0.001$; ns: not significant.

The indirect measurement of the chlorophyll content (Spad units) confirmed the biometric results (Figure 1). During the first 25 days after transplantation (DAT), all treatments did not show significant differences in the spad values even if an increasing trend could be observed for the bWB amended pots already from 22 DAT. From 27 to 29 DAT, bWB pots showed the highest spad values, followed by WB, while CTP escaroles had the lowest chlorophyll content. From 32 DAT until the end of the trial, plants amended with both biomasses resulted in the highest spad values, while CTP showed a slow decline of the chlorophyll content.

Lastly, the P, B, Mn, Fe, and Cu content of escarole leaves at the end of the trial, was also measured (Table 5). As observed for the available P of the corresponding soils, the application of WB and bWB significantly decreased the P concentration in escarole leaves by 63% and 59%, respectively, compared to those collected from the CTP pots. Similarly, the B content was 2 and 7 times higher in leaves from CTP compared to WBP and bWBP, respectively. Mn, Fe, and Cu concentration did not differ significantly among leaves of all treatments.

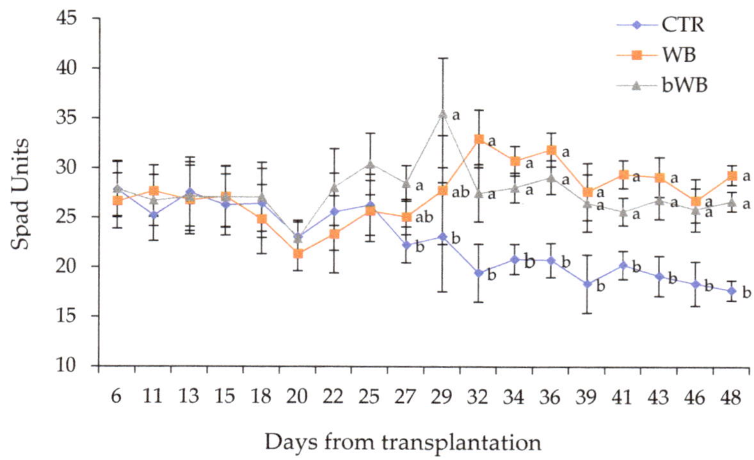

Figure 1. Effect of the biomasses on chlorophyll content of escaroles grown in control soil (CTA), soil amended with wasted bread (WB), and soil amended with bioprocessed wasted bread (bWB, treated with amylase and fermented with *Lactiplantibacillus plantarum* H64). [a–b] Different letters indicate significant differences among the data according to the HSD test. Vertical bars represent the standard deviation.

Table 5. Micronutrients and phosphorous content expressed as mg kg^{-1}, of escarole leaves grown in control soil (CTP), soil amended with wasted bread (WBP), and soil amended with bioprocessed wasted bread (bWBP, treated with amylase and fermented with *Lactiplantibacillus plantarum* H64).

Sample	B	Mn	Fe	Cu	P
CTP	15.45 ± 3.85 [a]	0.74 ± 0.25	14.77 ± 6.60	0.16 ± 0.03	358 ± 111 [a]
WBP	7.36 ± 0.96 [b]	0.76 ± 0.06	15.31 ± 2.93	0.15 ± 0.01	131 ± 50 [b]
bWBP	2.13 ± 2.52 [b]	1.06 ± 0.26	10.61 ± 3.24	0.13 ± 0.01	144 ± 29 [b]
	**	ns	ns	ns	*

Data are the means of three independent experiments ± standard deviations (n = 3). [a–b] Values in each column followed by a different letter are significantly different according to HSD.test. * Significant at $p \leq 0.05$; ** Significant at $p \leq 0.01$, ns: not significant.

4. Discussion

The long-term sustainability of food chains and the management of high levels of food loss and waste are among the challenges the global agri-food system have been facing in recent decades [24]. Bread, whose predicted production volume in 2021 was 209,874.8 million kilograms [25], represents a large portion of the global food waste, with economic and environmental repercussions [26]. The valorization of bakery waste as a food ingredient has been largely investigated recently, and different innovative biotechnological protocols have been proposed aiming at obtaining glucose syrup [27,28] or beer [26,29]. Recently, bioprocessing, e.g., enzymatic treatments and microbial fermentation, have been used to convert bread waste into valuable food ingredients, aiming at the improvement of the technological and sensory characteristics of the biomass, but also to the in situ enrichment of functional compounds such as dextran (with a positive impact on food texture) [30], antimicrobial compounds [31], and GABA [32].

Bioprocessed wasted bread, thanks to its suitability to be converted into a substrate for the growth of several microorganisms, was successfully used for the production of a medium for food industry starter cultivation [8]. Nevertheless, a major part of the wasted bread is no longer edible, not eligible for human consumption, and therefore disposed of as waste, thus representing an environmental issue due to the very high organic load. Only a small part of the wasted bread is employed for ethanol production or re-used as feed [7,26].

In this work, the potential of wasted bread to be used as a soil amendment was investigated. In addition to untreated wasted bread, bread biomass pretreated with enzymes and fermented with selected lactic acid bacteria was considered.

Overall, valorisation of food waste by conversion into products such as biofertilizers and biochar that can be added to the soil for increased nutrient inputs and fertility is gaining attention by the scientific community and industry [33]. Food waste fertilizers can be a relatively cheap source of nutrients compared to commercial inorganic fertilizer sources due to their large availability and the possibility of mass-scale and low-cost production. In addition to the nutrient role of the organic biomasses in soil, food waste can sometimes act as soil amendments, since they are able to affect the PGPM growth and survival, reduce pathogens, release nutrients, reduce leaching, increase water retention, and improve soil structure [34,35]. It was already observed that, as soil amendments, food waste-derived biomasses can increase plant yield and soil productivity [36].

The role of PGPM in soil fertility appears to be crucial; nevertheless, past research only focused on a few groups of common symbiotic rhizosphere microorganisms, such as rhizobia, *Bacillus*, *Pseudomonas*, and mycorrhizal fungi [37]. The functional role of other groups of potential PGPM, including LAB, has not been largely investigated [12], although such microorganisms could represent a genetic and metabolic resource for the development of biochemical solutions to pressing agricultural issues [12]. LAB are ubiquitous members of many plants, soil, and compost microbiomes, but little is known about the functional interactions between the LAB and their hosts.

LAB were shown to solubilize phosphate [38,39], likely through the production of organic acids, and it was also hypothesized that they can fix atmospheric nitrogen [39] or produce siderophores [38]. LAB could act as biocontrol agents; through the production of antimicrobial compounds, reactive oxygen species, and bacteriocins; by excluding pathogens by pre-emptively colonizing plant tissues vulnerable to infection and by altering the plant immune response [12]. Among LAB features, of interest not only from an agronomic but also from environmental purposes, the ability of *Lactiplantibacillus plantarum* to absorb Ni^{2+} and Cr^{2+} (from industrial wastewater) on the surface and inside their cells was proposed by Ameen et al. [40]. The superficial adsorption is possibly due to the electrostatic interaction of metals with the functional groups of the bacterial cell wall [41].

In this framework, the inoculum of wasted biomasses with properly selected LAB could guarantee the dominance of the LAB compared to other microbial groups, thanks to their capability to rapidly acidify the substrate and to produce antimicrobial compounds. In particular, the bioprocessed wasted bread harboured a very high population of the starter *L. plantarum* H64. The strain, previously selected for the ability to biosynthesize GABA in a matrix composed of wasted bread and wheat bran, allowed the repurpose of two of the main by-products of the cereal industry, promoting their application as a bread ingredient [32]. As expected, because of the starter carbohydrate metabolism, the bioprocessing enabled the production of organic acids which are in line with those previously reported in fermented surplus bread matrices [31,32]. On the contrary, proteolysis was not as pronounced if compared to common flour, which is explained by the fact that the proteases of the original flour, composing the bread dough, are degraded during the baking process. Indeed, unlike dairy LAB, most sourdough lactobacilli do not possess a cell-envelope-associated proteinase and depend on cereal-associated proteinase [42]. Hence, in surplus bread matrices, to ease the release of peptides available during LAB fermentation for their catabolism to amino acids or small bioactive sequences, the use of proteases should be considered, as previously reported [31,32]. Although a decrease in the total free amino acid content was observed in bioprocessed wasted bread, GABA content was 28% higher in bWB compared to WB. Additionally, one of the main advantages of the use of fermentation is the ability to prevent the proliferation of other microorganisms, either bacteria or molds, potentially spoiling bread. Indeed, a significant reduction of yeasts, mold, and *Enterobacteriaceae*, was observed after bioprocessing. This is an aspect particularly appealing in terms of the industrial application of bioprocessed biomass since it can guarantee a longer shelf-life of the amendment.

To better understand the principal changes occurring during cultivation, the biomasses were characterized for their main physicochemical properties. The relatively high EC values observed for WB and bWB are probably related to the presence of sodium chloride, commonly added to the bread formulation at 1–2% (*w*/*w*), however, bioprocessing slightly but significantly reduced the EC value of WB, most likely a consequence of sodium lactate formation in presence of a high concentration of lactic acid produced by LAB metabolism. Wasted bread bioprocessing also led to a slight but significant increase and decrease of OC and TN content, respectively. As a result, fermentation determined a positive balance between the C fixed in microbial biomass and the C lost in heterolactic fermentation as CO_2, whereas a major N loss as NH_3 through LAB catabolic pathways involving free amino acids could be responsible for the lower TN content in bWB compared to WB [43].

When the biomasses were used as amendments, the soil pH_{H2O} decreased because of the organic acids brought especially by bWB. The higher EC value of the biomasses reflected on that of treated soils, in fact, biomass mineralization could potentially release osmotically active compounds that could have contributed to the EC increase.

The use of biomasses raised the soil OC and TN content compared to the unamended soils. Among cultivated soils, CTP showed the lowest TN content at the end of the trial because of the uptake of nitrogen from the crop against no input. Since the physicochemical parameters (pH, EC, OC, and TN) showed the same trend in cultivated and uncultivated pots, it is safe to assume the biomasses, rather than the plants, were responsible for such changes. On the contrary, the availability of P was influenced by the crop since the absence of plants in amended pots resulted in a reduction of the P_{ava} content compared to all other treatments. The possible explanation for such behaviour is that the application of biomasses enhanced the microbial activity resulting in the immobilization of phosphate as microbial biomass and phytate, the dominant organic P form in soils, that accumulates due to the deficiency of hydrolytic enzymes and precipitates with metal ions [44]. In contrast, the presence of the plants produced a rhizosphere effect which provided phosphatases, responsible for the solubilization of organic P, and suitable organic acids that compete with phosphates for the sorption sites [10]. Indeed, among organic acids the most efficient in solubilizing soil P are the di- and tricarboxylic ones, mainly oxalic and citric acid, while *L. plantarum* H64 employed in the present study mainly produced monocarboxylic acids, such as lactic, acetic, and γ-aminobutyric acids. The lower pH and the higher organic acid content also led to a major availability of Mn, Fe, and Cu in amended soils. These elements, through the ligand exchange, were solubilized from their precipitated oxides, as a consequence of the biomass's addition as well as the soil microbial community and rhizosphere activities [9].

To evaluate whether soil improvements transmuted to beneficial changes in the plants, escarole growth and composition were monitored. Biometric parameters of escarole plants indicated that WB and bWB promoted plant growth. Even though during the first 25 DAT no significant differences were observed among treatments, probably because of the rooting and establishment of the plants, the following days revealed a trend. Soil amended with bWB was the first to prompt a better chlorophyll content compared to the other treatments because of its higher TN content and the larger supply of GABA that is correlated to several beneficial effects for plants. Indeed, GABA was found to (i) defend roots against pathogens, (ii) serve as an N reservoir, (iii) cryoprotect the tissues, and (iv) synthesize plant hormones [45]. In contrast, the control plants showed the lowest significant SPAD values from 27 DAT until the end of the trial presumably due to nutrient limitation.

The elemental composition of escarole leaves did not reflect the soil availability of the studied elements. In fact, CTP plant leaves showed a P content more than double that of WB and bWB possibly due to the immobilization of P in organic matter. Regardless, this result did not negatively influence the yield of the crop but rather the nutritional value of escarole. To avoid this inconvenience, transplanting should occur later than soil amendment to allow for better mineralization of the organic P. Regarding Mn, Fe, and Cu, their leaf content did not change among treatments although their soil content was higher in amended pots. It is hypothesized that their soil availability was already satisfied in the CTP pots since they are micronutrients, and/or those elements have been accumulated preferentially in the roots.

5. Conclusions

In our study, the application of wasted bread, raw or bioprocessed, resulted in higher escarole yield compared to the unfertilized control without any apparent phytotoxicity, thus confirming the possible re-utilization of such residual biomasses in agriculture as amendments. Although the effects could be transient, it is noteworthy that WB and bWB application resulted in a significantly higher soil OC and lower pH_{H2O} value, a feature that can ameliorate the cultivation of alkaline soils (typical of the Mediterranean area) through beneficial effects on the bioavailability of several nutrients. Nevertheless, such biomasses are not suitable for application in acidic soils (in which the excessive bioavailability of micronutrients and the release of Al from mineral weathering could be favoured).

It was previously reported that LAB promote growth in different crops, even though the underlying mechanisms behind this bio-stimulation remain unclear. The use of wasted bread fermented with *Lactiplantibacillus plantarum* resolved the transplantation stress sooner and further investigation is needed to study the effects of such pre-treatment on the standardization of biomass characteristics and its shelf-life. Additionally, WB can be subjected to contamination during storage making the effects of such biomass on soil quality unpredictable.

Even though further studies are necessary to fully exploit the potential of wasted bread as an amendment, the feasibility of its large-scale production is undeniable. Companies currently producing fertilizers could easily handle the collection of the bread from bakeries and large retailer networks thanks to the abundant and widespread availability of this wasted food product. Although a proper technological transfer is needed, the bioprocess proposed is cost-effective and implementable on an industrial scale. The supply to farmers might follow the current sale and distribution channels. In return, bakeries would not have to assume the disposal costs for managing bread waste (also no longer edible or reusable for feed purposes). It can be assumed that the entity of amendment treatment (approximately corresponding to 250 q/ha) could effectively offer a solution to food waste management and the economic and environmental sustainability of agricultural productions.

Author Contributions: Conceptualization and resources, C.G.R. and C.C. (Claudio Cocozza); formal analysis and data curation, C.C. (Claudio Cacace) and M.V.; writing—original draft preparation, C.C. (Claudio Cacace) and C.C. (Claudio Cocozza); writing—review and editing, M.V., C.G.R. and C.C. (Claudio Cocozza); supervision, G.B. All authors have read and agreed to the published version of the manuscript.

Funding: This research received no external funding.

Institutional Review Board Statement: Not applicable.

Informed Consent Statement: Not applicable.

Data Availability Statement: The data used to support the findings of this study are available from the corresponding author upon request.

Conflicts of Interest: The authors declare no conflict of interest.

References

1. De Mastro, F.; Cocozza, C.; Traversa, A.; Savy, D.; Abdelrahman, H.M.; Brunetti, G. Influence of crop rotation, tillage and fertilization on chemical and spectroscopic characteristics of humic acids. *PLoS ONE* **2019**, *14*, e0219099. [CrossRef] [PubMed]
2. Montemurro, F.; Maiorana, M.; Convertini, G.; Ferri, D. Alternative sugar beet production using shallow tillage and municipal solid waste fertilizer. *Agron. Sustain. Dev.* **2007**, *27*, 129–137. [CrossRef]

3. Abdeldaym, E.A.; Traversa, A.; Cocozza, C.; Brunetti, G. Effects of a 2-Year application of different residual biomasses on soil properties and potato yield. *Clean Soil Air Water* **2018**, *46*, 180026. [CrossRef]
4. Hardgrove, S.J.; Livesley, S.J. Applying spent coffee grounds directly to urban agriculture soils greatly reduces plant growth. *Urban For. Urban Green.* **2016**, *18*, 1–8. [CrossRef]
5. Kallenbach, C.M.; Conant, R.T.; Calderón, F.; Wallenstein, M.D. A novel soil amendment for enhancing soil moisture retention and soil carbon in drought-prone soils. *Geoderma* **2019**, *337*, 256–265. [CrossRef]
6. Capone, R.; El Bilali, H.; Debs, D.; Bottalico, F.; Cardone, G.; Berjan, S.; Elmenofi, G.A.G.; Aboubdillah, A.; Charbel, L.; Ali Arous, S. Bread and bakery products waste in selected mediterranean arab countries. *Am. J. Food Nutr.* **2016**, *4*, 40–50.
7. Melikoglu, M.; Webb, C. Use of waste bread to produce fermentation products. In *Food Industry Wastes*; Kosseva, M., Webb, C., Eds.; Academic Press: San Diego, CA, USA, 2013; pp. 63–76.
8. Verni, M.; Minisci, A.; Convertino, S.; Nionelli, L.; Rizzello, C.G. Wasted bread as substrate for the cultivation of starters for the food industry. *Front. Microbiol.* **2019**, *11*, 293. [CrossRef]
9. Sposito, G. *The Chemistry of Soil*; Oxford University Press: Oxford, UK, 2008.
10. Brunetti, A.; Traversa, A.; De Mastro, F.; Cocozza, C. Short term effects of synergistic inorganic and organic fertilization on soil properties and yield and quality of plum tomato. *Sci. Hortic.* **2019**, *252*, 342–347. [CrossRef]
11. Cocozza, C.; Ercolani, G.L. Siderophore production and associated characteristics in rhizosphere and non-rhizosphere fluorescent pseudomonads. *Ann. Microbiol.* **1997**, *47*, 17–28.
12. Lamont, J.R.; Wilkins, O.; Bywater-Ekegärd, M.; Smith, D.L. From yogurt to yield: Potential applications of lactic acid bacteria in plant production. *Soil Biol. Biochem.* **2017**, *111*, 1–9. [CrossRef]
13. AACC. *Approved Methods of the American Association of Cereal Chemistry*, 11th ed.; AACC: St. Paul, MN, USA, 2010.
14. Weiss, W.; Vogelmeier, C.; Gorg, A. Electrophoretic characterization of wheat grain allergens from different cultivars involved in bakers' asthma. *Electrophoresis* **1993**, *14*, 805–816. [CrossRef]
15. Rizzello, C.G.; Nionelli, L.; Coda, R.; De Angelis, M.; Gobbetti, M. Effect of sourdough fermentation on stabilisation, and chemical and nutritional characteristics of wheat germ. *Food Chem.* **2017**, *119*, 1079–1089. [CrossRef]
16. Church, F.C.; Swaisgood, H.E.; Porter, D.H.; Catignani, G.L. Spectrophotometric assay using o-phthaldialdehyde for determination of proteolysis in milk and isolated milk proteins. *J. Dairy Sci.* **1983**, *66*, 1219–1227. [CrossRef]
17. De Pasquale, I.; Verni, M.; Verardo, V.; Gómez-Caravaca, A.M.; Rizzello, C.G. Nutritional and functional advantages of the use of fermented black chickpea flour for semolina-pasta fortification. *Foods* **2021**, *10*, 182. [CrossRef]
18. Trinchera, L.; Leita, P.; Sequi, P. *Metodi di Analisi per i Fertilizzanti*; Ministero delle Politiche Agricole Alimentari e Forestali: Rome, Italy, 2006.
19. Ciavatta, C.; Antisari, L.V.; Sequi, P. Determination of organic carbon in soils and fertilizers. *Commun. Soil Sci. Plant Anal.* **1989**, *20*, 759–773. [CrossRef]
20. Olsen, S.R. *Estimation of Available Phosphorus in Soils by Extraction with Sodium Bicarbonate (No. 939)*; US Department of Agriculture: Washington, DC, USA, 1954.
21. IUSS Working Group WRB. World Reference Base for Soil Resources. In *International Soil Classification System for Naming Soils and Creating Legends for Soil Maps. World Soil Resources Reports No. 106*; FAO: Rome, Italy, 2015.
22. Soil Survey Staff. *Keys to Soil Taxonomy*, 12th ed.; USDA-Natural Resources Conservation Service: Washington, DC, USA, 2014.
23. Walkley, A.; Black, I.A. An examination of the Degtjareff method for determining soil organic matter, and a proposed modification of the chromic acid titration method. *Soil Sci.* **1934**, *37*, 29–38. [CrossRef]
24. FAO. *Developing Sustainable Food Value Chains—Guiding Principles*; FAO: Rome, Italy, 2014.
25. Statista. 2021. Available online: https://www.statista.com/outlook/40050100/100/bread/worldwide (accessed on 10 December 2021).
26. Brancoli, P.; Bolton, K.; Eriksson, M. Environmental impacts of waste management and valorisation pathways for surplus bread in Sweden. *Waste Manag.* **2020**, *117*, 136–145. [CrossRef]
27. Kwan, T.H.; Ong, K.L.; Haque, M.A.; Kwan, W.H.; Kulkarni, S.; Lin, C.S.K. Valorisation of food and beverage waste via saccharification for sugars recovery. *Biores. Technol.* **2018**, *255*, 67–75. [CrossRef]
28. Riaukaite, J.; Basinskiene, L.; Syrpas, M. Bioconversion of waste bread to glucose fructose syrup as a value-added product. In Proceedings of the FOODBALT 2019 13th Baltic Conference on Food Science and Technology "Food, Nutrition, Well-Being", Jelgava, Latvia, 2–3 May 2019; pp. 120–124.
29. Dlusskaya, E.; Jänsch, A.; Schwab, C.; Gänzle, M.G. Microbial and chemical analysis of a kvass fermentation. *Eur. Food Res. Technol.* **2008**, *227*, 261–266. [CrossRef]
30. Immonen, M.; Maina, N.H.; Wang, Y.; Coda, R.; Katina, K. Waste bread recycling as a baking ingredient by tailored lactic acid fermentation. *Int. J. Food Microbiol.* **2020**, *327*, 108652. [CrossRef]
31. Nionelli, L.; Wang, Y.; Pontonio, E.; Immonen, M.; Rizzello, C.G.; Maina, H.N.; Katina, K.; Coda, R. Antifungal effect of bioprocessed surplus bread as ingredient for bread-making: Identification of active compounds and impact on shelf-life. *Food Control* **2020**, *118*, 107437. [CrossRef]
32. Verni, M.; Vekka, A.; Immonen, M.; Katina, K.; Rizzello, C.G.; Coda, R. Biosynthesis of γ-aminobutyric acid by lactic acid bacteria in surplus bread and its use in bread-making. *J. Appl. Microbiol.* **2021**, *00*, 1–15. [CrossRef]
33. O'Connor, J.; Hoang, S.A.; Bradney, L.; Dutta, S.; Xiong, X.; Tsang, D.C.; Ramadass, K.; Vinu, A.; Kirkham, M.B.; Bolan, N.S. A review on the valorisation of food waste as a nutrient source and soil amendment. *Environ. Pollut.* **2021**, *272*, 115985. [CrossRef]

34. Sogn, T.A.; Dragicevic, I.; Linjordet, R.; Krogstad, T.; Eijsink, V.G.; Eich-Greatorex, S. Recycling of biogas digestates in plant production: NPK fertilizer value and risk of leaching. *Int. J. Recycl. Org. Waste Agric.* **2018**, *7*, 49–58. [CrossRef]
35. Tampio, E.; Salo, T.; Rintala, J. Agronomic characteristics of five different urban waste digestates. *J. Environ. Manag.* **2016**, *169*, 293–302. [CrossRef]
36. Barzee, T.J.; Edalati, A.; El-Mashad, H.; Wang, D.; Scow, K.; Zhang, R. Digestate biofertilizers support similar or higher tomato yields and quality than mineral fertilizer in a subsurface drip fertigation system. *Front. Sustain. Food Syst.* **2019**, *3*, 58. [CrossRef]
37. Vessey, J.K. Plant growth promoting rhizobacteria as biofertilizers. *Plant Soil* **2003**, *255*, 571–586. [CrossRef]
38. Shrestha, A.; Kim, B.S.; Park, D.H. Biological control of bacterial spot disease and plant growth-promoting effects of lactic acid bacteria on pepper. *Biocontrol Sci. Technol.* **2014**, *24*, 763–779. [CrossRef]
39. Giassi, V.; Kiritani, C.; Kupper, K.C. Bacteria as growth-promoting agents for citrus rootstocks. *Microbiol. Res.* **2016**, *190*, 46–54. [CrossRef]
40. Ameen, F.A.; Hamdan, A.M.; El-Naggar, M.Y. Assessment of the heavy metal bioremediation efficiency of the novel marine lactic acid bacterium, *Lactobacillus plantarum* MF042018. *Sci. Rep.* **2020**, *10*, 314. [CrossRef]
41. Kargar, S.H.M.; Shirazi, N.H. *Lactobacillus fermentum* and *Lactobacillus plantarum* bioremediation ability assessment for copper and zinc. *Arch. Microbiol.* **2020**, *202*, 1957–1963. [CrossRef] [PubMed]
42. Gänzle, M.; Gobbetti, M. Physiology and biochemistry of lactic acid bacteria. In *Handbook on Sourdough Biotechnology*; Gobbetti, M., Gänzle, M., Eds.; Springer: Boston, MA, USA, 2013; pp. 183–216.
43. Fernández, M.; Zúñiga, M. Amino acid catabolic pathways of lactic acid bacteria. *Crit. Rev. Microbiol.* **2006**, *32*, 155–183. [CrossRef] [PubMed]
44. Gerke, J. Phytate (Inositol hexakisphosphate) in soil and phosphate acquisition from inositol phosphates by higher plants. A review. *Plants* **2015**, *4*, 253–266. [CrossRef] [PubMed]
45. Molina-Rueda, J.J.; Pascual, M.B.; Pissarra, J.; Gallardo, F. A putative role for γ-aminobutyric acid (GABA) in vascular development in pine seedlings. *Planta* **2015**, *241*, 257–267. [CrossRef]

foods

MDPI

Article

Exploring the DPP-IV Inhibitory, Antioxidant and Antibacterial Potential of Ovine "Scotta" Hydrolysates

Roberto Cabizza [1], Francesco Fancello [1], Giacomo Luigi Petretto [2], Roberta Addis [2], Salvatore Pisanu [3], Daniela Pagnozzi [3], Antonio Piga [1] and Pietro Paolo Urgeghe [1,*]

1 Dipartimento di Agraria, Università degli Studi di Sassari, Viale Italia 39/A, 07100 Sassari, Italy; rcabizza@uniss.it (R.C.); fancello@uniss.it (F.F.); pigaa@uniss.it (A.P.)
2 Dipartimento di Chimica e Farmacia, Università degli Studi di Sassari, Via Muroni 23/A, 07100 Sassari, Italy; gpetretto@uniss.it (G.L.P.); raddis@uniss.it (R.A.)
3 Porto Conte Ricerche SRL, Loc Tramariglio, 07041 Alghero, Italy; pisanu@portocontericerche.it (S.P.); pagnozzi@portocontericerche.it (D.P.)
* Correspondence: paolou@uniss.it; Tel.: +39-0792-29378

Abstract: The aim of this work was to valorize the by-product derived from the ricotta cheese process (scotta). In this study, ovine scotta was concentrated by ultrafiltration and then subjected to enzymatic hydrolyses using proteases of both vegetable (4% E:S, 4 h, 50 °C) and animal origin (4% E:S, 4 h, 40 °C). The DPP-IV inhibitory, antioxidant, and antibacterial activities of hydrolysates from bromelain (BSPH) and pancreatin (PSPH) were measured in vitro. Both the obtained hydrolysates showed a significantly higher DPP-IV inhibitory activity compared to the control. In particular, BSPH proved to be more effective than PSPH (IC_{50} 8.5 ± 0.2 vs. 13 ± 1 mg mL^{-1}). Moreover, BSPH showed the best antioxidant power, while PSPH was more able to produce low-MW peptides. BSPH and PSPH hydrolysates showed a variable but slightly inhibitory effect depending on the species or strain of bacteria tested. BSPH and PSPH samples were separated by gel permeation chromatography (GPC). LC-MS/MS analysis of selected GPC fractions allowed identification of differential peptides. Among the peptides 388 were more abundant in BSPH than in the CTRL groups, 667 were more abundant in the PSPH group compared to CTRL, and 97 and 75 of them contained sequences with a reported biological activity, respectively.

Keywords: ovine scotta; bioactive peptides; bromelain; pancreatin; dipeptidyl peptidase IV inhibition; ovine second whey cheese; enzymatic hydrolysis

Citation: Cabizza, R.; Fancello, F.; Petretto, G.L.; Addis, R.; Pisanu, S.; Pagnozzi, D.; Piga, A.; Urgeghe, P.P. Exploring the DPP-IV Inhibitory, Antioxidant and Antibacterial Potential of Ovine "Scotta" Hydrolysates. *Foods* **2021**, *10*, 3137. https://doi.org/10.3390/foods10123137

Academic Editor: Manuel Castillo Zambudio

Received: 23 November 2021
Accepted: 14 December 2021
Published: 17 December 2021

Publisher's Note: MDPI stays neutral with regard to jurisdictional claims in published maps and institutional affiliations.

Copyright: © 2021 by the authors. Licensee MDPI, Basel, Switzerland. This article is an open access article distributed under the terms and conditions of the Creative Commons Attribution (CC BY) license (https://creativecommons.org/licenses/by/4.0/).

1. Introduction

"Scotta", also called "ricotta cheese exhaust whey" (RCEW) or "second cheese whey" (SCW) is the residual liquid by-product of the ricotta cheese production process, obtained by thermal flocculation of whey proteins, and heating the whey at a temperature of 85–90 °C and 78–85 °C for bovine or buffalo and ovine or goat whey, respectively [1]. Whey composition, the treatments performed during ricotta production process (i.e., adding milk, whey protein extraction method), and ricotta yield (depending on the temperature of protein coagulation, pH, and ionic strength of the whey) are the main factors that affect the physicochemical characteristics of scotta [2].

In Italy, scotta is mainly produced from bovine and ovine milk and, in less quantity, from buffalo and goat milk. During 2019 about 900,000 tons [3] of whey were transformed into ricotta cheese in Italy, representing about 16% of total whey produced, giving rise to more than 750,000 tons of scotta.

The production of scotta from ovine whey is mainly concentrated in Sardinia, which hosts about 22% of Italian dairy ewes [4] and where is produced more than 65% of Italian ovine milk. The Sardinian dairy system produces about 320,000 tons of ovine milk [5], which is almost completely destined for cheese production. About 12,000 tons of ricotta are produced in Sardinia, with a potential production of more than 250,000 tons of scotta. The disposal of scotta poses serious environmental concerns due to its high biochemical oxygen (BOD) and chemical oxygen demand (COD) [6–8]. Therefore, its valorization is becoming crucial for the dairy industry.

For many decades whey, and especially scotta, were underused or treated as waste, due to poor knowledge of their valuable components and the unavailability of adequate technologies to valorize such components. Over the years, several approaches to its valorization have been developed [9–11], since scotta still retains significant amount of useful compounds from whey, such as lactose, minerals, oligosaccharides, vitamins, proteins, soluble peptides, and free amino acids [9]. Scotta has been employed using biotechnological approaches as a growth substrate for some selected Lactic Acid Bacteria (LAB) for the production of lactic acid [12], or yeasts such as *Chlorella protothecoides* for the production of carotenoids, chlorophyll [13], and bioethanol [6,14], as ingredient to fortify ricotta, including bioactive peptides with antioxidant and anti-tyrosinase activities [15], bioactive peptides with angiotensin-I-converting enzyme (ACE)—inhibitory [16] biodegradable bioplastic [17], fermented drink [18] and hydrogen [19,20].

Furthermore, the interest in the application of enzymes of animals or plant sources applied to food matrices has grown during over the years [21,22], with the aim of valorizing by-products and reducing environmental impact [23]. The hydrolysis of whey proteins is a widespread practice, and the hydrolyzed whey protein (WPH) has had an important impact as functional or nutraceutical ingredient. WPH is produced by enzymatic hydrolysis of whey proteins, mainly from whey protein concentrates (WPC) or isolates (WPI), which leads to an increase in solubility and digestibility, reducing whey protein allergenic properties. Peptides obtained by enzymatic hydrolysis of whey proteins have shown interesting biological activities, with potential benefit for human health [24,25]; however, scientific studies specifically addressed to the valorization of scotta proteins and derived peptides are still poor [15].

Recently, several authors studied the potential of whey protein hydrolysates as inhibitors against dipeptidyl peptidase-4 (DPP-IV) [26–29]. DPPI-IV inhibitors play a key role in the treatments of type 2 diabetes (T2D), a worldwide diffused disease that affects 415 million people and in Italy accounts for more than 3 million patients, i.e., about 5% of the population [30].

The DPP-IV inhibitors adopted as oral antidiabetic agents act by promoting glucose homeostasis through the inhibition of the enzyme DPP-IV involved in the mechanism of degradation of glucose-dependent insulinotropic polypeptide (GIP) and glucagon-like peptide-1 (GLP-1), two key glucoregulatory hormones. DPP-IV inhibitors can reduce glucagon levels and at the same time stimulate insulin release.

The antioxidant potential of peptides derived from milk and whey proteins is well-reviewed [31]. However, up to now, just two works have focused on the antioxidant activity of scotta. Sommella et al. [11] reported bovine scotta peptides with antioxidant activity derived from αs1-casein and β-casein. Monari et al. [15] tested different proteases on bovine scotta, observing in vitro the antioxidant potential of the obtained hydrolysates.

The antibacterial activity of biopeptides encrypted into the protein fraction of ovine milk and whey has been already shown [32–34]. Several peptides of the C-terminal region of the ovine αs2-casein were shown to possess antibacterial activity against Gram-positive and Gram-negative bacteria, with the first group of bacteria more susceptible than the second group [33]. El-Zahar et al. [32] found that the peptic hydrolysates of whey protein inhibited, in a dose-dependent manner, the growth of *Escherichia coli*, *Bacillus subtilis*, and *Staphylococcus aureus*. Regarding scotta, recent work [11,15] showed the presence of several antibacterial peptides in the protein fraction of scotta.

Regarding the available proteases, bromelain and pancreatin were previously applied in the hydrolysis of dairy products and by-products to produce peptides with DPP-IV inhibitory [35] and antioxidant activity [15]. In addition, bromelain was reported to release antibacterial peptides from goat milk and whey [36].

While the literature has been mainly focused on the bioactive properties of peptides obtained from milk and whey proteins, the available data on scotta are limited to the bovine source [9] and, to the best of our knowledge, the present study is the first conducted on the ovine scotta as a potential source of bioactive peptides. Conversely, this matrix could represent an important source of such peptides, due to its higher amount of nitrogen, compared with bovine whey and scotta. The research conducted in this field could lead both to an adequate valorization of this by-product and an improvement of the sustainability of the dairy farms. Therefore, the aim of the present study was to evaluate the possibility of producing peptides with DPP-IV inhibitory, antioxidant and antibacterial activities from ultrafiltered ovine scotta by enzymatic hydrolysis with bromelain (BSPH), and pancreatin (PSPH), and evaluate their biological activity by in vitro tests. Furthermore, an in-depth characterization of peptide mixtures obtained from the hydrolysis with the two enzymes by gel permeation chromatography (GPC) and LC-MS/MS was performed.

2. Materials and Methods

2.1. Chemicals and Reagents

Analytical grade chemicals were obtained from Carlo Erba (Milano, Italy). Bromelain was obtained from Nutraceutica (Bologna, Italy) (2400 GDU g^{-1}) and from Enzyme Development Corporation (New York, NY, USA). Pancreatin from porcine pancreas (4 USP) was purchased from Sigma-Aldrich (St. Louis, MI, USA).

Synthetic peptides used for Gel Permeation Chromatography (GPC) calibration were bought from Sigma-Aldrich (St. Louis, MI, USA) (i.e., bovine serum albumin (BSA), β-Lactoglobulin (β-Lg), α-lactalbumin (α-La), aprotinin, bacitracin, tetrapeptide (Leu-Trp-Met-Arg), Asp-Glu, Tyr).

2.2. Scotta Concentrate Preparation

Ovine scotta was freshly collected from a dairy plant located in north Sardinia (Italy) after the manufacturing of ovine Ricotta cheese, and immediately refrigerated to 4 °C, delivered to the lab and stored at −20 °C. Scotta was thawed to 20 °C immediately before the concentration step. The mean chemical composition, determined according to the literature [37] was as follows: pH 6.19 ± 0.11; total solids, 6.73 ± 0.28% (*w*/*w*); fat, 0.05 ± 0.02% (*w*/*w*); total nitrogen (TN), 0.14 ± 0.01% (*w*/*w*); nitrogen soluble in water (NS), 0.08 ± 0.05% (*w*/*w*); non-protein nitrogen (NPN), 0.07 ± 0.05% (*w*/*w*); ash, 0.34 ± 0.23% (*w*/*w*).

Scotta (5 L) underwent a preliminary continuous skimming at 15,000× *g* using a lab-scale cream separator (TLE 100, Tecnolatte, Lodi, Italy), then was consecutively filtered through 5, 1.2, and 0.65 µm on conventional cartridge filters (Sartopure PP3 Midicap, Sartorius, Goettingen, Germany) with a surface area of 0.21, 0.15, and 0.15 m^2 respectively, fed by a SartoJet Membrane Pump (Sartorius, Goettingen, Germany). The 0.65 µm filtered skimmed scotta underwent tangential filtration at 20 °C through a 10 kDa Hydrosart membrane (Sartocon Slice Cassette, Sartorius, Goettingen, Germany), keeping a constant transmembrane pressure of 0.5 bar, and monitoring the removed permeate weight, which was precalculated in order to obtain a nitrogen concentration factor of 4×. The retentate was then reconstituted to the original weight by adding ultrapure (UP) water and subsequently underwent a diafiltration step using the same membrane, in order to reduce the mineral fraction and lactose. The collected retentate was sampled at the end of the process for total nitrogen (NT) determination and then stored at −20 °C until the subsequent hydrolysis steps.

2.3. Enzymatic Hydrolysis of Retentate Scotta Samples

The retentate obtained in Section 2.2, with a total nitrogen of 0.50 ± 0.01% (*w*/*w*), was split between two experiments to be hydrolyzed with bromelain (BSPH) and pancreatin (PSPH), respectively. For each experiment, three hydrolyses on 50 g (*n* = 3) of retentate were performed. Lab-scale enzymatic reactions were performed for 4 h. Enzymatic hydrolysis were performed, as recommended by the manufacturer, at 50 °C for bromelain and at 40 °C for pancreatin. The enzyme-substrate (E:S) ratio was fixed at 4% (enzyme weight to protein weight) for both experimental groups. Further three 50 g retentate control samples (CTRL) underwent the same procedure without enzyme addition, setting the temperature and duration to 50 °C and 4 h, respectively. The enzymatic reactions were performed in a water bath at constant temperature (±0.05 °C) and continuous magnetic stirring at 500 rpm using an AREX-6 Digital PRO Hot Plate Stirrer (Velp Scientific, Bohemia, NY, USA) equipped with a VTF EVO digital thermoregulator. All the reactions were stopped by heating the mixtures at 90 °C for 10 min to inactivate the proteases. Afterward, the mixtures were centrifuged twice at 14,000× *g* (Neya 16 R, Remi Elektrotechnki LTD, Vasai, India) for 15 min at 4 °C, and the precipitate was discarded. The obtained supernatants of bromelain scotta protein hydrolysate (BSPH), pancreatin scotta protein hydrolysate (PSPH) and of the control were freeze-dried (Labconco, Kansas City, MO, USA) and stored at −20 °C until further analysis, such as DPP-IV inhibition, antioxidant capacity and an antibacterial assay, as well as gel permeation chromatography (GPC) and LC-MS/MS characterization.

2.4. DPP-IV Inhibitory Activity

A DPP-IV drug discovery kit was used to measure the ability of hydrolysates to inhibit DPP-IV activity (Enzo Life Sciences Inc., Farmingdale, New York, NY, USA). The assays were conducted according to the manufacturer's instructions. Briefly, the kit contained human recombinant DPP-IV enzyme, a chromogenic substrate (H-Gly-Pro-pNA, MW = 328.8, 10 mM in DMSO), a calibration standard (p-nitroaniline, MW = 138, in assay buffer), an inhibitor as positive control (P32/98, MW = 260.4, 1 mM in DMSO), and an assay buffer (50 mM Tris, pH 7.5). All the reagents of the kit were stored at −70 °C; the analyses were conducted at room temperature. The freeze-dried protein hydrolysates were dispersed in ultrapure water in concentrations from 0.78 to 12.5 mg mL^{-1}. Assays were performed at 37 °C, in a 96-well microplate provided by the manufacturer and the reading was performed in a microplate reader every minute for a total of 30 min at λ 405 nm. Finally, absorbance values were plotted against time, and the "best fit" lines for data points and slope of the curves were obtained. Two technical replicates for each sample were performed. The % of inhibition was calculated with the formula:

$$\% \text{ activity remaining (with inhibitor)} = (\text{slope of inhibitor sample/control slope}) \times 100$$

The obtained data were analyzed by a one-way analysis of variance (ANOVA) using a Statgraphics Centurion XVI for Windows software package (version 16.2.04; Statpoint Technologies, Inc. Warrenton, Virginia, VA, USA). Fisher's least significant differences (LSD) test was applied to assess the difference between each pair of means ($p < 0.05$).

2.5. ABTS Radical Scavenging Activity

Antioxidant capacity was evaluated by colorimetric assay measuring the activity of the sample to scavenge the radical ABTS according to the method described by Petretto et al. [38]. The ABTS radical scavenging activity is based on the production of the radical cation (ABTS·+), prepared by reacting ABTS and potassium persulfate (2.45 mM) to reach a final concentration of 7 mM. Briefly, the solution obtained was kept in the dark at 25 °C for 12–16 h before the analysis. The ABTS radical solution was properly diluted with ethanol 70% to obtain an absorbance (λ = 734 nm) of 0.7 ± 0.02. The freeze-dried protein hydrolysates were dispersed in ultrapure water at concentrations ranging from 0.78 to 12.5 mg mL^{-1}. The reduction of radical ABTS was monitored at the start and after 50 min from the beginning of the reaction. Two technical replicates for each sample were performed. The antioxidant power of samples was expressed as a percentage of inhibition, and an IC_{50} value was calculated from the regression curve plotting different concentrations of hydrolysates against the percentage of activity, and expressed as the mean ± SD. Data were analyzed as described in Section 2.4.

2.6. Antibacterial Assays

The antibacterial activity of scotta hydrolysates was tested against six bacterial strains (see Table 1) belonging to three different species, namely, *Listeria monocytogenes* (four strains), *Staphylococcus aureus* and *Salmonella bongori*. The strains, stored at −80 °C, were thawed and precultured in Brain Heart infusion broth medium (BHI, WVR, Milano, Italy) for 24 h at 37 °C. Overnight cultures were then used to prepare microbial inoculation used for the test. One milliliter of overnight culture was centrifuged at 14,000× *g* for 2 min, then the pellets were resuspended in saline solution until reaching 0.2 optical density (OD) (~8 log10 CFU mL^{-1}).

Table 1. List of microorganisms, medium and culture condition for testing the antimicrobial activity of enzymatic hydrolysate of Scotta [1].

Tested Organisms	Source	Medium	Temperature and Time of Incubation
Staphylococcus aureus 20,231 DSMZ	DSMZ	BHI	37 °C × 24 h
Listeria monocytogenes B *Listeria monocytogenes* C	DAFS	BHI	37 °C × 24 h
Listeria monocytogenes E	DAFS	BHI	37 °C × 24 h
Listeria monocytogenes 20,600 DSMZ	DSMZ	BHI	37 °C × 24 h
Salmonella bongori 13,772 DSMZ	DSMZ	BHI	37 °C × 24 h

[1] DSMZ, Deutsche SammLung von Mikroorganismen und Zellkulturen, German Collection of Microorganism of Cell Cultures; DAFES, Collection of Microorganisms of Dipartimento di Agraria of the University of Sassari, Section of Food and Environmental Science.

The lyophilized hydrolysates (BSPH, PSPH) and non-hydrolysate (CTRL) were weighed and dissolved in Brain Heart Infusion broth (BHI), giving a final concentration of 100 mg mL^{-1}. The solutions were then filter-sterilized on a 0.22 µm filter (Sartorius). Aliquots of 100 µL of filtered BSPH, PSPH, CTRL, and BHI without hydrolysates (BHI-WH) as positive control were dispensed on 96-wells microtiter plates and inoculated with 5 µL of the bacterial suspension as previous prepared. Four wells for each strain and for each solution (BSPH, PSPH, CTRL and BHI-WH) were set up. The antibacterial assay was performed separately on separate microtiter plates for each sample and for each batch.

As blank samples, 100 µL BHI-hydrolysate and BHI-WH solutions before incubation were used. The microtiter plates were then incubated at 37 °C for 24 h and growth was measured automatically every 30 min at OD600 using a SPECTROstar nano microplate spectrophotometer reader (BMG Labtech, Ortenberg, Germany). Each growth curve was fitted by the primary model of Baranyi and Roberts [39] wrapped in DMFit Excel add-in [40], that was utilized also to evaluate the maximum specific growth rate (µ), the duration of lag phase (λ) according to Petretto et al. [38]. The 1000XOD absorbance values were log transformed to calculate the growth parameters with DMFIT add-in. Analysis of variance (ANOVA) was performed separately for each bacterial strain tested, using as factor the four treatments: BSPH, PSPH, CTRL, and BHI-WH to evaluate the influence of the two hydrolysates on the values of maximum specific growth rate (µmax) and lag phase (λ). When a significant effect was observed ($p < 0.05$), the differences between means were separated using the Tukey-Kramer multiple comparisons test. SPSS software, version 22, was used to conduct the statistical analyses.

2.7. Gel Permeation Chromatography

The freeze-dried scotta hydrolysates and control samples were reconstituted in water at 6 mg mL^{-1} and filtered on a 0.2 µm filter, then analyzed by Gel Permeation high performance liquid Chromatography (GPC) using an Agilent 1260 HPLC system equipped with a DAD detector. The separation was performed at 25 °C in isocratic mode with a mobile phase composed of 70:30 ACN:H_2O with 0.1% TFA, flowing continuously at

0.5 mL min^{-1} through a Phenomenex Yarra SEC-2000 column (300 × 7.8 mm; pore size 3 μm). Samples were filtered through 0.2 μm nylon filter and 20 μL were injected. The analytical signal was acquired for 30 min at 214 nm. A calibration of molecular weights (MW) was obtained by acquiring the retention volume of the following pure standards covering a MW range from 181.19 to 66,500 Da: Tyr, Asp-Glu, Leu-Trp-Met-Arg, bacitracin, aprotinin, α-La, β-Lg, BSA. The obtained linear model was adopted to determine the MW distribution of the hydrolysates. The results were expressed as relative abundance by summing the areas of the peaks detected at different molecular weight (obtained by the calibration curve) ranges (1 kDa, 1–5 kDa, 5–10 kDa and >10 kDa), as previously reported by [41]. Data were analyzed as described in Section 2.4.

A semi-preparative GPC step was further performed on the above-described samples by injecting 100 μL in the same conditions above described. The fraction corresponding to a calculated MW between 5000 and 330 Da was collected for further LC-MS/MS analysis.

2.8. LC-MS/MS Analysis

Fractions obtained from nine different semi-preparative GPC runs (three of CTRL, three of BSPH and three of PSPH hydrolysates) were brought to dryness and reconstituted in 0.2% formic acid. The peptide mixture concentration was estimated by measuring absorbance at 280 nm with a NanoDrop 2000 spectrophotometer (Thermo Scientific, San Jose, CA, USA), using dilutions of the MassPREP *E. coli* Digest Standard (Waters, Milford, MA, USA) to generate a calibration curve. Peptide concentration was adjusted to 1 μg μL^{-1}. Two technical replicates for each sample were performed.

LC-MS/MS analyses were carried out using a Q Exactive mass spectrometer (Thermo Scientific) interfaced with an UltiMate 3000 RSLCnano LC system (Thermo Scientific). After loading, peptide mixtures (4 μg per run) were concentrated and desalted on a trapping precolumn (Acclaim PepMap C18, 75 μm × 2 cm nanoViper, 3 μm, 100 Å, Thermo Scientific), using 0.2% formic acid at a flow rate of 5 μL min^{-1}. The peptide separation was performed at 35 °C using a C18 column (EASY-Spray column, 50 cm × 75 μm ID, PepMap C18, 3 μm, Thermo Scientific), using a linear gradient of 245 min from 5% to 37.5% of eluent B (0.1% formic acid in 80% acetonitrile) in eluent A (0.1% formic acid), at a flow rate of 250 nL min^{-1}. MS data were acquired using a data-dependent Top12 method dynamically choosing the most abundant precursor ions from the survey scan, under direct control of the Xcalibur software (version 1.0.2.65 SP2, Thermo Fisher Scientific), where a full-scan spectrum (from 300 to 1700 m/z) was followed by tandem mass spectra (MS/MS). The instrument was operated in positive mode with a spray voltage of 1.8 kV and a capillary temperature of 275 °C. Survey and MS/MS scans were performed in the Orbitrap with resolution of 70,000 and 17,500 at 200 m/z, respectively. The automatic gain control was set to 1,000,000 ions and the lock mass option was enabled on a protonated polydimethylcyclosiloxane background ion as an internal recalibration for accurate mass measurements. The dynamic exclusion was set to 30 s. Higher Energy Collisional Dissociation (HCD), performed at the far side of the C-trap, was used as fragmentation method by applying a 25 eV value for normalized collision energy. An isolation width of m/z 2.0. Nitrogen was used as the collision gas.

Peptide identification was performed using Proteome Discoverer (version 1.4; Thermo Scientific), with Sequest-HT as the search engine for protein identification, according to the following criteria: Database: custom, obtained by merging *Bos Taurus* and *Ovis aries* downloaded from UniprotKB, (release 2019_01); Precursor mass tolerance: 10 ppm; Fragment mass tolerance: 0.02 Da; Dynamic modification: methionine oxidation, Asparagine/Glutamine, Arginine deamidation, Serine/Threonine/Tyrosine phosphorylation), and Percolator for peptide validation (FDR < 1% based on peptide q-value). Results were filtered in order to keep only Rank 1 peptides, and protein grouping was allowed according to the maximum parsimony principle. Protein abundance was expressed by means of the normalized spectral abundance factor (NSAF). NSAF was calculated to evaluate the relative abundance of each protein and peptide, and "log ratio (log R)" values ($\log_2$ NSAF group a/NSAF group b) were obtained to estimate the fold changes of peptides between experimental groups expressed as base 2 on a logarithmic scale [42,43]. In this approach, the spectral counts of each peptide were divided by its length and normalized to the average number of spectral counts in a given analysis. In order to eliminate discontinuity due to SpC = 0, a correction factor, set to 0.01, was used.

Peptides showing log ratio >1.5 or <−1.5 were considered as differentially abundant between groups. A two-tailed *t*-test was applied, using in house software to evaluate the statistical significance of differences between groups. The differentially abundant peptides were then evaluated using the "profiles of potential biological activity" analysis, available on BIOPEP [44] to find within them any sequence with known DPP-IV inhibitory, antioxidant and antibacterial activity.

3. Results and Discussion

3.1. DPP-IV Inhibitory, Antioxidant Activity and GPC Profile of the Selected Hydrolysates

The DPP-IV inhibitory and ABTS activity of the obtained hydrolysates and of the control samples are showed in Table 2. The P32/98 positive control showed a IC_{50} of $1.395 \pm 0.007 \times 10^{-3}$ mg mL^{-1}. In contrast, the retentate control samples (CTRL) did not show a measurable inhibition. BSPH had a significantly higher DPP-IV inhibitory activity compared to PSPH (p = 0.0381) The obtained DPP-IV inhibitory activity was lower than that previously described for milk and whey hydrolysates obtained from WPC, WPI of different species [45–50]. However, the observed activity was just about one order of magnitude less than that measured on hydrolysates obtained from pure β-lactoglobulin [41]. In addition, a DPP-IV inhibitory activity on hydrolysates from scotta has never been measured before, leading to the consideration of this matrix as a candidate substrate for the industrial production of DPP-IV inhibitory peptides.

Table 2. DPP-IV and antioxidant activity of hydrolysates, and control [1].

Run	BSPH	PSPH	CTRL
DPP-IV IC_{50} (mg mL^{-1})	$8.5^{b} \pm 0.2$	$13^{a} \pm 1$	n.d.
ABTS IC_{50} (mg mL^{-1})	$0.79^{b} \pm 0.03$	$0.87^{ab} \pm 0.01$	$1.06^{a} \pm 0.18$

[1] Values are mean ± standard deviation (n = 3). Within rows, values with the same letter do not differ significantly from each other according to LSD test (p < 0.05). n.d.: absence of inhibition.

Further, the obtained data confirmed that enzymatic hydrolysis is a suitable way to increase the radical scavenging ability of sheep milk by-products [51]. In fact, BSPH showed a higher antioxidant activity compared to the control. Despite that, the hydrolysates did not differ regarding this property. A similar pattern was found by Monari et al. [15] for bovine scotta hydrolysates, which showed that the antioxidant activity of hydrolysates obtained using bromelain and pancreatin enzymes, did not differ significantly, even using different E:S ratios, both in unconcentrated bovine scotta and retentates.

Figure 1 shows the peptide distribution in terms of relative abundance according to the molecular weight obtained by gel permeation chromatography (GPC), and the comparison among BSPH, PSPH and CTRL.

As expected, most of BSPH and PSPH components were low molecular weight peptides (<1 kDa), whilst the high and medium molecular weight peptides (>10 kDa, 5–10 kDa, and 1–5 kDa) were more abundant in the control samples, which conversely showed a very low contribution of components with MW < 1 kDa (2.82 ± 0.21%). The amount of the 1–5 kDa fraction, even lower than the control, was significantly higher in BSPH (31 ± 0.9%) than in PSPH (23.56 ± 0.02%). Conversely, pancreatin in our system was more effective in producing peptides with low MW (74 ± 1%), compared to bromelain (66 ± 1%). Since pancreatin contains a mixture of proteases including chymotrypsin, trypsin and elastase [52], we suppose that in our system it exerted a more generalized proteolytic behavior than bromelain, which conversely has been reported to be less effective in producing free amino acids [15]. The specificity of bromelain may be responsible of the higher DPP-IV inhibitory activity measured in our hydrolysates.

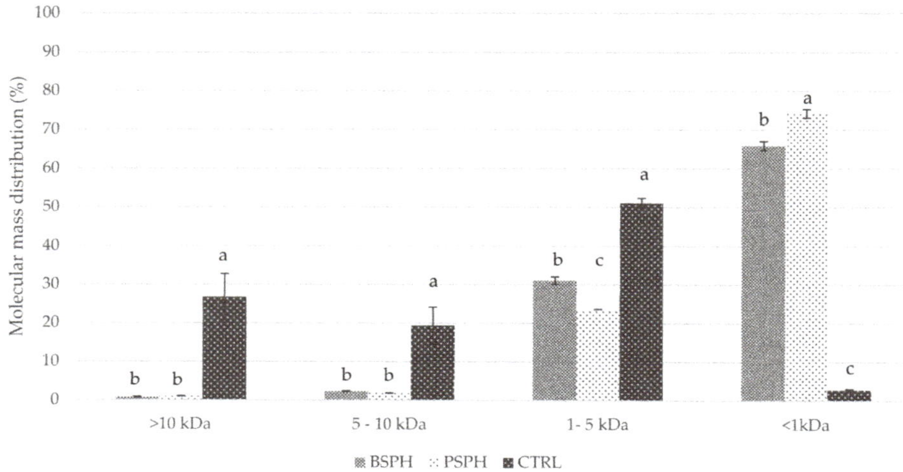

Figure 1. Distribution of the relative abundance (%) according to molecular weight obtained by gel permeation chromatography (GPC), and the comparison among BSPH, PSPH and CTRL. Values (n = 3) with the same letter do not differ significantly from each other according to LSD test ($p < 0.05$).

3.2. Antibacterial Activities of Hydrolysates

The enzymatic hydrolysates obtained from ovine scotta tested did not show a complete inhibitory or bactericidal effect on the target microorganisms at the concentration tested (100 mg mL^{-1}). However, the hydrolysates showed a variable but slightly inhibitory effect depending on the species or strain of bacteria tested. In particular, PSPH decreased significantly ($p < 0.001$), the maximum specific growth rate (μmax) respect to control being half the values in all bacteria tested except for *Salmonella bongori* 13,772 DSMZ strain, where the difference of its growth rate respect to control was not significant (see Figure 2A–F). The influence of BSPH was strain-dependent, decreasing significantly the μmax of *Listeria monocytogenes* B, *Listeria monocytogenes* D and *Staphylococcus aureus* 20,231 DSMZ, whereas the BSPH did not influence the μmax of *L. monocytogenes* 20,600 DSMZ, *L. monocytogenes* B and *Salmonella bongori* 13,772 DSMZ. An intriguing result was obtained with the CTRL, that decreased the μmax of *Listeria monocytogenes* B, *L. monocytogenes* C and *L. monocytogenes* E. All three strains were isolated from ovine ricotta cheese. Regarding the effect of scotta hydrolysates tested, no effect was observed on the lag time (λ) of *L. monocytogenes* C, *L. monocytogenes* E, *S. aureus* 20,231 DSMZ and *S. bongori* 13,772 DSMZ. An opposite effect of PSPH on lag time with respect to μmax was observed on *L. monocytogenes* 20,600 DSMZ (λ: 3.96 h) and *L. monocytogenes* B (λ: 1.42 h) strains, with a lag time for each bacterial strains that did not differ significantly from the control (3.83 h and 3.39 h for *L. monocytogenes* 20,600 DSMZ and *L. monocytogenes* B control respectively), but was significantly different from the other two treatments (BSPH and CTRL). Overall, all treatments reduced the bacterial density, confirming that scotta hydrolysates negatively influenced the growth of the bacteria tested. Contrasting with our results, Lestari et al. [36] revealed a strong antimicrobial activity of goat milk protein hydrolysate by using bromelain as a hydrolyzing agent. Indeed, the minimum inhibitory concentrations of these hydrolysates against *S. aureus* and *Escherichia coli* were below 100 ppm. Bovine β-LG and α-LA were previously treated with pancreatin, and the resulting hydrolysates were shown to possess antimicrobial activity [53].

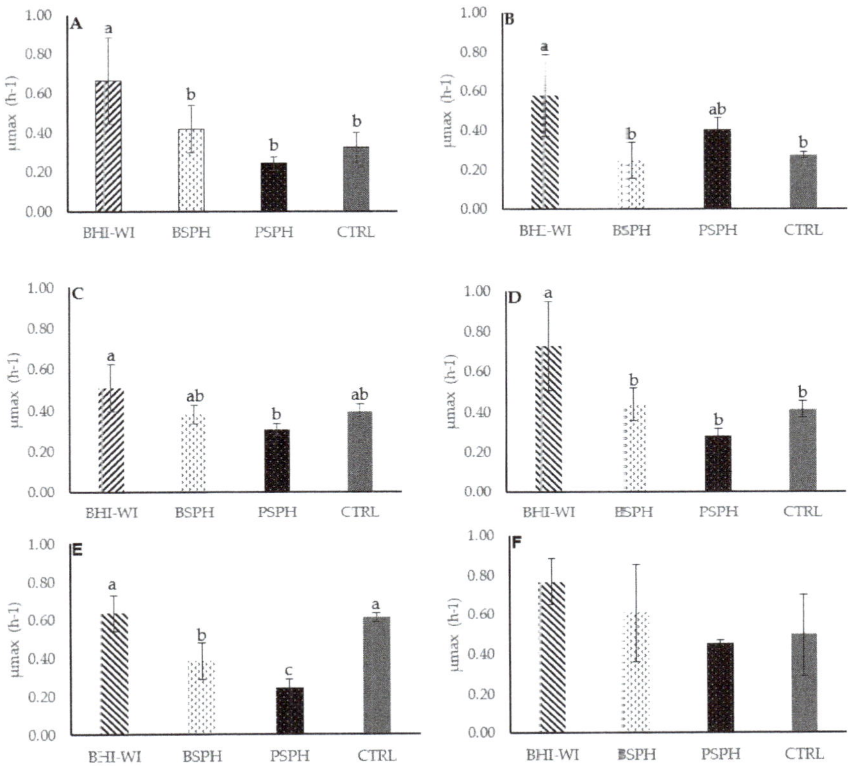

Figure 2. Effect of the different scotta-hydrolysates at concentration of 100 mg mL^{-1} on the maximum growth rate (μmax) of bacteria strains target. BHI-WH, Brain Heart infusion broth medium without hydrolysates; BSPH, Bromelain filter sterilized hydrolysate; PSPH, Pancreatin filter sterilized hydrolysate; CTRL, Scotta not hydrolysate filter sterilized). (Panel (**A–F**): *Listeria monocytogenes* B (**A**); *L. monocytogenes* C (**B**); *L. monocytogenes* 20,600 DSMZ (**C**); *L. monocytogenes* E (**D**); *Staphylococcus aureus* 20,231 DSMZ (**E**), *Salmonella bongori* 13,772 DSMZ (**F**)). Different lowercase letters above the bar indicate statistically significant differences between different treatments ($p < 0.001$).

3.3. LC-MS/MS Analysis of Scotta Hydrolysates

The LC-MS/MS analysis of the GPC fraction of the hydrolysates (BSPH, PSPH and CTRL) described in Section 2.8 allowed acquisition of 12,000 spectra from each run. A total of 58 ± 7 proteins and 547 ± 24 peptides were identified in BSPH, while 75 ± 6 proteins and 559 ± 33 peptides were identified in PSPH and further 83 ± 15 proteins and 1506 ± 381 peptides were identified in the CTRL samples (see Supplementary Materials, Sheet S1).

Considering BSPH vs. CTRL, a total of 29 proteins showed significant differences ($p \leq 0.05$). Twenty-one of them were more abundant in BSPH, while eight were more abundant in CTRL (see Supplementary Materials, Sheet S2). The differential analysis of BSPH vs. CTRL highlighted 751 significant peptides ($p \leq 0.05$). Of these, 388 were more abundant in BSPH samples, whilst 363 were more abundant in CTRL samples (see Supplementary Materials, Sheet S3).

Considering the available literature, differential peptides in BSPH were investigated to find sequences with reported DPP-IV inhibition, and antioxidant and antibacterial activity. The candidate peptides were further evaluated using the tool "profiles of potential biological activity" analysis, available on BIOPEP [44].

This approach highlighted 97 differential peptides containing at least one of the following sequences with known biological activity: LPQNI, VLGP, VLVLDTDYK, IPAVF, IPA, LKPTPEG, YPVEPF, YQEPVLGPVR, YVEEL, LDTDYKK, IDALNENK, KVAGT, AASDISLLDAGSAPLR, and ALK (see Table 3). All these peptides were attributable to β-lactoglobulin protein (P67976), except for LPQNI, VLGP, YPVEPF, YQEPVLGPVR derived from β-casein protein (P11839), and the tripeptide IPA originating from k-casein (P02669). In the following text and in the tables the active sequences contained in the identified peptides will be highlighted with bold characters.

The differential peptides within the DPP-IV sequence showed a length ranging from eight to twenty-eight amino acid residues. The shortest peptide was **LDTDYKK**Y from β-lactoglobulin with an estimated MW of 1044.52 Da (Log R = 2.38), whilst the longest was AIPPKKDQDKTE**IPA**INTIASAEPTVHS released from k-casein with an estimated MW of 3051.55 Da (Log R = 2.78).

Considering PSPH vs. CTRL, a total of 32 proteins showed a significant difference ($p \leq 0.05$). Twenty-five of them were more abundant in PSPH samples, while seven were more abundant in CTRL samples (see Supplementary Materials, Sheet S4). The peptide differential analysis of PSPH vs. CTRL highlighted 667 significant peptides ($p \leq 0.05$). Of these, 294 were differential in PSPH, and 373 were more abundant in the CTRL profile (see Supplementary Materials, Sheet S5).

The "profiles of potential biological activity" analysis on BIOPEP revealed 75 differential peptides containing at least one of the sequences previously observed in BSPH (see Table 4). The sequences with DPP-IV inhibitory activity were encrypted in peptides of 9 to 22 amino acid residues. In detail, K**IDALNENK** and **ALK**ALPMHI appeared the shortest peptides originating from β-lactoglobulin with an estimated MW of 1044.56 Da (Log R = 2.84) and MW of 992.59 Da (Log R = 2.03) respectively. Furthermore, the longer peptide identified was KDQDKTE**IPA**INTIASAEPTVH deriving from k-casein, with an estimated MW of 2884.55 Da (Log R =5.07).

Venn diagrams were used to evaluate the number of differential peptides, more abundant in BSPH vs. CRTL and PSPH vs. CRTL (Figure 3A,B, respectively), with potential biological activities.

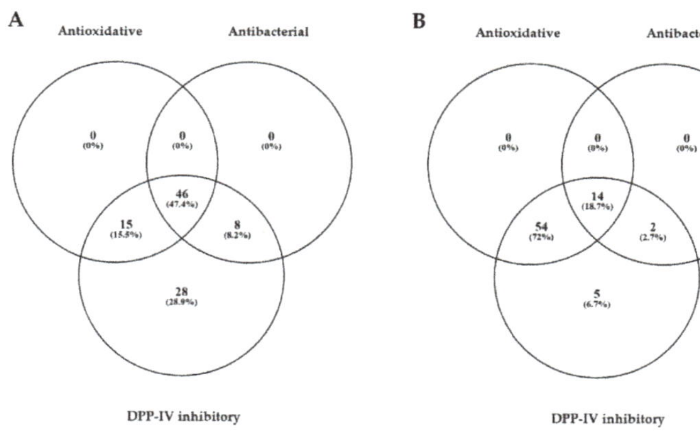

Figure 3. Distribution of the differential peptides more abundant in BSPH vs. CRTL (**A**) and PSPH vs. CRTL (**B**), according to their putative biological activities (DPP-IV inhibition, antioxidative and antibacterial properties).

Venn diagrams highlighted that none of the peptides contained sequences with only antibacterial or antioxidant known activity (Figure 3A,B). Interestingly, the 72% of the peptides in PSPH compared to the CTRL, contained sequences associated with antioxidant and DPP-IV inhibitory activity (Figure 3B). Moreover, a 5.6-fold higher number of peptides containing sequences with only DPP-IV inhibitory activity was found in BSPH vs. CTRL compared to PSPH vs. CTRL.

Furthermore, the differential analysis of BSPH vs. PSPH showed 82 proteins in total and 29 differentials ($p \leq 0.05$). Among them, 21 were more abundant in BSPH, while eight were more abundant in PSPH (see Supplementary Materials, Sheet S6). The peptide analysis indicated 1181 peptides, and 752 were significantly differential ($p \leq 0.05$). Among them, 388 peptides were more abundant in BSPH, while 364 were more abundant in PSPH (see Supplementary Materials, Sheet S7).

A total of 208 differential peptides contained sequences with known DPP-IV inhibitory activity (see Table 5).

Figure 4 groups differential peptides (BSPH vs. PSPH) generated by the same protein and dividing the components according to reported biological activity. Histograms show, for each protein (β-casein, k-casein, and β-lactoglobulin), the peptides more abundant in BSPH or in PSPH, respectively.

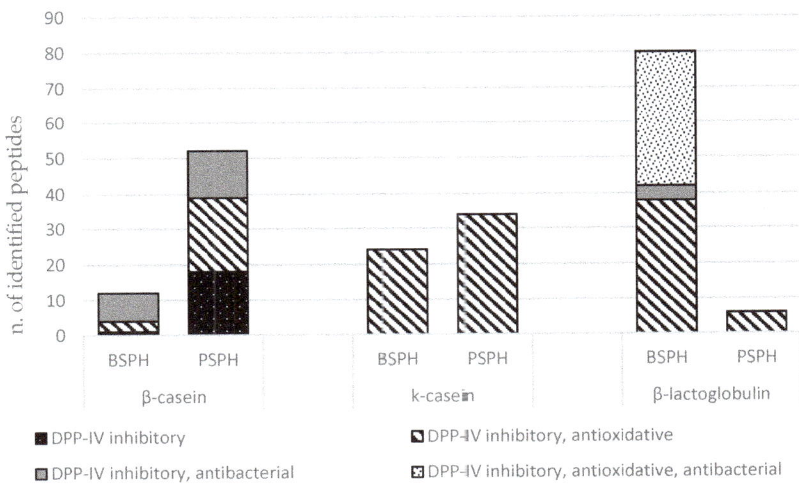

Figure 4. Number of differential peptides (BSPH vs. PSPH) grouped by the proteins and biological activities.

Sixty-four differential peptides were derived from β-casein, 52 of which were more abundant in PSPH, and 12 were more abundant in BSPH (Figure 4). Fifty-eight differential peptides were originated from k-casein 24, of which were more abundant in BSPH and 34 were more abundant in PSPH.

Interestingly, the number of differential peptides derived from β-lactoglobulin was differently distributed between BSPH and PSPH In fact, a total of 86 differential peptides derived from β-lactoglobulin were identified, 80 of them more abundant in BSPH and six in PSPH. This result could help to interpret the higher DPP-IV inhibitory activity showed in vitro by BSPH compared to PSPH. Moreover, none of the identified peptides derived from α-lactalbumin. Power et al. [50], found in silico a three-fold higher content of peptide sequences with potential DPP-IV inhibitory activity in β-lactoglobulin compared to α-lactalbumin. Furthermore, Tulipano et al. [54] found by in silico analysis that bovine β-lactoglobulin was a better source of DPP-IV inhibitory peptides compared with α-lactalbumin after treatment with digestive proteases.

Table 3. Analysis of differential peptides of BSPH vs. CTRL (Log R ≥ 1.5 and Log R ≤ −1.5).

ID Protein	Identified Peptide	Log_2 BSPH vs. CTRL	Activity	Reference
β-casein	GPIPNS**LPQNI**LPLT$^{79–94}$	3.82	Antioxidative; DPP-IV inhibitory;	[55,56]
	GPIPNS**LPQNI**LPLTQ$^{79–95}$	3.78		
β-casein	YQEP**VLGP**VR$^{206–215}$	2.1	Antioxidative; DPP-IV inhibitory;	[57–65]
β-lactoglobulin	**VLVLDTDYK**$^{110–118}$	2.21	Antioxidative; Antibacterial; DPP-IV inhibitory;	[66–73]
	VLVLDTDYKK$^{110–119}$	3.05		
	VLVLDTDYKKY$^{110–120}$	3.42		
	VLVLDTDYKKYL$^{110–121}$	2.72		
	VLVLDTDYKKYLL$^{110–122}$	2.17		
	K**VLVLDTDYK**KY$^{109–120}$	1.67		
	ENK**VLVLDTDYK**K$^{107–119}$	2.48		
	ENK**VLVLDTDYK**KY$^{107–120}$	3.45		
	ENK**VLVLDTDYK**KYL$^{107–121}$	2.18		
	DALENK**VLVLDTDYK**K$^{104–119}$	1.90		
	DALENK**VLVLDTDYK**KY$^{104–120}$	2.43		
	IDALENK**VLVLDTDYK**K$^{103–119}$	2.33		
	IDALENK**VLVLDTDYK**KY$^{103–120}$	3.05		
β-lactoglobulin	**IPAVF**KIDALNENK$^{96–109}$	2.43	Antioxidative; Antibacterial; DPP-IV inhibitory;	[26,67,68,71,74–76]
	TK**IPAVF**KIDALNENK$^{94–109}$	1.83		
k-casein	DQDKTE**IPA**INTIASAEPTVHS$^{134–155}$	5.40	DPP-IV inhibitory;	[26,41,75,77,78]
	AIPPKKDQDKTE**IPA**INT$^{128–145}$	5.39		
	E**IPA**INTIASAEPTVHS$^{139–155}$	5.28		
	IPPKKDQDKTE**IPA**INTIA$^{129–147}$	4.69		
	IPAINTIASAEPTVHS$^{140–155}$	4.62		
	IPPKKDQDKTE**IPA**IN$^{129–144}$	4.49		
	IPPKKDQDKTE**IPA**INT$^{129–145}$	4.46		
	PPKKDQDKTE**IPA**INTIAS$^{130–148}$	3.96		
	AIPPKKDQDKTE**IPA**INTIA$^{128–147}$	3.90		
	AIPPKKDQDKTE**IPA**IN$^{128–144}$	3.84		
	KDQDKTE**IPA**INT$^{132–145}$	3.59		
	KDQDKTE**IPA**INTIA$^{132–147}$	3.46		
	E**IPA**INTIASAEPTVH$^{139–154}$	3.37		
	DQDKTE**IPA**INTIAS$^{134–148}$	3.29		
	IPAINTIASAEPTVH$^{140–154}$	3.22		
	DQDKTE**IPA**INTIASAEPTVH$^{144–154}$	3.14		
	TE**IPA**INTIASAEPTVHS$^{138–155}$	2.99		
	PPKKDQDKTE**IPA**InTIASAEP$^{130–151}$	2.87		
	AIPPKKDQDKTE**IPA**INTIASAEPTVHS$^{128–155}$	2.78		
	DQDKTE**IPA**INTI$^{134–146}$	2.59		
	KDQDKTE**IPA**INTIASAEPTVHS$^{133–155}$	2.33		
	PPKKDQDKTE**IPA**INTIA$^{130–147}$	2.31		
	KDQDKTE**IPA**IN$^{133–144}$	1.84		
	DQDKTE**IPA**INTIA$^{134–147}$	1.50		
β-lactoglobulin	VYVEE**LKPTPEG**$^{59–70}$	2.10	Antioxidative; DPP-IV inhibitory;	[31]
	E**LKPTPEG**NLEILLQ$^{63–77}$	2.03		
	EE**LKPTPEG**NL$^{62–72}$	1.99		
	VYVEE**LKPTPEG**NL$^{59–72}$	1.74		
	VYVEE**LKPTPEG**NLE$^{59–73}$	1.69		
	EE**LKPTPEG**NLEILL$^{62–76}$	1.67		

Table 3. *Cont.*

ID Protein	Identified Peptide	Log_2 BSPH vs. CTRL	Activity	Reference
β-casein	EMPFPK**YPVEPFT**$^{123–135}$	4.23	Antibacterial; DPP-IV inhibitory;	[27,65,79–81]
	MPFPK**YPVEPF**TE$^{124–136}$	4.01		
	EMPFPK**YPVEPF**TE$^{123–136}$	3.96		
	MPFPK**YPVEPFT**$^{124–135}$	3.90		
	MPFPK**YPVEPF**TES$^{124–137}$	3.39		
	EMPFPK**YPVEPF**TES$^{123–137}$	3.15		
	MPFPK**YPVEPF**$^{124–134}$	1.93		
	EMPFPK**YPVEPF**$^{123–134}$	1.90		
β-casein	**YQEPVLGPVR**$^{208–212}$	2.10	DPP-IV inhibitory; Antioxidative;	[58,60,62,64]
β-lactoglobulin	**VYVEEL**KPTPEG$^{59–70}$	2.10	DPP-IV inhibitory; Antioxidative; Antibacterial	[24,82,83]
	VYVEELKPTPEGNL$^{59–72}$	1.74		
	VYVEELKPTPEGNLE$^{59–73}$	1.69		
β-lactoglobulin	ENKVLV**LDTDYKK**Y$^{107–120}$	3.45	Antioxidative; Antibacterial; DPP-IV inhibitory;	[44,84]
	VLV**LDTDYKK**Y$^{111–120}$	3.42		
	LDTDYKKYL$^{113–121}$	3.18		
	VLV**LDTDYKK**$^{112–120}$	3.05		
	IDALNENKVLV**LDTDYKK**Y$^{102–120}$	3.05		
	LV**LDTDYKK**Y$^{111–120}$	2.94		
	VLV**LDTDYKK**YL$^{112–121}$	2.72		
	V**LDTDYKK**Y$^{112–120}$	2.62		
	LV**LDTDYKK**YL$^{111–121}$	2.52		
	LDTDYKKYLL$^{113–122}$	2.48		
	ENKVLV**LDTDYKK**$^{107–119}$	2.48		
	DALNENKVLV**LDTDYKK**Y$^{103–120}$	2.43		
	LDTDYKKY$^{113–120}$	2.38		
	IDALNENKVLV**LDTDYKK**$^{102–119}$	2.33		
	LV**LDTDYKK**$^{111–119}$	2.23		
	ENKVLV**LDTDYKK**YL$^{107–121}$	2.18		
	VLV**LDTDYKK**YLL$^{112–122}$	2.17		
	DALNENKVLV**LDTDYKK**$^{103–119}$	1.90		
	KVLV**LDTDYKK**Y$^{109–120}$	1.67		
β-lactoglobulin	IPAVFK**IDALNENK**$^{96–109}$	3.09	DPP-IV inhibitory; Antioxidative; Antibacterial;	[44,85]
	IDALNENKVLVLDTDYKKY$^{102–120}$	3.05		
	IDALNENKVL$^{102–111}$	2.60		
	TKIPAVFK**IDALNENK**$^{94–109}$	2.46		
	IDALNENKVLVLDTDYKK$^{102–119}$	2.33		
	KIPAVFK**IDALNENK**$^{95–109}$	2.30		
	K**IDALNENK**$^{101–109}$	2.21		
	VFK**IDALNENK**$^{99–109}$	1.99		
	IDALNENKV$^{102–110}$	1.96		
β-lactoglobulin	LDIQ**KVAGT**WHS$^{27–39}$	1.92	DPP-IV inhibitory; Antioxidative;	[86]
β-lactoglobulin	**AASDISLLDAQSAPLR**$^{43–59}$	2.57	DPP-IV inhibitory;	[68,71]
	LAMA**ASDISLLDAQSAPLR**$^{39–59}$	2.37		
	MAASDISLLDAQSAPLR$^{42–59}$	1.88		
	AASDISLLDAQSAPLRV$^{43–59}$	1.53		
β-lactoglobulin	TPEVDNEALEKFDK**ALK**$^{143–159}$	4.03	DPP-IV inhibitory; Antioxidative;	[76]
	TPEVDNEALEKFDK**ALK**A$^{142–160}$	2.92		
	DNEALEKFDK**ALK**$^{147–159}$	2.38		
	EVDNEALEKFDK**ALK**$^{145–159}$	2.01		

The active sequences contained in longer peptides are highlighted with bold characters.

Table 4. Analysis of differential peptides of PSPH vs. CTRL (Log R ≥ 1.5 and Log R ≤ −1.5).

ID Protein	Identified Peptide	Log$_2$ PSPH vs. CTRL	Activity	Reference
β-casein	TGPIPNS**LPQNI**LPL$^{78–92}$	2.53	DPP-IV inhibitory; Antioxidative;	[55,56]
β-casein	QEP**VLGP**VRGPFPI$^{207–220}$	4.06	DPP-IV inhibitory;	[57–65]
	QEP**VLGP**VRGPFP$^{207–219}$	3.44		
	YQEP**VLGP**VRGPFPI$^{206–220}$	2.45		
	LYQEP**VLGP**VRGPFPI$^{205–220}$	1.91		
	EP**VLGP**VRGPFPI$^{208–220}$	1.85		
β-lactoglobulin	KIDALNENK**VLVLDTDYK**$^{101–118}$	2.83	DPP-IV inhibitory; Antioxidative; Antibacterial;	[66–73]
	KIDALNENK**VLVLDTDYKK**$^{101–119}$	2.50		
	VLVLDTDYKKY$^{112–120}$	2.20		
	IDALNENK**VLVLDTDYKK**$^{102–119}$	2.04		
	VLVLDTDYKKYL$^{112–121}$	1.93		
	KIDALNENK**VLVLDTDYK**KY$^{101–120}$	1.57		
β-lactoglobulin	**IPAVF**KIDALNENK$^{96–109}$	1.76	DPP-IV inhibitory; Antioxidative; Antibacterial;	[26,67,68,71,74–76]
k-casein	KDQDKTE**IPA**INTIASAEPT$^{133–152}$	5.99	DPP-IV inhibitory; Antioxidative;	[26,41,75,77,78]
	KDQDKTE**IPA**INT$^{133–144}$	5.85		
	TE**IPA**INTIASAEPTVH$^{138–154}$	5.17		
	KDQDKTE**IPA**INTIA$^{133–146}$	5.13		
	KDQDKTE**IPA**INTIASAEPTVH$^{133–154}$	5.07		
	DQDKTE**IPA**INTIASAEPT$^{134–152}$	5.05		
	DQDKTE**IPA**INTIASAEPTVH$^{134–154}$	5.04		
	KDQDKTE**IPA**IN$^{133–143}$	4.39		
	KDQDKTE**IPA**INTIAS$^{133–147}$	3.96		
	KDQDKTE**IPA**I$^{133–142}$	3.92		
	E**IPA**INTIASAEPTVH$^{139–154}$	3.91		
	KDQDKTE**IPA**INTI$^{133–145}$	3.90		
	DQDKTE**IPA**INTIAS$^{134–147}$	3.65		
	DQDKTE**IPA**INTIA$^{134–146}$	3.42		
	DQDKTE**IPA**INTI$^{134–145}$	3.32		
	MAIPPKKDQDKTE**IPA**$^{127–142}$	2.95		
	AIPPKKDQDKTE**IPA**IN$^{128–144}$	2.70		
	AIPPKKDQDKTE**IPA**INTIA$^{128–147}$	2.64		
	PPKKDQDKTE**IPA**IN$^{130–144}$	2.32		
	MAIPPKKDQDKTE**IPA**INTIA$^{127–147}$	2.09		
	AIPPKKDQDKTE**IPA**$^{128–142}$	1.68		
β-lactoglobulin	VEE**LKPTPEG**NLE$^{61–73}$	3.51	DPP-IV inhibitory; Antioxidative;	[31]
	VEE**LKPTPEG**NLEI$^{61–74}$	3.35		
	VEE**LKPTPEG**NLEILLQK$^{61–78}$	2.97		
	VEE**LKPTPEG**NLEIL$^{61–76}$	2.96		
	YVEE**LKPTPEG**NLE$^{60–73}$	2.89		
	VEE**LKPTPEG**DLE	2.64		
	VYVEE**LKPTPEG**N$^{59–71}$	2.64		
	VYVEE**LKPTPEG**NLE$^{59–73}$	2.63		
	YVEE**LKPTPEG**N$^{60–70}$	2.61		
	YVEE**LKPTPEG**NLEI$^{59–74}$	2.45		
	YVEE**LKPTPEG**NLEILLQK$^{59–78}$	2.28		
	YVEE**LKPTPEG**NLEIL$^{59–75}$	2.27		
	VYVEE**LKPTPEG**NLEILLQK$^{58–78}$	2.02		
	VEE**LKPTPEG**NL$^{60–72}$	1.93		
	RVYVEE**LKPTPEG**NLEILLQK$^{58–78}$	1.87		
	VYVEE**LKPTPEG**NL$^{58–72}$	1.76		
β-casein	EMPFPK**YPVEPF**$^{129–134}$	1.93	DPP-IV inhibitory; Antibacterial;	[27,65,79–81]

Table 4. *Cont.*

ID Protein	Identified Peptide	Log2 PSPH vs. CTRL	Activity	Reference
β-casein	**YQEPVLGPVR**GPFPI$^{208-217}$	2.45	DPP-IV inhibitory; Antioxidative;	[58,60,62,64]
	LYQEPVLGPVRGPFPI$^{206-215}$	1.91		
β-lactoglobulin	**VYVEEL**KPTPEGN$^{59-71}$	2.64	DPP-IV inhibitory; Antioxidative;	[24,82,83]
	VYVEELKPTPEGNLE$^{59-73}$	2.63		
	VYVEELKPTPEGNLEILLQK$^{59-78}$	2.02		
	R**VYVEEL**KPTPEGNLEILLQK$^{58-78}$	1.87		
	VYVEELKPTPEGNL$^{59-72}$	1.76		
β-lactoglobulin	KIDALNENKVLV**LDTDYKK**$^{101-119}$	2.50	DPP-IV inhibitory; Antibacterial; Antioxidative;	[44,84]
	VLV**LDTDYKK**Y$^{112-120}$	2.20		
	V**LDTDYKK**YL$^{112-121}$	2.12		
	IDALNENKVLV**LDTDYKK**$^{102-119}$	2.04		
	VLV**LDTDYKK**YL$^{112-121}$	1.93		
	KIDALNENKVLV**LDTDYKK**Y$^{101-120}$	1.57		
β-lactoglobulin	**KIDALNENKV**$^{101-110}$	3.39	DPP-IV inhibitory; Antioxidative;	[44,85]
	KIDALNENK$^{101-109}$	2.84		
	KIDALNENKVLVLDTDYK$^{101-118}$	2.83		
	KIDALNENKVLVLDTDYKK$^{101-119}$	2.50		
	IDALNENKVLVLDTDYKK$^{102-119}$	2.04		
	IPAVFK**IDALNENK**$^{96-109}$	1.76		
	KIDALNENKVLVLDTDYKKY$^{100-120}$	1.57		
β-lactoglobulin	GLDIQ**KVAGTW**H$^{27-38}$	1.73	DPP-IV inhibitory; Antioxidative;	[86]
β-lactoglobulin	SLAM**AASDISLLDAQSAPLR**V$^{39-59}$	2.56	DPP-IV inhibitory; Antibacterial;	[68,71]
	SLAM**AASDISLLDAQSAPLR**$^{39-58}$	2.21		
β-lactoglobulin	**ALK**ALPMHI$^{157-165}$	2.03	DPP-IV inhibitory; Antioxidative;	[76]

The active sequences contained in longer peptides are highlighted with bold characters.

Table 5. Analysis of differential peptides of BSPH vs. PSPH (Log R ≥ 1.5 and Log R ≤ −1.5).

ID Protein	Identified Peptide	Log2 BSPH vs. PSPH	Activity
β-casein	GPIPNS**LPQNI**LPLT$^{79-93}$	3.82	DPP-IV inhibitory; Antioxidative;
	GPIPNS**LPQNI**LPLTQ$^{79-94}$	3.78	
	LVYPFTGPIPNS**LPQNI**LPLTQTPVVVPPFLQPEIMGVPK$^{73-112}$	−1.54	
	S**LPQNI**LPLTQTPVVVPPFLQPEIMGVPKVKET$^{72-116}$	−1.62	
	TGPIPNS**LPQNI**LPLTQTPVVVPPFLQPEIMGVPKVKETMVPKH$^{78-121}$	−1.69	
	S**LPQNI**LPLTQTPVVVPPFLQPEIMGVPKVKETMVPKH$^{72-121}$	−1.91	
	S**LPQNI**LPLTQTPVVVPPFLQPEIMGVPKVK$^{72-114}$	−1.92	
	S**LPQNI**LPLTQTPVVVPPFLQPEIMGVPK$^{72-120}$	−1.98	
	FTGPIPNS**LPQNI**LPLTQTPVVVPPFLQPEIMGVPKVKETMVPKH$^{77-121}$	−2.44	
	FTGPIPNS**LPQNI**LPLTQTPVVVPPFLQPEIMGVPKVKETMVPK$^{77-120}$	−3.44	
	S**LPQNI**LPLTQTPVVVPPFLQPEIMGVPKVKETMVPK$^{72-120}$	−3.47	
	TGPIPNS**LPQNI**LPLTQTPVVVPPFLQPEIMGVPKVKETMVPK$^{78-120}$	−3.53	
β-casein	YQEP**VLGP**VR$^{206-215}$	2.10	DPP-IV inhibitory;
	YQEP**VLGP**VRGPFP$^{206-219}$	−1.61	
	VLPVPQKAVPQRDMPIQAFLLYQEP**VLGP**VRGPFP$^{185-219}$	−1.64	
	LSLSQPKVLPVPQKAVPQRDMPIQAFLLYQEP**VLGP**V$^{178-214}$	−1.71	
	AVPQRDMPIQAFLLYQEP**VLGP**VRGPFPI$^{192-220}$	−1.75	
	SLSQPKVLPVPQKAVPQRDMPIQAFLLYQEP**VLGP**VRGPFPILV$^{179-222}$	−1.77	
	AVPQRDMPIQAFLLYQEP**VLGP**VRGPFP$^{192-219}$	−1.80	
	EP**VLGP**VRGPFPIIV$^{208-222}$	−1.89	
	EP**VLGP**VRGPFPILV$^{208-222}$	−1.89	
	EP**VLGP**VRGPFPI$^{208-220}$	−2.01	
	FLLYQEP**VLGP**VRGPFP$^{203-219}$	−2.10	

Table 5. *Cont.*

ID Protein	Identified Peptide	Log2 BSPH vs. PSPH	Activity
	VLPVPQKAVPQRDMPIQAFLLYQEP**VLGP**VRGPFPILV$^{185-222}$	−2.14	
	VLPVPQKAVPQRDMPIQAFLLYQEP**VLGP**VRGPFPI$^{185-220}$	−2.24	
	YQEP**VLGP**VRGPFPIIV$^{206-222}$	−2.31	
	YQEP**VLGP**VRGPFPILV$^{206-222}$	−2.31	
	VLPVPQKAVPQRDMPIQAFLLYQEP**VLGP**VRGPFPIL$^{185-221}$	−2.33	
	EP**VLGP**VRGPFPII$^{208-221}$	−2.59	
	EP**VLGP**VRGPFPIL$^{208-221}$	−2.59	
	EP**VLGP**VRGPFP$^{208-219}$	−2.98	
β-lactoglobulin	ENK**VLVLDTDYK**KY$^{107-118}$	3.45	DPP-IV inhibitory; Antioxidative; Antibacterial;
	VL**VLDTDYK**KY$^{110-120}$	3.42	
	VLVLDTDYKK$^{112-119}$	3.05	
	IDALNENK**VLVLDTDYK**KY$^{102-120}$	3.05	
	VLVLDTDYKKYL$^{112-120}$	2.72	
	ENK**VLVLDTDYK**K$^{107-119}$	2.48	
	DALNENK**VLVLDTDYK**KY$^{103-120}$	2.43	
	IDALNENK**VLVLDTDYK**K$^{102-119}$	2.33	
	VLVLDTDYK$^{112-118}$	2.21	
	ENKVL**VLDTDYK**KYL$^{107-121}$	2.18	
	VLVLDTDYKKYLL$^{112-120}$	2.17	
	DALNENK**VLVLDTDYK**K$^{103-119}$	1.90	
	K**VLVLDTDYK**KY$^{109-120}$	1.67	
β-lactoglobulin	**IPAVF**KIDALNENK$^{106-109}$	3.09	DPP-IV inhibitory; Antioxidative; Antibacterial;
	TK**IPAVF**KIDALNENK$^{104-109}$	2.46	
	K**IPAVF**KIDALNENK$^{105-109}$	2.30	
k-casein	DQDKTE**IPA**INTIASAEPTVHS$^{134-155}$	5.40	DPP-IV inhibitory; Antioxidative;
	AIPPKKDQDKTE**IPA**INT$^{128-145}$	5.39	
	E**IPA**INTIASAEPTVHS$^{139-155}$	5.28	
	IPPKKDQDKTE**IPA**INTIA$^{129-147}$	4.69	
	IPAINTIASAEPTVHS$^{140-155}$	4.62	
	IPPKKDQDKTE**IPA**IN$^{129-144}$	4.49	
	IPPKKDQDKTE**IPA**INT$^{129-145}$	4.46	
	PPKKDQDKTE**IPA**INTIAS$^{130-148}$	3.96	
	AIPPKKDQDKTE**IPA**INTIA$^{128-147}$	3.90	
	AIPPKKDQDKTE**IPA**IN$^{128-144}$	3.84	
	KDQDKTE**IPA**INT$^{133-145}$	3.59	
	KDQDKTE**IPA**INTIA$^{133-147}$	3.46	
	E**IPA**INTIASAEPTVH$^{139-154}$	3.37	
	DQDKTE**IPA**INTIAS$^{134-148}$	3.29	
	IPAINTIASAEPTVH$^{140-154}$	3.22	
	DQDKTE**IPA**INTIASAEPTVH$^{134-154}$	3.14	
	TE**IPA**INTIASAEPTVHS$^{138-155}$	2.99	
	PPKKDQDKTE**IPA**INTIASAEP$^{130-151}$	2.87	
	AIPPKKDQDKTE**IPA**INTIASAEPTVHS$^{128-155}$	2.78	
	DQDKTE**IPA**INTI$^{134-146}$	2.59	
	KDQDKTE**IPA**INTIASAEPTVHS$^{133-155}$	2.33	
	PPKKDQDKTE**IPA**INTIA$^{130-147}$	2.31	
	KDQDKTE**IPA**IN$^{133-144}$	1.84	
	DQDKTE**IPA**INTIA$^{134-147}$	1.50	
	FMAIPPKKDQDKTE**IPA**INTIASAEPTVH$^{126-154}$	−1.53	
	MAIPPKKDQDKTE**IPA**INTIASAEPTVHSTPTTEAVVNAVDNP$^{127-169}$	−1.71	
	KTE**IPA**INTIASAEPTVH$^{137-154}$	−1.94	
	MAIPPKKDQDKTE**IPA**INTIASAEPTVHSTPTTEAVV$^{127-163}$	−2.01	
	IPPKKDQDKTE**IPA**INTIASAEPTVH$^{129-154}$	−2.18	
	MAIPPKKDQDKTE**IPA**INTIASAEPTVHSTP$^{127-157}$	−2.36	
	MAIPPKKDQDKTE**IPA**INTIASAEPTVHSTPTTEAVVNAV$^{127-166}$	−2.37	
	MAIPPKKDQDKTE**IPA**INTIASAEPTV$^{127-153}$	−2.39	
	MAIPPKKDQDKTE**IPA**INTIASAEP$^{127-151}$	−2.42	
	MAIPPKKDQDKTE**IPA**INT$^{127-144}$	−2.66	
	MAIPPKKDQDKTE**IPA**INTIASAEPTVHSTPTTEAVVNA$^{127-165}$	−2.77	
	AIPPKKDQDKTE**IPA**INTIASAEPTVH$^{128-154}$	−2.98	
	MAIPPKKDQDKTE**IPA**INTIASAEPTVHSTPTTEAVVNAVDNPE$^{127-170}$	−3.04	
	PPKKDQDKTE**IPA**INTIASAEPTVHSTPTTEAVVNAVDNPEA$^{129-169}$	−3.16	
	MAIPPKKDQDKTE**IPA**INTIASAEPTVHSTPTTEAVVNAVDNPEA$^{127-169}$	−3.19	
	PPKKDQDKTE**IPA**INTIASAEPTV$^{129-153}$	−3.42	

Table 5. *Cont.*

ID Protein	Identified Peptide	Log_2 BSPH vs. PSPH	Activity
	MAIPPKKDQDKTE**IPA**INTIASAEPTVHST$^{127-156}$	−3.52	
	MAIPPKKDQDKTE**IPA**INTIASAEPTVHSTPTTEA$^{127-161}$	−3.69	
	MAIPPKKDQDKTE**IPA**INTIASAEPTVHSTPTT$^{127-159}$	−3.69	
	MAIPPKKDQDKTE**IPA**INTIASAEPTVHSTPTTEAVVN$^{127-164}$	−3.73	
	MAIPPKKDQDKTE**IPA**INTIASAEPT$^{127-152}$	−3.90	
	MAIPPKKDQDKTE**IPA**INTIASAEPTVHSTPTTEAV$^{127-162}$	−3.95	
	MAIPPKKDQDKTE**IPA**IN$^{127-144}$	−3.96	
	MAIPPKKDQDKTE**IPA**INTIAS$^{127-148}$	−4.01	
	MAIPPKKDQDKTE**IPA**INTIASAEPTVHSTPTTEAVVNAVDN$^{127-168}$	−4.04	
	MAIPPKKDQDKTE**IPA**INTIASAEPTVHSTPTTEAVVNAVDNPEASS$^{127-173}$	−4.11	
	MAIPPKKDQDKTE**IPA**INTIASAEPTVHSTPTTEAVVNAVDNPEAS$^{127-172}$	−4.28	
	MAIPPKKDQDKTE**IPA**INTIASAEPTVHSTPTTE$^{127-160}$	−4.40	
	MAIPPKKDQDKTE**IPA**INTIASAEPTVHSTPTTEAVVNAVD$^{127-167}$	−4.64	
	MAIPPKKDQDKTE**IPA**INTIASA$^{127-149}$	−4.80	
	MAIPPKKDQDKTE**IPA**INTIASAEPTVHSTPT$^{127-158}$	−4.81	
	MAIPPKKDQDKTE**IPA**INTIASAE$^{127-150}$	−4.82	
	MAIPPKKDQDKTE**IPA**INTIASAEPTVHS$^{127-155}$	−5.60	
	MAIPPKKDQDKTE**IPA**INTIASAEPTVH$^{127-154}$	−6.28	
β-lactoglobulin	VYVEE**LKPTPEG**$^{59-70}$	2.10	
	E**LKPTPEG**NLEILLQ$^{63-77}$	2.03	
	EE**LKPTPEG**NL$^{62-72}$	1.99	
	VYVEE**LKPTPEG**NL$^{59-72}$	1.74	
	VYVEE**LKPTPEG**NLE$^{59-73}$	1.69	
	EE**LKPTPEG**NLEILL$^{62-76}$	1.67	
β-casein	EMPFPK**YPVEPFT**$^{122-135}$	4.23	DPP-IV inhibitory; Antibacterial;
	MPFPK**YPVEPF**TE$^{123-136}$	4.01	
	EMPFPK**YPVEPF**TE$^{122-136}$	3.96	
	MPFPK**YPVEPFT**$^{123-135}$	3.90	
	MPFPK**YPVEPF**TES$^{123-137}$	3.39	
	EMPFPK**YPVEPF**TES$^{122-137}$	3.15	
	MPFPK**YPVEPF**$^{123-134}$	1.93	
	EMPFPK**YPVEPF**$^{122-134}$	1.90	
	VKETMVPKHKEMPFPK**YPVEPF**TESQSLTLTDVE$^{113-156}$	−1.50	
	HKEMPFPK**YPVEPF**TESQ$^{121-138}$	−1.57	
	HKEMPFPK**YPVEPF**TESQSLTLTDVEKLH$^{121-149}$	−1.59	
	HKEMPFPK**YPVEPF**TESQSLT$^{121-141}$	−1.64	
	HKEMPFPK**YPVEPF**TESQSLTLTDVE$^{121-146}$	−1.73	
	HKEMPFPK**YPVEPF**TESQSLTLTDVEKLHLPLPLVQ$^{121-156}$	−1.74	
	HKEMPFPK**YPVEPF**TESQS$^{121-138}$	−1.75	
	HKEMPFPK**YPVEPF**TESQSL$^{121-139}$	−1.90	
	VKETMVPKHKEMPFPK**YPVEPF**TESQSL$^{113-140}$	−2.55	
	HKEMPFPK**YPVEPF**TESQSLTLTDVEK$^{121-147}$	−2.66	
	VKETMVPKHKEMPFPK**YPVEPF**TESQS$^{113-139}$	−2.67	
	EMPFPK**YPVEPF**TESQSLTLTDVEKLHLPLP$^{122-153}$	−2.75	
	HKEMPFPK**YPVEPF**TESQSLTLTDVEKLHLPLP$^{121-153}$	−3.66	
β-casein	**YQEPVLGPVR**$^{206-215}$	2.10	DPP-IV inhibitory; Antioxidative;
	YQEPVLGPVRGPFP$^{206-219}$	−1.61	
	VLPVPQKAVPQRDMPIQAFLL**YQEPVLGPVR**GPFP$^{185-219}$	−1.64	
	AVPQRDMPIQAFLL**YQEPVLGPVR**GPFPI$^{192-220}$	−1.75	
	SLSQFKVLPVPQKAVPQRDMPIQAFLL**YQEPVLGPVR**GPFPILV$^{178-222}$	−1.77	
	AVPQRDMPIQAFLL**YQEPVLGPVR**GPFP$^{192-219}$	−1.80	
	FLL**YQEPVLGPVR**GPFP$^{203-219}$	−2.10	
	VLPVPQKAVPQRDMPIQAFLL**YQEPVLGPVR**GPFPILV$^{185-222}$	−2.14	
	VLPVPQKAVPQRDMPIQAFLL**YQEPVLGPVR**GPFPI$^{185-220}$	−2.24	
	YQEPVLGPVRGPFPIIV$^{206-222}$	−2.31	
	YQEPVLGPVRGPFPILV$^{206-222}$	−2.31	
	VLPVPQKAVPQRDMPIQAFLL**YQEPVLGPVR**GPFPIL$^{185-220}$	−2.33	
β-lactoglobulin	**VYVEEL**KPTPEG$^{59-70}$	2.10	DPP-IV inhibitory; Antibacterial; Antioxidative;
	VYVEELKPTPEGNL$^{59-72}$	1.74	
	VYVEELKPTPEGNLE$^{59-73}$	1.69	

Table 5. *Cont.*

ID Protein	Identified Peptide	Log_2 BSPH vs. PSPH	Activity
β-lactoglobulin	ENKVLV**LDTDYKK**Y$^{107-120}$	3.45	DPP-IV inhibitory; Antibacterial; Antioxidative;
	VLV**LDTDYKK**Y$^{110-120}$	3.42	
	LDTDYKKYL$^{113-121}$	3.18	
	VLV**LDTDYKK**$^{110-119}$	3.05	
	IDALNENKVLV**LDTDYKK**Y$^{102-120}$	3.05	
	LV**LDTDYKK**Y$^{111-120}$	2.94	
	VLV**LDTDYKK**YL$^{110-121}$	2.72	
	V**LDTDYKK**Y$^{112-120}$	2.62	
	LV**LDTDYKK**YL$^{111-121}$	2.52	
	LDTDYKKYLL$^{113-122}$	2.48	
	ENKVLV**LDTDYKK**$^{107-119}$	2.48	
	DALNENKVLV**LDTDYKK**Y$^{103-120}$	2.43	
	LDTDYKKY$^{113-120}$	2.38	
	IDALNENKVLV**LDTDYKK**$^{102-119}$	2.33	
	LV**LDTDYKK**$^{111-119}$	2.23	
	ENKVLV**LDTDYKK**YL$^{107-121}$	2.18	
	VLV**LDTDYKK**YLL$^{109-122}$	2.17	
	DALNENKVLV**LDTDYKK**$^{103-119}$	1.90	
	KVLV**LDTDYKK**Y$^{109-120}$	1.67	
β-lactoglobulin	IPAVFK**IDALNENK**$^{106-109}$	3.09	DPP-IV inhibitory; Antioxidative;
	IDALNENKVLVLDTDYKKY$^{102-120}$	3.05	
	IDALNENKVL$^{102-111}$	2.60	
	TKIPAVFK**IDALNENK**$^{94-109}$	2.46	
	IDALNENKVLVLDTDYKK$^{102-119}$	2.33	
	KIPAVFK**IDALNENK**$^{95-109}$	2.30	
	K**IDALNENK**$^{101-109}$	2.21	
	VFK**IDALNENK**$^{99-109}$	1.99	
	IDALNENKV$^{102-110}$	1.96	
β-lactoglobulin	IPAVFK**IDALNENK**$^{106-109}$	3.09	DPP-IV inhibitory; Antioxidative;
	IDALNENKVLVLDTDYKKY$^{102-120}$	3.05	
	IDALNENKVL$^{102-111}$	2.60	
	TKIPAVFK**IDALNENK**$^{94-109}$	2.46	
	IDALNENKVLVLDTDYKK$^{102-119}$	2.33	
	KIPAVFK**IDALNENK**$^{95-109}$	2.30	
	K**IDALNENK**$^{101-109}$	2.21	
	VFK**IDALNENK**$^{99-109}$	1.99	
	IDALNENKV$^{102-110}$	1.96	
β-lactoglobulin	IPAVFK**IDALNENK**$^{106-109}$	3.09	DPP-IV inhibitory; Antioxidative;
	IDALNENKVLVLDTDYKKY$^{102-120}$	3.05	
	IDALNENKVL$^{102-111}$	2.60	
	TKIPAVFK**IDALNENK**$^{94-109}$	2.46	
	IDALNENKVLVLDTDYKK$^{102-119}$	2.33	
	KIPAVFK**IDALNENK**$^{95-109}$	2.30	
	K**IDALNENK**$^{101-109}$	2.21	
	VFK**IDALNENK**$^{99-109}$	1.99	
	IDALNENKV$^{102-110}$	1.96	
β-lactoglobulin	LDIQ**KVAGT**WHS$^{28-39}$	1.92	DPP-IV inhibitory; Antioxidative;
	IIVTQTMKGLDIQ**KVAGT**WH$^{19-38}$	−2.06	
β-lactoglobulin	**AASDISLLDAQSAPLR**$^{43-58}$	2.57	DPP-IV inhibitory; Antibacterial;
	LAMA**ASDISLLDAQSAPLR**$^{40-58}$	2.37	
	M**AASDISLLDAQSAPLR**$^{42-58}$	1.88	
	AASDISLLDAQSAPLRV$^{43-59}$	1.53	
β-lactoglobulin	TPEVDNEALEKFDK**ALK**$^{143-159}$	4.03	DPP-IV inhibitory; Antioxidative;
	TPEVDNEALEKFDK**ALK**A$^{143-160}$	2.92	
	DNEALEKFDK**ALK**$^{147-159}$	2.38	
	EVDNEALEKFDK**ALK**$^{145-159}$	2.01	
	NEALEKFDK**ALK**$^{148-159}$	−1.62	
	EALEKFDK**ALK**ALPMH$^{149-164}$	−1.77	
	NEALEKFDK**ALK**ALPMH$^{148-164}$	−2.37	
	NEALEKFDK**ALK**ALPMHIR$^{148-166}$	−2.65	
	EALEKFDK**ALK**ALPMHIR$^{149-166}$	−2.96	

The active sequences contained in longer peptides are highlighted with bold characters.

4. Conclusions

Due to its higher content of nutrients compared to other species, ovine scotta is a precious substrate that can be valorized through a multidisciplinary biotechnological approach with the aim of producing ingredients with specific biological activities. The enzymatic hydrolyses performed both with bromelain and pancreatin on retentate of scotta allowed enhancement of its DPP-IV inhibitory and antioxidant activities, bromelain being more promising in such an aim. Likewise, the antibacterial activity of hydrolysates slightly increased with respect to control, even if an inhibitory effect against some Listeria monocytogenes strains of the non-hydrolysates scotta was also noticed. LC-MS/MS analysis allowed identification among the experimental groups of several differential peptides that contain sequences with known activities among those here studied. Further studies are needed to optimize reaction conditions, in order to maximize such biological activities in relation to the specific objective.

Supplementary Materials: The following are available online at https://www.mdpi.com/2304-8158/10/12/3137/s1, Sheet S1: All identified peptides in BSPH, PSHP, and CTRL. Sheet S2: Differential proteins analysis of BSPH vs. CTRL. Sheet S3: Differential peptides analysis of BSPH vs. CTRL. Sheet S4: Differential proteins analysis of PSPH vs. CTRL. Sheet S5: Differential peptides analysis of PSPH vs. CTRL. Sheet S6: Differential proteins analysis of BSPH vs. PSPH. Sheet S7: Differential peptides analysis of BSPH vs. PSPH.

Author Contributions: Conceptualization, P.P.U.; methodology, P.P.U., R.C., F.F., G.L.P.; formal analysis, R.C., F.F., P.P.U., S.P.; investigation, R.C., F.F., R.A., S.P.; resources, P.P.U.; writing—original draft preparation, R.C., P.P.U., F.F., G.L.P.; writing—review and editing, P.P.U., R.C., G.L.P., F.F., D.P., S.P.; supervision, P.P.U. and A.P.; project administration, P.P.U.; funding acquisition, P.P.U. All authors have read and agreed to the published version of the manuscript.

Funding: This research was funded by Regione Autonoma della Sardegna-Legge Regionale 7 agosto 2007, n.7, Bando Capitale Umano ad Alta Qualificazione, annualità 2015 and by Fondo dell'Ateneo di Sassari per la ricerca 2020.

Data Availability Statement: The peptides identified in this study were analyzed for potential activity with the tool available on BIOPEP-UWM database at the following link: https://biochemia.uwm.edu.pl/en/biopep-uwm-2/.

Acknowledgments: The authors would like to thank Nutraceutica SRL and Enzyme Development Corporation for the supply of enzymes, and Caseificio F.lli Pinna (Thiesi, SS) for the scotta samples used in this study.

Conflicts of Interest: The authors declare no conflict of interest.

References

1. Salvatore, E.; Pes, M.; Falchi, G.; Pagnozzi, D.; Furesi, S.; Fiori, M.; Roggio, T.; Addis, M.F.; Pirisi, A. Effect of whey concentration on protein recovery in fresh ovine ricotta cheese. *J. Dairy Sci.* **2014**, *97*, 4686–4694. [CrossRef] [PubMed]
2. Pires, A.F.; Marnotes, N.G.; Rubio, O.D.; Garcia, A.C.; Pereira, C.D. Dairy by-Products: A Review on the Valorization of Whey and Second Cheese Whey. *Foods* **2021**, *10*, 1067. [CrossRef]
3. ISTAT Latte e Prodotti Lattiero Caseari: Prodotti per Tipo di Unità Produttiva. Available online: http://dati.istat.it/Index.aspx?QueryId=25267 (accessed on 19 September 2021).
4. Pulina, G.; Milán, M.J.; Lavín, M.P.; Theodoridis, A.; Morin, E.; Capote, J.; Thomas, D.L.; Francesconi, A.H.D.; Caja, G. Invited review: Current production trends, farm structures, and economics of the dairy sheep and goat sectors. *J. Dairy Sci.* **2018**, *101*, 6715–6729. [CrossRef] [PubMed]
5. ISTAT Latte e Prodotti Lattiero Caseari: Prodotti-Reg. Available online: http://dati.istat.it/Index.aspx?QueryId=25520 (accessed on 19 September 2021).
6. Sansonetti, S.; Curcio, S.; Calabrò, V.; Iorio, G. Optimization of ricotta cheese whey (RCW) fermentation by response surface methodology. *Bioresour. Technol.* **2010**, *101*, 9156–9162. [CrossRef]
7. Carvalho, F.; Prazeres, A.R.; Rivas, J. Cheese whey wastewater: Characterization and treatment. *Sci. Total Environ.* **2013**, *445-446*, 385–396. [CrossRef] [PubMed]
8. Mostafa, A.A. Treatment of cheese processing wastewater by physicochemical and biological methods. *Int. J. Microbiol. Res.* **2013**, *4*, 321–332.

9. Monti, L.; Donati, E.; Zambrini, A.V.; Contarini, G. Application of membrane technologies to bovine Ricotta cheese exhausted whey (scotta). *Int. Dairy J.* **2018**, *85*, 121–128. [CrossRef]
10. Pintado, M.E.; Macedo, A.C.; Malcata, F.X. Review: Technology, Chemistry and Microbiology of Whey Cheeses. *Food Sci. Technol. Int.* **2001**, *7*, 105–116. [CrossRef]
11. Sommella, E.; Pepe, G.; Ventre, G.; Pagano, F.; Conte, G.M.; Ostacolo, C.; Manfra, M.; Tenore, G.C.; Russo, M.; Novellino, E.; et al. Detailed peptide profiling of "Scotta": From a dairy waste to a source of potential health-promoting compounds. *Dairy Sci. Technol.* **2016**, *96*, 763–771. [CrossRef]
12. Secchi, N.; Giunta, D.; Pretti, L.; García, M.R.; Roggio, T.; Mannazzu, I.; Catzeddu, P. Bioconversion of ovine scotta into lactic acid with pure and mixed cultures of lactic acid bacteria. *J. Ind. Microbiol. Biotechnol.* **2012**, *39*, 175–181. [CrossRef]
13. Ribeiro, J.E.; Martini, M.; Altomonte, I.; Salari, F.; Nardoni, S.; Sorce, C.; Silva, F.L.D.; Andreucci, A. Production of Chlorella protothecoides biomass, chlorophyll and carotenoids using the dairy industry by-product scotta as a substrate. *Biocatal. Agric. Biotechnol.* **2017**, *11*, 207–213. [CrossRef]
14. Vincenzi, A.; Maciel, M.J.; Burlani, É.L.; Oliveira, E.C.; Volpato, G.; Lehn, D.N.; de Souza, C.F.V. Ethanol Bio-Production from ricotta cheese whey by several strains of the yeast Kluyveromyces. *Am. J. Food Technol.* **2014**, *9*, 281–291. [CrossRef]
15. Monari, S.; Ferri, M.; Russo, C.; Prandi, B.; Tedeschi, T.; Bellucci, P.; Zambrini, A.V.; Donati, E.; Tassoni, A. Enzymatic production of bioactive peptides from scotta, an exhausted by-product of ricotta cheese processing. *PLoS ONE* **2019**, *14*, e0226834. [CrossRef]
16. Pontonio, E.; Montemurro, M.; De Gennaro, G.V.; Miceli, V.; Rizzello, C.G. Anthypertensive Peptides from Ultrafiltration and Fermentation of the Ricotta Cheese Exhausted Whey: Design and Characterization of a Functional Ricotta Cheese. *Foods* **2021**, *10*, 2573. [CrossRef]
17. Raho, S.; Carofiglio, V.E.; Montemurro, M.; Miceli, V.; Centrone, D.; Stufano, P.; Schioppa, M.; Pontonio, E.; Rizzello, C.G. Production of the polyhydroxyalkanoate PHBV from ricotta cheese exhausted whey by haloferax mediterranei fermentation. *Foods* **2020**, *9*, 1459. [CrossRef]
18. Maragkoudakis, P.; Vendramin, V.; Bovo, B.; Treu, L.; Corich, V.; Giacomini, A. Potential use of scotta, the by-product of the ricotta cheese manufacturing process, for the production of fermented drinks. *J. Dairy Res.* **2016**, *83*, 104–108. [CrossRef] [PubMed]
19. Vasmara, C.; Marchetti, R. Initial pH influences in-batch hydrogen production from scotta permeate. *Int. J. Hydrogen Energy* **2017**, *42*, 14400–14408. [CrossRef]
20. Vasmara, C.; Pindo, M.; Micheletti, D.; Marchetti, R. Initial pH influences microbial communities composition in dark fermentation of scotta permeate. *Int. J. Hydrogen Energy* **2018**, *43*, 8707–8717. [CrossRef]
21. Mazorra-Manzano, M.A.; Ramírez-Suarez, J.C.; Yada, R.Y. Plant proteases for bioactive peptides release: A review. *Crit. Rev. Food Sci. Nutr.* **2018**, *58*, 2147–2163. [CrossRef] [PubMed]
22. Dinika, I.; Verma, D.K.; Balia, R.; Utama, G.L.; Patel, A.R. Potential of cheese whey bioactive proteins and peptides in the development of antimicrobial edible film composite: A review of recent trends. *Trends Food Sci. Technol.* **2020**, *103*, 57–67. [CrossRef]
23. Andler, S.M.; Goddard, J.M. Transforming food waste: How immobilized enzymes can valorize waste streams into revenue streams. *NPJ Sci. Food* **2018**, *2*, 19. [CrossRef] [PubMed]
24. Vargas-Bello-Pérez, E.; Márquez-Hernández, R.I.; Hernández-Castellano, L.E. Bioactive peptides from milk: Animal determinants and their implications in human health. *J. Dairy Res.* **2019**, *86*, 136–144. [CrossRef] [PubMed]
25. Minj, S.; Anand, S. Whey Proteins and Its Derivatives: Bioactivity, Functionality, and Current Applications. *Dairy* **2020**, *1*, 16. [CrossRef]
26. Tulipano, G.; Sibilia, V.; Caroli, A.M.; Cocchi, D. Whey proteins as source of dipeptidyl dipeptidase IV (dipeptidyl peptidase-4) inhibitors. *Peptides* **2011**, *32*, 835–838. [CrossRef]
27. Tagliazucchi, D.; Martini, S.; Shamsia, S.; Helal, A.; Conte, A. Biological activities and peptidomic profile of in vitro-digested cow, camel, goat and sheep milk. *Int. Dairy J.* **2018**, *81*, 19–27. [CrossRef]
28. Lacroix, I.M.E.; Li-Chan, E.C.Y. Dipeptidyl peptidase-IV inhibitory activity of dairy protein hydrolysates. *Int. Dairy J.* **2012**, *25*, 97–102. [CrossRef]
29. Nongonierma, A.B.; Fitzgerald, R.J. Dipeptidyl peptidase IV inhibitory and antioxidative properties of milk protein-derived dipeptides and hydrolysates. *Peptides* **2013**, *39*, 157–163. [CrossRef]
30. Ministero della Salute Diabete Mellito Tipo 2. Available online: https://www.salute.gov.it/portale/nutrizione/dettaglioContenutiNutrizione.jsp?lingua=italiano&id=5511&area=nutrizione&menu=croniche (accessed on 12 October 2021).
31. Corrochano, A.R.; Buckin, V.; Kelly, P.M.; Giblin, L. Invited review: Whey proteins as antioxidants and promoters of cellular antioxidant pathways. *J. Dairy Sci.* **2018**, *101*, 4747–4761. [CrossRef]
32. El-Zahar, K.; Sitohy, M.; Dalgalarrondo, M.; Choiset, Y.; Métro, F.; Haertlé, T.; Chobert, J.M. Purification and physicochemical characterization of ovine β-lactoglobulin and α-lactalbumin. *Food/Nahrung* **2004**, *48*, 177–183. [CrossRef]
33. López-Expósito, I.; Gómez-Ruiz, J.A.; Amigo, L.; Recio, I. Identification of antibacterial peptides from ovine αs2-casein. *Int. Dairy J.* **2006**, *16*, 1072–1080. [CrossRef]
34. Atanasova, J.; Ivanova, I. Antibacterial peptides from goat and sheep milk proteins. *Biotechnol. Biotechnol. Equip.* **2010**, *24*, 1799–1803. [CrossRef]
35. Lacroix, I.M.E.; Li-Chan, E.C.Y. Food-derived dipeptidyl-peptidase IV inhibitors as a potential approach for glycemic regulation—Current knowledge and future research considerations. *Trends Food Sci. Technol.* **2016**, *54*, 1–16. [CrossRef]

36. Lestari, P. Suyata Antibacterial activity of hydrolysate protein from Etawa goat milk hydrolysed by crude extract bromelain. *IOP Conf. Ser. Mater. Sci. Eng.* **2019**, *509*, 012111. [CrossRef]
37. Cabizza, R.; Rubattu, N.; Salis, S.; Pes, M.; Comunian, R.; Paba, A.; Addis, M.; Testa, M.C.; Urgeghe, P.P. Transfer of oxytetracycline from ovine spiked milk to whey and cheese. *Int. Dairy J.* **2017**, *70*, 12–17. [CrossRef]
38. Petretto, G.L.; Maldini, M.; Addis, R.; Chessa, M.; Foddai, M.; Rourke, J.P.; Pintore, G. Variability of chemical composition and antioxidant activity of essential oils between *Myrtus communis* var. Leucocarpa DC and var. Melanocarpa DC. *Food Chem.* **2016**, *197*, 124–131. [CrossRef]
39. Baranyi, J.; Roberts, T.A. A dynamic approach to predicting bacterial growth in food. *Int. J. Food Microbiol.* **1994**, *23*, 277–294. [CrossRef]
40. ComBase DMFit for Excel. Available online: https://www.combase.cc/index.php/en/8-category-en-gb/21-tools (accessed on 19 June 2021).
41. Le Maux, S.; Nongonierma, A.B.; FitzGerald, R.J. Peptide composition and dipeptidyl peptidase IV inhibitory properties of β-lactoglobulin hydrolysates having similar extents of hydrolysis while generated using different enzyme-to-substrate ratios. *Food Res. Int.* **2017**, *99*, 84–90. [CrossRef]
42. Old, W.M.; Meyer-Arendt, K.; Aveline-Wolf, L.; Pierce, K.G.; Mendoza, A.; Sevinsky, J.R.; Resing, K.A.; Ahn, N.G. Comparison of label-free methods for quantifying human proteins by shotgun proteomics. *Mol. Cell. Proteom.* **2005**, *4*, 1487–1502. [CrossRef] [PubMed]
43. Pisanu, S.; Cacciotto, C.; Pagnozzi, D.; Puggioni, G.M.G.; Uzzau, S.; Ciaramella, P.; Guccione, J.; Penati, M.; Pollera, C.; Moroni, P.; et al. Proteomic changes in the milk of water buffaloes (*Bubalus bubalis*) with subclinical mastitis due to intramammary infection by *Staphylococcus aureus* and by non-aureus staphylococci. *Sci. Rep.* **2019**, *9*, 15850. [CrossRef] [PubMed]
44. Minkiewicz, P.; Iwaniak, A.; Darewicz, M. BIOPEP-UWM Database of Bioactive Peptides: Current Opportunities. *Int. J. Mol. Sci.* **2019**, *20*, 5978. [CrossRef] [PubMed]
45. Liu, R.; Cheng, J.; Wu, H. Discovery of food-derived dipeptidyl peptidase IV inhibitory peptides: A review. *Int. J. Mol. Sci.* **2019**, *20*, 463. [CrossRef] [PubMed]
46. Nongonierma, A.B.; FitzGerald, R.J. Dipeptidyl peptidase IV inhibitory properties of a whey protein hydrolysate: Influence of fractionation, stability to simulated gastrointestinal digestion and food-drug interaction. *Int. Dairy J.* **2013**, *32*, 33–39. [CrossRef]
47. Song, J.J.; Wang, Q.; Du, M.; Ji, X.M.; Mao, X.Y. Identification of dipeptidyl peptidase-IV inhibitory peptides from mare whey protein hydrolysates. *J. Dairy Sci.* **2017**, *100*, 6885–6894. [CrossRef] [PubMed]
48. Konrad, B.; Anna, D.; Marek, S.; Marta, P.; Aleksandra, Z.; Józefa, C. The Evaluation of Dipeptidyl Peptidase (DPP)-IV, α-Glucosidase and Angiotensin Converting Enzyme (ACE) Inhibitory Activities of Whey Proteins Hydrolyzed with Serine Protease Isolated from Asian Pumpkin (*Cucurbita ficifolia*). *Int. J. Pept. Res. Ther.* **2014**, *20*, 483–491. [CrossRef] [PubMed]
49. Lacroix, I.M.E.; Li-Chan, E.C.Y. Inhibition of dipeptidyl peptidase (DPP)-IV and α-glucosidase activities by pepsin-treated whey proteins. *J. Agric. Food Chem.* **2013**, *61*, 7500–7506. [CrossRef]
50. Power, O.; Nongonierma, A.B.; Jakeman, P.; FitzGerald, R.J. Food protein hydrolysates as a source of dipeptidyl peptidase IV inhibitory peptides for the management of type 2 diabetes. *Proc. Nutr. Soc.* **2014**, *73*, 34–46. [CrossRef]
51. Corrêa, A.P.F.; Daroit, D.J.; Fontoura, R.; Meira, S.M.M.; Segalin, J.; Brandelli, A. Hydrolysates of sheep cheese whey as a source of bioactive peptides with antioxidant and angiotensin-converting enzyme inhibitory activities. *Peptides* **2014**, *61*, 48–55. [CrossRef]
52. Nongonierma, A.B.; FitzGerald, R.J. Enzymes Exogenous to Milk in Dairy Technology: Proteinases. *Encycl. Dairy Sci. Second Ed.* **2011**, *2*, 289–296. [CrossRef]
53. Biziulevičius, G.A.; Kislukhina, O.V.; Kazlauskaite, J.; Žukaite, V. Food-protein enzymatic hydrolysates possess both antimicrobial and immunostimulatory activities: A "cause and effect" theory of bifunctionality. *FEMS Immunol. Med. Microbiol.* **2006**, *46*, 131–138. [CrossRef]
54. Tulipano, G.; Faggi, L.; Nardone, A.; Cocchi, D.; Caroli, A.M. Characterisation of the potential of β-lactoglobulin and α-lactalbumin as sources of bioactive peptides affecting incretin function: In silico and in vitro comparative studies. *Int. Dairy J.* **2015**, *48*, 66–72. [CrossRef]
55. Uenishi, H.; Kabuki, T.; Seto, Y.; Serizawa, A.; Nakajima, H. Isolation and identification of casein-derived dipeptidyl-peptidase 4 (DPP-4)-inhibitory peptide LPQNIPPL from gouda-type cheese and its effect on plasma glucose in rats. *Int. Dairy J.* **2012**, *22*, 24–30. [CrossRef]
56. Otte, J.; Shalaby, S.M.; Zakora, M.; Pripp, A.H.; El-Shabrawy, S.A. Angiotensin-converting enzyme inhibitory activity of milk protein hydrolysates: Effect of substrate, enzyme and time of hydrolysis. *Int. Dairy J.* **2007**, *17*, 488–503. [CrossRef]
57. Sandré, C.; Gleizes, A.; Forestier, F.; Gorges-Kergot, R.; Chilmonczyk, S.; Léonil, J.; Moreau, M.-C.; Labarre, C. A peptide derived from bovine β-casein modulates functional properties of bone marrow-derived macrophages from germfree and human flora-associated mice. *J. Nutr.* **2001**, *131*, 2936–2942. [CrossRef] [PubMed]
58. Silva, S.V.; Malcata, F.X. Caseins as source of bioactive peptides. *Int. Dairy J.* **2005**, *15*, 1–15. [CrossRef]
59. Balthazar, C.F.; Pimentel, T.C.; Ferrão, L.L.; Almada, C.N.; Santillo, A.; Albenzio, M.; Mollakhalili, N.; Mortazavian, A.M.; Nascimento, J.S.; Silva, M.C.; et al. Sheep Milk: Physicochemical Characteristics and Relevance for Functional Food Development. *Compr. Rev. Food Sci. Food Saf.* **2017**, *16*, 247–262. [CrossRef]

60. Sowmya, K.; Bhat, M.I.; Bajaj, R.K.; Kapila, S.; Kapila, R. Buffalo Milk Casein Derived Decapeptide (YQEPVLGPVR) Having Bifunctional Anti-inflammatory and Antioxidative Features under Cellular Milieu. *Int. J. Pept. Res. Ther.* **2019**, *25*, 623–633. [CrossRef]
61. Nielsen, S.D.; Beverly, R.L.; Underwood, M.A.; Dallas, D.C. Release of functional peptides from mother's milk and fortifier proteins in the premature infant stomach. *PLoS ONE* **2018**, *13*, e0208204. [CrossRef]
62. Liu, H.; Tu, M.; Cheng, S.; Chen, H.; Wang, Z.; Du, M. An anticoagulant peptide from beta-casein: Identification, structure and molecular mechanism. *Food Funct.* **2019**, *10*, 886–892. [CrossRef]
63. Hao, X.; Yang, W.; Zhu, Q.; Zhang, G.; Zhang, X.; Liu, L.; Li, X.; Hussain, M.; Ni, C.; Jiang, X. Proteolysis and ACE-inhibitory peptide profile of Cheddar cheese: Effect of digestion treatment and different probiotics. *LWT* **2021**, *145*, 111295. [CrossRef]
64. Murray, N.M.; O'Riordan, D.; Jacquier, J.C.; O'Sullivan, M.; Holton, T.A.; Wynne, K.; Robinson, R.C.; Barile, D.; Nielsen, S.D.; Dallas, D.C. Peptidomic screening of bitter and nonbitter casein hydrolysate fractions for insulinogenic peptides. *J. Dairy Sci.* **2018**, *101*, 2826–2837. [CrossRef] [PubMed]
65. Nongonierma, A.B.; Fitzgerald, R.J. Structure activity relationship modelling of milk protein-derived peptides with dipeptidyl peptidase IV (DPP-IV) inhibitory activity. *Peptides* **2016**, *79*, 1–7. [CrossRef]
66. Jiang, B.; Zhang, X.; Yuan, Y.; Qu, Y.; Feng, Z. Separation of antioxidant peptides from pepsin hydrolysate of whey protein isolate by ATPS of EOPO Co-polymer (UCON)/Phosphate. *Sci. Rep.* **2017**, *7*, 13320. [CrossRef]
67. Mann, B.; Kumari, A.; Kumar, R.; Sharma, R.; Prajapati, K.; Mahboob, S.; Athira, S. Antioxidant activity of whey protein hydrolysates in milk beverage system. *J. Food Sci. Technol.* **2015**, *52*, 3235–3241. [CrossRef]
68. Pellegrini, A.; Dettling, C.; Thomas, U.; Hunziker, P. Isolation and characterization of four bactericidal domains in the bovine β-lactoglobulin. *Biochim. Biophys. Acta Gen. Subj.* **2001**, *1526*, 131–140. [CrossRef]
69. Elbarbary, H.A.; Ejima, A.; Sato, K. Generation of antibacterial peptides from crude cheese whey using pepsin and rennet enzymes at various pH conditions. *J. Sci. Food Agric.* **2019**, *99*, 555–563. [CrossRef]
70. Worsztynowicz, P.; Białas, W.; Grajek, W. Integrated approach for obtaining bioactive peptides from whey proteins hydrolysed using a new proteolytic lactic acid bacteria. *Food Chem.* **2020**, *312*, 126035. [CrossRef]
71. Chatterjee, A.; Kanawjia, S.K.; Khetra, Y.; Saini, P. Discordance between in silico & in vitro analyses of ACE inhibitory & antioxidative peptides from mixed milk tryptic whey protein hydrolysate. *J. Food Sci. Technol.* **2015**, *52*, 5621–5630. [CrossRef]
72. Conway, V.; Gauthier, S.F.; Pouliot, Y. Antioxidant activities of buttermilk proteins, whey proteins, and their enzymatic hydrolysates. *J. Agric. Food Chem.* **2013**, *61*, 364–372. [CrossRef]
73. Nongonierma, A.B.; FitzGerald, R.J. Strategies for the discovery and identification of food protein-derived biologically active peptides. *Trends Food Sci. Technol.* **2017**, *69*, 289–305. [CrossRef]
74. Lacroix, I.M.E.; Li-Chan, E.C.Y. Isolation and characterization of peptides with dipeptidyl peptidase-IV inhibitory activity from pepsin-treated bovine whey proteins. *Peptides* **2014**, *54*, 39–48. [CrossRef]
75. Silveira, S.T.; Martínez-Maqueda, D.; Recio, I.; Hernández-Ledesma, B. Dipeptidyl peptidase-IV inhibitory peptides generated by tryptic hydrolysis of a whey protein concentrate rich in β-lactoglobulin. *Food Chem.* **2013**, *141*, 1072–1077. [CrossRef]
76. Power, O.; Fernández, A.; Norris, R.; Riera, F.A.; FitzGerald, R.J. Selective enrichment of bioactive properties during ultrafiltration of a tryptic digest of β-lactoglobulin. *J. Funct. Foods* **2014**, *9*, 38–47. [CrossRef]
77. Bella, A.M.; Erickson, R.H.; Kim, Y.S. Rat intestinal brush border membrane dipeptidyl-aminopeptidase IV: Kinetic properties and substrate specificities of the purified enzyme. *Arch. Biochem. Biophys.* **1982**, *218*, 156–162. [CrossRef]
78. Ji, W.; Zhang, C.; Ji, H. Purification, identification and molecular mechanism of two dipeptidyl peptidase IV (DPP-IV) inhibitory peptides from Antarctic krill (*Euphausia superba*) protein hydrolysate. *J. Chromatogr. B Anal. Technol. Biomed. Life Sci.* **2017**, *1064*, 56–61. [CrossRef]
79. Ouertani, A.; Chaabouni, I.; Mosbah, A.; Long, J.; Barakat, M.; Mansuelle, P.; Mghirbi, O.; Najjari, A.; Ouzari, H.I.; Masmoudi, A.S.; et al. Two new secreted proteases generate a casein-derived antimicrobial peptide in Bacillus cereus food born isolate leading to bacterial competition in milk. *Front. Microbiol.* **2018**, *9*, 1148. [CrossRef]
80. Bounouala, F.Z.; Roudj, S.; Karam, N.E.; Recio, I.; Miralles, B. Casein Hydrolysates by Lactobacillus brevis and Lactococcus lactis Proteases: Peptide Profile Discriminates Strain-Dependent Enzyme Specificity. *J. Agric. Food Chem.* **2017**, *65*, 9324–9332. [CrossRef]
81. FitzGerald, R.J.; Cermeño, M.; Khalesi, M.; Kleekayai, T.; Amigo-Benavent, M. Application of in silico approaches for the generation of milk protein-derived bioactive peptides. *J. Funct. Foods* **2020**, *64*, 103636. [CrossRef]
82. Hernández-Ledesma, B.; Dávalos, A.; Bartolomé, B.; Amigo, L. Preparation of antioxidant enzymatic hydrolysates from α-lactalbumin and β-lactoglobulln. Identification of active peptides by HPLC-MS/MS. *J. Agric. Food Chem.* **2005**, *53*, 588–593. [CrossRef]
83. Almaas, H.; Eriksen, E.; Sekse, C.; Comi, I.; Flengsrud, R.; Holm, H.; Jensen, E.; Jacobsen, M.; Langsrud, T.; Vegarud, G.E. Antibacterial peptides derived from caprine whey proteins, by digestion with human gastrointestinal juice. *Br. J. Nutr.* **2011**, *106*, 896–905. [CrossRef]
84. Contreras, M. del M.; Hernández-Ledesma, B.; Amigo, L.; Martín-Álvarez, P.J.; Recio, I. Production of antioxidant hydrolyzates from a whey protein concentrate with thermolysin: Optimization by response surface methodology. *LWT Food Sci. Technol.* **2011**, *44*, 9–15. [CrossRef]

85. Demers-Mathieu, V.; Gauthier, S.F.; Britten, M.; Fliss, I.; Robitaille, G.; Jean, J. Antibacterial activity of peptides extracted from tryptic hydrolyzate of whey protein by nanofiltration. *Int. Dairy J.* **2013**, *28*, 94–101. [CrossRef]
86. Théolier, J.; Hammami, R.; Labelle, P.; Fliss, I.; Jean, J. Isolation and identification of antimicrobial peptides derived by peptic cleavage of whey protein isolate. *J. Funct. Foods* **2013**, *5*, 706–714. [CrossRef]

foods

MDPI

Article

Evaluating the Potential of the Defatted By-Product of *Aurantiochytrium* sp. Industrial Cultivation as a Functional Food

João Reboleira [1], Rafael Félix [1], Carina Félix [1], Marcelo M. R. de Melo [2], Carlos M. Silva [2], Jorge A. Saraiva [3], Narcisa M. Bandarra [4,5], Bárbara Teixeira [4,5], Rogério Mendes [4,5], Maria C. Paulo [6], Joana Coutinho [6] and Marco F. L. Lemos [1,*]

1 MARE—Marine and Environmental Sciences Centre, ESTM, Politécnico de Leiria, 2520-641 Peniche, Portugal; joao.reboleira@ipleiria.pt (J.R.); rafael.felix@ipleiria.pt (R.F.); carina.r.felix@ipleiria.pt (C.F.)
2 CICECO—Aveiro Institute of Materials, Department of Chemistry, University of Aveiro, Campus Universitário de Santiago, 3810-193 Aveiro, Portugal; marcelo.melo@ua.pt (M.M.R.d.M.); carlos.manuel@ua.pt (C.M.S.)
3 LAQV-REQUIMTE, Department of Chemistry, University of Aveiro, 3810-193 Aveiro, Portugal; jorgesaraiva@ua.pt
4 Division of Aquaculture and Upgrading, Portuguese Institute of the Sea and Atmosphere, Rua Alfredo Magalhães Ramalho, 6, 1495-006 Lisboa, Portugal; narcisa@ipma.pt (N.M.B.); barbara.p.b.teixeira@gmail.com (B.T.); rogerio@ipma.pt (R.M.)
5 CIIMAR, Interdisciplinary Centre of Marine and Environmental Research, University of Porto, Rua dos Bragas 289, 4050-123 Porto, Portugal
6 Depsiextracta Tecnologias e Biológicas, Lda., Zona Industrial do Monte da Barca Rua H, Lote 62, 2100-057 Coruche, Portugal; mariacastelo@depsiextracta.eu (M.C.P.); joanacoutinho@depsiextracta.eu (J.C.)
* Correspondence: marco.lemos@ipleiria.pt

Citation: Reboleira, J.; Félix, R.; Félix, C.; de Melo, M.M.R.; Silva, C.M.; Saraiva, J.A.; Bandarra, N.M.; Teixeira, B.; Mendes, R.; Paulo, M.C.; et al. Evaluating the Potential of the Defatted By-Product of *Aurantiochytrium* sp. Industrial Cultivation as a Functional Food. *Foods* **2021**, *10*, 3058. https://doi.org/10.3390/foods10123058

Academic Editors: Marco Poiana, Francesco Caponio and Antonio Piga

Received: 12 November 2021
Accepted: 4 December 2021
Published: 9 December 2021

Publisher's Note: MDPI stays neutral with regard to jurisdictional claims in published maps and institutional affiliations.

Copyright: © 2021 by the authors. Licensee MDPI, Basel, Switzerland. This article is an open access article distributed under the terms and conditions of the Creative Commons Attribution (CC BY) license (https://creativecommons.org/licenses/by/4.0/).

Abstract: While *Aurantiochytrium* sp. is an increasingly popular source of polyunsaturated fatty acids (PUFAs), its extraction generates high amounts of waste including the spent, defatted residue. The composition and bioactivities of this by-product could prove to be a major part of the sustainable valorisation of this organism within the framework of a circular economy. In this study, the defatted biomass of commercial *Aurantiochytrium* sp. was nutritionally characterised, and its amino acid profile was detailed. Additionally, the antioxidant and prebiotic potentials of an enzymatically digested sample of defatted *Aurantiochytrium* sp. were evaluated under a set of miniaturised in vitro assays. The nutritional profile of the spent *Aurantiochytrium* biomass revealed a protein and dietary-fibre rich product, with values reaching 26.7% and 31.0% for each, respectively. It also held high concentrations of glutamic and aspartic acid, as well as a favourable lysine/arginine ratio of 3.73. The digested samples demonstrated significant *Weissella cibaria* and *Bifidobacterium bifidum* growth-enhancing potential. Residual ferric reducing antioxidant power (FRAP) activity was likely attributed to antioxidant amino acids or peptides. The study demonstrated that some of the nutritional and functional potential that reside in the defatted *Aurantiochytrium* sp. waste encourages additional studies and the development of food supplements employing this resource's by-products under a biorefinery framework.

Keywords: spent biomass; prebiotic potential; enzymatic digestion; biorefinery; circular economy; by-products

1. Introduction

Aurantiochytrium sp. is a *Thraustochytrid* that has recently gained attention due to its high production of eicosapentaenoic acid and docosahexaenoic acid. They have emerged lately as an efficient economic alternative compared to other fish and microalgal oil sources by virtue of their simpler polyunsaturated fatty acids (PUFA) profiles and cost-effective culture conditions [1,2]. While the recovery of microbial oils avoids many of the problems associated with the traditional sources of PUFAs, the leftovers generated by such large-scale bioprocesses can still pose an environmental threat. A vast array of strategies and applications have been tested thus far in an attempt to add value to otherwise discarded microbial waste, with a large focus on the recycling of nutrients as a substrate for other economically feasible fermentations as well as the recovery of bioactive products [3–5]. Authors such as Medina (2015), Aida (2017), and Deshmukh (2021) have published distinct valorisation strategies applied to defatted microalgal or *Thraustochytrid* biomass, including the use as functional ingredients in biodegradable films, extraction of protein-rich antioxidant fractions, and direct use as nitrogen and phosphorous-rich additives to biofuel substrates [4,6,7]. The use of spent microalgal biomass has also been extended to livestock feed as either a soy or corn replacement as well as a nutritional supplement to traditional mixes [8].

Some of these applications fit under the designation of functional foods, which is a term that has acquired significant popularity in both social and scientific spheres. A recent definition given by Granato et al. (2020) states that functional foods, when regularly and efficaciously consumed as part of a diverse diet, can convey a positive effect on health beyond basic nutrition [9]. Said claims are regulated in most Western countries, limiting the classification to foods whose effects are verified via randomised, double-blind, and placebo-controlled clinical trials [10]. The potential held by functional foods in the prevention of many diseases deemed important in the 21st century, including obesity, type-2 diabetes, and several forms of cancer, has led to an enduring research interest over the course of the last 20 years [11–13]. The discovery of new functional foods is a large part of this effort, both as a way to find more bioactive ingredients and as a means to exploit new food resources in the form of new, high added-value products [13].

Due to the ever-growing demand for food of an expanding human population and their presence in an underexploited environment, marine food resources have found themselves under increased demand over recent years [14]. Among these, microalgae have confidently found their way into the niche market of food supplements. This was mostly due to exceptional amino-acid profiles in some cases comparable to terrestrial animal-sourced protein, as with the cyanobacteria *Arthrospira platensis* (commercial name "Spirulina") and the green algae *Chlorella* [6,15]. High dietary fibre content is also a highly desired feature in certain types of functional foods. These long-chain polysaccharides are incapable of being digested by the human digestive process and have been linked to numerous gastrointestinal health benefits [16,17]. Certain types of dietary fibre can be fermented by the gut microbiota, selectively promoting the growth of beneficial *Bifidobacterium, Lactobacillus, Bacillus, Streptococcus, Saccharomyces* and *Lactococcus* strains [18]. In turn, the proliferation of these strains has been associated with improvements to gut health via the suppression of pathogenic bacteria, improved gastro-intestinal flow, and short-chain fatty acid-driven immunomodulation, which are factors that are currently deemed essential in preventing intestinal and colonic cancers [18,19]. Carbohydrates that can reach the caeco-colon undigested and demonstrate the microbiota-enhancing effects described fall under the designation of prebiotics [20].

Given the recognition of how important gut health is in the prevention of serious human disease, the study of prebiotics is currently at the height of its development [18]. The complexities of digestion and the changes it incurs on the chemistry and function of dietary compounds has led researchers to develop increasingly elaborate in vitro models when assessing the prebiotic potential of foods and supplements [19]. While these advances have greatly improved the authenticity of the attributed label of prebiotics and led to great new insights on the bioavailability of certain nutrients, they are often difficult to reproduce without highly specialised, often custom-made equipment or access to clinical samples of human faeces [18,21]. Thus, quick assessments of the prebiotic potential of new foods are difficult to execute with these methodologies. Simpler, in vitro batch fermentations are still considered valuable screening tools for this very reason, despite their inadequate simulation of the digestion process [17].

With the removal of its lipid content, the spent biomass of *Aurantiochytrium* sp. still holds the potential of a nutritionally and functionally valuable food product. Considering that these organisms can accumulate approximately 50% of their weight in lipids, the defatted and dried remainder is a highly concentrated mixture of proteins and carbohydrates of exotic origin and whose nutritional and functional potential remains unexplored [22]. The present study seeks to confirm these claims via chemical analysis of the spent *Aurantiochytrium* biomass in addition to a screening of its antioxidant and prebiotic potentials.

2. Materials and Methods

2.1. Recovery of Spent Aurantiochytrium sp. Biomass

Aurantiochytrium sp. biomass under the commercial label Algamac 3050 was purchased from Pacific Trading–Aquaculture Ltd. (Dublin, Ireland). The biomass was supplied in vacuum-sealed plastic bags and as coarse flakes approximately 1.5 mm in length and 0.5 mm in thickness. Removal of the lipid fraction was carried out to simulate its industrial processing for the recovery of PUFAs using a lab-scale Soxhlet extraction apparatus. Samples of 5 g of biomass were loaded in a Soxhlet cartridge and extracted with n-hexane for 6 h. At the end of the extraction, the spent biomass was recovered from the cartridge solvent and dried overnight in a 50 °C oven.

2.2. Chemical Analysis of the Spent Aurantiochytrium sp. Biomass

Protein content was estimated using a LECO FP-528 DSP nitrogen analyser (LECO, St. Joseph, MI, USA). Fat content of both whole and spent *Aurantiochytrium* was determined according using the Bligh and Dyer technique with the modifications employed by Burja et al. (2007) applied to the standard method [23]. Ash and fibre content were determined according to the Association of Official Analytical Chemists (AOAC) standard methods 942.05 and 985.29, respectively [24,25]. A protein-rich fraction obtained from the defatted *Aurantiochytrium* sp. was prepared according to the procedure detailed by Vallabha et al. (2016), with some modifications [26]. These included a more prolonged extraction time (overnight) and the combination of all precipitated protein fractions. These were analysed using reverse-phase high-performance liquid chromatography following the method of Bidlingmeyer et al. (1984) after being subjected to a 24 h acid hydrolysis with 6N HCl with 0.1% phenol under vacuum and derivatisation with phenyl thiocarbamoyl [27].

2.3. Enzymatic Digestion of the Spent Aurantiochytrium sp. Biomass

A two-step simulated digestion of the defatted *Aurantiochytrium* sp. biomass was performed with a procedure adapted from Gawlik-Dziki et al. (2009) [28]. A solution of simulated saliva was prepared by dissolving 2.38 g Na_2HPO_4, 0.19 g KH_2PO_4, and 8 g NaCl in 1 L of distilled water and adjusting its pH to 6.75. Then, the solution was supplemented with 200 U of α-amylase (EC 3.2.1.1.). The simulated gastric digestion solution was an acidic (pH 1.2) 0.32% pepsin (porcine stomach mucosa, pepsin A, EC 3.4.23.1) dilution in 0.03 M NaCl. In 50 mL plastic centrifuge tubes, an approximate weight of 10 g of defatted *Aurantiochytrium* sp. flakes were mixed with 50 mL of simulated saliva and incubated in a 37 °C water bath for 10 min with occasional stirring using a steel spatula. Then, the slurry was brought to a pH of 1.2 using 5 M HCl, after which 50 mL of the simulated gastric solution were added. Then, a 120 min, 37 °C water bath incubation took place with occasional manual stirring. Afterwards, the digestion was halted via a short exposure to a 70 °C water bath (approximately 60 s), and the pH was brought up to 6.0 with a 1 M solution of $NaHCO_3$. Then, the entire digested slurry was treated as the digested defatted *Aurantiochytrium* sp. sample, from which 2 mL aliquots were gathered and stored at −20 °C prior to analysis.

2.4. In Vitro Prebiotic Potential Assay

The digested samples' potential to promote the growth of probiotic lactic acid bacteria was evaluated using a simplified and miniaturised method based on publications of

Wichienchot (2010) and Liu et al. (2016) and represented in Figure 1 [29,30]. The three bacterial strains selected for this assay include *Lactobacilus delbrueckii* subspecies *bulgaricus* DSMZ 20081, *Bifidobacterium bifidum* DSMZ 20456, and *Weissella cibaria* DSMZ 14295. Pre-cultures of *W. cibaria*, 48 h, 30 °C and pre-cultures of *B. bifidum* and *L. delbrueckii*, 72 h, 37 °C were prepared in MRS agar plates, with the latter two cultures having been maintained under anaerobic and oxygen-depleted atmospheres respectively using Mini Anaerocult A and C kits (Merck, Darmstadt, Germany). All pre-culture plates were prepared in triplicate as independent replicas. From these, 2.5×10^6 CFU/mL (*W. cibaria*) and 5.0×10^6 CFU/mL (*L. delbruecki* and *B. bifidum*) cellular suspensions were prepared in saline solution (0.85% NaCl; VWR). A set of master mixes were prepared encompassing all the necessary sample, blank, and control conditions in a 3:1:1 ratio of MRS medium, inoculum suspension, and sample, respectively. Digested *Aurantiochytrium* sp. samples were used without any further dilution, and digestion blanks provided a measure of growth induced by the enzymatic mixture (vehicle). Negative controls used the appropriate volume of saline solution instead of the digested sample. Then, 200 μL of each mixture were transferred to sterile round-bottom 96-well microplates and incubated for 72 h, with optical density (OD) readings occurring every 24 h at 600 nm using a microplate reader (EPOCH 2, BioTek Instruments, Winooski, VT, USA). Incubation of *W. cibaria* was conducted in aerobic conditions at 30 °C, while *B. bifidum* and *L. delbrueckii* microplates were enclosed in sealed bags under anaerobic and oxygen-depleted atmospheres at 37 °C as stated above.

A validation trial was performed using a 5.0×10^6 CFU/mL suspension of *L. delbrueckii* and inulin as a reference probiotic. Inulin concentrations ranging from 0.01 to 1% (*w/v*) were prepared in master mixes with identical MRS media and inoculum ratios as the main assay.

2.5. Antioxidant and Lipid Oxidation Protective Assays

The digested and defatted *Aurantiochytrium* sp. sample was subjected to a set of three antioxidant potential assays. Ferric-reducing antioxidant potential (FRAP) activity assay was performed according to Dudonné et al. (2009) with slight modifications to sample dilution rates [31]. First, 195 μL of ferric 2, 4, 6-tri (2-pyridyl)-s-triazine (TPTZ) along with 5 μL of either sample or iron sulphate standard were incubated for 30 min at 30 °C. The concentrations of the latter ranged from 20 to 1000 μM. A minimum of three independent assays were performed for each extraction condition tested. The 2,2-diphenyl-1-picrylhydrazyl (DPPH) radical reduction assay used a 96-well microplate-adapted protocol [32,33]. The working reagent was prepared by dissolving DPPH radical in absolute ethanol at a concentration of 0.1 mg/mL. The assay was conducted by pipetting 10 μL of each standard's or sample's concentration and of the digestion vehicle as control per well (8 wells each). In four wells, 190 μL of working reagent was added, and in the other four, 190 μL of ethanol was added. The plate was incubated in the dark for 60 min at room temperature, after which its absorbance (Abs) was read at 515 nm (EPOCH 2 microplate reader, BioTek® Instruments, Winooski, VT, USA). The amount of DPPH radical reduced by the standard per samples was calculated using a standard curve previously obtained, following the formula:

$$[DPPH](mM) = \frac{Abs_{Sample} - 0.0391}{5.1238}. \quad (1)$$

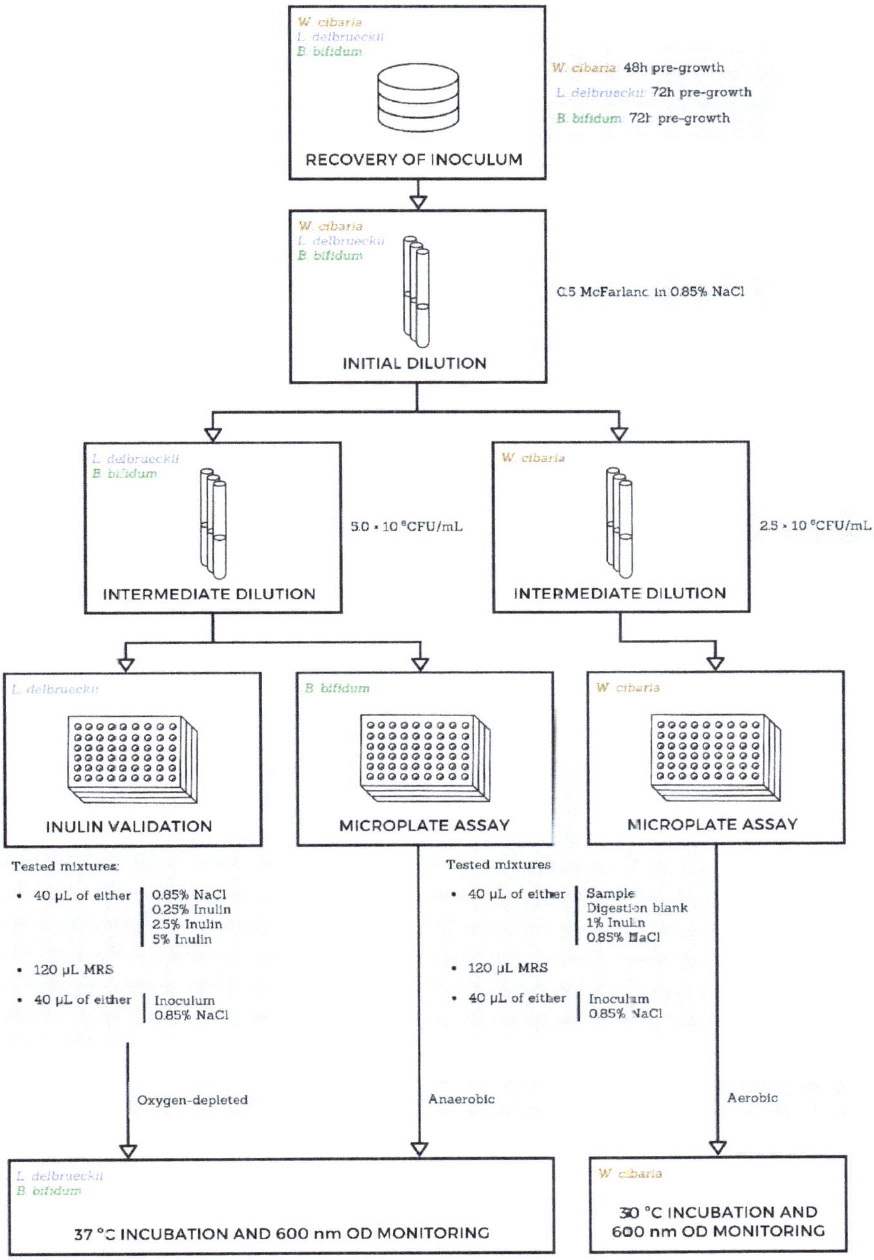

Figure 1. Flowchart representation of the miniaturised prebiotic potential assay employed in this study. Bacterial cultures were pre-grown in their respective optimal conditions for up to 72 h prior to the assay. The first step of the serial dilutions was performed identically for all cultures, with the following step adjusting for the required concentration. Inulin concentrations listed under "Tested mixtures" are higher than the tested concentrations as to account for the dilution occurring in the microplate well. Oxygen-depleted and anaerobic conditions were achieved using Merck's Mini Anaerocult A and C kits following manufacturer specifications. The incubations were prolonged for up to 72 h.

For each sample, Abs (515 nm) was calculated by subtracting the mean absorbance of the wells containing the sample and ethanol to the mean absorbance of the wells containing the sample and working reagent. The assay was performed in triplicate. The lipid peroxidation inhibitory potential (LPIP) was evaluated by a method adapted from Félix et al. (2020) and Yen and Hsieh (1998) [32–34]. The method was designed based on the auto-oxidation of a pure suspension of PUFAs in contact with air, in which case the only peroxides present are the lipid peroxides, which are quantifiable by the thiocyanate method. Briefly, a linoleic acid (LA) suspension was prepared (20 mM LA in Tween 20 at 5.6 mg/mL prepared in phosphate-buffered saline (PBS) at 20 mM, pH 7.1) and used as substrate. Then, 25 µL of extract at 1 mg/mL were pipetted onto a 2.0 mL microtube, in triplicate, and 125 µL of LA suspension and 100 µL of PBS (same as above) were added. As blanks, tubes with extract but with LA suspension's solvent (Tween 20 in PBS) instead of LA suspension were used (to determine peroxides native to the extract and subtract them from final result). As positive control (maximum peroxidation), 25 µL of extract vehicle (extract's solvent) and 125 µL of LA suspension were used (along with 100 µL of PBS), and as negative control, 25 µL of vehicle as 125 µL of LA suspension's solvent with 100 µL of PBS was used. All microtubes were incubated at 37 °C for 48 h in the dark and well capped. The experiment was performed in triplicate. Then, each tube was used to quantify peroxides by the thiocyanate method. From each tube, 20 µL (in triplicate) were sampled and added to a tube containing 940 µL of ethanol at 75% (v/v) and 20 µL of ammonium thiocyanate at 30% (w/v). Then, 20 µL of iron (II) chloride at 20 mM prepared in HCl at 3.5% (w/v) were added to each tube, and the mixture was properly homogenised using a vortex. Afterwards, each tube was used to read the absorbance in a microplate reader by pipetting 4 wells of 200 µL with the mixture. The absorbance was read at 500 nm, and the inhibitory potential was calculated:

$$LPIP\ (\%) = 100\frac{1 - Abs_{SampleBlank}}{Abs_{Pos.Ctrl} - Abs_{Neg.Ctrl}} \quad (2)$$

2.6. Statistical Analysis

All experiments were performed with at least three replicas and are presented as mean ± standard error. All graphical representations, descriptive statistics, one-way analysis of variance (ANOVA), and multiple comparisons tests (Tukey's honestly significant difference (HSD)) were all performed in GraphPad Prism v6.01 (GraphPad Software, Inc.; 2012, San Diego, CA, USA). Residual plots were used for all model assumptions, including normality, homoscedasticity, and independence. The type I error rate was at 0.05 for all statistical tests performed.

3. Results and Discussion

3.1. Chemical Analysis

Table 1 displays the macronutrient composition of both whole and defatted *Aurantiochytrium* sp. used in this study. A total lipid content of approximately 43% positions the samples used among the higher range of this parameter among other published results. Trovão et al. (2020) situated their set of *Aurantiochytrium* sp. within the 14 and 24% fat content range in their studies, while Ryu et al. (2013) achieved approximately 38.1% of lipidic weight growing *Aurantiochytrium* sp. in spent brewer's yeast [35,36]. Regardless, percentages as high as the ones presented here were previously achieved [37]. The protein content is within the expected values for cultured *Aurantiochytrium* sp., with authors such as Sami et al. (2013) and Moran et al. (2019) reporting this parameter at around 15% [38,39]. Fibre content was found to be about 31% of its dry weight, which is a percentage that stands far higher than most published results for *Aurantiochytrium* sp. Moran et al., 2019 found a maximum fibre content of around 3.4% in *Aurantiochytrium limacinum*. Such high values are not common even in other thraustochytrids [22].

Table 1. Macronutrient analysis performed on the whole *Aurantiochytrium* sp. (WA) and on its defatted counterpart (DA), shown as a percentage of dry weight. Each result is the average of at three measures ± standard error.

Sample ID	Lipid Content (g/100 g)	Protein Content (g/100 g)	Ash Content (g/100 g)	Fibre Content (g/100 g)
WA	42.7 ± 0.8	15.3 ± 0.7	10.7 ± 0.1	17.8 ± 1.5
DA	2.4 ± 0.8	26.7 ± 1.8	16.9 ± 0.9	31.0 ± 1.1

Looking at the macronutrient profile in isolation, *Aurantiochytrium* sp. in its defatted form already presents itself as a highly promising food product, with unusually high protein and dietary fibre contents. Similar profiles are found in marine organisms used as supplements, such as Spirulina and Chlorella [6]. With the loss of its lipid fraction, valued for its high PUFA content, the spent *Aurantiochytrium* sp. biomass may lose some of its nutritional richness. In turn, this depleted biomass is now much lighter in caloric content and thus much more compatible as a protein and dietary fibre supplement that is easily incorporated in a variety of diets.

3.2. Amino Acid Profile

The amino acid profile of the undigested spent *Aurantiochytrium* is shown in Table 2, listing the concentrations of 16 amino acids. Among these amino acids, eight essential amino acids were found. The major amino acids were glutamic acid (18 g/100 g) and aspartic acid (7.0 g/100 g), followed by serine, lysine, leucine, and proline. Cysteine, valine, and histidine were the among the least prevalent amino acids. The total amount of essential amino acids was higher than that of non-essential amino acids, with the high levels of leucine hinting at a potential use of this biomass in mid-workout energy snacks that aids muscle recovery and build-up [4]. While histidine levels were relatively low, they remain above the relative values of common plant or algae-based protein-rich supplements such as spirulina, soybean, and flaxseed. Histidine is a nutritionally essential amino acid that is also a precursor for several hormones (e.g., thyrotropin-releasing hormone), and critical metabolites affecting renal function, neurotransmission, gastric secretion, and the immune system [40]. The lysine/arginine/(Lys/Arg) ratio has been shown to positively affect the metabolic pathways of hypertension and have a positive effect on hypercholesterolemia, imparting lipidemic and atherogenic effects in rats even though the effects on humans were modest [26,41]. Although the exact mechanisms that lead to its positive effects are unknown, Yang et al. (2011) proposed that this amino acid ratio could limit the absorption rate of cholesterol [42]. The authors proposed either the slowdown of lipid absorption or promotion of 7 α-hydroxylase activity, which is a hepatic enzyme that limits the rate of cholesterol to bile acid conversion, as the mechanisms for this effect. In this study, the Lys/Arg ratio was found to be 3.73, which is quite similar to that of flaxseed (3.90) and a favourable ratio to be a useful protein ingredient in formulations intended to improve human health [40]. It should also be noted that the comparatively high amounts of glutamic and aspartic acid will likely contribute to an intensely umami flavour, which could grant any spent *Aurantiochytrium*-based supplement with desirable flavour-enhancing characteristics [43]. Such potential would need to be further investigated via sensory analysis.

Table 2. Amino acid profile of the defatted *Aurantiochytrium* sp. biomass in g/100 g of extracted protein and as a percentage of total detected amino acids. A collection of comparable values reported in the literature was included for both *Aurantiochytrium* sp. and other *Thraustochytrids*.

Amino Acid	*Aurantiochytrium* sp. Spent Biomass (g/100 g)	% of Total AA	*Aurantiochytrium* sp. from Literature (g/100 g)		References (*Auranti-ochytrium*)	Thraustochytrids from Literature (g/100 g)		References (Thraus-tochytrids)
			Min	Max		Min	Max	
Essential								
Alanine	2.2	4.0	0.8	3.9	[4,40]	0.53	7.5	[38,40,44]
Arginine	1.1	2.0	5.5	12.3		0.67	12.3	
Aspartic acid	7.0	12.7	5.6	7.1		2.91	14.7	
Glutamic acid	18	32.5	11.2	11.4		1.74	42.0	
Glycine	2.2	4.0	1.3	1.5		0.38	7.0	
Histidine	1.1	2.0	8.6	10.3		0.23	1.1	
Serine	4.2	7.6	2.6	3.2		0.46	10.8	
Threonine	1.8	3.3	0.3	0.8		0.38	1.5	
Non-essential								
Cysteine	0.79	1.4	0.3	0.4	[4,40]	0.14	1.4	[38,40,44]
Isoleucine	1.6	2.9	1.8	2.4		0.22	2.3	
Leucine	3.3	6.0	3.7	4.8		0.56	6.8	
Lysine	4.1	7.4	3.9	5.0		0.5	7.2	
Methionine	2.4	4.3	1.1	1.1		0.05	1.8	
Phenylalanine	1.4	2.5	2.1	2.7		0.36	3.7	
Proline	3.2	5.8	2.6	2.7		1.38	3.6	
Valine	0.93	1.7	2.7	3.4		0.34	4.0	

3.3. Antioxidant and Lipid Protective Activities

Table 3 lists the results of the three antioxidant assays performed on the digested defatted *Aurantiochytrium* sp. sample. There was no lipid oxidation prevention demonstrable using the LPIP method, which is unsurprising given the nature of the spent *Aurantiochytrium* sample. The majority of bioactive compounds, including those with antioxidant potential, whose presence in *Aurantiochytrium* sp. was previously reported, are lipophilic carotenoids and sterols [45]. These are expected to be mostly absent from the defatted biomass. Any residual activity, such as that which was detected in the FRAP assay, was likely caused by amino acids containing sulphur side chains, such as cysteine and methionine, or aromatic side chains, such as tyrosine, phenylalanine, and tryptophan [46,47]. The contribution of these effects to the overall antioxidant activity is heightened by the lipid-removal process. DPPH radical reduction activity was, similarly to the LPIP assay, indicative of a lack of lipophilic antioxidant compounds. In this instance, the ethanolic nature of the reaction medium results in the precipitation of most protein, and thus, any amino acid-driven activity is unrepresented [47].

Table 3. Antioxidant potential of the post-digestion defatted *Aurantiochytrium* sp. biomass, according to the DPPH radical reduction potential, ferric-reducing antioxidant potential (FRAP), and lipid peroxidation inhibitory potential (LPIP) assays. Each result is the average of at least three measures ± standard error.

Sample ID	DPPH (mM DPPH/mL)	FRAP (Fe(II) eq (mM)/mL)	LPIP (% of Ctrl)
Digested DA	0.025 ± 0.022	152.5 ± 6.2	162.1 ± 6.2

The presence of compounds with antioxidant activity in foods, regardless of their status as either functional or nutritious is, in most cases, highly desired. While the effects of dietary antioxidants in human health is still a contentious topic, their contribution as a positive factor in food preservation is generally well understood [48,49]. The presence of antioxidants in fatty foods, either as an ingredient or additive, is particularly desired, as these can greatly delay the loss in quality related to the oxidation of lipids [50]. While there are other compounds vulnerable to degradation under oxygen exposure, the defatted nature of the samples studied here means that lipid peroxidation phenomena are not as significant a concern for their long-term stability, and thus, a loss of antioxidant compounds may not be as sorely missed in a product based in defatted *Aurantiochytrium* [34]. Regardless, the presence of antioxidant proteins suggested by the FRAP activity assay could still provide a health benefit and warrants further research to determine their true chemical nature and concentrations.

3.4. Prebiotic Potential

3.4.1. Method Validation

The results of the prebiotic potential method validation trial are shown in Figure 2. The assay was successful in demonstrating noticeable changes in growth both with and without the presence of additives in its media. While the higher concentrations of the reference probiotic led to higher optical density readings after 72 h of incubation, it is interesting to note that this trend was not constant throughout the intermediate measures. Growth readings after 24 h suggested a preference for lower or null concentrations of inulin, which is likely associated with a breach in its maximum tolerable presence for this organism. Prolonged incubation revealed a reverse trend, with the growth of organisms under higher concentrations overtaking those previously mentioned. It is possible that inulin concentrations somewhere above 0.05%, together with the tested concentration of MRS media, provided an overabundance of soluble sugars, resulting in unfavourable osmotic pressures. These stressors were eventually overcome by the organism, which then made use of the higher abundance of nutrients to surpass the growth in the other conditions. Since these dynamics were easily verified in the conditions tested here, the testing of probiotic growth effects was carried on using this method.

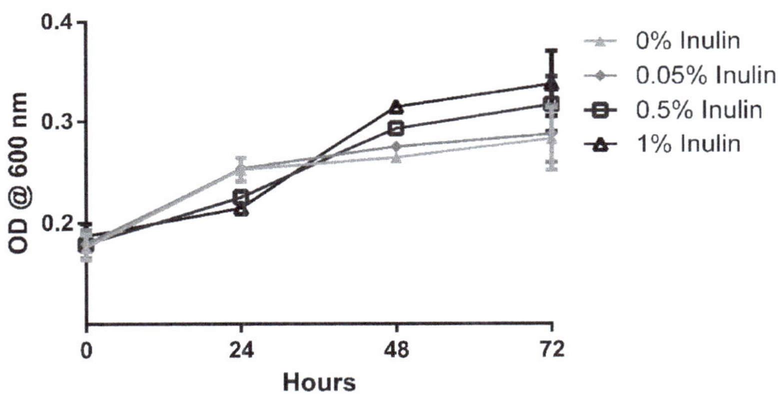

Figure 2. Growth of *Lactobacilus delbrueckii* under the effect of rising concentrations of inulin. Each point represents the average of at least three determinations ± standard error.

3.4.2. Growth Effect on Probiotic Cultures

Figure 3 shows the influence of the digested spent *Aurantiochytrium* solution (sample) and the digestion blank (vehicle) in the growth of *B. bifidum*, *W. cibaria*, and *L. delbrueckii* via a measure of optical density. For the first two microorganisms mentioned, the presence of the digested sample was favourable in promoting their growth, although the kinetics of each revealed differences that may influence the effectiveness of this sample as a prebiotic agent. The presence of digested defatted *Aurantiochytrium* solution had an immediate benefit on the growth of *W. cibaria*, as can be verified in the OD recorded after 24 h. This sudden growth appeared to consume nearly all the available substrate, and no further changes, worth noting, in cell density were observed over the course of the assay. *Weissela cibaria* has proven to be fastidious when compared to other lactic acid bacteria (LAB) [51]. Its accelerated growth kinetics and adaptability to harsh mediums has led to an increased interest in using it as a majority culture in sourdough starters [51]. *Bifidobacterium bifidum* revealed an equally favourable outcome for the probiotic growth enhancement potential of the digested defatted *Aurantiochytrium* solution, having reached 72 h of incubation with a significant DO difference between this sample and its digestion blank and growth control. In accordance with the features of a less fastidious anaerobic culture, the steepest increase in growth was observed between 24 and 48 h past medium inoculation, contrasting with the immediate spike in growth seen with *W. cibaria* immediately at 24 h. In contrast to the results commented so far, the growth of *L. delbrueckii* was unaffected by the presence of the digested defatted *Aurantiochytrium* sample or the digestion solution (vehicle). The scarcity of information regarding the non-lipid fraction of *Aurantiochytrium* sp. makes it difficult to point to specific causes for the selectivity of its prebiotic potential. Several studies testing the prebiotic potential of select compounds have also reported relatively lower growth of *L. delbrueckii*, but the authors did not establish likely causes [16,52,53]. Further studies of the non-lipidic fraction of *Aurantiochytrium* sp. might reveal hints about the selectivity of its prebiotic activity.

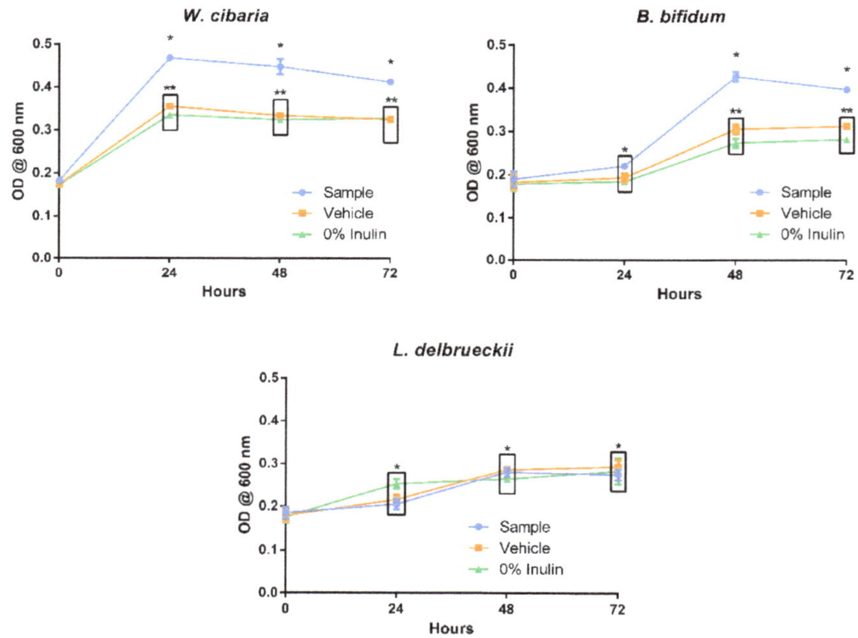

Figure 3. Growth of the probiotic strains *Bifidobacterium bifidum*, *Weissella cibaria*, and *Lactobacillus delbrueckii* under the presence of the digested defatted *Aurantiochytrium* sp. (Sample), the enzymatic digestion mixture (Vehicle), and the negative control saline solution (0% inulin). Each point represents the average of at least three determinations ± standard error. * and ** group measures are statistically identical within the same time point (either 24, 48, or 72 h).

The differences in kinetics between *W. cibaria* and *B. bifidum* are unlikely to reflect in a significant manner within in vivo conditions, given the extensive amount of biotransformations induced by digestion, as well as the influence of the remaining gut microbiota [19]. Selective or delayed growth of different gut bacteria is a hot topic of current-day prebiotic research, but the results shown here are more indicative of the differences between the growth characteristics of the studied strains rather than conclusive evidence of the selectivity of the substrate [54]. The results do show a significant increase in the growth of both *W. cibaria* and *B. bifidum* under the presence of digested spent *Aurantiochytrium* sp. Given the comparatively high amount of fibre content revealed by the chemical analysis, it is safe to say that this biomass, currently seen as industrial waste, may hold potential as a prebiotic supplement, with a stronger endorsement depending on further studies.

4. Conclusions

This study attempts an early exploration of functional food-related uses of spent *Aurantiochytrium* sp. biomass, which is a significant part of the industrial waste associated with the production of high added-value PUFAs from this source. Doing so could attribute this waste with value comparable to its lipid fraction and thus greatly increase the profitability and sustainability of all industrial exploitation of *Aurantiochytrium*. The chemical characterisations performed on the defatted biomass show a highly protein and dietary fibre-rich product that is simultaneously rid of most of its most caloric fraction. The amino acid profile revealed a fairly balanced distribution of essential variants, though it was not enough to warrant the use of this product as a protein supplement individually. High levels of glutamic and aspartic acid suggest it could be used as a source of flavour-enhancing umami compounds. The spent *Aurantiochytrium* was depleted of most of its lipophilic antioxidant compounds, but some residual activity was still registered by the FRAP assay, which was likely caused by antioxidant amino acids. The prebiotic potential assays revealed that after an in vitro enzymatic digestion, the sample held the capacity to enhance the growth of important probiotic strains. Together with its high dietary fibre content, these results point to a promising prebiotic supplement as one of the potential uses for the depleted biomass.

The results presented warrant a great deal of follow-up research, as many of its conclusions could be built upon a better understanding of the elemental composition of *Aurantiochytrium* sp., particularly of its non-lipid fractions. Additionally, the prebiotic potential demonstrated here is the result of a screening that is limited in scope and can be followed by either the testing of more strains and/or the use of a more in-depth digestion simulation and gut microbiota consortia from clinical samples. With these next steps fulfilled, a valuable application of the main by-products of *Aurantiochytrium* cultivation could be quickly implemented in both existing industries and upcoming biorefineries looking to maximise their profit and sustainability.

Author Contributions: Conceptualization, C.M.S., J.A.S., N.M.B., M.C.P., J.C., M.M.R.d.M. and M.F.L.L.; methodology, J.R., B.T. and R.M.; validation, R.F., C.F., N.M.B. and M.F.L.L.; investigation, J.R. and R.F.; resources, M.F.L.L.; writing—original draft preparation, J.R.; writing—review and editing, R.F., C.F., N.M.B. and M.F.L.L.; supervision, C.F. and M.F.L.L.; project administration, C.M.S., J.A.S., N.M.B., M.C.P., J.C. and M.F.L.L.; funding acquisition, C.M.S., J.A.S., N.M.B., M.C.P., J.C. and M.F.L.L. All authors have read and agreed to the published version of the manuscript.

Funding: This study had the support of Fundação para a Ciência e a Tecnologia (FCT) through the Strategic Project UID/MAR/04292/2020 to MARE, UIDB/50011/2020 & UIDP/50011/2020 to CICECO, and the University of Aveiro and FCT/MCT for the financial support for LAQV-REQUIMTE research Unit (FCT UIDB/50006/2020) through national founds, and, where applicable, co-financed by the FEDER, within the PT2020 Partnership Agreement This work was also funded by the project Algavalue—Valorização dos subprodutos do processo biotecnológico de produção de esqualeno e DHA pela microalga *Aurantiochytrium* sp. (POCI-01-0247-FEDER-017680) supported by COMPETE 2020, and Integrated Programme of SR&TD "SmartBioR" (reference Centro-01-0145-FEDER-000018) cofunded by Centro 2020 program, Portugal 2020, European Union, through the European Regional Development Fund.

Data Availability Statement: Data is available upon reasonable request.

Conflicts of Interest: The authors declare no conflict of interest.

References

1. Sun, X.M.; Xu, Y.S.; Huang, H. Thraustochytrid Cell Factories for Producing Lipid Compounds. *Trends Biotechnol.* **2020**, *65*, 102–113. [CrossRef]
2. Patel, A.; Rova, U.; Christakopoulos, P.; Matsakas, L. Mining of squalene as a value-added byproduct from DHA producing marine thraustochytrid cultivated on food waste hydrolysate. *Sci. Total Environ.* **2020**, *736*, 139691. [CrossRef] [PubMed]
3. Ou, L.; Thilakaratne, R.; Brown, R.C.; Wright, M.M. Techno-economic analysis of transportation fuels from defatted microalgae via hydrothermal liquefaction and hydroprocessing. *Biomass Bioenergy* **2015**, *72*, 45–54. [CrossRef]

4. Aida, T.M.; Maruta, R.; Tanabe, Y.; Oshima, M.; Nonaka, T.; Kujiraoka, H.; Kumagai, Y.; Ota, M.; Suzuki, I.; Watanabe, M.M.; et al. Nutrient recycle from defatted microalgae (Aurantiochytrium) with hydrothermal treatment for microalgae cultivation. *Bioresour. Technol.* **2017**, *228*, 186–192. [CrossRef] [PubMed]
5. Bellou, S.; Baeshen, M.N.; Elazzazy, A.M.; Aggeli, D.; Sayegh, F.; Aggelis, G. Microalgal lipids biochemistry and biotechnological perspectives. *Biotechnol. Adv.* **2014**, *32*, 1476–1493. [CrossRef] [PubMed]
6. Deshmukh, A.R.; Aloui, H.; Khomlaem, C.; Negi, A.; Yun, J.H.; Kim, H.S.; Kim, B.S. Biodegradable films based on chitosan and defatted Chlorella biomass: Functional and physical characterization. *Food Chem.* **2021**, *337*, 127777. [CrossRef]
7. Medina, C.; Rubilar, M.; Shene, C.; Torres, S.; Verdugo, M. Protein fractions with techno-functional and antioxidant properties from Nannochloropsis gaditana microalgal biomass. *J. Biobased Mater. Bioenergy* **2015**, *9*, 417–425. [CrossRef]
8. Gatrell, S.; Lum, K.; Kim, J.; Lei, X.G. Nonruminant nutrition symposium: Potential of defatted microalgae from the biofuel industry as an ingredient to replace corn and soybean meal in swine and poultry diets. *J. Anim. Sci.* **2014**, *92*, 1306–1314. [CrossRef]
9. Granato, D.; Barba, F.J.; Bursa'cbursa ´c, D.; Kovačevi ´c, K.; Lorenzo, J.M.; Cruz, A.G.; Putnik, P. Annual Review of Food Science and Technology Functional Foods: Product Development, Technological Trends, Efficacy Testing, and Safety. *Annu. Rev. Food Sci. Technol.* **2020**. [CrossRef]
10. Holdt, S.L.; Kraan, S. Bioactive compounds in seaweed: Functional food applications and legislation. *J. Appl. Phycol.* **2011**, *23*, 543–597. [CrossRef]
11. Ziemer, C.J.; Gibson, G.R. An overview of probiotics, prebiotics and synbiotics in the functional food concept: Perspectives and future strategies. *Proc. Int. Dairy J.* **1998**, *8*, 473–479. [CrossRef]
12. Kim, B.; Hong, V.M.; Yang, J.; Hyun, H.; Im, J.J.; Hwang, J.; Yoon, S.; Kim, J.E. A review of fermented foods with beneficial effects on brain and cognitive function. *Prev. Nutr. Food Sci.* **2016**, *21*, 297–309. [CrossRef] [PubMed]
13. Rai, A.K.; Pandey, A.; Sahoo, D. Biotechnological potential of yeasts in functional food industry. *Trends Food Sci. Technol.* **2019**, *83*, 129–137. [CrossRef]
14. Álvarez-Viñas, M.; Flórez-Fernández, N.; Torres, M.D.; Domínguez, H. Successful approaches for a red seaweed biorefinery. *Mar. Drugs* **2019**, *17*, 620. [CrossRef]
15. Imbimbo, P.; D'Elia, L.; Liberti, D.; Olivieri, G.; Monti, D.M. Towards green extraction methods from microalgae learning from the classics. *Appl. Microbiol. Biotechnol.* **2020**, *104*, 9067–9077. [CrossRef] [PubMed]
16. Adebola, O.O.; Corcoran, O.; Morgan, W.A. Synbiotics: The impact of potential prebiotics inulin, lactulose and lactobionic acid on the survival and growth of lactobacilli probiotics. *J. Funct. Foods* **2014**, *10*, 75–84. [CrossRef]
17. Pham, V.T.; Mohajeri, M.H. The application of in vitro human intestinal models on the screening and development of pre-And probiotics. *Benef. Microbes* **2018**, *9*, 725–742. [CrossRef]
18. Patel, A.K.; Singhania, R.R.; Awasthi, M.K.; Varjani, S.; Bhatia, S.K.; Tsai, M.L.; Hsieh, S.L.; Chen, C.W.; Dong, C. Di Emerging prospects of macro- and microalgae as prebiotic. *Microb. Cell Fact.* **2021**, *20*, 1–16. [CrossRef]
19. Minekus, M.; Alminger, M.; Alvito, P.; Ballance, S.; Bohn, T.; Bourlieu, C.; Carrière, F.; Boutrou, R.; Corredig, M.; Dupont, D.; et al. A standardised static in vitro digestion method suitable for food-an international consensus. *Food Funct.* **2014**, *5*, 1113–1124. [CrossRef]
20. Li, D.; Kim, J.M.; Jin, Z.; Zhou, J. Prebiotic effectiveness of inulin extracted from edible burdock. *Anaerobe* **2008**, *14*, 29–34. [CrossRef]
21. Zheng, L.X.; Chen, X.Q.; Cheong, K.L. Current trends in marine algae polysaccharides: The digestive tract, microbial catabolism, and prebiotic potential. *Int. J. Biol. Macromol.* **2020**, *151*, 344–354. [CrossRef] [PubMed]
22. de la Peña, M.R.; Teruel, M.B.; Oclarit, J.M.; Amar, M.J.A.; Ledesma, E.G.T. Use of thraustochytrid Schizochytrium sp. as source of lipid and fatty acid in a formulated diet for abalone Haliotis asinina (Linnaeus) juveniles. *Aquac. Int.* **2016**, *24*, 1103–1118. [CrossRef]
23. Burja, A.M.; Armenta, R.E.; Radianingtyas, H.; Barrow, C.J. Evaluation of fatty acid extraction methods for Thraustochytrium sp. ONC-T18. *J. Agric. Food Chem.* **2007**, *55*, 4795–4801. [CrossRef] [PubMed]
24. Tada, S.; Innami, S. A Simplified Modification of the AOAC Official Method for Determination of Total Dietary Fiber Using Newly Developed Enzymes. *J. AOAC Int.* **2007**, *90*, 217–224. [CrossRef] [PubMed]
25. Thiex, N.; Novotny, L.; Crawford, A. Determination of Ash in Animal Feed: AOAC Official Method 942.05 Revisited. *J. AOAC Int.* **2012**, *95*, 1392–1397. [CrossRef]
26. Vallabha, V.S.; Tapal, A.; Sukhdeo, S.V.; Govindaraju, K.; Tiku, P.K. Effect of arginine:lysine ratio in free amino acid and protein form on l-NAME induced hypertension in hypercholesterolemic Wistar rats. *RSC Adv.* **2016**, *6*, 73388–73398. [CrossRef]
27. Bidlingmeyer, B.A.; Cohen, S.A.; Tarvin, T.L. Rapid analysis of amino acids using pre-column derivatization. *J. Chromatogr. B Biomed. Sci. Appl.* **1984**, *336*, 93–104. [CrossRef]
28. Gawlik-Dziki, U.; Dziki, D.; Baraniak, B.; Lin, R. The effect of simulated digestion in vitro on bioactivity of wheat bread with Tartary buckwheat flavones addition. *LWT-Food Sci. Technol.* **2009**, *42*, 137–143. [CrossRef]
29. Wichienchot, S.; Jatupornpipat, M.; Rastall, R.A. Oligosaccharides of pitaya (dragon fruit) flesh and their prebiotic properties. *Food Chem.* **2010**, *120*, 850–857. [CrossRef]
30. Liu, Z.; Wang, W.; Huang, G.; Zhang, W.; Ni, L. In vitro and in vivo evaluation of the prebiotic effect of raw and roasted almonds (Prunus amygdalus). *J. Sci. Food Agric.* **2016**, *96*, 1836–1843. [CrossRef]

31. Dudonné, S.; Vitrac, X.; Coutiére, P.; Woillez, M.; Mérillon, J.M. Comparative study of antioxidant properties and total phenolic content of 30 plant extracts of industrial interest using DPPH, ABTS, FRAP, SOD, and ORAC assays. *J. Agric. Food Chem.* **2009**, *57*, 1768–1774. [CrossRef] [PubMed]
32. Yen, G.C.; Hsieh, C.L. Antioxidant Activity of Extracts from Du-zhong (Eucommia ulmoides) toward Various Lipid Peroxidation Models in Vitro. *J. Agric. Food Chem.* **1998**, *46*, 3952–3957. [CrossRef]
33. Félix, R.; Carmona, A.M.; Félix, C.; Novais, S.C.; Lemos, M.F.L. Industry-friendly hydroethanolic extraction protocols for grateloupia turuturu UV-shielding and antioxidant compounds. *Appl. Sci.* **2020**, *10*, 5304. [CrossRef]
34. Félix, R.; Valentão, P.; Andrade, P.B.; Félix, C.; Novais, S.C.; Lemos, M.F.L. Evaluating the in vitro potential of natural extracts to protect lipids from oxidative damage. *Antioxidants* **2020**, *9*, 231. [CrossRef]
35. Trovão, M.; Pereira, H.; Costa, M.; Machado, A.; Barros, A.; Soares, M.; Carvalho, B.; Silva, J.T.; Varela, J.; Silva, J.T. Lab-Scale Optimization of Aurantiochytrium sp. Culture Medium for Improved Growth and DHA Production. *Appl. Sci.* **2020**, *10*, 2500. [CrossRef]
36. Ryu, B.G.; Kim, K.; Kim, J.; Han, J.I.; Yang, J.W. Use of organic waste from the brewery industry for high-density cultivation of the docosahexaenoic acid-rich microalga, Aurantiochytrium sp. KRS101. *Bioresour. Technol.* **2013**, *129*, 351–359. [CrossRef]
37. Jakobsen, A.N.; Aasen, I.M.; Josefsen, K.D.; Strøm, A.R. Accumulation of docosahexaenoic acid-rich lipid in thraustochytrid Aurantiochytrium sp. strain T66: Effects of N and P starvation and O2 limitation. *Appl. Microbiol. Biotechnol.* **2008**, *80*, 297–306. [CrossRef] [PubMed]
38. Sami, R.; Lianzhou, J.; Yang, L.; Ma, Y.; Jing, J. Evaluation of fatty acid and amino acid compositions in okra (abelmoschus esculentus) grown in different geographical locations. *Biomed Res. Int.* **2013**, *2013*. [CrossRef] [PubMed]
39. Moran, C.A.; Morlacchini, M.; Keegan, J.D.; Fusconi, G. Increasing the Omega-3 Content of Hen's Eggs Through Dietary Supplementation with Aurantiochytrium limacinum Microalgae: Effect of Inclusion Rate on the Temporal Pattern of Docosahexaenoic Acid Enrichment, Efficiency of Transfer, and Egg Characteristics. *J. Appl. Poult. Res.* **2019**, *28*, 329–338. [CrossRef]
40. Tran, T.L.N.; Miranda, A.F.; Mouradov, A.; Adhikari, B. Physicochemical characteristics of protein isolated from thraustochytrid oilcake. *Foods* **2020**, *9*, 779. [CrossRef] [PubMed]
41. Vega-López, S.; Matthan, N.R.; Ausman, L.M.; Harding, S.V.; Rideout, T.C.; Ai, M.; Otokozawa, S.; Freed, A.; Kuvin, J.T.; Jones, P.J.; et al. Altering dietary lysine:arginine ratio has little effect on cardiovascular risk factors and vascular reactivity in moderately hypercholesterolemic adults. *Atherosclerosis* **2010**, *210*, 555–562. [CrossRef] [PubMed]
42. Yang, L.; Chen, J.; Xu, T.; Qiu, W.; Zhang, Y.; Zhang, L.; Xu, F.; Liu, H. Rice protein extracted by different methods affects cholesterol metabolism in rats due to its lower digestibility. *Int. J. Mol. Sci.* **2011**, *12*, 7594–7608. [CrossRef] [PubMed]
43. Temussi, P.A. The good taste of peptides. *J. Pept. Sci.* **2012**, *18*, 73–82. [CrossRef] [PubMed]
44. Yang, J.; Song, X.; Wang, L.; Cui, Q. Comprehensive analysis of metabolic alterations in Schizochytrium sp. strains with different DHA content. *J. Chromatogr. B Anal. Technol. Biomed. Life Sci.* **2020**, *1160*, 122193. [CrossRef] [PubMed]
45. De Melo, M.M.R.; Sapatinha, M.; Pinheiro, J.; Lemos, M.F.L.; Bandarra, N.M.; Batista, I.; Paulo, M.C.; Coutinho, J.; Saraiva, J.A.; Portugal, I.; et al. Supercritical CO_2 extraction of Aurantiochytrium sp. biomass for the enhanced recovery of omega-3 fatty acids and phenolic compounds. *J. CO_2 Util.* **2020**, *38*, 24–31. [CrossRef]
46. Apak, R.; Özyürek, M.; Güçlü, K.; Çapanoğlu, E. Antioxidant activity/capacity measurement. 1. Classification, physicochemical principles, mechanisms, and electron transfer (ET)-based assays. *J. Agric. Food Chem.* **2016**, *64*, 997–1027. [CrossRef] [PubMed]
47. Alam, M.N.; Bristi, N.J.; Rafiquzzaman, M. Review on in vivo and in vitro methods evaluation of antioxidant activity. *Saudi Pharm. J.* **2013**, *21*, 143–152. [CrossRef] [PubMed]
48. Di Lorenzo, C.; Colombo, F.; Biella, S.; Stockley, C.; Restani, P. Polyphenols and Human Health: The Role of Bioavailability. *Nutrients* **2021**, *13*, 273. [CrossRef] [PubMed]
49. Gordon, M.H. Significance of Dietary Antioxidants for Health. *Int. J. Mol. Sci.* **2011**, *13*, 173–179. [CrossRef] [PubMed]
50. Bolumar, T.; Andersen, M.L.; Orlien, V. Antioxidant active packaging for chicken meat processed by high pressure treatment. *Food Chem.* **2011**, *129*, 1406–1412. [CrossRef]
51. Ricciardi, A.; Parente, E.; Zotta, T. Modelling the growth of Weissella cibaria as a function of fermentation conditions. *J. Appl. Microbiol.* **2009**, *107*, 1528–1535. [CrossRef] [PubMed]
52. Chen, P.W.; Liu, Z.S.; Kuo, T.C.; Hsieh, M.C.; Li, Z.W. Prebiotic effects of bovine lactoferrin on specific probiotic bacteria. *BioMetals* **2017**, *30*, 237–248. [CrossRef] [PubMed]
53. Goderska, K. The antioxidant and prebiotic properties of lactobionic acid. *Appl. Microbiol. Biotechnol.* **2019**. [CrossRef] [PubMed]
54. Hirano, R.; Sakanaka, M.; Yoshimi, K.; Sugimoto, N.; Eguchi, S.; Yamauchi, Y.; Nara, M.; Maeda, S.; Ami, Y.; Gotoh, A.; et al. Next-generation prebiotic promotes selective growth of bifidobacteria, suppressing Clostridioides difficile. *Gut Microbes* **2021**, *13*, 1973835. [CrossRef]

foods

MDPI

Article

Compositional Features and Nutritional Value of Pig Brain: Potential and Challenges as a Sustainable Source of Nutrients

Jaruwan Chanted [1], Worawan Panpipat [1], Atikorn Panya [2], Natthaporn Phonsatta [2], Ling-Zhi Cheong [3] and Manat Chaijan [1,*]

1 Food Technology and Innovation Research Center of Excellence, School of Agricultural Technology and Food Industry, Walailak University, Nakhon Si Thammarat 80160, Thailand; Jaruwanchanted@gmail.com (J.C.); pworawan@wu.ac.th (W.P.)
2 Food Biotechnology Research Team, Functional Ingredients and Food Innovation Research Group, National Center for Genetic Engineering and Biotechnology (BIOTEC), National Science and Technology Development Agency, Bangkok 12120, Thailand; atikorn.pan@biotec.or.th (A.P.); natthaporn.pho@biotec.or.th (N.P.)
3 Zhejiang-Malaysia Joint Research Laboratory for Agricultural Product Processing and Nutrition, College of Food and Pharmaceutical Science, Ningbo University, Ningbo 315211, China; cheonglingzhi@nbu.edu.cn
* Correspondence: cmanat@wu.ac.th; Tel.: +66-7567-2384; Fax: +66-7567-2302

Citation: Chanted, J.; Panpipat, W.; Panya, A.; Phonsatta, N.; Cheong, L.-Z.; Chaijan, M. Compositional Features and Nutritional Value of Pig Brain: Potential and Challenges as a Sustainable Source of Nutrients. *Foods* **2021**, *10*, 2943. https://doi.org/10.3390/foods10122943

Academic Editors: Marco Poiana, Francesco Caponio and Antonio Piga

Received: 4 November 2021
Accepted: 29 November 2021
Published: 30 November 2021

Publisher's Note: MDPI stays neutral with regard to jurisdictional claims in published maps and institutional affiliations.

Copyright: © 2021 by the authors. Licensee MDPI, Basel, Switzerland. This article is an open access article distributed under the terms and conditions of the Creative Commons Attribution (CC BY) license (https://creativecommons.org/licenses/by/4.0/).

Abstract: The goal of this study was to establish the nutritional value and compositional properties of the brains of crossbred pigs (Landrace–Large white–Duroc (LLD)), in order to realize the zero-waste concept and increase the use of by-products in the sustainable meat industry. Fat (9.25% fresh weight (fw)) and protein (7.25% fw) were the principal dry matters of pig brain, followed by carbohydrate and ash. Phospholipid and cholesterol had a 3:1 ratio. Pig brain had a red tone (L^* = 63.88, a^* = 5.60, and b^* = 15.43) and a high iron content (66 mg/kg) due to a total heme protein concentration of 1.31 g/100 g fw. The most prevalent macro-element was phosphorus (14 g/kg), followed by potassium, sodium, calcium, and magnesium. Zinc, copper, and manganese were among the other trace elements discovered. The most prevalent nitrogenous constituents were alkali-soluble protein, followed by water-soluble protein, stromal protein, salt-soluble protein, and non-protein nitrogen. Essential amino acids were abundant in pig brain (44% of total amino acids), particularly leucine (28.57 mg/g protein), threonine, valine, and lysine. The total lipid, neutral, and polar lipid fractions of the pig brain had different fatty acid compositions. The largest amount was observed in saturated fatty acids (SFA), followed by monounsaturated fatty acids (MUFA) and polyunsaturated fatty acids (PUFA). Stearic acid and palmitic acid were the most common SFA. Oleic acid was the most prevalent MUFA, while docosahexaenoic acid was the most common PUFA. Thus, the pig brain can be used in food formulations as a source of nutrients.

Keywords: pork; fatty acid; amino acid; mineral; meat; by-product; sustainability

1. Introduction

Pork consumption is currently on the rise among the world's population. China, the European Union, the United States, Brazil, and Russia are the world's top five pork producers [1]. Thailand's pig production was estimated to be over 20.5 million heads in 2020, with pork consumption estimated to be around 1.3 million tons [1]. By-products, such as blood, bone, bristle, fat trimmings, viscera, and brain, are created as a result of increased pig consumption [2–4]. By-products constitute between 60% and 70% of the butchered carcass, with roughly 40% being edible and 20% being inedible [4]. Some of these by-products are widely used in many countries around the world in a variety of traditional dishes [3] and can have value effectively added using additional processes, such as thermal, chemical, centrifugation, washing, and combined processes to produce lard, flavor concentrate, plasma, red blood cell, gelatin, protein hydrolysates, and other products [4–7]. However, a range of factors, such as religion, culture, income, and personal taste, have an impact on the utilization of meat by-products. Depending on the country and local traditions, various meat by-products can be considered edible in certain locations but inedible in others. In reality, high-nutrient by-products such as liver, heart, blood, lung, spleen, kidney, brains, and tripe are used in the cuisines of some countries around the world [3]. Naturally, the nutritional makeup of each by-product is dependent on the animal type from which it is derived [3]. The basic composition and nutritional worth of these by-products should be assessed first, in order to find approaches to boost their value.

Although pig brain is an important by-product of slaughtering and pork processing, it has not yet been extensively used, particularly for human consumption. There is also no academic knowledge on how to increase the value of pig brain. Only traditional cookery, such as soup, gravy, stew, curry, and fried food, has been recognized as a primary manner of using the pig brain. When using pig brain as food, not only should the issue of safety, particularly of prion diseases, be considered, but the chemical compositions and nutritive value must also be assessed in order to provide nutritional information and pave the way for improved exploitation of the pig brain [8]. As a result, the goal of this study was to determine the nutritional value and compositional features of pig brain. The findings may be valuable in boosting pig brain intake and subsequent exploitation in the sustainable meat sector in order to achieve zero waste.

2. Materials and Methods

2.1. Chemicals

All chemicals and reagents used in this study, e.g., trichloroacetic acid (TCA), sodium dodecyl sulfate (SDS), potassium hydroxide (KOH), and potassium chloride (KCl), were acquired from Sigma-Aldrich (St. Louis, MO, USA).

2.2. Collection and Preparation of Pig Brains

Ten brains of crossbred pigs (Landrace–Large white–Duroc, LLD) at 4 months of age were collected from the Shaw Processing Food Co., Ltd. in Nakhon Si Thammarat, Thailand. The brains came from healthy pigs, approved by Thailand's Bureau of Livestock Standards and Certification. Within 1 h, the obtained samples were delivered to Walailak University's Food Technology and Innovation Laboratory in ice with a sample-to-ice ratio of 1:2 (*w/w*). The brains were then rinsed in cold water (4 °C), drained, and chopped with a Talsa Bowl Cutter K15e (The Food Machinery Co., Ltd., Kent, UK) to create a homogeneous composite sample. After vacuum packing (DZQ-400, Afapa Vacuum Equipment Co., Ltd., Shanghai, China), ground samples were kept at −80 °C for no more than 1 month before being utilized experimentally.

2.3. Proximate Composition

Moisture (A.O.A.C method number 950.46), crude protein (A.O.A.C method number 928.08, Kjeldahl factor of 6.25), fat (A.O.A.C method number 963.15), ash (A.O.A.C method number 920.153), and carbohydrate (calculated by the difference) were all investigated in the proximate composition of pig brain [9]. The results were expressed in grams per 100 g of fresh weight (fw).

2.4. Determination of Total Phospholipid and Total Cholesterol Contents

Bligh and Dyer's method [10] was used to extract lipid from the pig brain. Samples (25 g) were homogenized with 200 mL of a mixture of chloroform, methanol, and distilled water (1:2:1, *v/v/v*) at 9500 rpm for 2 min at 4 °C using an IKA Labortechnik homogenizer (Selangor, Malaysia). Then, 50 mL chloroform was added to the homogenate, and the mixtures were homogenized at the same speed for 1 min. After that, 25 mL distilled water was added, and the mixtures were homogenized at the same speed for 30 s. The homogenate was centrifuged at 3000× *g* for 15 min at 4 °C using an RC-5B plus centrifuge (Sorvall, Norwalk, CT, USA) and then transferred to a separating flask. The chloroform phase was drained into a 125 mL Erlenmeyer flask containing around 2–5 g of sodium sulfate, agitated well, and filtered through Whatman No. 4 filter paper (GE Healthcare Bio-Sciences Corp., Pittsburgh, PA, USA) into a round-bottom flask. A rotary evaporator (Model N-100, Eyela Ltd., Tokyo, Japan) was used to evaporate the solvent at 40 °C.

The total phospholipid content of the extracted oil was determined using a modified Stewart method [11]. Oil samples (20 µL) were dissolved in chloroform to obtain a final volume of 2 mL. Then, 1 mL of thiocyanate reagent (a mixture of 0.10 M ferric chloride hexahydrate and 0.40 M ammonium thiocyanate) was added. The lower layer was removed after 1 min of vigorous mixing and the absorbance at 488 nm was determined. Phosphatidylcholine (0–50 ppm) was used to create a standard curve. The total phospholipid content was measured in g/100 g fw.

The cholesterol content of the oil samples was determined using a modified version of Beyer and Jensen's method [12]. The oil sample (0.1–0.2 g) was saponified using 2% alcoholic KOH for 10 min. The unsaponified fraction was extracted with 2 × 10 mL hexane. The extracts were rinsed with 5 mL distilled water and dried at 45 °C in a W350 Memmert temperature-controlled water bath (Schwabach, Germany). The dried extract was resuspended in 3 mL of glacial acetic acid, and 2 mL coloring reagent was added. To prepare the coloring reagent, the stock reagent of 10% (*w/v*) $FeCl_3 \cdot 6H_2O$ in glacial acetic acid was made and then 1 mL of the stock reagent was diluted with 100 mL of concentrated H_2SO_4. The absorbance of the reaction mixture was read at 565 nm against a glacial acetic acid blank using a UV–Vis spectrophotometer (UV-1900, Shimadzu, Kyoto, Japan). A standard curve was prepared using cholesterol in glacial acetic acid at 0 to 120 mg/L. The total cholesterol content was measured in g/100 g fw.

2.5. Total Heme Protein Content and Color Measurement

Using Chaijan and Undeland's method [13], the total heme protein content was assessed and stated in g of hemoglobin per 100 g of sample. A sample and 3 volumes of 0.1 M phosphate buffer, pH 7 containing 5% SDS (*w/v*) was homogenized at 13,500 rpm for 20 s. The homogenate was heated in a water bath (85 °C) for 1 h and cooled under running tap water for 10 min. The solution was then centrifuged (5000× *g*/15 min/25 °C). The absorbance of the supernatant was read at 535 nm using a UV–Vis spectrophotometer with phosphate buffer as a blank. A standard curve of bovine hemoglobin (0–20 µM) was used.

A portable Hunterlab ColorFlex® EZ device (Hunter Assoc. Laboratory; Reston, VA, USA) was used to collect colorimetric data of the pig brain in triplicate. A white and black standard were used to calibrate the device. The measurement modes tristimulus L^* (lightness), a^* (redness/greenness), and b^* (yellowness/blueness) were chosen. According to Chen et al. [14], the redness index (a^*/b^*) was computed.

2.6. Mineral Composition

The mineral composition including phosphorus (P), potassium (K), sodium (Na), calcium (Ca), magnesium (Mg), iron (Fe), zinc (Zn), copper (Cu), manganese (Mn), and chromium (Cr) was determined [9]. Samples (4 g) were mixed with 4 mL of strong nitric acid and vigorously shaken for 5 min. The mixtures were heated on a hot plate until digestion was completed. The digested samples were transferred to a volumetric flask and filled to a capacity of 10 mL with deionized water. An inductively coupled plasma optical emission spectrophotometer was used to analyze the solution (PerkinElmer, Model 4300DV, Norwalk, CT, USA). The flow rates of argon to plasma, auxiliary, and nebulizer were kept at 15, 0.2, and 0.8 L/min, respectively. The sample's flow rate was set at 1.5 mL/min. The mineral content was determined and expressed in mg/kg.

2.7. Protein Fractionation

The protein composition of pig brain was fractionated according to Hashimoto et al. [15]. The non-protein nitrogenous (NPN) compound fraction, water-soluble protein fraction, salt-soluble protein fraction, alkali-soluble protein fraction, and stromal protein fraction were separated from the pig brain proteins due to their varied solubilities. Briefly, the ground sample (20 g) were homogenized in 200 mL of phosphate buffer (15.6 mM Na_2HPO_4, 3.5 mM KH_2PO_4), pH 7.5 with an IKA Labortechnik homogenizer. The homogenate was centrifuged (5000× g/15 min/4 °C) using an RC-5B plus centrifuge. The residue was homogenized and centrifuged again after being mixed with 200 mL of the same buffer. These two supernatants were combined, and TCA was added to obtain a final concentration of 5% (*w/v*). The resulting precipitate was collected by filtration and referred to as the water-soluble protein fraction. The filtrate was used as the NPN fraction. For the above residue, 10 vol of phosphate buffer (15.6 mM Na_2HPO_4, 3.5 mM KH_2PO_4) containing 0.45 mM KCl, pH 7.5 was added. The mixture was homogenized and centrifuged (5000× g/15 min/4 °C). The procedure was carried out twice more. Both supernatants were combined and used as the salt-soluble protein fraction. The precipitate obtained was added with 5 vol of 0.1 M NaOH and stirred for 10 h at 4 °C. The mixtures were then centrifuged (5000× g/15 min/4 °C). The supernatant was given as the alkali soluble protein fraction. The final precipitate was used as the stromal protein fraction. The nitrogen distribution was estimated after each fraction by the Kjeldahl method [9].

2.8. Amino Acid Profile and Fourier Transform Infrared (FTIR) Spectroscopy

According to Chinarak et al. [16], the amino acid profile of the pig brain was measured. In a 10 mL crimp seal glass vial, freeze-dried samples were combined with 5 mL of hydrolysis solution (6 M HCl, 5% thioglycolic acid, and 1% phenol) and tightly sealed. One 1 mL of the sample was centrifuged at 10,000× g for 10 min) after being hydrolyzed at 110 °C for 18 h. The supernatant (100 μL) was then neutralized with 1 M sodium carbonate. An aliquot of 25 μL was transferred to a 2 mL GC glass vial, which was then filled with 50 μL of 200 nM norleucine as an internal standard. After drying for around 1–2 h at 60 °C, 50 μL of dichloromethane was added and dried for another 30 min to eliminate any remaining water. Then, a derivatizing agent, *N*-tert-Butyldimethylsilyl-*N*-methyltrifluoroacetamide, with 1% *tert*-Butyldimethylchlorosilane (50 μL) and acetonitrile (50 μL), were mixed with the samples and subjected to incubate in a hot-air oven (100 °C/4 h). The sample (2 μL) was analyzed using a Shimadzu GCMS-TQ8050 NX (Kyoto, Japan) after cooling to room temperature.

A horizontal attenuated total reflectance (ATR) trough plate crystal cell (45° ZnSe; 80 mm long, 10 mm wide and 4 mm thick) (Pike Technology, Inc., Madison, WI, USA), equipped with a Bruker Model Vector 33 FTIR spectrometer (Bruker Co., Ettlingen, Germany), was used to perform FTIR analysis on the freeze-dried pig brain and pig brain lipid. In the mid-infrared region (500–4000 cm^{-1}), 16 scans, at a resolution of 4 cm^{-1}, were used to capture FTIR spectra at room temperature (26–29 °C). A reference air spectrum was acquired as a background. Analysis of spectral data was carried out using the OPUS 3.0 data collection software program [17].

2.9. Fatty Acid Profile of Total Lipid, Neutral Lipid Fraction, and Polar Lipid Fraction

According to Estefanell et al. [18], the neutral and polar fractions of total lipids were fractionated by adsorption chromatography on silica cartridges (Sep-pak; Waters S.A., MA, USA) using 30 mL chloroform and 20 mL chloroform/methanol (49:1, *v/v*) as neutral lipid solvents, followed by a 30 mL methanol rinse to yield the polar fraction. Fatty acid methyl esters (FAME) in the samples (total, neutral, and polar lipids) were determined using a gas chromatography/quadrupole time of flight (GC/Q-TOF) mass spectrometer (GC 7890B/MSD 7250, Agilent technologies, USA) coupled to the PAL auto sampler system (CTC Analytics AG, Switzerland). The MassHunter software was used to collect MS data (Version 10.0, Agilent Technologies, Santa Clara, CA, USA). The complete method and optimal condition can be obtained from the report of Chinarak et al. [16].

2.10. Statistical Analyses

All analyses were run in triplicate. An analysis of variance was performed on the data. Duncan's multiple-range test was used to compare the means. SPSS 17.0 for Windows (SPSS Inc., Chicago, IL, USA) was used to conduct the statistical analysis.

3. Results and Discussion

3.1. Proximate Composition

The approximate analysis of the pig brain is shown in Table 1. Moisture was the most abundant composition in the pig brain sample, accounting for 79.96% (fw), alike to other pork by-products [19]. Among the dry matter, fat was the most abundant composition (9.25% fw), followed by protein (7.25% fw), carbohydrate (2.21% fw), and ash (1.33% fw), respectively. Lipids and their intermediates are important parts of the brain's structure and function. Behind adipose tissue, the brain has the second largest lipid content, with lipids accounting for half of the brain's dry weight [20]. Krafft et al. [21] reported that brain tissue is composed of around 70–83% water, 7.5–8.5% protein (fw), and 5–15% lipid (fw). Brain tissue contained high content of lipid, which was mainly divided into three major groups, including neutral lipids, phospholipids, and sphingolipids [21]. Pig brain had considerably higher total lipid content than fresh pork loin (1.27–3.41% fw) [22] and pork by-products, such as heart, liver, lung, stomach small intestine, large intestine, spleen, uterus, and pancreas (0.28–7.18% fw) [19]. The protein content of the pig brain, on the other hand, was lower than that of fresh pork (19.80%, fw) [23] and other pork by-products, particularly viscera portions (8.45–22.05% fw), which have been reported in the literature [19]. The results revealed that the pig brain could be a rich source of nutrients, especially lipid and protein. The total protein and lipid content of the original material are both important factors in deciding how to use and recover by-products [24,25].

Table 1. Chemical composition of pig brain.

Compositions (g/100 g Fresh Weight)	Values
Moisture	79.96 ± 0.19
Fat	9.25 ± 0.32
Protein	7.25 ± 0.96
Carbohydrate	2.21 ± 0.42
Ash	1.33 ± 0.16
Total phospholipid	0.86 ± 0.00
Total cholesterol	0.30 ± 0.01
Total heme protein	1.31 ± 0.03

Values are given as mean ± standard deviation from triplicate determinations.

3.2. Total Phospholipids and Cholesterol Contents

Phospholipids are functional and structural components of cell membranes that are easily absorbed by the body and therefore perform their functions. Phospholipids are a dietary supply of lipoproteins, which play a significant role in lipid metabolism. Dietary phospholipid has been shown, in mammalian studies, to limit lipid deposition in the liver by blocking lipid absorption and oxidation [26,27]. Phospholipid is one of the most abundant lipids in brain tissue [21]. From this study, the phospholipid content of the oil extracted from pig brain was 0.86 g/100 g fw. The phospholipid content of crude lipid from pig brain was similar to that of alternative foods such as sago palm weewil (*Rhynchophorus ferregineus*) larvae (2.6–9.3 g/100 g lipid) [28], but it was lower than that of popular ingredients, such as fish oil (13.7–32.9 g/100 g lipid) [29] and krill oil (32.5 g/100 g lipid) [30]. Phospholipid is a natural surfactant that may be used to prepare emulsions and has high emulsifying characteristics in general [31]. Phospholipids can also act as an antioxidant [32]. Phospholipids can form associations with cholesterol in cell membranes [33]. Cholesterol is also a significant constituent of brain cell membranes [34]. The typical quantity of cholesterol found in an animal is around 2.1 g/kg of body weight [35]. Table 1 shows that the total cholesterol concentration of the pig brain was 0.30 g/100 g fw, resulting in a phospholipid to cholesterol ratio of around 3:1. Pig brain lipids had a lower cholesterol level (3.27 g/100 g lipid) than certain typical foods, such as cooked duck egg (6.4 g/100 g lipid) and raw squid (16.9 g/100 g lipid) [28]. Cholesterol is a lipid produced by the liver of animals and found in all animal-based foodstuffs, including eggs, red meat, and fish [36]. Cholesterol, which is one of the main structural components of cell membranes and is turned into hormones, is essential to good health at a normal level [37]. Hypercholesterolemia, cardiovascular disease, and coronary heart disease are all linked to a high-cholesterol diet [37]. The adult population should consume no more than 300 mg of cholesterol each day [28]. Although pig brain lipids are high in phospholipids, the cholesterol content should be considered when consuming pig brain. Separating cholesterol from pig brain oil could be an approach for increasing the oil's use as a functional ingredient.

3.3. Total Heme Protein and Color

The pig brain has a total heme protein concentration of 1.31 g/100 g fw. This was attributed to the presence of a blood vessel in the brain, which is evident in Table 1. Blood vessels, particularly capillaries, form a dynamic and intricate architecture that transports oxygen and nutrients to the brain [38]. As food, the presence of heme protein in form of blood may influence the storage stability, specifically, discoloration, rancidity development, and microbial growth [13,39,40]. Positive a^* (5.60) and b^* (15.43) values with a redness index of 0.36 indicated that the presence of heme protein caused the pig brain to be reddish brown in color (Table 2). The high moisture content in the pig brain (Table 1) was represented by a high L^* (63.88) value (Table 2). The elevated redness levels could be explained by the presence of heme proteins. The amount of moisture, membrane protein, connective tissue, lipid, and heme proteins in pig brain may affect its color. Moisture, membrane protein, and connective tissue can cause lightness; lipid can give yellowness; heme proteins (e.g., hemoglobin) can cause redness; and hemoglobin oxidation/denaturation can cause a yellow-brownish color.

Table 2. Color attributes and appearances of pig brain.

Attributes	Values
Color	
L^*	63.88 ± 0.25
a^*	5.60 ± 0.40
b^*	15.43 ± 0.63
Redness index	0.36 ± 6.95
Appearance	
Whole	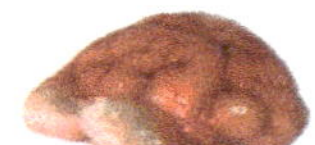
Ground	

Values are given as mean ± standard deviation from triplicate determinations. L^* = lightness, a^* = redness/greenness, b^* = yellowness/blueness.

3.4. Mineral Composition

As shown in Table 3, pig brain is a source of macro and trace elements. Minerals' value as dietary ingredients is based on more than just their nutritional and physiological functions. By stimulating or suppressing enzyme-catalyzed and other reactions, minerals contribute to food flavor, color, and texture [41]. P (14.0 g/kg) was found to be more abundant than the other macro-elements, followed by K (9.6 g/kg), Na (5.6 g/kg), Ca

(2.2 g/kg), and Mg (0.7 g/kg). Mineral concentrations in pig brain were higher than those seen in other pork by-products [19]. Free element, salt, phosphoproteins, and phospholipids are all examples of P. Pig brain phosphoproteins with calcium-binding capacity have been isolated [42]. K is required for the maintenance of cell membrane potentials, and a lack of it is linked to hypertension and an increased risk of cardiovascular disease [43]. The major extracellular osmolyte, serum Na, is the most essential predictor of serum osmolality, including the central nervous system of the brain [44]. The pig brain had a lower Ca concentration than the major Ca source in human diet—cows' milk [43]—but it could be another source of Ca. Mg is important for nerve transmission and muscle conduction from a neurological standpoint. It also protects neurons against excessive excitement, which can cause neuronal cell death, and has been linked to a variety of neurological illnesses [45].

Table 3. Mineral profile of pig brain.

Mineral	Contents (mg/kg)
Macro-element	
Phosphorous (P)	14,021.76 ± 97.20
Potassium (K)	9635.85 ± 132.18
Sodium (Na)	5624.04 ± 23.28
Calcium (Ca)	2196.99 ± 53.57
Magnesium (Mg)	699.39 ± 3.91
Trace-element	
Iron (Fe)	65.79 ± 0.66
Zinc (Zn)	37.90 ± 0.51
Copper (Cu)	12.42 ± 0.13
Manganese (Mn)	0.99 ± 0.01
Chromium (Cr)	nd

nd: not detected. Values are given as mean ± standard deviation from triplicate determinations.

Fe had the highest concentration of trace elements observed, followed by Zn, Cu, and Mn. Cr was not found in the pig brain (Table 3). Fe was found to be the most abundant trace element in pig visceral by-products, according to Seong et al. [19]. These trace minerals are necessary for human health, and a lack of them can cause nutritional deficiency symptoms [46]. Trace elements are molecular micronutrients that are needed in minute amounts but are crucial for the viability of many physiological functions in living tissues [41]. Pig brain Fe, Zn, and Cu levels were higher than those seen in other pork by-products [19]. Fe is the most abundant essential trace element in the human body, and it is a necessary micronutrient for brain development Normal brain growth, myelination, and neurotransmission all require Fe [47]. The overall amount of Fe in the body is around 3–5 g, with the majority of it in the blood and the rest in the liver, bone marrow, and muscles [41]. Fe is one of the key nutrients for blood's optimal function; anemia, especially in pregnant women and children, is caused by a lack of Fe [48]. The pigment level in the pig brain may be caused by high Fe concentrations. The prostate, portions of the eye, the muscle, brain, bones, kidneys, and liver all store zinc [49]. After Fe, Zn is the second most prevalent transition metal in organisms [41]. Cu is found in practically all physiological tissues and is mostly stored in the liver as well as the brain, heart, kidneys, and muscles [50]. Mn is a mineral that aids in the growth of the body, metabolism, and enzymatic defense mechanisms [51].

3.5. Distribution of Nitrogenous Constituents and Amino Acid Profile

Table 4 shows the distribution of nitrogenous constituents in pig brain. The ratio of NPN to protein N was roughly 1:28. Alkali-soluble protein accounted for 47.79% of the protein N, followed by water-soluble protein (21.60%), stroma (21.10%), and salt-soluble protein (9.49%). The presence of alkali-soluble protein in the membrane has been identified [52]. Proteins undergo modifications in alkaline conditions, which drive them apart by repulsion, allowing for associations with water and therefore solubilization [53,54]. In the cytosolic fluid of the brain, water-soluble and salt-soluble proteins may be found [55].

Complexins, for example, are a cytosolic protein family [56]. Collagen has been discovered in the brain as one of the principal stroma [57], particularly in the capillary walls [38,58].

Table 4. Distribution of nitrogenous constituents and amino acid compositions of pig brain.

Compositions	Values
Nitrogenous constituents (mg N/g fresh weight)	
Protein nitrogen	
Alkali-soluble protein	15.15 ± 0.40 (47.79%) *
Water-soluble protein	6.85 ± 0.00 (21.60%)
Stromal protein	6.69 ± 0.00 (21.10%)
Salt-soluble protein	3.01 ± 0.40 (9.49%)
Non-protein nitrogen (NPN)	1.15 ± 0.40
Amino acid compositions (mg/g protein)	
Essential amino acid (EAA)	
Leucine	28.57 ± 0.46
Threonine	25.50 ± 0.19
Valine	17.76 ± 0.17
Lysine	16.06 ± 0.02
Phenylalanine	12.89 ± 0.07
Isoleucine	12.57 ± 0.11
Histidine	8.70 ± 0.11
Methionine	7.56 ± 0.10
Non-essential amino acid (NEAA)	
Glutamic acid/glutamine	44.11 ± 0.40
Aspartic acid/asparagine	31.09 ± 0.32
Serine	18.90 ± 0.17
Arginine	18.38 ± 0.16
Glycine	15.72 ± 0.15
Alanine	15.45 ± 0.88
Proline	11.69 ± 0.06
Threonine	9.43 ± 0.09
Cysteine	1.13 ± 0.00
EAA (% of total amino acid)	43.86
NEAA (% of total amino acid)	56.14
Umami-taste active amino acid (UAA, % of total amino acid)	35.99
Functional amino acid (FAA, % of total amino acid)	67.41
Hydrophobic amino acid (% of total amino acid)	33.85

Values are given as mean ± standard deviation from triplicate determinations. * Values reported in the parenthesis represent the percentage respect the total protein nitrogen. The umami-taste active amino acid (UAA) index was calculated by adding the individual values of glutamine, asparagine, glycine, and alanine. The functional amino acid (FAA) index was calculated by adding the individual values of leucine, threonine, methionine, arginine, glycine, alanine, proline, and cysteine. The hydrophobic amino acid index was calculated by adding the individual values of leucine, valine, phenylalanine, isoleucine, alanine, proline, and cysteine.

The pig brain was found to be a good supply of essential amino acids (EAA), based on the findings (Table 4). The nutritional quality of food proteins is governed by the composition, quantity, and availability of EAA [59]. The total EAA of pig brain protein was around 44% of total amino acid, which was higher than the joint WHO/FAO/UNU expert consultation's recommendation of around 29% [60]. The most abundant EAA in pig brain was leucine (28.57 mg/g), followed by threonine, valine, lysine, isoleucine, phenylalanine, histidine, and methionine. Valine, phenylalanine, lysine, histidine, leucine, and isoleucine were the most common EAA in pig liver and heart [19]. EAA profiles of pig brain protein were similar to those of animal proteins such as milk, egg, fish, and meat [59,61].

The most prevalent non-essential amino acid (NEAA) was glutamic acid (44.11 mg/g), followed by aspartic acid, serine, glycine, alanine, proline, threonine, and cysteine. Glutamic acid, aspartic acid, glycine, and alanine are classified as umami-taste active amino acids (UAA), which play a role in the umami flavor [59,62,63]. The UAA level of pig brain was around 36% of total amino acid (Table 4). The UAA concentration of pig brain was similar to that of farm-raised sturgeon caviar (37.44–38.04%) and seaweeds (37.59–42.50%) [59], but it was higher than that of wild edible mushrooms (7–22%) [64]. As a result, the umami flavor of the pig brain could be detectable. Umami-relevant substances in pork *longissimus* and *biceps femoris* muscles include glutamic acid, total free amino acid, inosine monophosphate (IMP), and soluble oligopeptides [65] Functional amino acid (FAA) are amino acids found in living beings that have a role in and control critical metabolic pathways [66]. FAA can be EAA or NEAA and is composed of arginine, aspartic acid, cysteine, glutamic acid, glycine, leucine, methionine, proline, tryptophan, and tyrosine [66]. FAA aids in the treatment of some metabolic abnormalities and the modulation of the immune system [59,66]. The presence of FAA is significant due to their critical biological roles. In pig brain, FAA was detected in significant amounts (67.41%) (Table 4).

The amount and arrangement of hydrophobic and hydrophilic amino acids, particularly the interfacial characteristics, are critical to protein techno-functionality [63]. The hydrophobic amino acids made up 33.85% of the total amino acids in the pig brain. The ability of a protein to operate as an emulsifying agent should be linked to its hydrophobicity. Furthermore, the solubility profile of pig brain protein may be influenced by its amino acid composition. Protein solubility has been shown to be influenced by hydrophilic amino acids such as glutamic acid and aspartic acid [67]. As a result, pig brain can be exploited as a nutritious source of protein with potential technical applications.

3.6. FTIR Spectra

Figure 1 shows the FTIR spectra of whole pig brain and its lipid. The whole pig brain and its lipids have "fingerprint" spectra in the wavenumber range of 2700 to 3500 cm^{-1}, which are originated from stretch vibrations of CH, NH, and OH groups and related to their chemical compositions [21]. In the range of 400 to 1800 cm^{-1}, the spectra of whole pig brain and lipid fraction differed, indicating the presence of distinct chemical bonds and functional groups in the samples. In the whole pig brain, the various bands revealed the presence of water, protein, lipid, and carbohydrate. Some peaks may, however, be overlapping due to the combination of various compositions. According to Guillén and Cabo [68] and Lerma-García et al. [69], the presence of the vibration OH stretching of water allows the first band (3600–3200 cm^{-1}) to function. The peaks at 2850–2950 cm^{-1} correspond to the CH stretching bonds of methyl and methylene. The peaks at 1738 cm^{-1} are due to the presence of triglyceride functional groups (C=O stretching). The amide I peak at 1655 cm^{-1} (C=O stretching/hydrogen bonding paired with CN stretch and CCN deformation), the amide II peak at 1547 cm^{-1} (NH ending mixed with CN stretching), and the amide III peak at 1452 cm^{-1} (CN stretching and NH deformation) are all characteristic of proteins [70,71]. The amide A and amide B peaks were also identified at 3282 and 2922 cm^{-1}, respectively [63]. Finally, carbohydrates are responsible for the peaks in the 900–1200 cm^{-1} range [72], with the most conspicuous saccharide band about 1370 cm^{-1} [21].

In the spectra of lipids, due to the purity of the lipid and their lipid composition (e.g., triglyceride, phospholipid, cholesterol, sphingomyelin, and cerebrosides), the peaks in the range 400 to 1800 cm^{-1} were different from those of the original whole pig brain. Attributed to the existence of ester groups, Krafft et al. [21] found that the neutral lipid triacylglyceride showed a peak at 1729 cm^{-1}. The existence of phospholipids could explain the presence of a high peak at 835 cm^{-1}. It has been reported that phosphatidic acid, the parent component of phospholipids, has an 860 cm^{-1} band [21]. Several phospholipid varieties, including phosphatidylcholine (PC), phosphatidylethanolamine (PE), phosphatidylserine (PS), and phosphatidylinositol (PI), were discovered in the brain, each with a different spectrum, due to differences in functional head groups or fatty acid residues [21]. According to Krafft et al. [21], the cholesterol spectrum has multiple sharp bands ranging from 400 to 1200 cm^{-1}, with the most intense ones at 429, 548, 608, and 702 cm^{-1}. Furthermore, the peak at 1440 cm^{-1} is attributed to cholesterol CH distortion. In the spectra of sphingomyelin, choline bands are positioned at 718 and 875 cm^{-1}, the same as in PC, because sphingomyelin has a PC residue linked to the ceramide backbone [21].

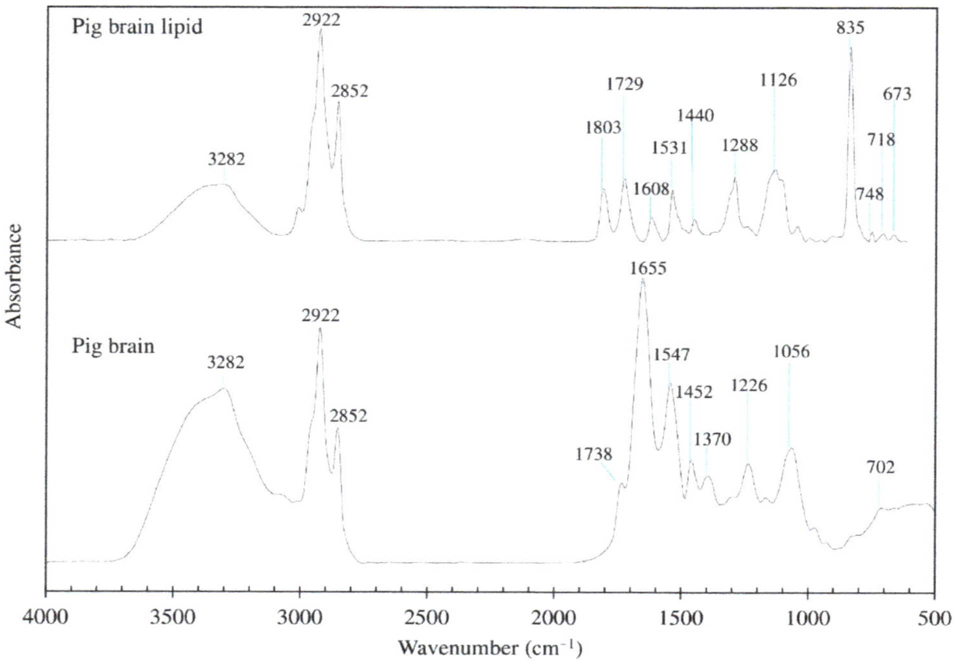

Figure 1. Fourier transform infrared (FTIR) spectra of the freeze-dried pig brain and pig brain lipid.

3.7. Fatty Acid Profiles

Lipids are important parts of the brain's structure and function. Essential fatty acids must be delivered into the brain from the circulation, despite the fact that some fatty acids can be produced de novo [73]. Unlike adipose tissue, which stores fatty acids predominantly as triglycerides, the brain is thought to create phospholipids for cell membranes primarily from acylated lipids [20]. The components of brain lipids can be separated into two categories: neutral lipid and polar lipid [21]. As a result, the fatty acid composition of total lipid was compared with that of neutral and polar lipid fractions. It was discovered that there is a difference in fatty acid profiles in pig brain between the total lipid and its fractions (Table 5). In general, the brain's fatty acid makeup is distinctive, with a high concentration of long-chain polyunsaturated fatty acids (PUFA), such as eicosapentaenoic acid and docosahexaenoic acid (DHA) [73]. Mas et al. [74] and Seong et al. [19] reported on the fatty acid content of several pork by-products and found that there was a considerable variance in the fatty acid composition.

From the results, saturated fatty acid (SFA) was found to have the highest content, followed by monounsaturated fatty acid (MUFA) and polyunsaturated fatty acid (PUFA) ($p < 0.05$). The most prevalent SFAs were stearic acid (C18:0) and palmitic acid (C16:0). The results were in agreement with Seong et al. [19], who reported that palmitic acid and stearic acid were the main SFAs found in pork by-products. The most prevalent MUFA was oleic acid (C18:1), while the most common PUFA was DHA ($p < 0.05$). DHA is a kind of n-3 PUFA that provides a variety of health effects. DHA was only discovered in the heart, liver, and spleen in a prior investigation by Seong et al. [19], with the liver having the greatest quantity (3.47%). Therefore, pig brain had a higher DHA content than other pork by-products.

Table 5. Fatty acid profile of total lipid, polar lipid fraction, and neutral lipid fraction of pig brain.

Fatty Acids (% of Total Fatty Acid)	Total Lipid	Polar Lipid Fraction	Neutral Lipid Fraction
Saturated fatty acid (SFA)			
Lauric acid (C12:0)	0.06 ± 0.09 b	0.17 ± 0.29 a	0.21 ± 0.33 a
Myristic acid (C14:0)	0.45 ± 0.31 b	0.57 ± 0.52 a	0.38 ± 0.13 b
Pentadecanoic acid (C15:0)	0.07 ± 0.02 b	0.10 ± 0.06 a	0.06 ± 0.02 b
Palmitic acid (C16:0)	14.30 ± 0.87 b	22.24 ± 10.37 a	15.49 ± 3.92 b
Heptadecanoic acid (C17:0)	0.84 ± 0.11 a	0.57 ± 0.08 b	0.52 ± 0.14 c
Stearic acid (C18:0)	17.92 ± 0.89 c	24.40 ± 14.61 a	23.12 ± 2.28 b
Arachidic acid (C20:0)	0.25 ± 0.04 a	0.23 ± 0.00 b	0.25 ± 0.06 a
Heneicosanoic acid (C21:0)	0.04 ± 0.04 c	0.08 ± 0.03 a	0.06 ± 0.03 b
Behenic acid (C22:0)	0.13 ± 0.07 c	0.21 ± 0.02 a	0.16 ± 0.11 b
Lignoceric acid (C24:0)	0.12 ± 0.07 b	0.24 ± 0.05 a	0.16 ± 0.30 b
Total SFA	34.18 ± 1.69 c	48.81 ± 2.27 a	40.43 ± 3.21 b
Monounsaturated fatty acid (MUFA)			
Palmitoleic acid (C16:1 *n*-7)	1.09 ± 0.27 a	0.91 ± 0.58 b	0.72 ± 0.20 c
Elaidic acid (C18:1 *n*-9 trans)	0.06 ± 0.04 c	0.30 ± 0.10 a	0.10 ± 0.12 b
Cis-9-octadecenoic acid (C18:1 *n*-9)	20.40 ± 1.17 c	25.28 ± 13.38 b	27.35 ± 5.38 a
Cis-11-eicosenoic acid (C20:1 *n*-11)	2.02 ± 0.43 a	1.01 ± 1.11 c	1.59 ± 0.32 b
Total MUFA	23.60 ± 1.71 c	27.50 ± 10.15 b	29.76 ± 0.16 a
Polyunsaturated fatty acid (PUFA)			
Cis-9,12-octadecadienoic acid (C18:2 *n*-6)	1.41 ± 0.27 a	0.82 ± 0.80 c	1.13 ± 0.31 b
Cis-9,12,15-octadecatrienoic acid (C18:3 *n*-3)	0.06 ± 0.03 a	nd	0.07 ± 0.10 a
Cis-6,9,12-octadecatrienoic acid (C18:3 *n*-6)	0.05 ± 0.01 a	0.07 ± 0.47 a	0.05 ± 0.07 a
Cis-11, 14-eicosadienoic acid (C20:1 *n*-6)	0.26 ± 0.01 a	0.14 ± 0.04 c	0.18 ± 0.03 b
Cis-8, 11, 14-eicosatrienoic acid (C20:3 *n*-6)	1.57 ± 0.65 a	0.88 ± 2.33 b	0.93 ± 0.48 b
Cis-5, 8, 11, 14, 17-eicosapentaenoic acid (C20:5 *n*-3, EPA)	0.06 ± 0.35 a	nd	nd
Cis-4,7,10,13,16,19-docosahexaenoic acid (C22:6 *n*-3, DHA)	20.05 ± 1.14 a	9.13 ± 10.21 c	11.78 ± 5.78 b
Total PUFA	23.46 ± 1.78 a	11.06 ± 8.41 c	14.13 ± 2.45 b
Total *n*-3 PUFA	20.16 ± 1.15 a	9.07 ± 6.84 c	11.84 ± 2.27 b
Total *n*-6 PUFA	3.29 ± 0.95 a	1.93 ± 2.78 c	2.30 ± 0.72 b
Ratio *n*-3/*n*-6	6.12 ± 0.17 a	4.70 ± 0.21 c	5.15 ± 0.10 b

nd: not detected. Values are given as mean ± standard deviation from triplicate determinations. Different letters in the same row indicate significant differences ($p < 0.05$).

MUFAs and PUFAs have long been thought to be beneficial fats, especially when used to replace saturated fats in the diet. This helps to keep blood cholesterol levels in the normal range and has been connected to a lower risk of coronary heart disease [75]. Total *n*-3 PUFA to total *n*-6 PUFA ratios in polar lipid fraction, neutral lipid fraction, and total lipid were 4.72, 5.16, and 6.12, respectively. The pig brain contains a considerable number of important *n*-3 fatty acids, as seen by the relatively high *n*-3/*n*-6 PUFA ratio. Because of their high bioavailability, *n*-3 fatty acid-containing phospholipids have a considerable preventive effect against cardiovascular disease, antioxidant activity, memory improvement, and other possible health advantages [76].

4. Conclusions

Important nutrients, such as phospholipids, proteins, essential amino acids (leucine, threonine, valine, lysine, phenylalanine, isoleucine, histidine, and methionine), DHA, and minerals (e.g., P, K, Ca, Mg, Fe, Zn, and Cu), were found in the pig brain. The pig brain in this experiment also included a significant amount of UAA, which can be employed to improve the sensory qualities of pig brain-based products. As a result, the pig brain could be used as a source of nutrients in food preparations. On the other hand, the pig brain exhibited high quantities of SFA and cholesterol, which should be taken into account. To properly investigate the usage of pig brain for human nutrition, such components may need to be removed. To achieve the zero-waste goal, the cholesterol and SFA fractions, on the other hand, may be used in cosmetics in the future. Using the optimal refining approach, the nutritious protein with techno-functionality from pig brain may be isolated, along with the lipid separation for food sustainability.

Author Contributions: Conceptualization, J.C., W.P., L.-Z.C. and M.C.; methodology, J.C., W.P., A.P., L.-Z.C. and M.C.; software, J.C. and N.P.; validation, W.P. and M.C.; investigation, J.C.; resources, W.P., A.P. and M.C.; data curation, J.C. and N.P.; writing—original draft preparation, J.C., W.P. and M.C.; writing—review and editing, J.C., W.P., L.-Z.C. and M.C.; supervision, M.C. and W.P.; funding acquisition, M.C. and W.P. All authors have read and agreed to the published version of the manuscript.

Funding: This research was supported by National Research Council of Thailand (NRCT) under the Research and Researchers for Industries (RRI) program, Thailand [grant number PHD61I0056].

Institutional Review Board Statement: Not applicable.

Informed Consent Statement: Not applicable.

Data Availability Statement: Not applicable.

Acknowledgments: We would like to thank Food Technology and Innovation Research Center of Excellence, Walailak University, for facility support.

Conflicts of Interest: The authors declare no conflict of interest.

References

1. Department of Livestock. *Statistics Livestock in Thailand;* Bureau of Livestock Development and Development, Ministry of Agriculture and Cooperatives: Bangkok, Thailand, 2020.
2. Ockerman, H.W.; Basu, L. By-products. In *Encyclopedia of Meat Sciences*, 1st ed.; Jensen, W.K., Devine, C., Dikeman, M., Eds.; Elsevier Academic Press: Amsterdam, The Netherlands, 2004; pp. 104–112.
3. Toldrá, F.; Aristoy, M.-C.; Mora, L.; Reig, M. Innovations in value-addition of edible meat by-products. *Meat Sci.* **2012**, *92*, 290–296. [CrossRef]
4. Campbell, R.E.; Kenney, P.B. Edible By-products from the Production and Processing of Muscle Foods. In *Muscle Foods*, 1st ed.; Campbell, R.E., Kenney, P.B., Eds.; Springer: Singapore, 1994; pp. 79–105.
5. Nollet, L.M.L.; Toldrá, F. Introduction of offal meat: Definitions, regions, cultures, generalities. In *Handbook of Analysis of Edible Animal By-Products*, 1st ed.; Nollet, L.M.L., Toldrá, F., Eds.; CRC Press: Boca Raton, FL, USA, 2011; pp. 3–11.
6. Estévez, M.; Ventanas, J.; Cava, R.; Puolanne, E. Characterisation of a traditional Finnish liver sausage and different types of Spanish liver pâtés: A comparative study. *Meat Sci.* **2005**, *71*, 657–669. [CrossRef]
7. Santos, E.M.; González-Fernández, C.; Jaime, I.; Rovira, J. Physicochemical and sensory characterization of Morcilla de Burgos, a traditional Spanish blood sausage. *Meat Sci.* **2003**, *65*, 893–898. [CrossRef]
8. Collinge, J. Prion diseases of humans and animals: Their causes and molecular basis. *Annu. Rev. Neurosci.* **2001**, *24*, 519–550. [CrossRef] [PubMed]
9. AOAC. *Official Methods of Analysis*, 16th ed.; Association of Official Analytical Chemists: Washington, DC, USA, 2000.
10. Bligh, E.G.; Dyer, W.J. A rapid method of total lipid extraction and purification. *Can. J. Biochem. Physiol.* **1959**, *37*, 911–917. [CrossRef]
11. Stewart, J.C.M. Colorimetric determination of phospholipids with ammonium ferrothiocyanate. *Anal. Biochem.* **1980**, *104*, 10–14. [CrossRef]
12. Beyer, R.S.; Jensen, L.S. Overestimation of the cholesterol content of eggs. *J. Agric. Food Chem.* **1989**, *37*, 917–920. [CrossRef]
13. Chaijan, M.; Undeland, I. Development of a new method for determination of total haem protein in fish muscle. *Food Chem.* **2015**, *173*, 1133–1141. [CrossRef]
14. Chen, H.-H.; Chiu, E.-M.; Huang, J.-R. Color and gel-forming properties of horse mackerel (*Trachurus japonicus*) as related to washing conditions. *J. Food Sci.* **1997**, *62*, 985–991. [CrossRef]
15. Hashimoto, K.; Watabe, S.; Kono, M.; Shiro, K. Muscle protein composition of sardine and mackerel. *Bull. Ja. Soc. Sci. F.* **1979**, *45*, 1435–1441. [CrossRef]
16. Chinarak, K.; Panpipat, W.; Summpunn, P.; Panya, A.; Phonsatta, N.; Cheong, L.-Z.; Chaijan, M. Insights into the effects of dietary supplements on the nutritional composition and growth performance of sago palm weevil (*Rhynchophorus ferrugineus*) larvae. *Food Chem.* **2021**, *363*, 130279. [CrossRef] [PubMed]
17. Chaijan, M.; Panpipat, W. Feasibility of a pH driven method for maximizing protein recovery of over-salted albumen. *Food Biosci.* **2018**, *24*, 89–94. [CrossRef]
18. Estefanell, J.; Mesa-Rodríguez, A.; Ramírez, B.; La Barbera, A.; Socorro, J.; Hernandez-Cruz, C.M.; Izquierdo, M.S. Fatty acid profile of neutral and polar lipid fraction of wild eggs and hatchlings from wild and captive reared broodstock of *Octopus vulgaris*. *Front. Physiol.* **2017**, *8*, 453. [CrossRef] [PubMed]

19. Seong, P.N.; Park, K.M.; Cho, S.H.; Kang, S.M.; Kang, G.H.; Park, B.Y.; Moon, S.S.; Van Ba, H. Characterization of edible pork by-products by means of yield and nutritional composition. *Korean J. Food Sci. Anim. Resour.* **2014**, *34*, 297–306. [CrossRef] [PubMed]
20. Hamilton, J.A.; Hillard, C.J.; Spector, A.A.; Watkins, P.A. Brain Uptake and Utilization of Fatty Acids, Lipids and Lipoproteins: Application to Neurological Disorders. *J. Mol. Neurosci.* **2007**, *33*, 2–11. [CrossRef]
21. Krafft, C.; Neudert, L.; Simat, T.; Salzer, R. Near infrared Raman spectra of human brain lipids. *Spectrochim. Acta A* **2005**, *61*, 1529–1535. [CrossRef]
22. Pérez-Palacios, T.; Ruiz, J.; Martín, D.; Muriel, E.; Antequera, T. Comparison of different methods for total lipid quantification in meat and meat products. *Food Chem.* **2008**, *110*, 1025–1029. [CrossRef] [PubMed]
23. Ma, J.; Sun, D.-W.; Pu, H.; Wei, Q.; Wang, X. Protein content evaluation of processed pork meats based on a novel single shot (snapshot) hyperspectral imaging sensor. *J. Food Eng.* **2019**, *240*, 207–213. [CrossRef]
24. Mamelona, J.; Santi-Louis, R.; Pelletier, E. Nutritional composition and antioxidant properties of protein hydrolysates pre-pared from echinoderm byproducts. *J. Food Sci. Technol.* **2010**, *45*, 147–154. [CrossRef]
25. Pereira, P.M.D.C.C.; Vicente, A.F.D.R.B. Meat nutritional composition and nutritive role in the human diet. *Meat Sci.* **2013**, *93*, 586–592. [CrossRef] [PubMed]
26. Buang, Y.; Wang, Y.-M.; Cha, J.-Y.; Nagao, K.; Yanagita, T. Dietary phosphatidylcholine alleviates fatty liver induced by orotic acid. *Nutrition* **2005**, *21*, 867–873. [CrossRef]
27. Rossmeisl, M.; Medrikova, D.; van Schothorst, E.; Pavlisova, J.; Kuda, O.; Hensler, M.; Bardova, K.; Flachs, P.; Stankova, B.; Vecka, M.; et al. Omega-3 phospholipids from fish suppress hepatic steatosis by integrated inhibition of biosynthetic pathways in dietary obese mice. *Biochim. Biophys. Acta (BBA) Mol. Cell Biol. Lipids* **2014**, *1841*, 267–278. [CrossRef] [PubMed]
28. Chinarak, K.; Chaijan, M.; Panpipat, W. Farm-raised sago palm weevil (Rhynchophorus ferrugineus) larvae: Potential and challenges for promising source of nutrients. *J. Food Compos. Anal.* **2020**, *92*, 103542. [CrossRef]
29. Zhong, Y.; Madhujith, T.; Mahfouz, N.; Shahidi, F. Compositional characteristics of muscle and visceral oil from steelhead trout and their oxidative stability. *Food Chem.* **2007**, *104*, 602–608. [CrossRef]
30. Albalat, A.; Nadler, L.E.; Foo, N.; Dick, J.R.; Watts, A.J.R.; Philp, H.; Neil, D.M.; Monroig, O. Lipid Composition of Oil Extracted from Wasted Norway Lobster (Nephrops norvegicus) Heads and Comparison with Oil Extracted from Antarctic Krill (Euphasia superba). *Mar. Drugs* **2016**, *14*, 219. [CrossRef]
31. Lu, F.S.H.; Nielsen, N.S.; Baron, C.P.; Jensen, L.H.S.; Jacobsen, C. Physico-chemical properties of marine phospholipid emul-sions. *J. Am. Oil Chem. Soc.* **2012**, *89*, 2011–2024. [CrossRef]
32. Pan, Y.; Tikekar, R.V.; Nitin, N. Effect of antioxidant properties of lecithin emulsifier on oxidative stability of encapsulated bioactive compounds. *Int. J. Pharm.* **2013**, *450*, 129–137. [CrossRef] [PubMed]
33. Ohvo-Rekilä, H.; Ramstedt, B.; Leppimäki, P.; Slotte, J.P. Cholesterol interactions with phospholipids in membranes. *Prog. Lipid Res.* **2002**, *41*, 66–97. [CrossRef]
34. Petrov, A.M.; Kasimov, M.R.; Zefirov, A.L. Brain Cholesterol Metabolism and Its Defects: Linkage to Neurodegenerative Diseases and Synaptic Dysfunction. *Acta Naturae* **2016**, *8*, 58–73. [CrossRef]
35. Dietschy, J.M. Central nervous system: Cholesterol turnover, brain development and neurodegeneration. *Biol. Chem.* **2009**, *390*, 287–293. [CrossRef]
36. Ma, H. Cholesterol and human health. *Nat. Sci.* **2004**, *2*, 17–21.
37. Naviglio, D.; Gallo, M.; Le Grottaglie, L.; Scala, C.; Ferrara, L.; Santini, A. Determination of cholesterol in Italian chicken eggs. *Food Chem.* **2012**, *132*, 701–708. [CrossRef]
38. Carrier, M.; Guilbert, J.; Lévesque, J.-P.; Tremblay, M.-È.; Desjardins, M. Structural and Functional Features of Developing Brain Capillaries, and Their Alteration in Schizophrenia. *Front. Cell. Neurosci.* **2020**, *14*, 595002. [CrossRef]
39. Abdollahi, M.; Marmon, S.; Chaijan, M.; Undeland, I. Tuning the pH-shift protein-isolation method for maximum hemo-globin-removal from blood rich fish muscle. *Food Chem.* **2016**, *212*, 213–224. [CrossRef] [PubMed]
40. Chaijan, M.; Panpipat, W.; Banerjee, R.; Verma, A.; Siddiqui, M. Mechanism of Oxidation in Foods of Animal Origin. In *Natural Antioxidants*, 1st ed.; Banerjee, R., Verma, A.K., Siddiqui, M.W., Eds.; CRC Press: Waretown, NJ, USA, 2016; pp. 1–37.
41. Bhattacharya, P.T.; Misra, S.R.; Hussain, M. Nutritional aspects of essential trace elements in oral health and disease: An ex-tensive review. *Scientifica* **2016**, *5464373*. [CrossRef] [PubMed]
42. Wolff, D.J.; Siegel, F.L. Purification of a Calcium-binding Phosphoprotein from Pig Brain. *J. Biol. Chem.* **1972**, *247*, 4180–4185. [CrossRef]
43. Wu, R.A.; Ding, Q.; Yin, L.; Chi, X.; Sun, N.; He, R.; Luo, L.; Ma, H.; Li, Z. Comparison of the nutritional value of mysore thorn borer (Anoplophora chinensis) and mealworm larva (Tenebrio molitor): Amino acid, fatty acid, and element profiles. *Food Chem.* **2020**, *323*, 126818. [CrossRef] [PubMed]
44. Kengne, F.G.; Decaux, G. Hyponatremia and the Brain. *Kidney Int. Rep.* **2018**, *3*, 24–35. [CrossRef]
45. Kirkland, A.; Sarlo, G.L.; Holton, K.F. The Role of Magnesium in Neurological Disorders. *Nutrients* **2018**, *10*, 730. [CrossRef]
46. Tapiero, H.; Tew, K.D. Trace elements in human physiology and pathology: Zinc and metallothioneins. *Biomed. Pharmacother.* **2003**, *57*, 399–411. [CrossRef]
47. Ekstrand, B.; Scheers, N.; Rasmussen, M.K.; Young, J.F.; Ross, A.B.; Landberg, R. Brain foods-the role of diet in brain perfor-mance and health. *Nutr. Rev.* **2021**, *79*, 693–708. [CrossRef]

48. Benoist, B. Iron-deficiency anemia: Reexamining the nature and magnitude of the public health problem. *J. Nutr.* **2001**, *131*, 564S–703S.
49. Pfeiffer, C.C.; Braverman, E.R. Zinc, the brain and behavior. *Biol. Psychiatry* **1982**, *17*, 513–532. [PubMed]
50. Osredkar, J.; Sustar, N. Copper and zinc, biological role and significance of copper/zinc imbalance. *J. Clin. Toxicol.* **2011**, *S3*, 1–18. [CrossRef]
51. Aschner, J.L.; Aschner, M. Nutritional aspects of manganese homeostasis. *Mol. Asp. Med.* **2005**, *26*, 353–362. [CrossRef]
52. Gurd, J.W.; Evans, W.H.; Perkins, H.R. Chemical characterization of the proteins and glycoproteins of mouse liver plasma membranes solubilized by sequential extraction with aqueous and organic solvents. *Biochem. J.* **1972**, *126*, 459–466. [CrossRef]
53. Panpipat, W.; Chaijan, M. Biochemical and physicochemical characteristics of protein isolates from bigeye snapper (*Priacanthus tayenus*) head by-product using pH shift method. *Turk. J. Fish. Aquat. Sci.* **2016**, *16*, 41–50.
54. Panpipat, W.; Chaijan, M. Functional properties of pH-shifted protein isolates from bigeye snapper (Priacanthus tayenus) head by-product. *Int. J. Food Prop.* **2017**, *20*, 596–610. [CrossRef]
55. Bernier, I.; Tresca, J.-P.; Jollès, P. Ligand-binding studies with a 23 kDa protein purified from bovine brain cytosol. *Biochim. Biophys. Acta* **1986**, *871*, 19–23. [CrossRef]
56. McMahon, H.T.; Missler, M.; Li, C.; Südhof, T.C. Complexins: Cytosolic proteins that regulate SNAP receptor function. *Cell* **1995**, *83*, 111–119. [CrossRef]
57. Seppänen, A.; Suuronen, T.; Hofmann, S.C.; Majamaa, K.; Alafuzoff, I. Distribution of collagen XVII in the human brain. *Brain Res.* **2007**, *1158*, 50–56. [CrossRef] [PubMed]
58. Yokoi, K.; Kojic, M.; Milosevic, M.; Tanei, T.; Ferrari, M.; Ziemys, A. Capillary-wall collagen as a biophysical marker of nanotherapeutic permeability into the tumor microenvironment. *Cancer Res.* **2014**, *74*, 4239–4246. [CrossRef] [PubMed]
59. Machado, M.; Machado, S.; Pimentel, F.B.; Freitas, V.; Alves, R.C.; Oliveira, M.B.P.P. Amino acid profile and protein quality assessment of macroalgae produced in an integrated multi-trophic aquaculture system. *Foods* **2020**, *9*, 1382. [CrossRef]
60. Joint WHO/FAO/UNU Expert Consultation. *Protein and Amino Acid Requirements in Human Nutrition*; World Health Organization Technical Report Series; World Health Organization: Geneva, Switzerland, 2007; pp. 1–265.
61. Kulma, M.; Kouřimská, L.; Homolková, D.; Božik, M.; Plachý, V.; Vrabec, V. Effect of developmental stage on the nutritional value of edible insects. A case study with Blaberus craniifer and Zophobas morio. *J. Food Compos. Anal.* **2020**, *92*, 103570. [CrossRef]
62. Fan, S.H.; Liu, T.T.; Wan, P.; Zhu, Q.; Xia, N.; Wang, Q.Z.; Chen, D.W. Enrichment of the umami-taste-active amino acids and peptides from crab sauce using ethanol precipitation and anion-exchange resin. *J. Food Process. Preserv.* **2021**, *45*, e15390. [CrossRef]
63. Chaijan, M.; Chumthong, K.; Kongchoosi, N.; Chinarak, K.; Panya, A.; Phonsatta, N.; Cheong, L.; Panpipat, W. Characterisation of pH-shift-produced protein isolates from sago palm weevil (*Rhynchophorus ferrugineus*) larvae. *J. Insects Food Feed.* **2021**, 1–12. [CrossRef]
64. Sun, L.; Liu, Q.; Bao, C.; Fan, J. Comparison of free total amino acid compositions and their functional classifications in 13 wild edible mushrooms. *Molecules* **2017**, *22*, 350. [CrossRef]
65. Sasaki, K.; Motoyama, M.; Mitsumoto, M. Changes in the amounts of water-soluble umami-related substances in porcine longissimus and biceps femoris muscles during moist heat cooking. *Meat Sci.* **2007**, *77*, 167–172. [CrossRef]
66. Wu, G. Functional amino acids in nutrition and health. *Amino Acids* **2013**, *45*, 407–411. [CrossRef]
67. Trevino, S.R.; Scholtz, J.M.; Pace, C.N. Amino acid contribution to protein solubility: Asp, Glu, and Ser contribute more fa-vorably than the other hydrophilic amino acids in RNase Sa. *J. Mol. Biol.* **2007**, *366*, 449–460. [CrossRef]
68. Guillén, M.D.; Cabo, N. Infrared spectroscopy in the study of edible oils and fats. *J. Sci. Food Agric.* **1997**, *75*, 1–11. [CrossRef]
69. Lerma-García, M.J.; Gori, A.; Cerretani, L.; Simó-Alfonso, E.F.; Caboni, M.F. Classification of Pecorino cheeses produced in Italy according to their ripening time and manufacturing technique using Fourier transform infrared spectroscopy. *J. Dairy Sci.* **2010**, *93*, 4490–4496. [CrossRef] [PubMed]
70. Nagarajan, M.; Benjakul, S.; Prodpran, T.; Songtipya, P.; Kishimura, H. Characteristics and functional properties of gelatin from splendid squid (Loligo formosana) skin as affected by extraction temperatures. *Food Hydrocoll.* **2012**, *29*, 389–397. [CrossRef]
71. Riaz, T.; Zeeshan, R.; Zarif, F.; Ilyas, K.; Muhammad, N.; Safi, S.Z.; Rahim, A.; Rizvi, S.A.A.; Rehman, I.U. FTIR analysis of natural and synthetic collagen. *Appl. Spectrosc. Rev.* **2018**, *53*, 703–746. [CrossRef]
72. Wolkers, W.F.; Oliver, A.E.; Tablin, F.; Crowe, J.H. A Fourier-transform infrared spectroscopy study of sugar glasses. *Carbohydr. Res.* **2004**, *339*, 1077–1085. [CrossRef] [PubMed]
73. Bruce, K.D.; Zsombok, A.; Eckel, R.H. Lipid processing in the brain: A key regulator of systemic metabolism. *Front. Endocrinol.* **2017**, *8*, 60. [CrossRef] [PubMed]
74. Mas, G.; Llavall, M.; Coll, D.; Roca, R.; Díaz, I.; Oliver, M.; Gispert, M.A.; Realini, C.E. Effect of an elevated monounsaturated fat diet on pork carcass and meat quality traits and tissue fatty acid composition from York-crossed barrows and gilts. *Meat Sci.* **2011**, *89*, 419–425. [CrossRef]
75. Degirolamo, C.; Rudel, L.L. Dietary monounsaturated fatty acids appear not to provide cardioprotection. *Curr. Atheroscler. Rep.* **2010**, *12*, 391–396. [CrossRef]
76. Bao, Z.; Zhang, P.; Chen, J.; Gao, J.; Lin, S.; Sun, N. Egg yolk phospholipids reverse scopolamine–induced spatial memory deficits in mice by attenuating cholinergic damage. *J. Funct. Foods* **2020**, *69*, 103948. [CrossRef]

foods

MDPI

Article

Functionalization of a Vegan Mayonnaise with High Value Ingredient Derived from the Agro-Industrial Sector

Alessandra De Bruno, Rosa Romeo, Antonio Gattuso, Amalia Piscopo * and Marco Poiana

Department of AGRARIA, University Mediterranea of Reggio Calabria, Vito, 89124 Reggio Calabria, Italy; alessandra.debruno@unirc.it (A.D.B.); rosa.romeo@unirc.it (R.R.); antonio.gattuso@unirc.it (A.G.); mpoiana@unirc.it (M.P.)

* Correspondence: amalia.piscopo@unirc.it; Tel.: +39-965-1694366

Abstract: This work aimed to evaluate the antioxidant effect determined by the addition of phenolic extract on the oxidative stability and quality of vegan mayonnaise. Two different antioxidant extracts containing 100 mg L^{-1} of hydroxytyrosol and obtained by olive mill wastewater were used in the preparation. After preliminary studies, already evaluated in other works, on hydrophilic and lipophilic food matrices, the results of this study could contribute to understanding the effects of the enrichment on emulsified food systems with phenolic extracts. The functionalized mayonnaise samples were monitored up to 45 days of storage at 10 °C in comparison with a control sample for microbiological, physicochemical, antioxidant, sensory properties and for oxidative stability. The results achieved through this work showed the efficacy of the use of phenolic extract as ingredients for its positive effect on chemical properties of mayonnaise. The adding extracts lead to the increase of oxidative stability with an induction period higher (about 24 h) than the control sample (about 12 h).

Keywords: antioxidant activity; mayonnaise; olive mill wastewater; oxidative stability; phenolic extract

Citation: De Bruno, A.; Romeo, R.; Gattuso, A.; Piscopo, A.; Poiana, M. Functionalization of a Vegan Mayonnaise with High Value Ingredient Derived from the Agro-Industrial Sector. *Foods* **2021**, *10*, 2684. https://doi.org/10.3390/foods10112684

Academic Editor: Theodoros Varzakas

Received: 24 September 2021
Accepted: 1 November 2021
Published: 3 November 2021

Publisher's Note: MDPI stays neutral with regard to jurisdictional claims in published maps and institutional affiliations.

Copyright: © 2021 by the authors. Licensee MDPI, Basel, Switzerland. This article is an open access article distributed under the terms and conditions of the Creative Commons Attribution (CC BY) license (https://creativecommons.org/licenses/by/4.0/).

1. Introduction

Food industries and, particularly, the olive oil industry produce large quantities of by-products that can be a serious environmental problem. Olive mill wastewater could be an economic and natural source of antioxidants due to its high content of phenolic compounds with a wide array of biological activities [1,2]. Scientific researches have shown that their recovery is important at the environmental level for the reduction of pollution; and in food technologies for different aims such as: nutritious, functional agents and for shelf life extension. Afkhami et al. [3] have studied the orange juice enriched with encapsulated polyphenolic extract of lime waste; Romeo et al. [4,5] studied the using the phenolic extract obtained by olive mill wastewater for the enrichment of hydrophilic model system and the application of natural antioxidants in a lipid system (oil).

Mayonnaise represents one of the most widely consumed sauces in the world [6]: it is a semisolid oil-in-water emulsion, prepared traditionally with egg yolk and 60–80% of oil. The presence of egg in mayonnaise formulation is important both for emulsion and for the taste and color but it is a critical point for the health aspect due to the cholesterol amount [7]. Nowadays an increasing number of people is following a vegetarian or flexitarian diet to prevent cardiovascular diseases resulting from bad nutrition. Many scientific works have been carried out on the possibility of the egg's removal in mayonnaise and replace them with soya, wheat, and milk proteins [8]. For example, soya milk and sunflower oil are used to formulate a vegan mayonnaise. As for all the foods with a high oil content, mayonnaise is susceptible to deterioration due to autooxidation of the unsaturated fats that can negatively affect physicochemical and sensorial attributes of food [9]. Lipid oxidation in mayonnaise causes the development of potentially toxic reaction products, undesirable off-flavors and, simultaneously, it decreases the shelf life of mayonnaise [10,11]. In order to manage these problems, various strategies can be used for avoiding or reduce oxidative processes. One of the common ways to delay lipid oxidation is the use of antioxidants. The efficacy of an antioxidant is influenced by different factors, such as its interaction with other ingredients and its ability to be located at the interface where oxidation takes place.

Generally, synthetic and commercial antioxidants, such as butylated hydroxy toluene (BHT), butylated hydroxy anisole (BHA) and ethylene diamine tetraacetic acid (EDTA), are widely used in mayonnaise to prevent rancidity. However, today the substitution of chemical ingredients with natural ingredients is highly appreciated by the consumer for the health effect and it shows also a great potential for improving food stability against lipid oxidation. Natural antioxidants can act as retarders, when they protect target lipids from oxidation initiators or hinder the propagation phase, the so-called chain-breaking antioxidants [12,13]. For it, the use of plant extracts, rich in antioxidant constituents, such as polyphenols, is an increasing trend in the food industry because they are an alternative to synthetic compounds with reducing and antimicrobial effect [14,15]. The olive oil production generates a considerable amount of olive oil mill waste, rich in organic compounds, mainly phenols. Only a small fraction of phenolic components is transferred to olive oil (1–2%) while the remaining portion is lost in olive oil by-products [16]. This work aimed to evaluate the effect of phenolic extracts obtained by olive mill wastewater in physicochemical and antioxidant characteristics of a vegan mayonnaise during storage period.

2. Materials and Methods

2.1. Phenolic Extract Preparation

Olive of Ottobratica cv were processed by a three-phase centrifugation apparatus in an olive oil mill located in the province of Reggio Calabria. The obtained olive Mill Wastewater (OMWW) were transferred in Food Technologies laboratory of the Mediterranean University of Reggio Calabria (Italy) where were submitted to two different extraction methods.

Method A*:* was carried out following the method reported by Romeo et al. [4]. An aliquot of OMWW was acidified to pH 2 with HCl and washed three times with hexane (1:1, *v*/*v*) in order to remove the lipid fraction. After shaken and centrifuged (Nüve, Ankara, Turkey) the extraction procedure was carried out by means of ethyl acetate three times and the solvent was recovered in a separating funnel (1:4 *v*/*v*). The ethyl acetate was separated and evaporated using a rotary vacuum evaporator at 25 °C. Finally, the dry residues were again dissolved in 100 mL of water, filtered using PTFE 0.45 μm (diameter 15 mm) syringe filter. The obtained sample, named PE_A, was then stored at 4 °C until subsequent analyses.

Method B*:* an aliquot of OMWW was acidified to pH 2 with citric acid. After 30 min of shaken and 5 min of centrifugation (6000 rpm) the sample was filtrated with a paper filter and concentrated in an oven at 50 °C. The final residue characterized by gelatinous consistency was extracted with water (1:5, *w*/*v*) in an ultrasound system (Sonoplus Ultrasonic homogenisers, Series 2000.2, HD 2200.2. BANDELIN, Ultraschall seit 1955) for 30 min. Finally, the obtained extract was filtered using PTFE 0.45 μm (diameter 15 mm) syringe filter. The sample, named PE_B, was then stored at 4 °C until subsequent analyses.

2.2. Mayonnaise Preparation

A schematic overview of the vegan mayonnaise production process is reported in Figure 1. The main ingredients used for the formulation were: soya milk, sunflower oil, lemon juice, salt and phenolic extracts (PE_A and PE_B). All ingredients were mixed using a lab-scale mixer (Bimby TM31, Vorwerk, Wuppertal, Germany) in a three-step process in order to maintain a closely packed emulsion. 1st step: soya milk and salt were mixed (1.100 $g \cdot min^{-1}$, 1 min, 37 °C); 2nd step: sunflower oil and lemon juice was slowly added under continuous mixing (2000 $g \cdot min^{-1}$, 3 min) until a mayonnaise emulsion had been formed; 3rd step: the PE_A and PE_B amounts corresponding to 100 mg L^{-1} of Hydroxytyrosol (respectively, 50 and 45 g) were incorporating to the mixture (300 $g \cdot min^{-1}$, 1 min). Mayonnaise samples were stored in capped containers and refrigerated at 10 °C for storage. All analyses were performed at 0, 15, 30 and 45 days of storage. The two enriched mayonnaises, $EMPE_A$ and $EMPE_B$, were compared with a sample without PE, named control.

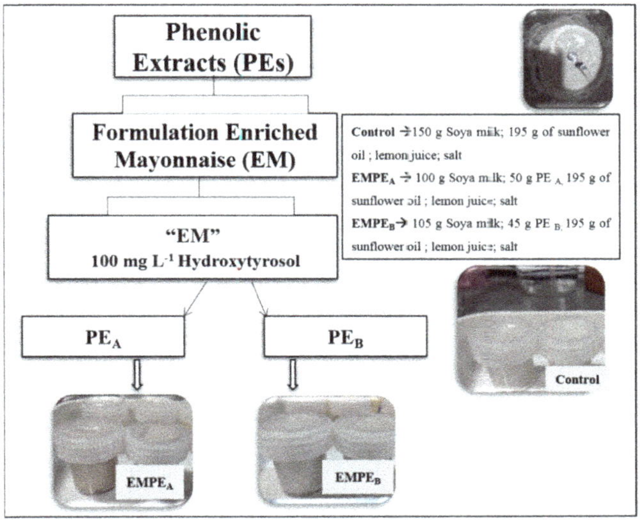

Figure 1. Schematic overview of formulation of mayonnaise.

2.3. Antioxidant Characterization of Phenolic Extracts

The main antioxidant parameters, such as: total phenol content (TPC), ABTS and DPPH assays, were performed spectrophotometrically following the method described by De Bruno et al. [16], with some modifications.

For TPC analysis, 0.1 mL of the phenolic extracts (PE_A and PE_B), were placed in a 25 mL volumetric flask and mixed with 20 mL of deionized water and 0.625 mL of the Folin Ciocalteau reagent. After 3 min, 2.5 mL of a saturated solution of Na_2CO_3 (20%) were added. The content was mixed and diluted to volume with deionized water. Thereafter, the mixture was incubated for 12 h at room temperature and in the dark. The absorbance of the samples was measured at 725 nm against a blank using a double-beam ultraviolet-visible spectrophotometer (Agilent 8453 UV–Vis, Germany) and compared with a gallic acid calibration curve (concentration between 1 and 10 mg L^{-1}). The results were expressed as mg of GAE 100 mL^{-1}.

For DPPH assay, 10 μL of PE extracts (PE_A and PE_B) were added to 2990 μL of a 6×10^{-5} M of methanol solution of DPPH (2.2-diphenyl-1-picrylhydrazyl, Carlo Erba, MI, Italy) in a cuvette and left in the dark for 30 min (till stabilization). The decrement of absorbance was determined by a spectrophotometer at 515 nm against methanol as blank and at the temperature of 20 °C.

For ABTS assay, 10 μL of PE extracts (PE_A and PE_B) were added to 2990 μL of ABTS reaction mixture (2.20-azino-bis (3-ethylbenzothiazoline-6-sulfonic acid), and the absorbance was measured after 6 min at 734 nm against ethanol as blank by a spectrophotometer. For both the assays, the radical scavenging activity was plotted against Trolox concentration (from 1.5 to 24 μM) and the results were expressed as μmol Trolox mL^{-1} of PE.

The quantification of the main phenolic compounds was carried out following the method described by Romeo et al. [4], through a UHPLC-DAD analysis. 5 μL of PE was injected in a UHPLC system that consisted of an UHPLC PLATINblue (Knauer, Berlin, Germany) equipped with a binary pump system using a Knauer blue orchid column C18 (1.8 μm, 100 × 2 mm) coupled with a PDA-1 (Photo Diode Array Detector) PLATINblue (Knauer, Berlin, Germany). The mobile phases were (A) water acidified with acetic acid (pH 3.10) and (B) acetonitrile; the gradient elution program consisted of 0–3 min, 95% A;

3–15 min, 95–60% A; 15–15.5 min, 60–0% A. Finally, returning to the initial conditions was achieved during analysis keeping the column at 30 °C. External standards (concentration between 1 and 100 mg L^{-1}) were used for the quantification and the results were expressed as mg 100 mL^{-1}.

2.4. Physicochemical, Microbiological and Antioxidant Evaluation of Enriched Mayonnaise Samples (EM)

2.4.1. Physicochemical Analysis

The pH of mayonnaise samples was measured at 25 °C using a digital pH meter (Crison Basic 20, Spain) according to AOAC [17]. Total acidity (% oleic acid) was performed according to Official and standard methods (AOCS, [18–20]). The moisture content (MC, %) was tested in an Electronic Moisture Analyser MA37 (Sartorius, Goettingen, Germany). The analysis was performed using 5 g of the sample at 105 °C. The color analysis was evaluated using a reflection colorimeter (Minolta CR 300, Japan) with reference to a CIE L*a*b* coordinates by using a D65 illuminant. Each sample was homogeneously distributed into a glass vessel and the color was recorded at 10 different points.

2.4.2. Microbiological Analysis

The viable populations of the principal groups of microorganisms were determined by plate inoculation and incubation at 32 °C up to 3 days before counting the colonies in the following selective media: total mesophilic bacteria in Plate Count Agar (Plate Count Agar, Conda-Pronadisa, Spain), lactic acid bacteria in MRS Agar (LAB) (Oxoid), yeasts and moulds in OGYA (Oxoid).

2.4.3. Oxidative Stability in Accelerated Storage Test

To investigate the effect of PE extracts in delaying or inhibiting of fat oxidation, mayonnaise samples with and without extract were subjected to high oxidative stress in OXITEST reactor. Oxitest analysis allows to detect the time necessary to reach an end point of oxidation that corresponds to a detectable rancidity or a rapid change in the oxidation rate. An oxidation Test Reactor (VELP Scientifica, Usmate Velate, MB, Italy) was used in order to evaluate the opposition to fat oxidation. This method is recognized by AOCS International Standard Procedure (Cd 12c–16) for the determination of oxidation stability of food, fats, and oils (AOAC, [21]). The analysis consists of monitoring the oxygen uptake of the reactive constituent of food samples to determine the oxidative stability under conditions of accelerated oxidation. Briefly, 5 g of oil sample were distributed homogenously in a hermetically sealed titanium chamber; oxygen was purged into the chamber up to a pressure of 6 bar. The reactor temperature was set at 90 °C. These reaction working conditions allow obtaining the sample Induction Period (IP) within a short time. The OXITEST allows to measure the modification of absolute pressure inside the two chambers and, through the OXISoftTM Software (Version 10002948 Usmate Velate, MB, Italy), automatically generates the IP expressed as hours by the graphical method.

2.4.4. Analysis of Antioxidant Compounds

The extraction of antioxidant compounds from Mayonnaise samples (EM) and the evaluation of antioxidant parameters were carried out following the method reported by Romeo et al. [5], opportunely modified. Two grams of EM were added with 2 mL of methanol: water (70:30) and 2 mL of hexane and mixed with a Vortex for 10 min. The hydro-alcoholic phase was separated from the oil phase in a refrigerated centrifuge apparatus (NF 1200R, Nüve, Ankara, Turkey) at 5000 rpm, 4 °C for 10 min. Hydro-alcoholic extracts were recovered with a syringe, filtered through a 0.45 µm nylon filter, diameter 15 mm (Thermo Fischer Scientific, Waltham, MA, USA), and utilized for the phenolic compounds quantification and antioxidant activity.

For the total phenolic determination in EM, an aliquot of the diluted extract was mixed with 0.300 mL of Folin reagent and 0.25 mL of deionized water and, after 4 min, with 2.4 mL of an aqueous solution of Na_2CO_3 (5%). The mixture was maintained in a 40 °C water bath for 20 min and TPC was determined at 750 nm. The results were expressed as mg of gallic acid equivalent kg^{-1} of Mayonnaise. The total antioxidant capacity assays (DPPH and ABTS) and the determination of the main bioactive phenolic compounds in EM samples were analysed with the same methods reported in Section 2.3, with some modifications. For DPPH and ABTS assays, the radical scavenging activity was expressed as µmol Trolox 100 g^{-1} of EM; while the individual phenolic compounds were expressed as mg kg^{-1} EM.

2.4.5. Sensory Evaluation

Sensory characteristics including colour, flavour "taste and odour", consistency, appearance, overall acceptability was evaluated in EM. The test was performed by a panel of 8 judges (males and females) from 25 to 50 years old, recruited among researchers and technicians of the Food Science and Technology Unit of Reggio Calabria University with previous experience in sensory analysis. The judges were trained before the sessions to identify the attributes to be evaluated Sensory data were elaborated by calculating the median of results.

2.5. Statistical Analysis

Results of the present study were expressed as mean ± SD of three measurements (n = 3). Appropriate test statistics, Multivariate and One-way ANOVA with Tukey's post-hoc test, and t-test were at $p < 0.05$ were performed by SPSS Software (Version 15.0, SPSS Inc., Chicago, IL, USA).

3. Results

3.1. Characterization of Phenolic Extracts

The main antioxidant parameters evaluated on the two phenolic extracts (PE_A and PE_B) were reported in Table 1. Significant differences were noted between the extracts, particularly for TPC and ABTS assays with higher results in PE_A (TPC: 7895 mg GAE L^{-1} PE; ABTS: 28,604 µmol TE mL^{-1} PE) respect to PE_B (TPC: 7258 mg GAE L^{-1} PE; ABTS: 25,716 µmol TE mL^{-1} PE).

Table 1. Antioxidant parameters and individual Phenolic Compounds of phenolic extracts.

Antioxidant Properties	PE_A	PE_B	Sign
DPPH	1156 ± 18	1071 ± 9	*
ABTS	25,716 ± 35	28,604 ± 18	**
TPC	7895 ± 8	7258 ± 14	**
Hydroxytyrosol	739 ± 1	837 ± 4	**
Tyrosol	152 ± 2	148 ± 0.3	ns
Chlorogenic Acid	17 ± 0.1	16 ± 0.3	ns
Vanillic Acid	39 ± 0.0	40 ± 0.5	ns
Caffeic Acid	25 ± 0.2	21 ± 0.3	**
p-coumaric Acid	64 ± 0.3	61 ± 0.0	**
Oleuropein	28 ± 0.8	75 ± 0.3	**

Note: The data are presented as means ± SD. Student's t test performed between the two phenolic extracts (PE_A and PE_B): * significant difference at $p < 0.05$; ** significant difference at $p < 0.01$. ns not significant. µmol TE mL^{-1} PE for ABTS and DPPH and mg 100 mL^{-1} PE for TPC and single phenolics.

Considering the partition coefficient of wastewaters phenols mixture, the extraction with ethyl acetate by different steps allows retaining most phenolic compounds soluble in the organic phase, as reported by Soberón et al. [22]. This explains the results observed in PE_A. On the other hand, PE_B was obtained by water extraction, so it was characterized by phenols insoluble in the organic phase. The only problem linked to the extract PE_A could be represented by the typology of solvents used for the extraction, namely hexane and ethyl acetate, but the extractive procedure was carefully carried out with the aim to use the obtained antioxidant as functional ingredients in food matrixes. For this reason, the solvent has been totally evaporated at the end of the extraction and the solutes had to be recovered with water. To verify if traces of solvent (hexane and Ethyl acetate) persisted in the phenolic extract before the food application, we analyzed the headspace of the hydrophilic phase (through a GC-MS) and have proved absent.

Similarly, differences in antioxidant activity of the phenolic extracts may be ascribed to the polarity of extracting solvents and thus to the chemical characteristics of extracted compounds [23]. The antioxidant activity measured by ABTS assay showed a higher value for both extracts compared to DPPH assay. Likewise, Bibi Sadeer et al. [24] investigated that ABTS cationic radical showed high solubility in organic and aqueous media, thus it is capable to screen the activity of both lipophilic and hydrophilic compounds. In contrast, DPPH radical dissolves in an organic medium reacting only with lipophilic phenolics. The principal phenolics in the extracts were hydroxytyrosol (PE_A: 759 and PE_B: 837 mg 100 mL^{-1}) and tyrosol (PE_A: 152 and PE_B: 148 mg 100 mL^{-1}), in agreement with literature [5,25]. Di Mattia et al. [26] reported that tyrosol and hydroxytyrosol are effective in preventing primary and secondary oxidation in o/w emulsion ensuring the oxidative stability during storage.

3.2. *Qualitative and Quantitative Characterization of Enriched Vegan Mayonnaise (EM)*

3.2.1. Physicochemical Aspects

The colour of enriched mayonnaise was evaluated after the phenolic enrichment, considering that it is the main parameters which affect the consumer's choice (Table 2).

Table 2. Colour parameters of mayonnaise during storage period (days).

Parameters	Time	Control	$EMPE_A$	$EMPE_B$	Sign.
L*	0	89.03 ± 0.39 ^{aA}	82.10 ± 0.47 ^{bB}	76.26 ± 0.20 ^{cB}	**
	15	88.93 ± 0.54 ^{aAB}	82.33 ± 0.33 ^{bB}	76.29 ± 0.16 ^{cB}	**
	30	88.37 ± 0.37 ^{aB}	82.47 ± 0.29 ^{bB}	76.45 ± 0.36 ^{cB}	**
	45	87.59 ± 0.67 ^{aC}	83.12 ± 0.21 ^{bA}	76.82 ± 0.29 ^{cA}	**
	Sign.	**	**	**	
a*	0	−0.29 ± 0.07 ^{c}	3.41 ± 0.03 ^{bA}	3.83 ± 0.03 ^{aA}	**
	15	−0.28 ± 0.09 ^{c}	3.44 ± 0.05 ^{bA}	3.79 ± 0.08 ^{aA}	**
	30	−0.27 ± 0.05 ^{c}	3.17 ± 0.10 ^{bB}	3.48 ± 0.17 ^{aB}	**
	45	−0.25 ± 0.09 ^{c}	2.59 ± 0.05 ^{bC}	3.30 ± 0.05 ^{aC}	**
	Sign.	ns	**	**	
b*	0	10.78 ± 0.12 ^{cB}	12.86 ± 0.11 ^{bC}	14.71 ± 0.07 ^{aC}	**
	15	10.68 ± 0.26 ^{cB}	12.89 ± 0.12 ^{bBC}	14.77 ± 0.12 ^{aC}	**
	30	10.89 ± 0.04 ^{cB}	12.98 ± 0.08 ^{bB}	15.13 ± 0.17 ^{aB}	**
	45	11.16 ± 0.28 ^{cA}	14.11 ± 0.06 ^{bA}	16.57 ± 0.19 ^{aA}	**
	Sign.	**	**	**	

Note: The data are presented as means ± SD. Means within a row with different letters are significantly different by Tukey's post hoc test. Abbreviation: ns, not significant. ** Significance at $p < 0.01$. Small letters show differences among the different samples and capital letters show differences for the single sample during the storage period.

Moreover, the monitoring of its colour was considered crucial to verify the formation of compounds following an oxidative deterioration. The replacement of ingredients compared with the traditional formulation of mayonnaise, leads to a physical and chemical variation, and can have an effect on colour of the final products [27], in particular in this study which involved the use of brown extracts. The addition of phenolic extracts (PEs) and the storage time promoted a significant variation of colour parameters ($p < 0.05$) of enriched mayonnaise. Lightness decreased after phenolic extracts (PEs) addition, more with PE_B, whereas vegan mayonnaises denoted higher a* and b* parameters after the enrichment. Storage time leads an increase of yellowness parameter and a decrease of redness parameter showing a trend opposite to that proved by Altunkaya et al. [28]. In contrast, no variations were observed for redness between the first and the last day of storage for Control sample. Previous research has proved that colour parameters, in particular lightness, are related to fat droplet sizes [29]. Probably, the modification of fat droplet size that occurred following the addition of phenolic compounds may produce the colour detected changes [30].

Food safety and quality are important to consumers. As it is well known, the pH, acidity values and moisture content play an important role in chemical and microbiological stability of fat foods. For this, in order to evaluate the potential application of PEs, all samples were subjects to chemical and microbiological analysis. The pH values of mayonnaise samples analysed at 1st time ranged from 2.92 to 5.01 (Control: 5.01 > EMPE$_A$: 2.92 > EMPE$_B$: 3.74), therefore, the addition of PE allows an acidification of the enriched samples. slightly lower pH value (Control: 4.97 > EMPE$_A$: 2.97 > EMPE$_B$: 3.85) were observed at the end of storage period (45 days) according to Rasmy et al. [31]. The two enriched samples showed a decrease of the TA during the time of storage (Figure 2), instead the control sample showed increase of acidity. The highest measured value was in the sample EMPE$_A$ (9.41 g Oleic acid 100 g^{-1} mayonnaise). The high acidity value is consistent with pH of PE (pH 2), that induced a decrease of emulsion pH and an increase of TA value.

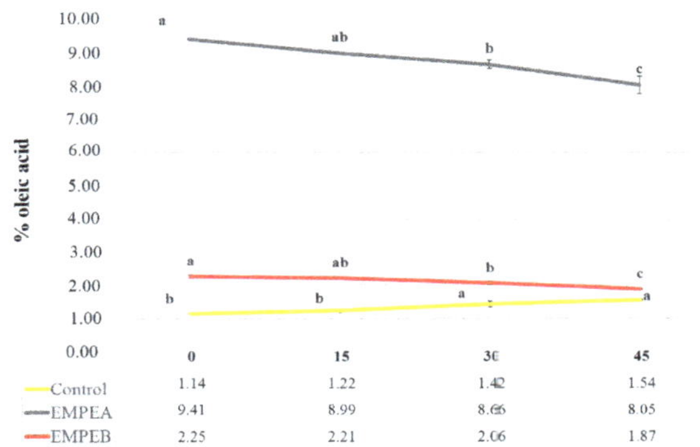

Figure 2. Changes in Total Acidity in the samples during storage time. Different letters show differences for $p < 0.05$.

The highest moisture content value was determined in EMPE$_A$ sample (40.43%) at 1st time, this value decreased during the storage period, indeed at the 45th day was of about 35.95% (Figure 3). In addition, the Control sample showed a significant variation of the moisture content from 38.38 to 30.88%. EMPE$_B$ showed instead a slight, no significant variation over time (35.18 to 35.72%).

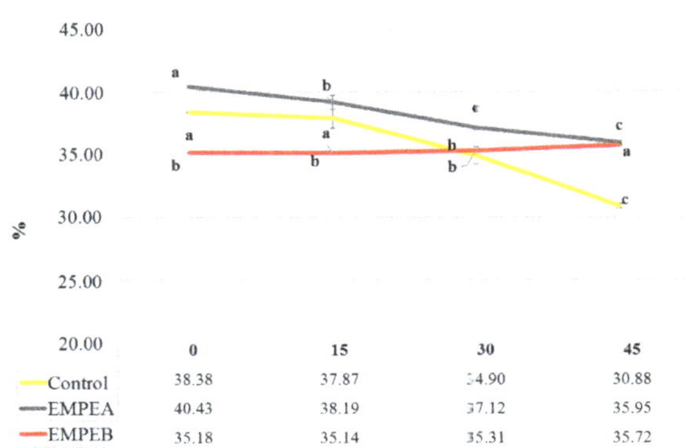

Figure 3. Changes in Moisture content in the samples during storage time. Different letters show differences for $p < 0.05$.

3.2.2. Microbiological Parameters

All samples were evaluated for mesophilic aerobic, yeast, moulds and lactic bacteria count. The results detected in all samples were below quantification limits (<1 cfu/mL, data not shown) for 45 days, accordance with that reported also by Martillanes et al. [29]. Probably, the pH conditions of the mayonnaise samples prevented the growth of food spoilage microorganisms [10], acting as an antimicrobial agent.

3.2.3. Sensory Parameters

The enrichment with PE, leads to a variation of sensory parameters compared to the control sample (Figure 4). In general, all the tested samples showed differences for descriptors among them, except for saltness. Flavour, bitterness, spreadability were affected by the addition of PE; in particular, the increase of the perception of bitterness meant such as acid and pungent taste, was linked probably to the high acidity of the PE_B extract (pH 2), as well as to the amount of oleuropein occurred in the enriched samples, acknowledged as responsible of bitter tasting. The natural proteins present in soya milk, determine the formation of the emulsion. Enriched Mayonnaise resulted less consistent than the control sample, probably due to the partial substitution of soya milk with PE. As described by Giacintucci et al. [32], the incorporation of PE in fact modifies the dispersion degree of emulsion with consequences on hardness, consistency end elasticity of samples. Overall, although the sensory evaluation reveals that the addition of PE interferes with the main sensorial attributes, the overall acceptability of $EMPE_B$ can be considered good compared to the control sample.

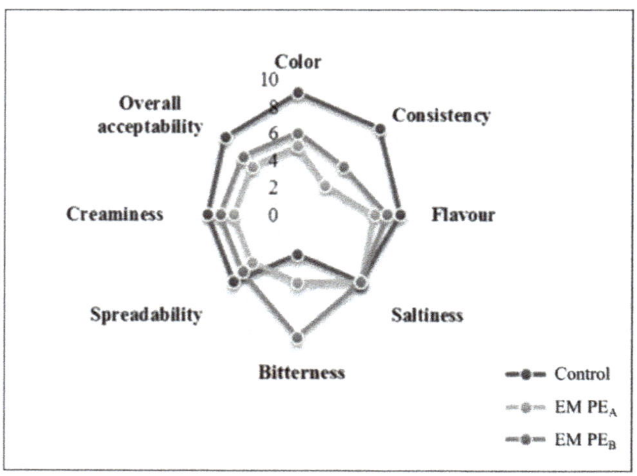

Figure 4. Spider plot of sensory attributes of mayonnaise samples.

3.2.4. Oxidative Stability and Antioxidant Activity of EM

The rate of lipid oxidation in an emulsion is influenced by several factors, including the molecular structure of lipids, heat, light, physical characteristics of emulsion droplets and processing conditions [33]. As it can see in Figure 5, at 1st day of production, enriched samples showed the longer induction period ($EMPE_A$: 25:15 h and $EMPE_B$: 23:57 h) compared to the control sample (13:05 h). Although, after 45 days of storage lower induction periods were observed in all samples, PEs seems to exert a protective role on thermal oxidative stability of emulsions. At the end of storage, the resistance to rancidity was found to be of 33% and 58% higher rather than the Control for $EMPE_A$ and $EMPE_B$, respectively.

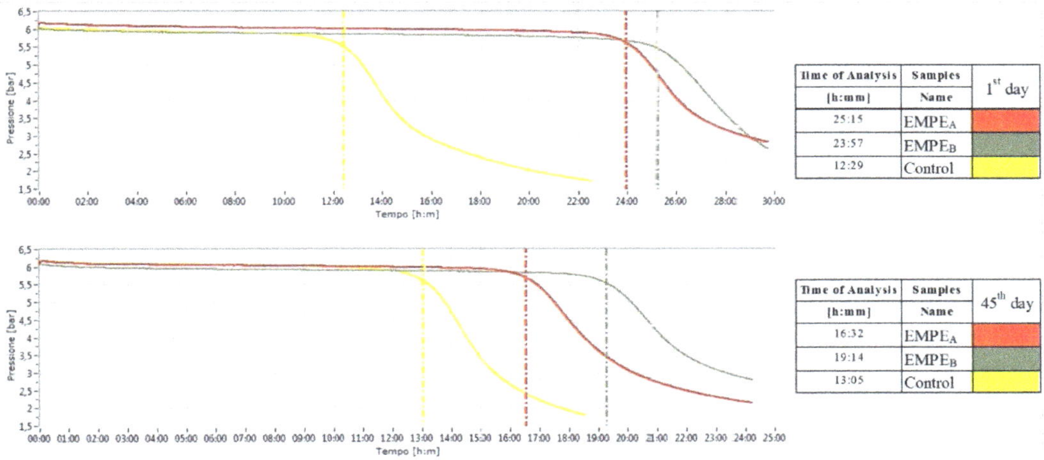

Figure 5. Oxidation curves at 1st and 45th day of storage: red (EMPE$_A$), dark grey (EMPE$_B$), yellow (Control).

Likewise, Raikos, [34], showed that the addition of natural antioxidant can be a reliable strategy to improve the resistance to lipid oxidation of fat emulsion. In a previous article written by Paradiso et al. [35], it was reported that the catechol structure characterizing, e.g., hydroxytyrosol and oleuropein exerts a marked inhibiting activity towards oxidation in emulsion. To verify the actual formulation effect on the inhibition of rancidity processes, chromatographic quantification and antioxidant evaluation were performed. UHPLC analysis showed that the main compounds were transferred from extracts to emulsion. The highest content of hydroxytyrosol was detected in EMPE$_A$ (82.75 mg kg^{-1}) while similar content of tyrosol was quantified in the two EMPE (EMPE$_A$ 19.28 and EMPE$_B$ 18.12 $\pm$ 0.16 mg kg^{-1}). Even though, after 45 days, a significant decrease of Hydroxytyrosol was detected equal to 46% for EMPEA sample and 41% for EMPE$_B$. It is conceivable that the concentration of bioactive compounds was still relevant in term of antioxidant efficiency. However, it is important to point out that mayonnaise is a multiphase system. In this regard, the polarity antioxidant of different compounds which in turn affects their partition into the different phases, play a key role in antioxidant real effectiveness [36].

In view of the above, multiple assays, TPC, DPPH, and ABTS were performed to allow a full insight into the antioxidant capacity of extracts. As reported in Table 3, the scavenging effect of PE$_3$ extracts against ABTS radical cation showed the same trend of TPC. Either way, no significant variations were observed during storage ($p > 0.05$). Only a decrease of 9% in ABTS$^+$ results were observed for samples enriched with PE$_A$ extract. The addition of PE$_B$ extract had a radical scavenging potential against DPPH radical: 134 µmol TE 100 g^{-1} after 1st day while. 78 µmol TE 100 g^{-1} were instead measured in MPE$_A$ at the same storage time. Nevertheless, ANOVA data elaboration reveals a significant effect of storage time on EMPE$_B$. After 45 days a decrement of 25% of antioxidant activity was detected for EMPE$_B$ while no significant variation was observed for EMPE$_A$. The highest results of antioxidant activity were showed by ABTS assay, particularly in EMPE$_B$ samples with values greater than 610 µmol TE 100 g^{-1}. The results obtained from different assays can be correlated to the polarity of compounds present in the food matrix (hydrophilic or lipophilic) for this reason, the antioxidant efficiency has responded better with ABTS test.

Table 3. Variation of the mayonnaise antioxidant parameters at 1st and 45th day of storage.

	Time (Day)	$EMPE_A$	$EMPE_B$	Sign.
DPPH	1st	78 ± 14	134 ± 11	**
	45th	74 ± 11	100 ± 1	*
	Sign.	**n.s.**	**	
ABTS	1st	463 ± 50	613 ± 74	**
	45th	590 ± 38	752 ± 146	**ns**
	Sign.	**	**ns**	
TPC	1st	323 ± 8	413 ± 18	**
	45th	353 ± 18	404 ± 28	*
	Sign.	*	**ns**	
Hydroxytyrosol	1st	82.75 ± 1.06	66.63 ± 0.18	**
	45th	44.00 ± 0.17	39.21 ± 0.08	**
	Sign.	**	**	
Tyrosol	1st	19.28 ± 0.39	18.12 ± 0.16	**ns**
	45th	12.61 ± 0.07	10.32 ± 0.04	**
	Sign.	**	**	
Clorogenic Acid	1st	4.61 ± 0.02	1.93 ± 0.06	**
	45th	3.18 ± 0.04	1.26 ± 0.01	**
	Sign.	**	**	
Vanillic Acid	1st	0.43 ± 0.02	1.66 ± 0.08	**
	45th	0.37 ± 0.02	0.34 ± 0.02	**ns**
	Sign.	**ns**	**	
Caffeic Acid	1st	2.95 ± 0.08	2.25 ± 0.06	*
	45th	2.18 ± 0.18	1.90 ± 0.08	**ns**
	Sign.	*	*	
***p*−cumaric Acid**	1st	6.50 ± 0.11	7.06 ± 0.08	*
	45th	2.55 ± 0.06	4.36 ± 0.07	**
	Sign.	**	**	
Oleuropein	1st	32.45 ± 0.20	32.55 ± 0.64	**ns**
	45th	28.74 ± 0.13	25.16 ± 0.04	**
	Sign.	**	**	

Note: The data are presented as means ± SD. Abbreviation: ns, not significant. ** Significance at $p < 0.01$; * Significance at $p < 0.05$. μmol TE 100 g^{-1} PE for ABTS and DPPH and mg kg^{-1} for TPC and single phenolics.

4. Conclusions

Based on the results, the use of different extracts is a valuable choice to improve the qualitative characteristics of O/W emulsions. The specific phenolic composition of extracts plays a key role in the nutritional parameters of vegan mayonnaise. The concentration of hydroxytyrosol and tyrosol transferred in the samples allows slowing down undesirable oxidation process improving the shelf life of products. In addition, the parameters related to the antioxidant capacity of extracts, TPC, DPPH and ABTS assays, evidenced that the enrichment could have potential health—properties for consumers. Even so, all antioxidant assays indicated that the phenolic extracts had high antioxidant activity and for this reason could be considered suitable for use as a high value−added ingredient.

Finally, the results obtained with the use of the phenolic extract (PE_B), recovered with the use of only water as a solvent for mayonnaise formulation, opens the way to green methodologies in the recovery of added value molecules from wastewaters. An improvement of the recipe for vegan mayonnaise preparation, with the aim to increase the acceptability of the consumer, could be acquired by adding some ingredients that allow improving the color and taste at an aromatic level.

Author Contributions: Conceptualization, A.D.B. and R.R.; methodology, A.D.B., A.G. and R.R; software, A.D.B. and R.R; validation, A.P. and M.P.; formal analysis, A.D.B., R.R. and A.G.; investigation, A.D.B., A.G. and R.R.; resources, M.P.; data curation, A.D.B. and A.P.; writing—original draft preparation, A.D.B. and R.R.; writing—review and editing, A.P and M.P.; visualization, A.D.B., A.P. and M.P.; supervision, A.P. and M.P.; project administration, M.P. All authors have read and agreed to the published version of the manuscript.

Funding: This research was funded by the grant of MIUR (Ministry of Education, University and Research), PRIN 2017JTNK78 "Good-by-Waste".

Data Availability Statement: The data presented in this study are available on request from the corresponding author.

Conflicts of Interest: The authors declare no conflict of interest.

References

1. Caporaso, N.; Formisano, D.; Genovese, A. Use of phenolic compounds from olive mill wastewater as valuable ingredients for functional foods. *Crit. Rev. Food Sci. Nutr.* **2017**, *58*, 2829–2841. [CrossRef]
2. Giuffrè, A.M.; Piscopo, A.; Sicari, V.; Poiana, M. The effects of harvesting on phenolic compounds and fatty acids content in virgin olive oil (cv Roggianella). *Riv. Ital. Sostanze Grasse* **2012**, *87*, 14–23.
3. Afkhami, R.; Goli, M.; Keramat, J. Functional orange juice enriched with encapsulated polyphenolic extract of lime waste and hesperidin. *Int. J. Food Sci. Technol.* **2018**, *53*, 634–643. [CrossRef]
4. Romeo, R.; De Bruno, A.; Imeneo, V.; Piscopo, A.; Poiana, M. Evaluation of enrichment with antioxidants from olive oil mill wastes in hydrophilic model system. *J. Food Process. Preserv.* **2019**, *43*, e14211. [CrossRef]
5. Romeo, R.; De Bruno, A.; Imeneo, V.; Piscopo, A.; Poiana, M. Impact of Stability of Enriched Oil with Phenolic Extract from Olive Mill Wastewaters. *Foods* **2020**, *9*, 856. [CrossRef] [PubMed]
6. Romeo, R.; De Bruno, A.; Piscopo, A.; Brenes, M.; Poiana, M. Effects of Phenolic Enrichment on Antioxidant Activity of Mayonnaise. *Chem. Eng. Trans.* **2021**, *87*, 127–132.
7. Anton, M.; Martinet, V.; Dalgararrondo, M.; Beaumal, V.; David-Briand, E.; Rabesona, H. Chemical and structural characterisation of low-density lipoproteins purified from hen egg yolk. *Food Chem.* **2003**, *83*, 175–183. [CrossRef]
8. Karshenas, M.; Goli, M.; Zamindar, N. The effect of replacing egg yolk with sesame–peanut defatted meal milk on the physicochemical, colorimetry, and rheological properties of low-cholesterol mayonnaise. *Food Sci. Nutr.* **2018**, *6*, 824–833. [CrossRef] [PubMed]
9. Raikos, V.; Hayes, H.; Ni, H. Aquafaba from commercially canned chickpeas as potential egg replacer for the development of vegan mayonnaise: Recipe optimisation and storage stability. *Int. J. Food Sci. Tech.* **2020**, *55*, 1935–1942. [CrossRef]
10. Ghorbani Gorji, S.; Smyth, H.E.; Sharma, M.; Fitzgerald, M. Lipid oxidation in mayonnaise and the role of natural antioxidants: A review. *Trends Food Sci. Technol.* **2016**, *56*, 88–102. [CrossRef]
11. Azhagu Saravana Babu, P.; Vajiha Aafrin, B.; Archana, G.; Sabina, K.; Sudharsan, K.; Radha Krishnan, K.; Babuskin, S.; Sivarajan, M.; Sukumar, M. Polyphenolic and phytochemical content of Cucumis sativus seeds and study on mechanism of preservation of nutritional and quality outcomes in enriched mayonnaise. *Int. J. Food Sci. Tech.* **2016**, *51*, 1417–1424. [CrossRef]
12. Ahmad, M.; Kaloo, Z.A.; Ganai, B.A.; Ganaie, H.A.; Singh, S. Phytochemical screening of Meconopsis aculeata royle an important medicinal plant of Kashmir Himalaya: A perspective. *Res. J. Phytochem* **2016**, *10*, 1–9. [CrossRef]
13. Yang, Y.; Song, X.; Xiaonan, S.; Qi, B.; Wang, Z.; Li, Y.; Jiang, L. Rosemary extract can be used as a synthetic antioxidant to improve vegetable oil oxidative stability. *Ind. Crop Prod.* **2016**, *80*, 141–147. [CrossRef]
14. Farhadi, K.; Esmaeilzadeh, F.; Hatam, M.; Forough, M.; Molaie, R. Determination of phenolic compounds content and antioxidant active in skin, pulp, seed, cane and leaf of five native grape cultivars in We Azerbaijan province, Iran. *Food Chem.* **2016**, *199*, 847–855. [CrossRef]
15. Samec, D.; Maretic, M.; Lugaric, I.; Mesic, A.; Sondi, B.A.; Duralija, B. Assessment of the differences in the physical, chemical and phytochemical properties of four strawberry cultivars using principal component analysis. *Food Chem.* **2016**, *194*, 828–834. [CrossRef]
16. De Bruno, A.; Romeo, R.; Piscopo, A.; Poiana, M. Antioxidant quantification in different portions obtained during olive oil extraction process in an olive oil press mill. *J. Sci. Food Agric.* **2020**, *101*, 1119–1126. [CrossRef]
17. Association of Official Analytical Chemistry. *Official Methods of Analysis of AOAC International*, 17th ed.; Author: Gaithersburg, MD, USA, 2005.
18. AOCS. Method Ca 5a 40. In *Official Methods and Recommended Practices of the American Oil Chemists' Society*, 6th ed.; AOCS Press: Champaign, IL, USA, 2017.
19. AOCS. Method Cd 8–53. In *Official Methods and Recommended Practices of the American Oil Chemists' Society*, 6th ed.; AOCS Press: Champaign, IL, USA, 2017.
20. AOCS. Method Ch 5-91. In *Official Methods and Recommended Practices of the American Oil Chemists' Society*; AOCS Press: Champaign, IL, USA, 1989.

21. AOCS. Method Cd 12c-16. In *Official Methods and Recommended Practices of the American Oil Chemists' Society*, 6th ed.; AOCS Press: Champaign, IL, USA, 2017.
22. Soberón, L.F.; Carelli, A.A.; González, M.T.; Ceci, L.N. Method for phenol recovery from "alperujo": Numerical optimization and predictive model. *Eur. Food Res. Technol.* **2019**, *245*, 1641–1650. [CrossRef]
23. Brahmi, F.; Mechri, B.; Dabbou, S.; Dhibi, M.; Hammami, M. The efficacy of phenolics compounds with different polarities as antioxidants from olive leaves depending on seasonal variations. *Ind Crop Prod.* **2012**, *38*, 146–152. [CrossRef]
24. Bibi Sadeer, N.; Montesano, D.; Albrizio, S.; Zengin, G.; Mahomoodally, M.F. The Versatility of Antioxidant Assays in Food Science and Safety-Chemistry- Applications, Strengths, and Limitations. *Antioxidants* **2020**, *9*, 709. [CrossRef] [PubMed]
25. Wang, Z.; Wang, C.; Yuan, J.; Zhang, C. Adsorption characteristics of adsorbent resins and antioxidant capacity for enrichment of phenolics from two-phase olive waste. *J. Chromatogr. B Anal. Technol. Biomed. Life Sci.* **2016**, *1040*, 38–46. [CrossRef]
26. Di Mattia, C.; Paradiso, V.M.; Andrich, L.; Giarnetti, L.; Caponio, F.; Pittia, P. Effect of Olive Oil Phenolic Compounds and Maltodextrins on the Physical Properties and Oxidative Stability of Olive Oil O/W Emulsions. *Food Biophys.* **2014**, *9*, 396–405. [CrossRef]
27. Flamminii, F.; Di Mattia, C.D.; Sacchetti, G.; Neri, L.; Mastrocola, D.; Pittia, P. Physical and Sensory Properties of Mayonnaise Enriched with Encapsulated Olive Leaf Phenolic Extracts. *Foods* **2020**, *9*, 997. [CrossRef] [PubMed]
28. Altunkaya, A.; Hedegaard, R.V.; Harholt, J.; Brimer, L.; Gokmene, V.; Skibsted, L.H. Oxidative stability and chemical safety of mayonnaise enriched with grape seed extract. *Food Funct.* **2013**, *4*, 1647–1653. [CrossRef]
29. Martillanes, S.; Rocha-Pimienta, J.; Gil, M.V.; Ayuso-Yuste, M.C.; Delgado-Adámez, J. Antioxidant and antimicrobial evaluation of rice bran (Oryza sativa L.) extracts in a mayonnaise-type emulsion. *Food Chem.* **2020**, *308*, 125633. [CrossRef] [PubMed]
30. Park, J.J.; Olawuyi, I.F.; Lee, W. Characteristics of low-fat mayonnaise using different modified arrowroot starches as fat replacer. *Int. J. Biol. Macromol.* **2020**, *153*, 215–223. [CrossRef] [PubMed]
31. Rasmy, N.M.; Hassan, A.A.; Foda, M.I.; El-Moghazy, M.M. Assessment of the antioxidant activity of sage (Salvia officinalis L.) extracts on the shelf life of mayonnaise. *World J. Dairy Food Sci.* **2012**, *7*, 28–40.
32. Giacintucci, V.; Di Mattia, C.; Sacchetti, G.; Neri, L.; Pittia, P. Role of olive oil phenolics in physical properties and stability of mayonnaise-like emulsions. *Food Chem.* **2016**, *213*, 369–377. [CrossRef]
33. Kiokias, S.; Varzakas, T.H.; Arvanitoyannis, I.S.; Labropoulos, A.E. Lipid Oxidation and Control of Oxidation. *Adv. J. Food Biochem.* **2009**, 383–408.
34. Raikos, V.; McDonagh, A.; Ranawana, V.; Duthie, G. Processed beetroot (Beta vulgaris L.) as a natural antioxidant in mayonnaise: Effects on physical stability, texture and sensory attributes. *Food Sci. Hum. Well.* **2016**, *5*, 191–198.
35. Paradiso, V.M.; Flamminii, F.; Pittia, P.; Caponio, F.; Di Mattia, C. Radical Scavenging Activity of Olive Oil Phenolic Antioxidants in Oil or Water Phase during the Oxidation of O/W Emulsions: An Oxidomics Approach. *Antioxidants* **2020**, *9*, 996. [CrossRef]
36. Sorensen, A.; Nielsen, N.; Decker, E.; Let, M.; Xu, X.; Jacobsen, C. The efficacy of compounds with different polarities as antioxidants in emulsions with omega-3 lipids. *J. Am. Oil Chem. Soc.* **2011**, *88*, 489–502. [CrossRef]

foods

Article

The Effect of Cold Press Chia Seed Oil By-Products on the Rheological, Microstructural, Thermal, and Sensory Properties of Low-Fat Ice Cream

Ilker Atik [1], Zeynep Hazal Tekin Cakmak [2], Esra Avcı [2] and Salih Karasu [2,*]

1 Food Technology Program, Afyon Vocational School, Afyon Kocatepe University, Afyonkarahisar 03200, Turkey; ilkeratik@hotmail.com
2 Department of Food Engineering, Faculty of Chemical and Metallurgical Engineering, Davutpasa Campus, Yildiz Technical University, Istanbul 34210, Turkey; zhazaltekin@gmail.com (Z.H.T.C.); esravci93@hotmail.com (E.A.)
* Correspondence: skarasu@yildiz.edu.tr; Tel.: +90-212-383-46-23

Abstract: This study aimed to investigate the utilization of cold-pressed chia-seed oil by-products (CSOB) in a low-fat ice cream formulation as a fat replacer and stabilizer. In the study, ice cream emulsion mixtures were formulated by using 0.2–0.4% xanthan gum (XG), 2.5–12.5% fat, and 1–3% CSOB. Optimization was performed using the response surface methodology (RSM) and full factorial central composite design (CCD) based on the flow behavior rheological properties of the emulsions obtained from 17 different experimental points. All of the emulsion samples showed non-Newtonian shear-thinning flow behavior. The consistency coefficient (K) values of the emulsion samples were found to be 4.01–26.05 Pas^n and were significantly affected by optimization parameters ($p < 0.05$). The optimum formulation was determined as 0.29% XG, 2.5% CSOB, 2.5% fat. The low-fat (LF-IC) and full-fat control samples (FF-IC) were compared to samples produced with an optimum formulation (CBLF-IC) based on the steady shear, frequency sweep, and 3-ITT (three interval thixotropy test) rheological properties, thermal properties, emulsion stability, light microscope images, and sensory quality. CBLF-IC showed similar rheological behavior to FF-IC. The mix of CBLF-IC showed higher emulsion stability and lower poly-dispersity index (PDI) value and fat globule diameters than those of FF-IC and LF-IC. The thermal properties of the samples were significantly affected by the addition of CSOB in an ice cream mix. CBLF-IC exhibited a lower temperature range (ΔT), enthalpy of fusion (ΔH_f), and freezing point temperature (T_f) than those of FF-IC and LF-IC. While CBLF-IC exhibited a higher overrun value than other samples, it showed similar sensory properties to the FF-IC sample. The results of this study suggested that CSOB could be used successfully in low-fat ice cream production. This study also has the potential to gain new perspectives for the evaluation of CSOB as a fat substitute in a low-fat ice cream.

Keywords: by-product; rheological properties; optimization; melting profile

Citation: Atik, I.; Tekin Cakmak, Z.H.; Avcı, E.; Karasu, S. The Effect of Cold Press Chia Seed Oil By-Products on the Rheological, Microstructural, Thermal, and Sensory Properties of Low-Fat Ice Cream. *Foods* **2021**, *10*, 2302. https://doi.org/10.3390/foods10102302

Academic Editors: Marco Poiana, Francesco Caponio and Antonio Piga

Received: 21 August 2021
Accepted: 22 September 2021
Published: 28 September 2021

Publisher's Note: MDPI stays neutral with regard to jurisdictional claims in published maps and institutional affiliations.

Copyright: © 2021 by the authors. Licensee MDPI, Basel, Switzerland. This article is an open access article distributed under the terms and conditions of the Creative Commons Attribution (CC BY) license (https://creativecommons.org/licenses/by/4.0/).

1. Introduction

Ice cream is a frozen milk-based dessert as a frozen aerated emulsion (O/W) containing partially combined fat globules, air bubbles, and ice crystals. Ice cream includes milk, milk cream, emulsifiers, sweeteners, stabilizers, and flavorings [1]. Ice cream generally contains high dairy or non-dairy fat (10–16%) [2,3]. As one of the most important ingredients in ice cream, milk fat interacts with other ingredients to improve texture, mouthfeel, creaminess, and overall sensation of lubricity [4]. Ice cream is a popular dessert, especially during hot weather, with an annual global consumption of around 2 L per person [5]. However, the demand for low-calorie foods has increased in recent years. Therefore, many studies have been carried out to develop new additives and new products in the food industry to meet this demand. In order to ensure the stability of ice cream, some special additives with stabilizer and emulsifier properties are required, and these additives are used in ice cream production.

Stabilizers are preferred in ice cream production because they provide specific and important functions such as increasing the viscosity and smoothness of the ice cream mix, improving aeration, reducing ice recrystallization, and the rate of structural collapse during melting [2,6]. Emulsifiers are substances that provide a fine dispersion of foods by reducing surface tension [7]. Emulsifiers minimize the formation of ice crystals by increasing the volume capacity of the ice cream and its resistance to melting. Stabilizers and emulsifiers also provide the ice cream with dryness and hardness, as well as a smooth structure and the desired oily feeling [8–10].

Chia (*Salvia hispanica* L.) seed, belonging to the *Lamiaceae* family, has a high amount of oil (25–38%), carbohydrates (26–41%), high dietary fiber (18–30%), and protein (19–23%) [11]. Chia seeds have emulsifying and stabilizing properties due to their high protein content and water-soluble branching polysaccharides. They have the potential to be used as a natural emulsifier and stabilizer in many food compositions [12,13]. Researchers have studied the effect of chia seed gel mucilage on ice cream as a stabilizer and emulsifier [14–18]. In our study, different from these studies, a cold press oil by-product was used in the preparation of ice cream. In addition, in our study, the by-product ratio was determined under optimum conditions, and comprehensive properties such as rheological, microstructural, emulsion stability, thermal and sensory properties in the low-fat and high-fat control samples were analyzed in this study. It was also used for the first time in this study for 3-ITT analysis and thermal loop rheology of ice cream samples. Our research paper contains differences from other studies both in terms of material and analysis methods.

Fat substitutes can be used to reduce the amount of fat in foods and replace it with oil-like substances and reduce the caloric value with new additives. Fat substitutes can provide many of the properties that fat gives to foods and be classified as carbohydrate-derived fat substitutes, protein-derived fat substitutes, synthetic fat substitutes, and fat compound substitutes. When these substances are used instead of fat in foods, the fat in the food can be partially or completely reduced, and the energy from fat is minimized but provides palatable products [19]. Recently, cold-pressed oil industry by-products have also been used as fat substitutes [20–22]. Cold-press oil by-products such as cold-press oils include a high amount of protein, carbohydrate, fiber contents, and nutritive components without any solvent trace, and these by-products can be used as a natural fat replacer, emulsifier, stabilizer, and a natural antioxidant source in food emulsions [23,24].

Chia-seed by-products derived from the cold-press oil industry have the potential to be utilized as a building strengthener and fat replacer due to their ability to develop relatively viscous solutions. Furthermore, Capitani et al. [25] reported that the residual meal/by-product obtained from the oil-extracting process of chia seeds was determined as a good source of total dietary fiber, consisting mostly of insoluble dietary fiber. Therefore, chia seed by-products can improve food texture, utilize a role as texturing and stabilizing agent due to high dietary fiber content. Akcicek and Karasu [20] used the cold-pressed chia seed oil by-products as a fat replacer in a low-fat salad dressing. However, chia seed oil by-products as a fat replacer, stabilizer, and emulsifier in a low-fat ice-cream formulation have not been investigated. Many studies are present in the literature about the development of low-fat ice cream formulations [7,26–30].

The main purpose of this study is to produce ice cream with an optimum feature of low-fat content by using cold-pressed chia seed oil by-products without causing any change in the rheological and melting properties of ice cream. In this study, the optimization process was carried out based on the rheological analysis results of the ice creams produced with different formulations. The rheological and melting properties of the ice creams produced after the optimization process were investigated. Thus, production parameters with optimum rheological and melting properties were determined.

2. Material and Methods

2.1. Material

In this study, CSOB, xanthan gum (XG), egg yolk powder (EYP), milk cream (35% milkfat), pasteurized milk (1.5% milk fat), sugar, and water were used for producing ice cream samples. CSOB, 8.07% fat, 56.10% total carbohydrate including fiber, 23.50% protein, 29.26% fiber, 5.21% ash, 6.85% moisture, was supplied from ONEVA Food Co. (Istanbul, Turkey). XG and egg yolk powder (EYP) were obtained from Sigma-Aldrich (Sigma Chemical Co., St. Louis, MO, USA), and the sugar, milk, and milk cream were purchased from the local market.

2.2. Methods

The study was carried out in two stages. In the first stage, the responses were obtained at 17 different points for ice cream samples through the Design Expert Software (Version 7; Stat-Easy Co., Minneapolis, MN, USA), and the formulation was optimized according to flow behavior characteristics. In the second stage, the dynamic rheological and melting profile properties of the ice cream samples produced according to the optimum formulation were examined and characterized.

2.2.1. Experimental Design

The formulation of ice cream samples was determined using the Design Expert Software (Version 7; Stat-Easy Co., Minneapolis, MN, USA). Response surface methodology (RSM) was carried out for the study of three factors, namely fat concentration (2.5–12.5 g/100 g) (X_1), CSOB concentration (1–3 g/100 g) (X_2), and gum concentration (0.2–04 g/100 g) (X_3), using an unblocked full factorial central composite design (CCD). Seventeen different experimental points were obtained by using Design Expert Software (Version 7; Stat-Easy Co., Minneapolis, MN, USA) to determine the optimum amount of fat, CSOB, and gum content. For the estimation of the error, the design consists of three of the factorial points. A quadratic model was fitted to the experimental data for each response. Model applicability was evaluated based on the R^2, R^2-adj, lack of fit, F, and *p*-values obtained from ANOVA. Response surface plots showing the effect of model parameters on the *K* value were built by Design Expert Software. The optimization was performed based on the highest desirability value. The formulation with the lowest fat content with a desirability value of 1 was determined as the optimum formulation. Three central points were used. Analysis of all points was performed in triplicate, and the results were reported as mean values and standard deviations. The formulation of the samples has been determined in accordance with the ice cream standard. Accordingly, 2.5% fat ratio representing low-fat ice cream ($0.5 \leq$ milk fat < 3), 7.5% fat ratio representing semi-fat ice cream ($3 \leq$ milk fat < 8), and 12.5% fat ratio representing fat ice cream ($8 \leq$ milk fat ≤ 13) were used. Based on these fat ratios, the amount of milk and milk cream to be used has been calculated. The percentage of fat, XG, and CSOB vary in formulations, and the amounts of sugar (12%) and EYP (3%) remain constant. Seventeen formulations obtained in this way are shown in Table 1.

2.2.2. Ice-Cream Preparation

Ice cream formulations consisted of 2.5–12.5% fat (derived from 1.5% milk fat and 98.5% milk cream), 12% sugar, 3% egg yolk powder (EYP) and 0.2–0.4% xanthan gum (XG), 1–3% CSOB. XG and milk were weighed according to the formulations determined in the first stage of ice cream production. XG was slowly added in order to dissolve it thoroughly and kept in a magnetic stirrer at 1000 rpm. Then, CSOB, which was weighted according to the formulation, was added and mixed. Then milk cream was added sequentially. The EYP and sugar were added and mixed. Then, the samples were homogenized with ultra turrax (Daihan. HG-15D) at 1000 rpm for 3 min. After this process, the ice cream mixture samples were kept at 0–4 °C for 2 days to mature. Afterward, ice cream was produced from the mixes using an ice cream machine (DeLonghi IL Gelataio ICK5000, Treviso, Italy). Ice cream samples were stored at −18 °C for analysis.

Table 1. Steady shear power-law parameters of the ice cream mixtures contained CSOB with ingredient compositions.

Samples	XG (%)	Fat (%)	CSOB (%)	K (Pas^n)	n	R^2
C1	0.30	7.5	3	17.23	0.26	1.00
C2	0.20	12.5	1	10.95	0.29	1.00
C3	0.40	2.5	3	11.61	0.27	0.99
C4	0.40	2.5	1	9.46	0.23	0.99
C5	0.30	12.5	2	14.83	0.32	1.00
C6	0.20	7.5	2	7.51	0.35	1.00
C7	0.30	2.5	2	8.28	0.24	0.99
C8	0.20	12.5	3	12.57	0.35	1.00
C9	0.30	7.5	1	9.38	0.32	1.00
C10	0.30	2.5	1	4.01	0.26	0.99
C11	0.40	12.5	3	26.05	0.28	1.00
C12	0.40	7.5	2	20.69	0.37	0.99
C13	0.30	7.5	2	15.06	0.38	1.00
C14	0.30	7.5	2	15.01	0.38	1.00
C15	0.40	12.5	1	17.52	0.30	1.00
C16	0.30	7.5	2	15.62	0.37	1.00
C17	0.20	2.5	3	5.30	0.33	0.99

XG: xanthan gum, CSOB: chia seed oil by-product, K: consistency coefficient, n: flow behavior index, R^2: regression coefficient corresponding to the power law model.

2.2.3. Analysis of the Ice Cream Mix

Rheological Analyzes

Flow behavior, dynamic, and 3-ITT rheological properties of ice cream mixes were determined using a stress and temperature-controlled rheometer (MCR 302, Anton Paar, Australia). A parallel plate probe (PP50, Anton Paar, Australia) was used for rheological measurement. All measurements were performed at 25 °C and duplicated for the accuracy of the results.

Flow Behavior Rheological Properties

Flow behavior rheological properties of ice cream samples prepared using CSOB were determined using a parallel plate probe (plate diameter 50 mm, gap size 0.5 mm) with a shear rate in the range 0.1–100 (s^{-1}). The measurement was carried out at a constant temperature of 25 °C, and 3 parallel studies were carried out for each sample. The data obtained from the rheological analysis were fitted to the power law model, and nonlinear regression was used to calculate model parameters;

$$\tau = K\gamma^n \quad (1)$$

In Equation (1), the τ value represents the shear stress (Pa), K the consistency coefficient (Pas^n), γ the shear rate (s^{-1}), and n the flow behavior index.

Dynamic Rheological Properties

Parallel plate configuration was used for the dynamic rheological analysis of ice cream samples. Initially, the amplitude sweep test was performed between 0.1% and 100% strain to determine the linear viscoelastic region, and according to the result, the frequency sweep test was studied in the frequency range of 0.1–10 Hz and angular velocity of 0.1–64 s^{-1} (ω). Elastic modulus (G') and viscose modulus (G'') corresponding to angular velocity and frequency values were determined. The parameters for dynamic rheological properties were found using the power law model and nonlinear regression;

$$G' = K'(\omega)^{n'} \quad (2)$$

$$G'' = K''(\omega)^{n''} \quad (3)$$

In Equations (2) and (3), the G' value represents storage modulus (Pa), G'' value loss modulus (Pa), ωangular velocity value (s^{-1}), K', K'' consistency coefficient values (Pas^n) and n', n'' flow behavior index values.

3-ITT

3-ITT rheological properties of ice cream samples that contained CSOB were determined as 0.5 s^{-1} as constant shear rate value and 150 s^{-1} as variable shear rate value. The linear viscoelastic region has been taken into consideration in the selection of the values, and the linear viscoelastic region of the samples ends at 50 s^{-1}. The ice cream samples were subjected to a very low shear rate (0.5 s^{-1}) for 100 s during the first time interval. In the second time interval, 150 s^{-1} was exposed to the determined cutting force for 40 s. In the third time interval, the dynamic rheological behavior of the ice cream in the second time interval was tested by subjecting the samples to the low shear rate level in the first time interval. For this purpose, the change in the viscoelastic solid structure (G') of the samples was observed. The behavior of samples produced using CSOB in the third time interval was modeled using a second-order structural kinetic model;

$$\left[\frac{G' - G_e}{G_0 - G_e}\right]^{1-n} = (n-1)kt - 1 \quad (4)$$

In the model, the G' value indicates the change in the storage module (Pa), G_0 indicates the initial storage module value (Pa) in the 3rd time interval, G_e represents the storage module at the moment when the product fully recovered, in other words, the storage module (Pa) at the moment when the product is fully balanced, and k is the thixotropic velocity constant.

Emulsion Stability Test

The emulsion stability of ice cream mix was determined by the thermal loop test previously described by Tekin et al. [31]. The ice cream mix samples were subjected to ten thermal cycles from 23 to 45 °C in a high-temperature stability test. An angular frequency (ω) and strain value were 10 Hz and 0.5%, respectively. The heating and cooling rates were set at 11 °C/min. The maximum points of all cycles were recorded by using the rheometer software (Rheoplus for MCR 301) using the internal loop. The relative structural change value (Δ) was calculated for G^* values by dividing the maximum value of each cycle by the value of the first cycle.

The Determination Particle Size

The size of the fat globules and zeta potential value of the samples was determined by the particle size measuring device (Nanosizer, Malvern Instruments, Worcestershire, UK). The samples were diluted 500-fold with ultrapure water before homogenization by stirring in an ultrasonic water bath for 1 min. The particle size of the samples was determined according to the dynamic light scattering technique [31].

2.2.4. Analysis of the Ice Cream Samples

Overrun Measurement

The volumes of ice cream mixtures (just before freezing) and ice cream samples (immediately after freezing) were weighed, and values were recorded [15]. Overrun values of the ice cream samples were calculated according to the formula described by Equation (5);

$$\text{Overrun } (\%) = \frac{W_2 - W_1}{W_1} \times 100 \tag{5}$$

where W_1, and W_2 represent the weight of a unit volume of ice cream mix and the weight of a unit volume of ice cream after freezing, respectively.

Thermal Properties of Ice Cream Samples

The thermal properties of ice cream samples were analyzed by a differential scanning calorimeter (DSC) by A DTA-DSC (differential scanning calorimetry) operating at atmospheric pressure (STA44gf3, Netzsch, Germany) according to the method reported by Hwang et al. [32]. Ice cream samples of 10 mg were placed in a pre-weighed aluminum sample pan, the pan was sealed using a Quick Press pan crimper (T_{zero}), and the thermal data were recorded from −20 to +50 °C in a nitrogen atmosphere with a heating rate of 1 °C/min. An empty pan was used as the reference. The flow rates of nitrogen gas for cooling were 50 mL/min. The onset temperatures (T_0), T_f, and ΔH_f of the transitions of ice formation and ice melting were determined. The onset temperatures are considered as the intersection of the tangent and baseline to the left side of the melting peak. Freezing points were determined by using the temperature of the steepest slope. The enthalpy of fusion was calculated by extrapolating the baseline under the peak by connecting the flat baseline before and after the melting peak and integrating the peak above the baseline. The amount of ice formed per gram of sample (freezable water) was determined by the method described by Soukoulis et al. [33] by dividing the melting enthalpy with the pure ice fusion latent heat (S = 334 J/g).

Sensory Analysis of Ice Cream

A trained panel of 10 members (graduate students and academic staff from the Food Engineering Department at Yildiz Technical University in Istanbul, Turkey) assessed the sensory characteristics of ice cream samples. Panel members were instructed on the ice cream samples prior to the commencement of the tests. The training consisted of a two-hour thorough presentation to the panelists on the purpose of the study and the characteristics of the samples. The panelists were asked to identify the optimal persimmon content for the required enhanced ice cream in terms of look and color, odor, taste and flavor, texture, melting resistance, and overall acceptability. Ice cream samples were evaluated using a scaling method of descriptive attributes for all parameters (1 = undesired, 9 = desired) [34].

Color Measurement

The color parameters of ice cream samples were measured using a colorimeter (CR-400 Konica, Minolta, Tokyo, Japan). The CIELAB coordinate system was used, and the L^*, a^* and b^* parameters were evaluated. L^*, a^* and b^* parameters represent whiteness/darkness, redness/greenness, and yellowness/blueness, respectively.

2.3. Statistical Analysis

Ice cream samples were produced in three replications, and three parallel measurements were performed from each replication. The mean and standard deviation values were presented. Statistical applications were carried out with the Statistica software package (Stat Soft Inc., Tulsa, UK). The Duncan multiple comparison test was used to compare the mean values of the parameters after optimization ($p < 0.05$). As a result of the applied rheological analysis, power law model parameters were calculated with the help of nonlinear regression analysis. The Statistica software program (Stat Soft Inc., Tulsa, UK) was used for nonlinear regression analysis.

3. Result and Discussion

3.1. Determination of Steady Shear Rheological Properties of Ice Cream Mix for the Formulation Optimization

In this study, the data of the steady shear rheological properties were used for the formulation optimization of ice cream mixtures. The flow curves of the ice cream mix obtained from 17 different trial points are shown in Figure 1. There was a decrease in the slope of the shear rate versus shear stress graphs of the ice cream mixes, indicating that the viscosity of all samples decreased with increasing shear rate. The reduction in viscosity can be explained by the structural breakdown of the intermolecular interaction [35]. The ice cream mixes showed shear-thinning flow characteristics, which is typical flow behavior for an ice cream mix. The shear-thinning behavior is an important factor in choosing the pump size for mixing [35].

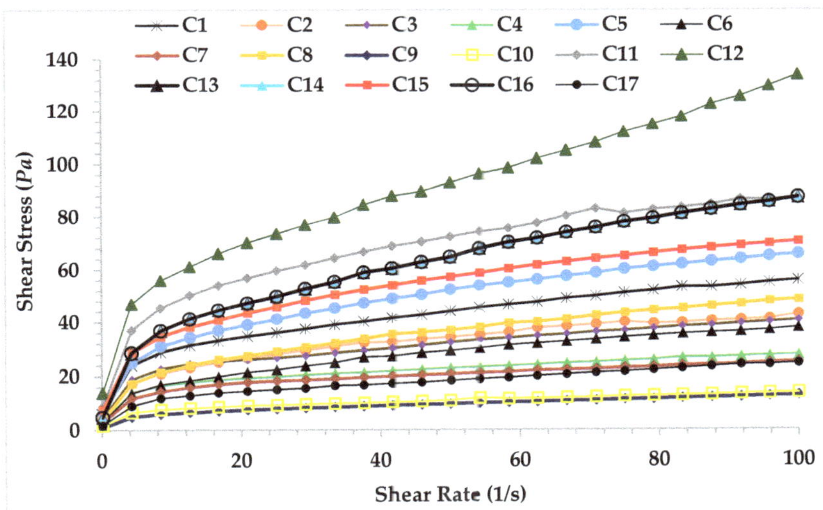

Figure 1. Steady shear rheological behavior of the ice cream mixes.

Power law model parameters (*K* and *n* values) and determination coefficient (R^2) calculated for 17 different points created with the trial design are shown in Table 1. R^2 values of the power law model were higher than 0.99. This shows that the power law model is suitable for determining the flow behavior properties of ice cream mixes. According to the ice cream formulations, *K* and *n* values differed and were found as 4.01–26.05 Pas^n and 0.23–0.38, respectively. *n* values lower than 1 indicated that all ice cream mixtures exhibited the non-Newtonian pseudoplastic flow behavior (Table 1). Dairy products generally exhibit shear-thinning (pseudoplastic) behavior with flow behavior indexes of $0 < n < 1$ [36]. As seen in Table 1, the sample −10 (0.3% XG, 2.5% fat, 1% CSOB) showed the lowest *K* value (4.01 Pas^n), while the sample-11 (0.4% XG, 12.5% fat, 3% CSOB) exhibited the highest *K* value (26.05 Pas^n). The *K* value of the samples increased with increasing CSOB, gum, and fat content. These results showed that increasing the *K* value with the increase in CSOB can improve the shear-thinning properties of ice cream mixes. The desired consistency values can be obtained by using CSOB even at low-fat and gum content. Thus, CSOB can be used for improving rheological properties in low-fat ice cream.

3.2. The Effect of Model Parameters on K and n Value and Determination Optimum Formulation

Figure 2 presents the impact of gum, fat, and CSOB in the formulation on *K* value. The increase in fat, gum, and CSOB, as shown in Figure 2, resulted in an increase in the *K* value of the ice cream mixes. The structure of CSOB, which contains polysaccharides with high water-holding ability, can explain this phenomenon. This polysaccharide structure offers excellent water retention and stabilizing capabilities. CSOB can be adsorbed at the interface area and has surface-active qualities, in addition to its stabilizer action. Because of these CSOB characteristics, the *K* value of the ice cream mixes increased. Furthermore, increasing the amount of gum resulted in a considerable rise in the *K* value, particularly in formulations including xanthan gum at a specific level. The increase in *K* value in both increases in fat, CSOB, and gum content can be explained by the synergistic impact of the ice cream mix's components.

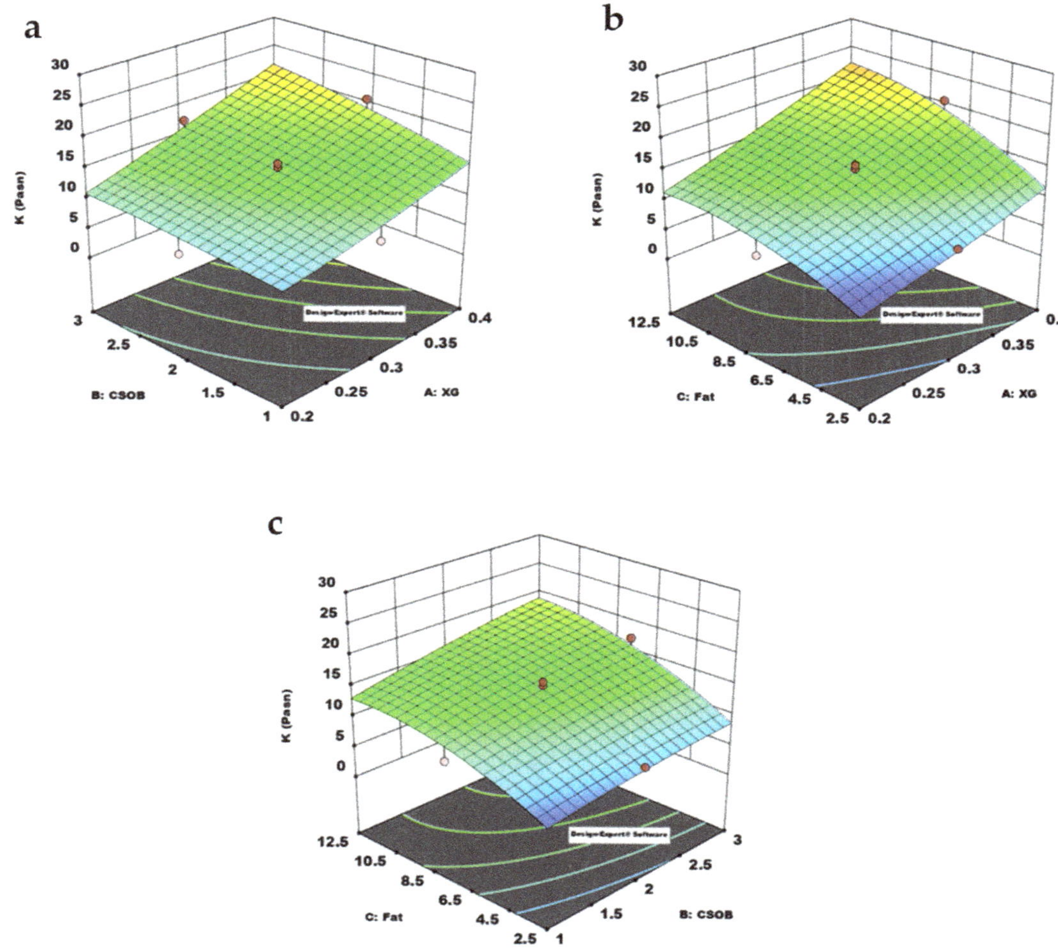

Figure 2. Response surface plot showing the effect of model parameters on the *K* value of ice cream mixes. (**a**): XG-CSOB, (**b**): XG-Fat, (**c**): CSOB-Fat (A: XG (xanthan gum), B: CSOB (chia seed oil by-product), C: fat (milk fat), *K*: consistency coefficient).

Table 2 showed that the response surface method (RSM) and the quadratic model were used to describe the influence of formulation components on the flow behavior parameters

(K, n) of ice-cream mixtures. As seen in Table 2, the ANOVA analysis of variance was used to statistically evaluate the influence of dependent variables on the K value of the ice cream mixes.

Table 2. Quadratic model parameter's corresponding K value.

Source	Sum of Squares	df	Mean Squares	F-Value	p-Value	
Model	529.67	9	58.85	56.29	<0.0001	significant
A-XG	280.46	1	280.46	268.24	<0.0001	
B-CSOB	86.69	1	86.69	82.91	<0.0001	
C-Fat	138.80	1	138.80	132.75	<0.0001	
AB	11.93	1	11.93	11.41	0.0118	
AC	2.29	1	2.29	2.19	0.1824	
BC	0.9184	1	0.9184	0.8783	0.3799	
A^2	1.91	1	1.91	1.83	0.2183	
B^2	0.0066	1	0.0066	0.0063	0.9391	
C^2	7.74	1	7.74	7.41	0.0297	
Residual	7.32	7	1.05			
Lack of Fit	7.09	5	1.42	12.18	0.0776	not significant
R^2					0.9864	
Adjusted R^2					0.9688	
Predicted R^2					0.8918	
Adeq Precision					30.5150	

XG: xanthan gum, CSOB: chia seed oil by-product, K: consistency coefficient.

The R^2 and Adjusted R^2 values of the model used were determined as 0.9864 and 0.9618, respectively. The differences between Adjusted R^2 and predicted R^2 was lower than 0.2, and the lack of fit value was found as insignificant. These results indicated that the quadratic model could be successfully used to describe the effect of formulation on the K value of the samples. The p-value of the model was lower than 0.05, indicating that the model terms significantly affected the K value. In this model, A, B, C, AB, C^2 are significant model terms. The linear effect of all independent variables was significant. The interaction and model terms A and B, and the quadratic effect of C was also be found as significant.

Steady shear rheological analyzes showed that the increase in the amount of CSOB in the formulation resulted in a significant increase in the K values of the samples. Chia seeds affect fat binding and gel-forming properties due to the functional properties of dietary fiber. Olivos-Lugo et al. [37] reported that chia seeds are high in dietary fiber (34.6%), oil contents (32.2%), and protein (24.6%). Therewith, Akcicek and Karasu [20] suggested that CSOB could be used as a fat replacer in a low-fat salad dressing. Considering the properties of CSOB in this study, it is understood that it has a stabilizer feature due to its polysaccharide content and that the proteins it contains can be adsorbed in the interface area and have emulsifier quality. Proteins show surface-active properties and decrease the interfacial tension, which is predicted to cause an increase in consistency coefficient. The stabilizer feature comes from the branched polysaccharide structure in its content, and these polysaccharides can hold water [12]. Based on these properties, CSOB causes an increase in consistency coefficient (K) and can be used in the production of low-fat ice cream. With the increase in the K value of the mixtures, the n value decreases and shows pseudoplastic behavior, which is the typical flow behavior characteristic of ice cream. Due to the solid particles in CSOB, it significantly increases the K value by affecting the viscosity of the ice cream mixture. The fat, emulsifier, stabilizer, and CSOB used in ice cream mixtures significantly increase the K consistency coefficient of the mixtures. The main reason for this is the synergistic interaction between the components. The K value of the full-fat control sample was used to determine the optimum formulation. The sample with the highest desirability value and minimum fat content was selected as the optimum formulation. The optimum formulation was determined as 2.5% fat, 0.29% XG, 2.51% CSOB.

3.3. Rheological Properties of Optimum and Control Ice Cream Mixtures

The steady shear, frequency sweep, and thixotropic properties of the samples produced based on optimum formulation were compared with the values of the full-fat and low-fat control sample. The flow curves of optimum and control ice cream samples were given in Figure 3. The ice cream mixtures performed a pseudoplastic (shear-thinning) rheological behavior; that is, the shear stress increased with increasing shear rate (Figure 3). Pseudoplastic rheological behavior is the typical behavior to characterize the flow properties of most ice cream mixtures [38,39]. The optimum and high-fat control samples showed similar viscosity curves. The power law model was used to determine the consistency coefficient (*K*) and flow behavior index (*n*) values of the optimal and control ice cream mixtures. Table 3 represents the power law model parameters (*K* and *n* values) as well as the determination coefficient (R^2). *K* values varied from 3.46 to 5.66 Pas^n, *n* values from 0.30 to 0.33, and R^2 values higher than 0.99. While the *K* value of the full-fat control samples and the *K* value of the sample containing CSOB were found to be statistically insignificant, the *K* value of both samples was found to be significantly higher than the low-fat control. This result showed that an ice cream mix similar to the consistency properties of full-fat ice cream could be produced with the use of CSOB.

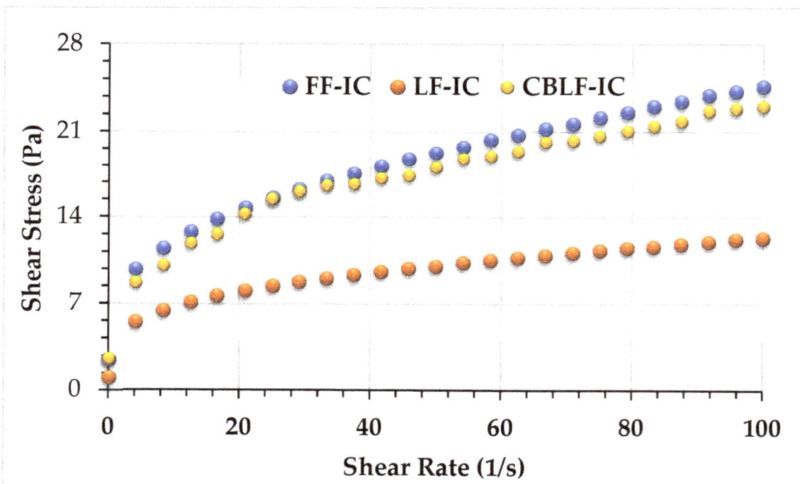

Figure 3. Steady shear rheological behavior of ice cream mixes. FF-IC: full-fat ice cream, LF-IC: low-fat ice cream, CBLF-IC: low-fat ice cream with chia seed oil by-product.

The dynamic rheological properties of ice creams produced from optimum and high-fat and low-fat control ice cream formulations were investigated. The frequency sweep test can simulate the liquid behavior of ice cream samples during chewing in the mouth [40], which helps to comprehensively evaluate the impact of CSOB addition on ice cream quality. Increasing G' and G'' values of samples with increasing frequency is evidence of gel-like behavior in ice cream samples [41]. As seen in Figure 4, the value of G' of all samples was higher than the value of G'', indicating that the solid character of all ice cream samples is more dominant than the liquid character. As seen in Figure 4, the G' value of FF-IC (12.5% fat) and CBLF-IC (2.5% fat) samples were almost at the same level, which explained that 10% fat can be compensated with 2.51% CSOB. As can be seen from the graph, the LF-IC instance has the lowest G' and the lowest G''. The G' and G'' values of the LF-IC (2.5% fat, 0.35% XG) were lower than the G' and G'' values of the CBLF-IC (2.5% fat, 0.29% XG, 2.51% CSOB). As it can also be seen in Table 3, the CBLF-IC has an elastic modulus value similar to FF-IC with full-fat content (12.5%). Aziz et al. [42] investigated the effect of adding okra gum on the rheological, textural, and melting properties of low-fat ice cream samples. It was stated that G' values were higher than G'' values for all ice cream samples. Substituting the fat content in ice cream with okra gum increased the viscous modulus (G'') values of the samples. Previous studies on viscoelastic properties were consistent with the results of our study in terms of G' and G'' values. Synergetic interactions between CSOB and ice cream ingredients can lead to improved food quality and expanded food applications due to enhanced functional properties. It may also have commercial potential for cost reduction.

Table 3. Steady shear, dynamic, and 3-ITT rheological model parameters of FF-IC, LF-IC, and CBLF-IC.

Rheological Analysis		**Samples**		
		FF-IC	**LF-IC**	**CBLF-IC**
Steady shear $\sigma = K \times \gamma^n$	K (Pas^n)	5.66 ± 0.07 [a]	3.46 ± 0.03 [b]	5.53 ± 0.09 [a]
	n	0.32 ± 0.01 [b]	0.36 ± 0.01 [a]	0.30 ± 0.02 [b]
	R^2	>0.99	>0.99	>0.99
Frequency $G'' = K'' \times (\omega)^{n'}$	K'	23.41 ± 0.18 [a]	3.09 ± 0.06 [b]	23.09 ± 0.24 [a]
	n'	0.268 ± 0.002 [b]	0.535 ± 0.003 [a]	0.268 ± 0.002 [b]
	R^2	0.993	0.979	0.9878
$G'' = K'' \times (\omega)^{n''}$	K''	8.46 ± 0.12 [a]	2.46 ± 0.02 [b]	8.15 ± 0.35 [a]
	n''	0.212 ± 0.001 [c]	0.288 ± 0.002 [a]	0.268 ± 0.003 [b]
	R^2	0.979	0.983	0.978
3-ITT	G_0'	21.12 ± 0.08 [a]	8.42 ± 0.04 [c]	18.86 ± 0.05 [b]
	G_e'	38.73 ± 0.15 [a]	14.81 ± 0.08 [c]	26.34 ± 0.11 [b]
	k	0.027 ± 0.001 [a]	0.011 ± 0.000 [b]	0.025 ± 0.001 [a]
	G_e' / G_0'	1.834	1.760	1.397
	$k \times 1000$	27.00	11.10	25.33
	R^2	0.983	0.995	0.996
ζ-potential (mV)		-39.13 ± 1.041 [a]	-32.03 ± 1.464 [b]	-28.90 ± 1.058 [c]
d_{32} (μm)		1.305 ± 0.044 [a]	0.877 ± 0.024 [b]	0.494 ± 0.012 [c]
PdI		0.741 ± 0.127 [a]	0.493 ± 0.167 [a]	0.534 ± 0.174 [a]

FF-IC: full-fat ice cream, LF-IC: low-fat ice cream, CBLF-IC: low-fat ice cream with chia seed oil by-product. Different lowercase letters in the same line indicate a statistical difference ($p < 0.05$).

The dynamic rheological parameters (K', K'', n', and n'' values) were also calculated by using the power law model (Table 3). The R^2 values of the model were found in the range of 0.97–0.98. As can be seen in Table 3, the K' and K'' values of the samples were in the range of 3.09–23.41 and 2.46–8.46, respectively; the values of n' and n'' were found in the range of 0.268–0.535 and 0.212–0.288, respectively. The K' values were higher than the K'' values for all samples. Accordingly, all of the ice cream samples showed a viscoelastic solid character. When CSOB was added to low-fat ice cream, the K' and K'' values increased when compared to the low-fat ice cream sample.

All ice cream samples in the third interval exhibited thixotropic behavior (Figure 5). This result indicated that all ice cream samples could recover their viscoelastic character after high sudden deformation during food processing, such as homogenization or pumping. For ice cream combinations, this flow behavior is desirable. Akcicek and Karasu [20] investigated the thixotropic behavior of salad dressing samples stabilized by chia seed oil by-products and found that recoverable characteristics in the third interval are similar to our findings. The current investigation found that CSOB enhanced the thixotropic behavior of ice cream samples following rapid deformation.

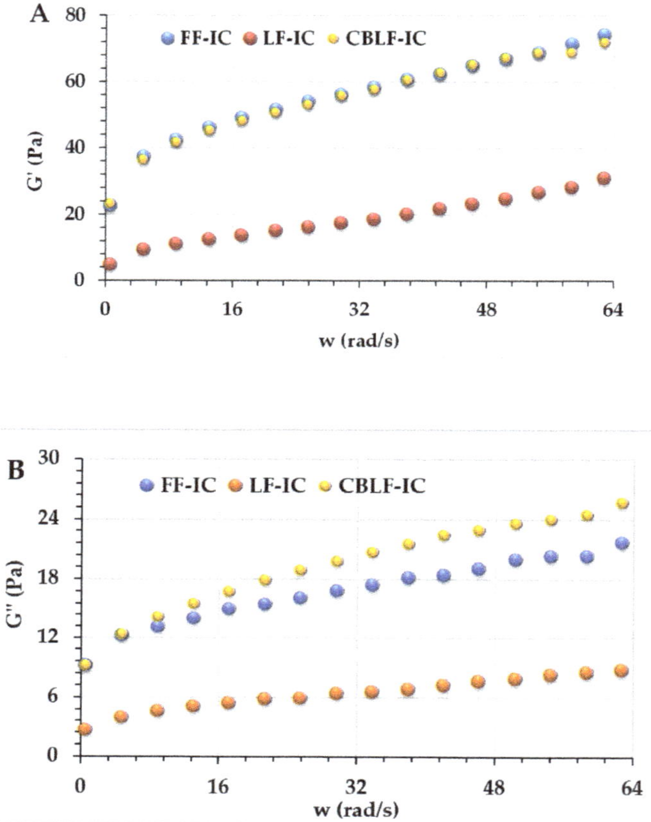

Figure 4. Dynamic rheological properties of the ice cream mixes. (**A**): Storage modulus (G') vs. ω, (**B**): loss modulus (G'') vs. ω. FF-IC: full-fat ice cream, LF-IC: low-fat ice cream, CBLF-IC: low-fat ice cream with chia seed oil by-product.

Parameters (G_0', G_e', k, G_0'/G_e') were obtained with the second-order structural kinetic model. G_0', G_e', G_0'/G_e', $k \times 1000$, and R^2 values were 8.42–21.12, 14.81–38.73, 1.760–1.834, 11.10–27.00, and 0.995–0.996, respectively. FF-IC showed the highest G_0', G_e', G_0'/G_e', and k × 1000 values, indicating that the full-fat control sample (FF-IC) showed the highest thixotropic behavior. The amount of fat in an O/W emulsion has a significant impact on its rheological characteristics. Therefore, the decrease in the fat content of the ice cream samples causes weak rheological properties, especially the recoverable character, as the fat content of the ice cream samples decreases. Although CBLF-IC has low-fat content (2.5%), FF-IC and CBLF-IC showed similar thixotropic behavior and viscoelastic solid character so the higher recoverable behavior obtained with CSOB addition can be explained by more intermolecular interactions by the formation of small aggregates of hydrocolloid. These rheological properties indicate that CSOB can be utilized to enhance the rheological properties and thixotropy of low-fat ice cream samples.

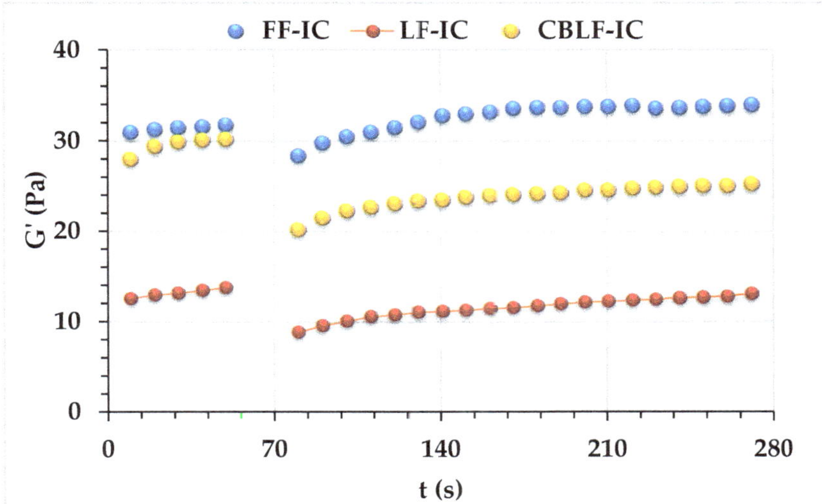

Figure 5. 3-ITT rheological properties of ice cream mixes. FF-IC: full-fat ice cream, LF-IC: low-fat ice cream, CBLF-IC: low-fat ice cream with chia seed oil by-product.

3.4. Emulsion Stability and Microstructure Properties of Ice Cream Mixes

The emulsion stability of the ice cream mixes was determined by the thermal loop test. The thermal loop test could be used as a fast method to measure emulsion stability by subjecting to different numbers of thermal cycles. The temperature fluctuations during processing, production, storage, and transportation stages were simulated in this test [31]. The structural or morphological changes due to the applied thermal stress are determined by the change of modules (G^*, G') from cycle to cycle. Figure 6 shows the change in the G^* value after applied thermal stress. As can be seen, after 10 applied thermal cycles, a slight increase in the G^* value of all samples was observed. This insignificant change indicates that all samples are resistant to thermal stress and have high emulsion stability. The percent change (ΔG^*) in the G^* values of the samples was calculated as 14.37, 10.10, 8.37 for FF-IC, LF-IC, and CBLF-IC, respectively. This may indicate that the sample containing CSOB may show higher stability than other samples.

The fat particle size (d_{32}), PDI value, and zeta potential values of the samples were found as 0.494–1.305 μm, (−39.13)–(−28.90) mV and 0.493–0.741, respectively (Table 3). As can be seen, the sample containing CSOB exhibited lower particle size and PDI value. These results are consistent with the thermal loop test. A decrease was observed in the zeta potential values of the samples as the water ratio increased. However, all samples have a sufficient zeta potential value. Figure 7 shows the milk fat particle distribution. The high-fat control sample and the sample containing CSOB have homogeneous droplet distribution and low-fat droplet size. These results indicated that the use of CSOB could have a positive effect on the fat droplet size and distribution, and emulsion stability in ice cream.

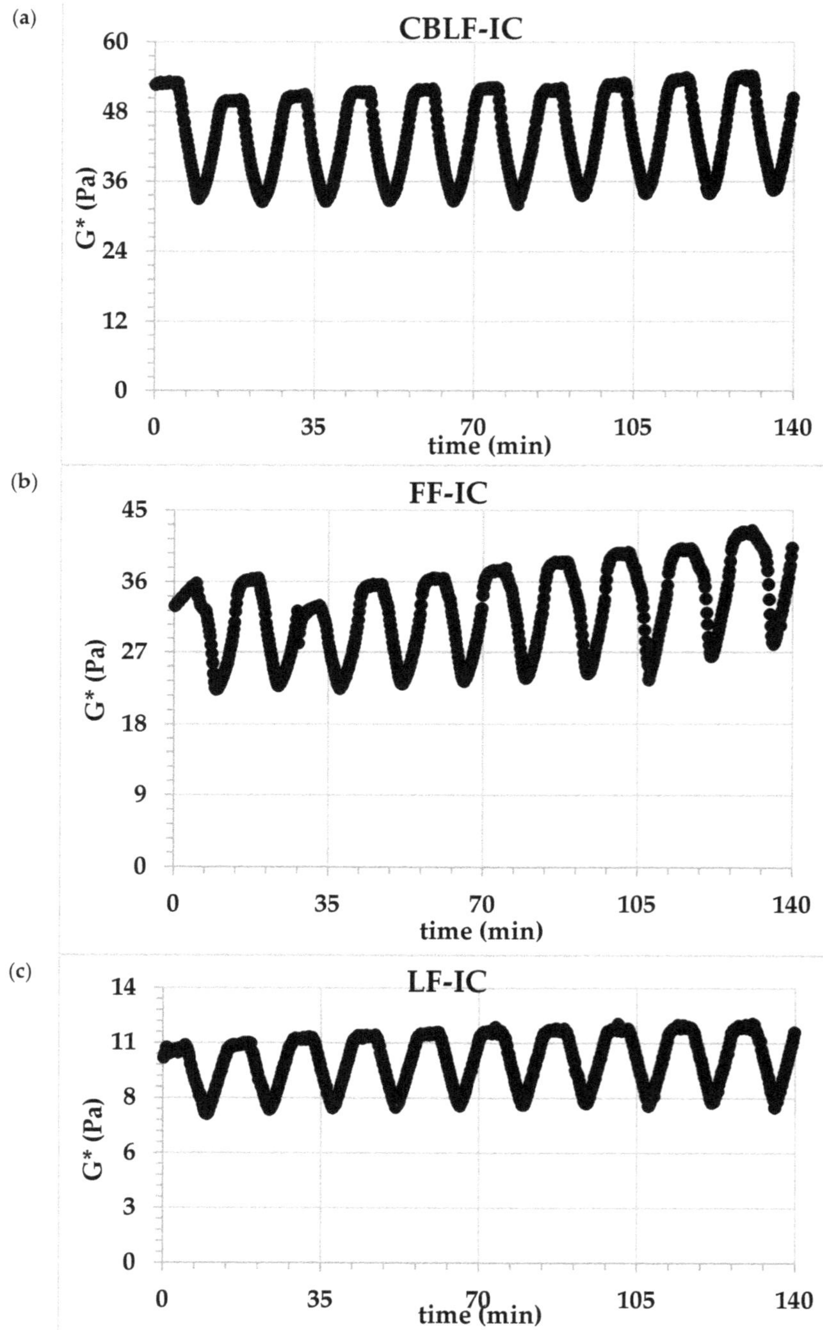

Figure 6. The change in the G^* value after applying the thermal loop. (**a**) CBLF-IC: low-fat ice cream with chia seed oil by-product, (**b**) FF-IC: full-fat ice cream, (**c**) LF-IC: low-fat ice cream.

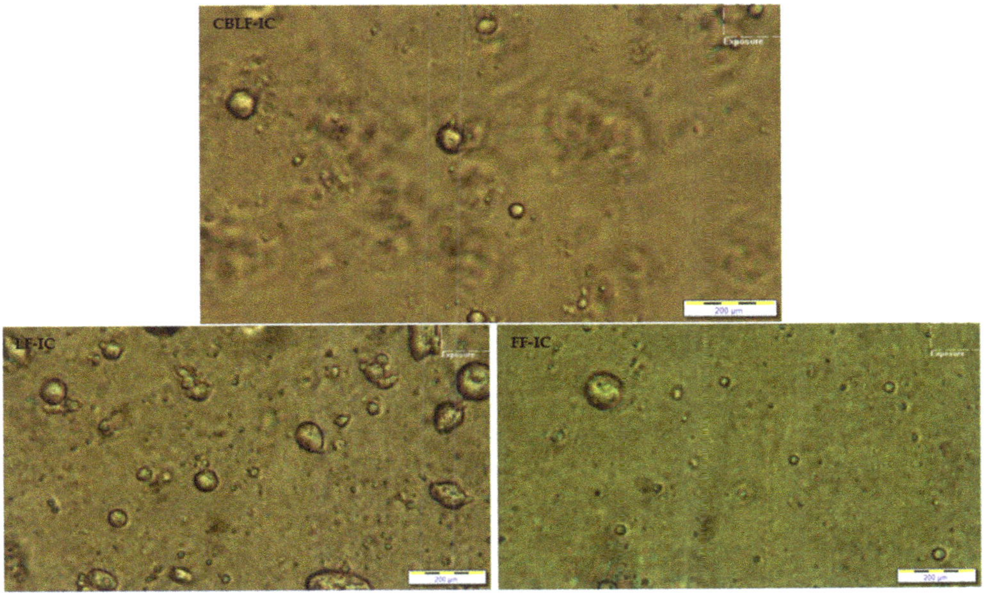

Figure 7. Light microscopy pictures of the ice cream mixtures. FF-IC: full-fat ice cream, LF-IC: low-fat ice cream, CBLF-IC: low-fat ice cream with chia seed oil by-product. Samples were characterized at room temperature at 20× magnification.

3.5. *Quality Parameters of Ice Cream*

3.5.1. Thermal Properties of Ice Cream

The thermal properties of the ice cream samples are presented in Table 4. The freezing points (T_f) of FF-IC, LF-IC, and CBLF-IC were obtained as −3.66, −3.35, and −3.99 °C, respectively. The freezing points of the ice cream samples were significantly differed ($p < 0.05$). The freezing point temperature of ice cream is closely related to the serum phase concentration and the soluble biopolymer concentration. Generally, the T_f value decreases as the serum phase concentration increases or the molecular weight of the soluble biopolymers decreases [43,44]. Soukoulis, Lebesi and Tzia [33] investigated the effect of different fiber contents on the thermal properties of ice cream. Similar to our study, a significant decrease was observed in the T_f values of ice creams when 2% wheat and oat fiber were added. Researchers explained this result by increasing the serum phase concentration due to the high water holding capacity of wheat and oat fiber. They reported an increase in the T_f value with the addition of apple fiber and inulin, which have higher soluble fiber content. The high water holding capacity of the insoluble fibers in the CSBO content may have caused a decrease in the T_f value. In addition, the use of less XG in the sample containing CSOB may have caused a decrease in the less soluble biopolymer substance in the serum phase, thus significantly reducing the T_f value. The higher T_f value obtained from LF-IC could be due to a decrease in solid concentration by reducing fat contents.

Table 4. Thermal properties of the control ice cream mixtures and optimum ice cream mixture contained CSOB.

Sample	T_0 (°C)	T_{end} (°C)	T_f (°C)	ΔT (°C)	ΔH_f (J/g)	W_f (%)	Overrun (%)
FF-IC	-11.67 ± 0.05^{C}	5.39 ± 0.04^{A}	-3.66 ± 0.02^{B}	17.06 ± 0.05^{A}	166.01 ± 2.68^{B}	49.72 ± 0.28^{B}	75.41 ± 0.18^{B}
LF-IC	-11.0 ± 0.15^{B}	4.95 ± 0.01^{B}	-3.35 ± 0.05^{A}	15.95 ± 0.08^{B}	199.32 ± 2.45^{A}	59.51 ± 1.05^{A}	71.23 ± 0.31^{C}
CBLF-IC	-10.18 ± 0.04^{A}	3.23 ± 0.02^{C}	-3.98 ± 0.03^{C}	14.16 ± 0.10^{C}	146.85 ± 1.25^{C}	43.95 ± 0.74^{C}	86.31 ± 0.62^{A}

FF-IC: full-fat ice cream, LF-IC: low-fat ice cream, CBLF-IC: low-fat ice cream with chia seed oil by-product. Different letters in the same column indicate a statistical difference ($p < 0.05$).

The melting resistance of ice cream represents the ability of ice cream to resist melting when exposed to high temperatures. The heating system in the DSC provides the temperature that causes the formation of an endothermic peak. The melting enthalpy (ΔH) is calculated by integrating with the area of the endothermic peak and is the amount of energy leaving the system. It occurs as a change in the overall internal energy of food [45]. In our study, the melting enthalpy values of the samples were found as 166.0, 199.3, and 146.8 J/g for the FF-IC, LF-IC, and CBLF-IC, respectively. The freezable water amount is a critical parameter affecting ice melting enthalpy in ice cream. Melting enthalpy is less in ice creams with lower freezable water content [46]. The enthalpy value of the sample CBLF-IC (containing CSOB) was found to be lower than that of FF-IC and LF-IC (the full-fat and low-fat control samples, respectively). These results can be explained by the interaction between water and CSOB. The addition of CSOB to ice creams mix could increase the amount of bound water and reduced the amount of freezable water thanks to its water-binding capacity, which may lead to a decrease in enthalpy. In addition, as the fat reduction in ice cream samples was balanced by adding water, it caused high ice formation in ice cream. However, the CSOB produced a balanced effect by chemically binding the free water, preventing excessive ice crystal formation. The lower enthalpy value of ice cream containing CSOB than FF-IC despite higher water content could be explained by the higher water retention capacity and lowering amount of freezable water.

ΔT values were found as 14.16, 15.95, and 17.05 for the samples CBLF-IC, LF-IC, and FF-IC, respectively. The sample containing CSOB had the lowest ΔT value than control samples. The temperature range (ΔT) could be used as an indicator of the uniformity of the size distribution of ice crystals. Therefore, a narrow melting temperature range indicates a more homogeneous distribution of ice crystals melting in a smaller temperature range [47]. The enrichment of ice creams in terms of polysaccharides and protein with CSOB facilitates the formation of tiny ice crystals, contributing to the improvement of texture perception and stability of ice crystals during cold storage.

3.5.2. Overrun Properties of Ice Cream

The overrun values are shown in Table 4. There was a significant difference between the overrun values of the samples. The sample containing CSOB exhibited the highest overrun value, while the low-fat control sample exhibited the lowest overrun value. The increase in the overrun value with the addition of CSOB may be due to the protein and high molecular carbohydrates in the CSOB content. Proteins play an important role in increasing foam stability thanks to their emulsification properties [48]. With the addition of CSOB to the formulation, both the protein content is gained, and the high molecular carbohydrate ratio is increased. Thus, an increase in the overrun value was observed with the addition of CSOB. Akalın et al. [49] reported that dietary fibers obtained from orange, apple, and wheat provided a significant increase in overrun values compared to the control sample. On the other hand, Mansour et al. [50] reported that the addition of datary fiber powder significantly reduced the overrun value of the ice cream samples. Some researchers have suggested that there is a relationship between the overrun value and the rheological properties [6,26]. Samakradhamrongthai et al. [51] reported that the overrun value increased with increased ice cream mix viscosity. The higher overrun with increasing viscosity could be explained by a more efficient breakdown of the incorporated air cells to a smaller air cell size of ice cream mix during freezing [51,52]. Similarly, the increase in the overrun value with the addition of CSOB in our study can be explained by the increase in the consistency index of the ice cream mix.

3.5.3. Sensory and Color Quality of Ice Cream

Sensory scores of ice cream samples are shown in Table 5. As can be seen, a significant difference was observed between the sensory scores of FF-IC and LF-IC. Whole-fat ice cream showed the highest value in all criteria. Low-fat ice cream containing CSOB (CBLF-IC), on the other hand, showed similar sensory properties to the full-fat sample (FF-IC), except for the color and appearance criteria. With the addition of CSOB, the change in color values compared to the full-fat sample (FF-IC) is expected. There was no significant difference in the overall acceptability criteria of the CSOB-containing sample (CBLF-IC) and the full-fat control sample (FF-IC). This indicated that the color difference observed with the addition of CSOB could not adversely affect the consumability of ice cream. Eskandari and Akbari [53] and Akın et al. [54] reported that the addition of dietary fiber and other fat replacers did not cause a negative change in sensory scores of ice cream, similar to our study. In another study [51], the reduced fat ice cream prepared with inulin showed an acceptable sensory score, similar to our study. These results indicated that with the addition of CSOB, the tested quality properties of ice cream could be improved without adversely affecting the sensory properties and that low-fat ice cream could be produced in a similar way to achieve the quality properties of full-fat ice cream. L^*, a^* and b^* color values of the samples are presented in Table 6. L^*, a^* and b^* values of the samples were found as 77.08–83.42, 4.23–4.93 and 13.63–15.41, respectively. As can be seen, a significant difference was detected between the color values of the samples. The highest L^* value was measured from the FF-IC sample, while the lowest L^* value was obtained from the CBLF-IC sample. The high L^* value of the FF-IC sample can be explained by the higher fat content than the other two samples. In addition, with the introduction of CSOB into the ice cream formulation, a decrease in L^*, a^* and b^* values was observed. This result shows that CBLF-IC will cause a significant change in the color values of the ice cream samples.

Table 5. Sensory analysis of ice cream samples.

Sample	Color and Appearance	Icy Structure and Consistency	Foreign Taste and Smell	Cream Taste	Melting Resistance	General Acceptance
FF-IC	7.90 ± 1.10 A	7.01 ± 1.10 A	7.20 ± 1.40 A	7.00 ± 2.26 A	7.15 ± 2.08 A	7.44 ± 1.50 A
LF-IC	6.75 ± 1.25 B	5.01 ± 1.73 B	6.10 ± 1.73 A,B	5.70 ± 1.51 A	5.13 ± 1.45 B	5.72 ± 1.85 B
CBLF-IC	6.21 ± 0.92 B	6.00 ± 1.42 A,B	5.50 ± 2.51 A,B	6.40 ± 1.25 A	6.11 ± 1.10 A,B	6.12 ± 1.35 A,B

FF-IC: full-fat ice cream, LF-IC: low-fat ice cream, CBLF-IC: low-fat ice cream with chia seed oil by-product. Different letters in the same column indicate a statistical difference ($p < 0.05$).

Table 6. Color properties of the control ice cream mixtures and optimum ice cream mixture containing CSOB.

Sample	L^*	a^*	b^*
FF-IC	83.42 ± 0.23 A	4.93 ± 0.02 A	15.41 ± 0.08 A
LF-IC	79.07 ± 0.43 B	4.81 ± 0.12 A	15.11 ± 0.15 A
CBLF-IC	77.08 ± 0.58 C	4.23 ± 0.15 B	13.63 ± 0.01 B

FF-IC: full-fat ice cream, LF-IC: low-fat ice cream, CBLF-IC: low-fat ice cream with chia seed oil by-product. Different letters in the same column indicate a statistical difference ($p < 0.05$).

4. Conclusions

The consumption of chia seed oil has been increasing in recent years due to its polyunsaturated fatty acids, especially linolenic acid. As a result of the production of cold- pressed chia seed oil, waste rich in carbohydrates, fiber and protein emerges. Evaluation of this waste is an important issue for cold press manufacturers. In this study, the potential for the use of CSOB in the production of low-fat ice cream was investigated. A significant increase was observed in the K value of the ice cream mix samples with the increase in CSOB in the formulation. The low-fat sample (CBLF-IC) containing CSOB, produced as a result of the optimization, showed similar rheological properties to the full-fat sample (FF-IC). Emulsion stability, fat globule size and distribution, thermal properties, and overrun analyzes indicated that the quality of ice cream mix and ice cream could be improved with the addition of CSOB. According to the results of the sensory analysis, the addition of CSOB did not cause a significant decrease in the sensory qualities of ice cream. This study has shown that CSOB can be used successfully in the production of low-fat ice cream and can provide a new perspective for the evaluation of this by-product with low economic value.

Author Contributions: Conceptualization, S.K. and Z.H.T.C.; methodology, S.K.; software, E.A. and Z.H.T.C.; Funding acquisition, I.A.; investigation, E.A. and Z.H.T.C.; data curation, E.A. and Z.H.T.C.; writing—original draft preparation, Z.H.T.C. and E.A.; writing—review and editing, I.A. and S.K.; visualization, Z.H.T.C. and S.K.; supervision, S.K. All authors have read and agreed to the published version of the manuscript.

Funding: This research received no external funding.

Data Availability Statement: The data presented in this study are available on request from the corresponding author.

Conflicts of Interest: The authors declare no conflict of interest.

References

1. Karaman, S.; Toker, Ö.S.; Yüksel, F.; Çam, M.; Kayacier, A.; Dogan, M. Physicochemical, bioactive, and sensory properties of persimmon-based ice cream: Technique for order preference by similarity to ideal solution to determine optimum concentration. *J. Dairy Sci.* **2014**, *97*, 97–110. [CrossRef] [PubMed]
2. Marshall, R.T.; Goff, H.D.; Hartel, R.W. *Ice Cream*; Springer Science & Business Media: Amsterdam, The Netherlands, 2003.
3. Akbari, M.; Eskandari, M.H.; Davoudi, Z. Application and functions of fat replacers in low-fat ice cream: A review. *Trends Food Sci. Technol.* **2019**, *86*, 34–40. [CrossRef]
4. Giese, J. Fats, oils, and fat replacers. *Food Technol.* **1996**, *50*, 77–84.
5. Góral, M.; Kozłowicz, K.; Pankiewicz, U.; Góral, D.; Kluza, F.; Wójtowicz, A. Impact of stabilizers on the freezing process, and physicochemical and organoleptic properties of coconut milk-based ice cream. *LWT* **2018**, *92*, 516–522. [CrossRef]
6. BahramParvar, M.; Goff, H.D. Basil seed gum as a novel stabilizer for structure formation and reduction of ice recrystallization in ice cream. *Dairy Sci. Technol.* **2013**, *93*, 273–285. [CrossRef]
7. Baer, R.J.; Wolkow, M.D.; Kasperson, K.M. Effect of Emulsifiers on the Body and Texture of Low Fat Ice Cream1. *J. Dairy Sci.* **1997**, *80*, 3123–3132. [CrossRef]
8. Arbuckle, W.S. *The Little Ice Cream Book*; Springer: Amsterdam, The Netherlands, 1981.
9. Zhang, Z.; Goff, H. On fat destabilization and composition of the air interface in ice cream containing saturated and unsaturated monoglyceride. *Int. Dairy J.* **2005**, *15*, 495–500. [CrossRef]
10. Lal, S.N.; O'Connor, C.J.; Eyres, L. Application of emulsifiers/stabilizers in dairy products of high rheology. *Adv. Colloid Interface Sci.* **2006**, *123*, 433–437. [CrossRef]
11. Mohd Ali, N.; Yeap, S.K.; Ho, W.Y.; Beh, B.K.; Tan, S.W.; Tan, S.G. The promising future of chia, *Salvia hispanica* L. *J. Biomed. Biotechnol.* **2012**, *2012*, 171956. [CrossRef]
12. Capitani, M.I.; Corzo-Rios, L.; Chel-Guerrero, L.; Betancur-Ancona, D.; Nolasco, S.M.; Tomás, M.C. Rheological properties of aqueous dispersions of chia (*Salvia hispanica* L.) mucilage. *J. Food Eng.* **2015**, *149*, 70–77. [CrossRef]
13. Avila-De La Rosa, G.; Alvarez-Ramirez, J.; Vernon-Carter, E.; Carrillo-Navas, H.; Pérez-Alonso, C. Viscoelasticity of chia (*Salvia hispanica* L.) seed mucilage dispersion in the vicinity of an oil-water interface. *Food Hydrocoll.* **2015**, *49*, 200–207. [CrossRef]
14. Chavan, V.; Gadhe, K.; Dipak, S.; Hingade, S. Studies on extraction and utilization of chia seed gel in ice cream as a stabilizer. *J. Pharmacogn. Phytochem.* **2017**, *6*, 1367–1370.
15. Campos, B.E.; Ruivo, T.D.; da Silva Scapim, M.R.; Madrona, G.S.; Bergamasco, R.d.C. Optimization of the mucilage extraction process from chia seeds and application in ice cream as a stabilizer and emulsifier. *LWT Food Sci. Technol.* **2016**, *65*, 874–883. [CrossRef]
16. Feizi, R.; Goh, K.K.; Mutukumira, A.N. Effect of chia seed mucilage as stabiliser in ice cream. *Int. Dairy J.* **2021**, *120*, 105087. [CrossRef]
17. Ürkek, B. Effect of using chia seed powder on physicochemical, rheological, thermal, and texture properties of ice cream. *J. Food Process. Preserv.* **2021**, *45*, e15418. [CrossRef]
18. Chaves, M.A.; Piati, J.; Malacarne, L.T.; Gall, R.E.; Colla, E.; Bittencourt, P.R.; de Souza, A.H.; Gomes, S.T.; Matsushita, M. Extraction and application of chia mucilage (*Salvia hispanica* L.) and locust bean gum (*Ceratonia siliqua* L.) in goat milk frozen dessert. *J. Food Sci. Technol.* **2018**, *55*, 4148–4158. [CrossRef]
19. Rothwell, J. Sugars and other sweeteners for ice cream and other frozen desserts. In Proceedings of the International Symposium, Athens, Greece, 18–19 September 1997.
20. Akcicek, A.; Karasu, S. Utilization of cold pressed chia seed oil waste in a low-fat salad dressing as natural fat replacer. *J. Food Process. Eng.* **2018**, *41*, e12694. [CrossRef]
21. Tekin-Cakmak, Z.H.; Karasu, S.; Kayacan-Cakmakoglu, S.; Akman, P.K. Investigation of potential use of by-products from cold-press industry as natural fat replacers and functional ingredients in a low-fat salad dressing. *J. Food Process. Preserv.* **2021**, *8*, e15388.
22. Tekin, Z.H.; Karasu, S. Cold-pressed flaxseed oil by-product as a new source of fat replacers in low-fat salad dressing formulation: Steady, dynamic and 3-ITT rheological properties. *J. Food Process. Preserv.* **2020**, *44*, e14650. [CrossRef]

23. Aksoy, F.S.; Tekin-Cakmak, Z.H.; Karasu, S.; Aksoy, A.S. Oxidative stability of the salad dressing enriched by microencapsulated phenolic extracts from cold-pressed grape and pomegranate seed oil by-products evaluated using OXITEST. *Food Sci. Technol.* **2021**. [CrossRef]
24. Karaman, S.; Karasu, S.; Tornuk, F.; Toker, O.S.; Gecgel, U.; Sagdic, O.; Ozcan, N.; Gül, O. Recovery potential of cold press byproducts obtained from the edible oil industry: Physicochemical, bioactive, and antimicrobial properties. *J. Agric. Food Chem.* **2015**, *63*, 2305–2313. [CrossRef] [PubMed]
25. Capitani, M.I.; Spotorno, V.; Nolasco, S.M.; Tomás, M.C. Physicochemical and functional characterization of by-products from chia (*Salvia hispanica* L.) seeds of Argentina. *LWT Food Sci. Technol.* **2012**, *45*, 94–102. [CrossRef]
26. Akalın, A.S.; Karagözlü, C.; Ünal, G. Rheological properties of reduced-fat and low-fat ice cream containing whey protein isolate and inulin. *Eur. Food Res. Technol.* **2008**, *227*, 889–895. [CrossRef]
27. Tiwari, A.; Sharma, H.K.; Kumar, N.; Kaur, M. The effect of inulin as a fat replacer on the quality of low-fat ice cream. *Int. J. Dairy Technol.* **2015**, *68*, 374–380. [CrossRef]
28. Yilsay, T.Ö.; Yilmaz, L.; Bayizit, A.A. The effect of using a whey protein fat replacer on textural and sensory characteristics of low-fat vanilla ice cream. *Eur. Food Res. Technol.* **2006**, *222*, 171–175. [CrossRef]
29. Tekin, E.; Sahin, S.; Sumnu, G. Physicochemical, rheological, and sensory properties of low-fat ice cream designed by double emulsions. *Eur. J. Lipid Sci. Technol.* **2017**, *119*, 1600505. [CrossRef]
30. Ismail, E.; Al-Saleh, A.; Metwalli, A. Effect of inulin supplementation on rheological properties of low-fat ice cream. *Life Sci. J.* **2013**, *10*, 1742–1746.
31. Tekin, Z.H.; Avci, E.; Karasu, S.; Toker, O.S. Rapid determination of emulsion stability by rheology-based thermal loop test. *LWT* **2020**, *122*, 109037. [CrossRef]
32. Hwang, J.-Y.; Shyu, Y.-S.; Hsu, C.-K. Grape wine lees improves the rheological and adds antioxidant properties to ice cream. *LWT Food Sci. Technol.* **2009**, *42*, 312–318. [CrossRef]
33. Soukoulis, C.; Lebesi, D.; Tzia, C. Enrichment of ice cream with dietary fibre: Effects on rheological properties, ice crystallisation and glass transition phenomena. *Food Chem.* **2009**, *115*, 665–671. [CrossRef]
34. Aloğlu, H.Ş.; Özcan, Y.; Karasu, S.; Çetin, B.; Sağdiç, O. Influence of transglutaminase treatment on the physicochemical, rheological, and melting properties of ice cream prepared from goat milk. *Mljekarstvo* **2018**, *68*, 126–138. [CrossRef]
35. Sharma, M.; Singh, A.K.; Yadav, D.N. Rheological properties of reduced fat ice cream mix containing octenyl succinylated pearl millet starch. *J. Food Sci. Technol.* **2017**, *54*, 1638–1645. [CrossRef]
36. Norton, I.T.; Spyropoulos, F.; Cox, P. *Practical Food Rheology: An Interpretive Approach*; John Wiley & Sons: London, UK, 2010.
37. Olivos-Lugo, B.; Valdivia-López, M.; Tecante, A. Thermal and physicochemical properties and nutritional value of the protein fraction of Mexican chia seed (*Salvia hispanica* L.). *Food Sci. Technol. Int.* **2010**, *16*, 89–96. [CrossRef]
38. Aime, D.; Arntfield, S.; Malcolmson, L.; Ryland, D. Textural analysis of fat reduced vanilla ice cream products. *Food Res. Int.* **2001**, *34*, 237–246. [CrossRef]
39. Karaca, O.B.; GÜVEN, M.; Yasar, K.; Kaya, S.; Kahyaoglu, T. The functional, rheological and sensory characteristics of ice creams with various fat replacers. *Int. J. Dairy Technol.* **2009**, *62*, 93–99. [CrossRef]
40. Zhang, H.; Chen, J.; Li, J.; Wei, C.; Ye, X.; Shi, J.; Chen, S. Pectin from Citrus Canning Wastewater as Potential Fat Replacer in Ice Cream. *Molecules* **2018**, *23*, 925. [CrossRef] [PubMed]
41. Kurt, A.; Kahyaoglu, T. Rheological properties and structural characterization of salep improved by ethanol treatment. *Carbohydr. Polym.* **2015**, *133*, 654–661. [CrossRef]
42. Aziz, N.S.; Sofian-Seng, N.-S.; Yusop, S.M.; Kasim, K.F.; Razali, N.S.M. Functionality of okra Gum as a novel carbohydrate-based fat replacer in ice cream. *Food Sci. Technol. Res.* **2018**, *24*, 519–530. [CrossRef]
43. Hartel, R. Crystallization in Foods. In *Handbook of Industrial Crystallization*, 2nd ed.; Myerson, A.S., Ed.; Elsevier: Amsterdam, The Netherlands, 2002; pp. 287–304.
44. Fuangpaiboon, N.; Kijroongrojana, K. Sensorial and physical properties of coconut-milk ice creammodified with fat replacers. *Maejo Int. J. Sci. Technol.* **2017**, *11*, 133–147.
45. Roos, Y.H. Mapping the Different States of Food Components Using State Diagrams. In *Modern Biopolymer Science*; Kasapis, S., Norton, I.T., Ubbink, J.B., Eds.; Academic Press: San Diego, CA, USA, 2009; pp. 261–276.
46. Pintor-Jardines, A.; Arjona-Roman, J.L.; Totosaus-Sanchez, A.; Severiano-Perez, P.; Gonzalez-Gonzalez, L.R.; Escalona-Buendia, H.B. The influence of agave fructans on thermal properties of low-fat, and low-fat and sugar ice cream. *LTW Food Sci. Technol.* **2018**, *93*, 679–685. [CrossRef]
47. Alvarez, V.B.; Wolters, C.L.; Vodovotz, Y.; Ji, T. Physical Properties of Ice Cream Containing Milk Protein Concentrates. *J. Dairy Sci.* **2005**, *88*, 862–871. [CrossRef]
48. Aykan, V.; Sezgin, E.; Guzel-Seydim, Z. Use of fat replacers in the production of reduced-calorie vanilla ice cream. *Eur. J. Lipid Sci. Technol.* **2008**, *110*, 516–520. [CrossRef]
49. Akalın, A.S.; Kesenkas, H.; Dinkci, N.; Unal, G.; Ozer, E.; Kınık, O. Enrichment of probiotic ice cream with different dietary fibers: Structural characteristics and culture viability. *J. Dairy Sci.* **2018**, *101*, 37–46. [CrossRef] [PubMed]
50. Mansour, A.I.A.; Ahmed, M.A.; Elfaruk, M.S.; Alsaleem, K.A.; Hammam, A.R.A.; El-Derwy, Y.M.A. A novel process to improve the characteristics of low-fat ice cream using date fiber powder. *Food Sci. Nutr.* **2021**, *9*, 2836–2842. [CrossRef]

51. Samakradhamrongthai, R.S.; Jannu, T.; Supawan, T.; Khawsud, A.; Aumpa, P.; Renaldi, G. Inulin application on the optimization of reduced-fat ice cream using response surface methodology. *Food Hydrocoll.* **2021**, *119*, 106873. [CrossRef]
52. VanWees, S.R.; Rankin, S.A.; Hartel, R.W. The microstructural, melting, rheological, and sensorial properties of high-overrun frozen desserts. *J. Texture Stud.* **2020**, *51*, 92–100. [CrossRef]
53. Eskandari, M.H.; Akbari, M. The effect of inulin on the physicochemical properties and sensory attributes of low-fat ice cream. *Int. Dairy J.* **2016**, *57*, 52. [CrossRef]
54. Akın, M.B.; Akın, M.S.; Kırmacı, Z. Effects of inulin and sugar levels on the viability of yogurt and probiotic bacteria and the physical and sensory characteristics in probiotic ice-cream. *Food Chem.* **2007**, *104*, 93–99. [CrossRef]

MDPI
St. Alban-Anlage 66
4052 Basel
Switzerland
Tel. +41 61 683 77 34
Fax +41 61 302 89 18
www.mdpi.com

Foods Editorial Office
E-mail: foods@mdpi.com
www.mdpi.com/journal/foods

www.ingramcontent.com/pod-product-compliance
Lightning Source LLC
Chambersburg PA
CBHW061811210926
51850CB00006B/673
9783036559490